D1233968

6/8/89

The Numerical Analysis
of Ordinary
Differential Equations

The Numerical Analysis of Ordinary Differential Equations

Runge–Kutta and General Linear Methods

J. C. BUTCHER

Department of Computer Science
University of Auckland

A Wiley–Interscience Publication

JOHN WILEY & SONS

Chichester · New York · Brisbane · Toronto · Singapore

Copyright © 1987 by John Wiley & Sons Ltd.

Library of Congress Cataloging-in-Publication Data:

Butcher, J. C. (John Charles), 1933–
  The numerical analysis of ordinary
differential equations.
  'A Wiley–Interscience publication.'
  Bibliography: p.
  Includes index.
  1. Differential equations—Numerical solutions.
2. Runge–Kutta formulas.
I. Title.
QA372.B9   1987   515.3′52   86-5600

ISBN 0 471 91046 5

British Library Cataloguing in Publication Data:

Butcher, J. C.
  The numerical analysis of ordinary
  differential equations: Runge–Kutta
  and general linear methods.
  1. Differential equations—Numerical
  solutions
  I. Title
  515.3′52   QA372

ISBN 0 471 91046 5

Printed and bound in Great Britain

ki nga rakau me nga putiputi o Aotearoa

—

Preface

Ordinary differential equations are at the heart of our perception of the physical universe. For this reason numerical methods for their solution are central tools for obtaining quantitative information on physical behaviour.

The advice given to Alice to start at the beginning, to keep on going until she got to the end and then to stop is a remarkably simple summary of the step-by-step approach to the solution of initial value problems. Although it is not necessarily the best way to read this book there is, at least at chapter level, a reasonably natural progression. Chapter 1 contains introductory sections on a number of mathematical topics. These range from differential calculus on a vector space to graphs and combinatorics. It is perhaps surprising that aspects of these two diverse parts of mathematics together provide the necessary background for investigations of the order of Runge–Kutta and other methods. Although many readers will wish to jump right over this elementary work, they might find it convenient to refer back to it to put later parts of the book into context.

Chapter 2 is intended to provide an introduction to numerical methods for solving initial value problems by dealing with particular classes of methods as generalizations of the method of Euler. In the case of Runge–Kutta methods, the introduction is not placed very far ahead of what it introduces. In the case of linear multistep methods, the introduction must stand more or less alone for the time being. It is intended that a sequel to this book will contain a more detailed study of linear multistep methods. In addition to its role as an introduction to the later parts of the book, chapter 2 can be thought of as a self-contained elementary course on numerical methods for ordinary differential equations.

In chapter 3, Runge–Kutta methods are studied in considerable detail and, if there is a centrepiece of this book, this is it. There exists a great wealth of theoretical knowledge concerning Runge–Kutta methods but even the small part of this that has been presented here might be thought not to be justified by the role of these methods in practical computation. The author, however, disputes this view on the two grounds that the potential efficacy of Runge–

Kutta methods in their own right has not yet been fully realized and that Runge–Kutta methods provide a theoretical framework for understanding general linear methods where the potential may be even greater.

It is to general linear methods that chapter 4 is devoted. These provide a means of studying a wide range of interesting methods in a unified manner. The theory for this large class of methods can be applied immediately to linear multistep methods and some of the conclusions arrived at through this path provide an alternative insight to these conclusions arrived at through the traditional approach.

In each of chapters 3 and 4, much of the emphasis is theoretical. In the sequel to this book, now in the early planning stage, practical issues concerning the design of efficient differential equation software will be considered in more detail. This will deal with linear multistep methods as well as with the methods the present volume emphasises.

Finally, this book contains an extensive bibliography of the subject up to the year 1982. While it cannot be claimed to be complete, it is as close to this as could be achieved within practical limitations. The intention has been that journal articles and conference proceedings should be included but that theses and technical reports should not. When this dividing line has been breached, it has been because of my ignorance of the nature of various publications which have been referred to by other authors or because review journals have acknowledged the more formal nature of some report series by treating them for review purposes in the same way as journal papers. One or two important series of conference proceedings are included within the alphabetical list of authors and editors and their works. For example, entries for relevant members of the Lecture Notes in Mathematics may be found under L.

The transliteration rules for names originally in Cyrillic have been based on those now used in *Mathematical Reviews* and by the British Standards Institute. This has been slightly simplified in that both щч and щ are transliterated in the same way as shch (rather than щч as sh.ch). A number of arbitrary decisions on the spelling of names have been taken. For example, Обрешковъ, sometimes spelt Обречков, has been transliterated as Obreshkov rather than the more common Romanization of Obrechkov. It has also been decided to use translated titles except where the original language was English, French, German or Italian. It is intended that the bibliography from 1983 onwards should be continued in the sequel to this volume.

To accommodate the small number of works referred to in the text but which fall outside the scope of the main bibliography, a list of additional references is supplied. The nature of a particular reference in the text can be distinguished by the style of brackets used around the year of publication. For example, a paper published in 1982 would be referred to as (1982) if it is in the bibliography and as [1982] if it is an additional reference.

The numbering system used within the text of the book has been chosen to be at the same time hierarchical and concise. For example, (314a), (314b), . . . are equations within subsection 314 which is in section 31 which is in chapter

3 and definition 435A and theorem 435B are to be found in subsection 435 in section 43 in chapter 4. Figures and tables inherit the numbers of their subsections with the addition of a distinguishing prime (') in the few cases where this becomes necessary. A similar numbering system is used for algorithms. Since ALGOL is no longer widely used as a publication language, algorithms have not geen given in the traditional manner as formal procedure declarations but rather as sequences of Pascal statements. It is hoped in this way to achieve a style of presentation which is easy to read as it stands but which, at the same time, can be readily translated into FORTRAN or other languages.

I wish to acknowledge my great debt to the many people who have assisted me through discussions and critical readings, in typing and iterated retyping, in library work concerned with the bibliography and in many other ways. My decision not to name them individually is an indication of how many collaborators I have had and that any line drawn below those who are named would undervalue the assistance of the others.

J. C. BUTCHER
Auckland, New Zealand
November 1985

Contents

1
Mathematical and computational introduction

10 MATHEMATICAL PRELIMINARIES

100 Vectors, matrices and norms.

In dealing with systems of differential equations, especially with equations derived from physical problems, the use of vectors is essential. Thus, rather than regarding the system

$$y_1'(x) = f_1(x, y_1(x), y_2(x), \ldots, y_n(x)),$$

$$y_2'(x) = f_2(x, y_1(x), y_2(x), \ldots, y_n(x)),$$

$$\vdots \qquad \vdots \qquad \vdots$$

$$y_n'(x) = f_n(x, y_1(x), y_2(x), \ldots, y_n(x)),$$

as a family of n equations in n dependent variables $y_1, y_2, \ldots, y_n$, it is convenient to regard it as a single equation

$$y'(x) = f(x, y(x))$$

in a single vector-valued variable y with components at a particular value of x given by

$$y(x) = \begin{bmatrix} y_1(x) \\ y_2(x) \\ \vdots \\ y_n(x) \end{bmatrix} = [y_1(x), y_2(x), \ldots, y_n(x)]^T,$$

where ^{T} denotes transpose.

The linear structures in the set of real numbers $\mathbb{R}$ associated with addition and subtraction generalize in a straightforward way to corresponding structures in the set of n-dimensional vectors of real numbers $\mathbb{R}^n$. Not so straightforward are the generalizations of multiplication of a number by a constant

1

and the absolute value of a number. The natural generalizations of these are respectively multiplication of a vector by a matrix and the norm of a vector.

If $v = [v_1, v_2, \ldots, v_n]^T \in \mathbb{R}^n$ and $A : \mathbb{R}^n \to \mathbb{R}^m$ is a linear function, then A is characterized by the $m \times n$ matrix elements given by

$$
A = \begin{bmatrix}
a_{11} & a_{12} & \cdots & a_{1n} \\
a_{21} & a_{22} & \cdots & a_{2n} \\
\vdots & \vdots & & \vdots \\
a_{m1} & a_{m2} & \cdots & a_{mn}
\end{bmatrix}
$$

and the elements of $Av \in \mathbb{R}^m$ are given by

$$
Av = \begin{bmatrix}
a_{11}v_1 + a_{12}v_2 + \cdots + a_{1n}v_n \\
a_{21}v_1 + a_{22}v_2 + \cdots + a_{2n}v_n \\
\vdots & \vdots & & \vdots \\
a_{m1}v_1 + a_{m2}v_2 + \cdots + a_{mn}v_n
\end{bmatrix}.
$$

In particular, if $m = 1$, the set of matrices such as A is known as the 'dual space' and is sometimes denoted by $(\mathbb{R}^n)^*$. The elements of $(\mathbb{R}^n)^*$ are conveniently written as transposes of corresponding elements of $\mathbb{R}^n$. If $u^T \in (\mathbb{R}^n)^*$ then $u^T v \in \mathbb{R}$ is the scalar product of u and v.

Another important special case is $m = n$ for which A maps members of $\mathbb{R}^n$ to members of $\mathbb{R}^n$. The multiplication of a vector v by a real number c is equivalent to multiplication by the 'scalar matrix':

$$
\begin{bmatrix}
c & 0 & \cdots & 0 \\
0 & c & \cdots & 0 \\
\vdots & \vdots & & \vdots \\
0 & 0 & \cdots & c
\end{bmatrix}
$$

for which the (i, j) element is $c\delta_{ij}$ where the 'Kronecker delta' is

$$
\delta_{ij} = \begin{cases} 1, & i = j, \\ 0, & i \neq j. \end{cases}
$$

The unit matrix I, which leaves a vector v unchanged by multiplication, is the scalar matrix with $c = 1$.

It is convenient to write $e_i \in \mathbb{R}^n$ as the ith 'natural basis vector', for $i = 1, 2, \ldots, n$. That is, the jth component of e_i equals δ_{ij}.

To generalize the absolute value of a number, that is the distance between zero and the number, we make use of norms. The norm of v, written as $\| v \|$, measures the magnitude of v, that is the distance between the zero vector and v. Although the fixed notation $\|\cdot\|$ is usually adopted, there is a wide variety of possible choices of the functional form of $\|\cdot\|$. In fact, any function from

$\mathbb{R}^n$ to $\mathbb{R}$ may be used as a norm as long as it satisfies the conditions

$$\| 0 \| = 0 \tag{100a}$$

$$\| v \| > 0 \qquad \text{for all } v \in \mathbb{R}^n \backslash \{0\} \tag{100b}$$

$$\| cv \| = | c | \cdot \| v \| \qquad \text{for all } c \in \mathbb{R}, \; v \in \mathbb{R}^n, \tag{100c}$$

$$\| u + v \| \leq \| u \| + \| v \| \qquad \text{for all } u, v \in \mathbb{R}^n. \tag{100d}$$

Note that 0 denotes either the zero vector in a vector space or the zero number, depending on the context.

The reason that it is not always necessary to be specific about which norm is being used is that many facts expressed in terms of one norm imply the corresponding facts in which a different norm is used. For example, suppose that $x_1, x_2, \ldots$ is a sequence of vectors which converges to a limit ξ in the sense that the sequence of real numbers $\| x_1 - \xi \|, \| x_2 - \xi \|, \ldots$ converges to zero. Then this statement will also be true if any other norm is used instead. In fact, if $\| \cdot \|$ and $\| \cdot \|'$ are two different norms then there exist positive constants α, β such that, for all x,

$$\alpha \| x \|' \leq \| x \| \leq \beta \| x \|'. \tag{100e}$$

In applications to numerical analysis, the three most convenient norms are usually denoted by $\| \cdot \|_1$, $\| \cdot \|_2$ and $\| \cdot \|_\infty$ and referred to as the l_1, l_2 and l_∞ norms respectively. For a vector $v = [v_1, v_2, \ldots, v_n]^\mathrm{T}$, they are defined by

$$\| v \|_1 = \sum_{i=1}^{n} | v_i |$$

$$\| v \|_2 = \left(\sum_{i=1}^{n} v_i^2 \right)^{1/2}$$

$$\| v \|_\infty = \max_{i=1}^{n} | v_i |.$$

The l_∞ norm is particularly useful in that if two vectors u, v are related by $\| u - v \| \leq d$ then an equivalent statement is that each component of u differs from the corresponding component of v by no more than d.

If $A : \mathbb{R}^n \to \mathbb{R}^m$, then it is easy to see that $\| Av \| / \| v \|$ is bounded for $v \neq 0$, where different types of norms could be used in $\mathbb{R}^n$ and $\mathbb{R}^m$. We define $\| A \|$ as the supremum of this quantity. In particular, if $m = n$ and the same norm is used in $\| Av \|$ and in $\| v \|$ then the norm defined by $\| A \| = \sup_{v \neq 0} \| Av \| / \| v \|$ is referred to as the 'norm subordinate to' the norm used in $\mathbb{R}^n$. In the case of l_1, l_2 and l_∞, these names and notations carry over to the corresponding subordinate norms. Formulae for these subordinate norms are

$$\| A \|_1 = \max_{j=1}^{n} \sum_{i=1}^{n} |a_{ij}|,$$

$$\| A \|_2 = \rho(A^{\mathrm{T}}A)^{\frac{1}{2}},$$

$$\| A \|_\infty = \max_{i=1}^{n} \sum_{j=1}^{n} |a_{ij}| = \| A^{\mathrm{T}} \|_1,$$

where ρ is the 'spectral radius' function which we will also use for other purposes in subsection 104.

In the case that $m = 1$, so that linear functions on $\mathbb{R}^n$ to $\mathbb{R}$ are members of the dual vector space, the norms of these linear functions have an interesting relation with the norms of the corresponding members of $\mathbb{R}^n$. In fact, $\| u^{\mathrm{T}} \|_1 = \| u \|_\infty$, $\| u^{\mathrm{T}} \|_2 = \| u \|_2$, $\| u^{\mathrm{T}} \|_\infty = \| u \|_1$ in the sense that

$$\sup_{v \neq 0} \frac{|u^{\mathrm{T}}v|}{\| v \|_1} = \| u \|_\infty$$

$$\sup_{v \neq 0} \frac{|u^{\mathrm{T}}v|}{\| v \|_2} = \| u \|_2$$

$$\sup_{v \neq 0} \frac{|u^{\mathrm{T}}v|}{\| v \|_\infty} = \| u \|_1.$$

101 Multilinear functions.

Generalizing the idea of linear functions on a vector space we consider a collection of $s+1$ vector spaces $\mathbb{R}^{n_1}, \mathbb{R}^{n_2}, \ldots, \mathbb{R}^{n_s}, \mathbb{R}^m$ and functions from $\mathbb{R}^{n_1} \times \mathbb{R}^{n_2} \times \cdots \times \mathbb{R}^{n_s}$ to $\mathbb{R}^m$ which are linear in each of its s arguments separately. This means that, if A is such an s-linear function, then for each $i = 1, 2, \ldots, s$ and for $x_1 \in \mathbb{R}^{n_1}, x_2 \in \mathbb{R}^{n_2}, \ldots, x_i, y_i \in \mathbb{R}^{n_i}, \ldots, x_s \in \mathbb{R}^{n_s}$ and $c \in \mathbb{R}$,

$$A(x_1, x_2, \ldots, x_i + y_i, \ldots, x_s)$$
$$= A(x_1, x_2, \ldots, x_i, \ldots, x_s) + A(x_1, x_2, \ldots, y_i, \ldots x_s) \quad (101a)$$

$$A(x_1, x_2, \ldots, cx_i, \ldots, x_s)$$
$$= cA(x_1, x_2, \ldots, x_i, \ldots, x_s). \quad (101b)$$

Note that if $s = 2$ then such a function is referred to as a bilinear function and, in general, as a multilinear function.

Let $a^i_{j_1 j_2 \ldots j_s}$ denote component i in $A(e_{j_1}, e_{j_2}, \ldots, e_{j_s})$, where $e_{j_1}, e_{j_2}, \ldots, e_{j_s}$ are basis vectors. Then, because of (101a) and (101b), we can write

$$A(x_1, x_2, \ldots, x_s) = \sum_{i=1}^{m} \sum_{j_1=1}^{n_1} \sum_{j_2=1}^{n_2} \cdots \sum_{j_s=1}^{n_s} a^i_{j_1 j_2 \ldots j_s} x_1^{j_1} x_2^{j_2} \ldots x_s^{j_s} e_i, \quad (101c)$$

where

$$x_1 = [x_1^1, x_1^2, \ldots, x_1^{n_1}]^T, \qquad x_2 = [x_2^1, x_2^2, \ldots, x_2^{n_2}]^T,$$

$$\ldots, \qquad x_s = [x_s^1, x_s^2, \ldots, x_s^{n_s}]^T.$$

We can now estimate $\| A(x_1, x_2, \ldots, x_s) \|_\infty$ using (101c). We find that

$$\| A(x_1, x_2, \ldots, x_s) \|_\infty \le \max_{i=1}^{m} \sum_{j_1=1}^{n_1} \sum_{j_2=1}^{n_2} \ldots \sum_{j_s=1}^{n_s}$$

$$| a_{j_1 j_2 \ldots j_s}^i | \| x_1 \|_\infty \| x_2 \|_\infty \ldots \| x_s \|_\infty$$

so that, if $x_1, x_2, \ldots, x_s$ are all non-zero, then $\| A(x_1, x_2, \ldots, x_s) \|_\infty /$ $(\| x_1 \|_\infty \| x_2 \|_\infty \ldots \| x_s \|_\infty)$ is bounded and the supremum of this is defined as the infinity norm of A. Because of the existence of the bounds in (100e), $\| A(x_1, x_2, \ldots, x_s) \| / (\| x_1 \| \| x_2 \| \ldots \| x_s \|)$ will always be bounded for non-zero $x_1, x_2, \ldots, x_s$, for an arbitrary collection of the $s + 1$ norms occurring in this expression. Thus $\| A \|$ will always exist as the supremum of this.

102   Calculus in vector spaces.

For a function $f : X \to Y$, where X and Y are vector spaces, we consider the difference between the value of f at two different points x and x_0 in an open set $U \subseteq X$, and we ask how this difference might be related to $x - x_0$ in a limiting case as $\| x - x_0 \| \to 0$. We distinguish three levels of dependence of $f(x) - f(x_0)$ on $x - x_0$ known respectively as continuity, Lipschitz continuity and differentiability.

The meaning of continuity of the function f at x_0 is that $\| f(x) - f(x_0) \| \to 0$ as $\| x - x_0 \| \to 0$. That is, given any positive number, ε, we can choose a positive number δ such that wherever x satisfies $\| x - x_0 \| < \delta$, then $\| f(x) - f(x_0) \| < \varepsilon$. Lipschitz continuity at x_0 is the special case of continuity at x_0 in which, for some $r > 0$, a constant L can be chosen so that

$$\| f(x) - f(x_0) \| \le L \| x - x_0 \| \tag{102a}$$

for all x satisfying $\| x - x_0 \| \le r$. This implies continuity since we can choose δ such that $\varepsilon \ge L\delta$. If (102a) holds for all x and $x_0 \in X$, we have a global Lipschitz condition.

Differentiability of f at x_0 requires that $f(x) - f(x_0)$ should approximate a linear function applied to the vector $x - x_0$. That is, if A is this linear function, then $\| f(x) - f(x_0) - A(x - x_0) \|$ should tend to zero more rapidly than $\| x - x_0 \|$ as $\| x - x_0 \| \to 0$. This linear function A is the derivative of f at x_0 and generalizes this concept in one-dimensional calculus. The notation $f'(x_0)$ will be carried over from real variable calculus. Note that the linear function $f'(x_0)$, if it exists, is unique since if there were two such linear functions A and B then

$$\| [f(x) - f(x_0) - A(x - x_0)] - [f(x) - f(x_0) - B(x - x_0)] \|$$

would tend to zero more rapidly than $\| x - x_0 \|$ so that $\| (A - B)(x - x_0) \| / \| x - x_0 \| \to 0$ as $x \to x_0$. By writing $x - x_0 = tv$, where t is real and v is a non-zero constant vector such that $\| (A - B)v \| \geq \frac{1}{2} \| A - B \| \, \| v \|$ and letting $t \to 0$, we find that $\| A - B \| = 0$, implying that $A = B$. The formal definition of $f'(x_0)$ is as follows.

DEFINITION 102A

The function $f: X \to Y$ is *differentiable* at x_0 if there exists a linear function $f'(x_0): X \to Y$ such that, for any $\varepsilon > 0$, there exists $\delta > 0$ such that if $\| x - x_0 \| < \delta$ then

$$\| f(x) - f(x_0) - f'(x_0)(x - x_0) \| \leq \varepsilon \| x - x_0 \|.$$

If f is differentiable at all points in an open set U, we write $f' : U \to L(X, Y)$ as the function taking every point to its derivative. Note that here $L(X, Y)$ denotes the vector space of linear functions on X to Y.

Various elementary results from real calculus carry over to vector space calculus. In particular, the chain rule on the differentiability of the composition of differentiable functions is expressed in the following result.

THEOREM 102B.

If $f: U \to Y$ is differentiable at $x_0 \in U \subseteq X$ and $g: V \to Z$ is differentiable at $f(x_0) \in V \subseteq Y$, then $g \circ f$ is differentiable at x_0 and furthermore $(g \circ f)'(x_0) = g'(f(x_0))(f'(x_0))$.

Proof.

$$\| (g \circ f)(x) - (g \circ f)(x_0) - [g'(f(x_0))(f'(x_0))] (x - x_0) \|$$
$$= \| (g(f(x)) - g(f(x_0)) - g'(f(x_0))(f(x) - f(x_0)))$$
$$+ g'(f(x_0))(f(x) - f(x_0) - f'(x_0)(x - x_0)) \|$$
$$\leq \| g(f(x)) - g(f(x_0)) - g'(f(x_0))(f(x) - f(x_0)) \|$$
$$+ \| g'(f(x_0)) \| \cdot \| f(x) - f(x_0) - f'(x_0)(x - x_0) \|$$

and the result follows immediately from definition 102A and the continuity of f. ∎

In the same way that f' is defined, so it is possible to define second derivatives by regarding f' as a function on an open set $U \subseteq X$ to $L(X, Y)$, which is itself a vector space with norm defined in terms of the norms used with the spaces X and Y.

DEFINITION 102C.

If f' exists on $U \subseteq X$, then f is *twice differentiable* at $x_0 \in U$ if there exists a linear function $f''(x_0) : X \to L(X, Y)$ such that, for all $\varepsilon > 0$, there exists $\delta > 0$ such that if $\| x - x_0 \| < \delta$ then

$$\| f'(x) - f'(x_0) - f''(x_0)(x - x_0) \| \le \varepsilon \| x - x_0 \|. \tag{102b}$$

In this definition, $f''(x_0) \in L(X, L(X, Y))$ where this linear space may be identified in a simple way with the set $L_2(X, Y)$ of bilinear functions on $X \times X$ to Y. If $A \in L(X, L(X, Y))$ then the corresponding $\bar{A} \in L_2(X, Y)$ is defined by $A(u)(v) = \bar{A}(u, v)$ for all $u, v \in X$. It can be shown that $\| A \| = \| \bar{A} \|$ and we can use this identification of $L(X, L(X, Y))$ with $L_2(X, Y)$ to replace (102b) by

$$\| f'(x)(k) - f'(x_0)(k) - f''(x_0)(x - x_0, k) \| \le \varepsilon \| x - x_0 \| \cdot \| k \| \tag{102c}$$

where this is to hold for all $k \in X$.

In just the same way we can introduce third, fourth, ... derivatives according to the following recursive definition.

DEFINITION 102D.

Suppose that $f^{(s-1)}$ exists on an open set $U \subseteq X$. Then f is said to be s *times differentiable* at $x_0 \in U$ if there exists an s-linear function $f^{(s)}(x_0) \in L_s(X, Y)$, called the sth *derivative* of f at x_0, such that for all $\varepsilon > 0$, there exists $\delta > 0$ such that if $\| x - x_0 \| < \delta$ then the inequality

$$\| f^{(s-1)}(x)(k_1, k_2, \ldots, k_{s-1}) - f^{(s-1)}(x_0)(k_1, k_2, \ldots, k_{s-1})$$

$$- f^{(s)}(x_0)(x - x_0, k_1, k_2, \ldots, k_{s-1}) \|$$

$$\le \varepsilon \| x - x_0 \| \| k_1 \| \| k_2 \| \cdots \| k_{s-1} \| \tag{102d}$$

holds for all $k_1, k_2, \ldots, k_{s-1} \in X$.

Note that the notation $L_2(X, Y)$ has been extended to $L_s(X, Y)$, the set of s-linear functions on X^s to Y. For detailed properties of the quantities defined in this subsection, the reader is referred to Dieudonné [1969].

103 Partial derivatives.

Even though we have been emphasizing the interpretation of a function $f : \mathbb{R}^n \to \mathbb{R}^m$ as a single entity, we will now revert to the alternative interpretation of f as a collection of m scalar-valued functions. It will be convenient to denote these functions using superscripts rather than subscripts so that they become $f^1, f^2, \ldots, f^m$. Note that, using the natural basis for the dual space $(\mathbb{R}^m)^*$, we can write f^i as the composition $f^i = e_i^T \circ f$. If the derivative $f'(x)$ exists it may be interpreted as a collection of $m \times n$ scalars $f^i_j(x)$, where $i = 1, 2, \ldots, m$ and $j = 1, 2, \ldots, n$. In terms of the natural basis $d_1, d_2, \ldots, d_n$

for $\mathbb{R}^n$, we have

$$f_j^i(x) = e_i^T f'(x) d_j. \tag{103a}$$

Another well-established notation is to write

$$f_j^i(x) = \frac{\partial f^i(x^1, x^2, \ldots, x^n)}{\partial x^j}, \tag{103b}$$

where

$$x = [x^1, x^2, \ldots, x^n]^T.$$

For higher derivatives, we write, by analogy with (103a),

$$f_{j_1 j_2 \cdots j_s}^i(x) = e_i^T f^{(s)}(x)(d_{j_1}, d_{j_2}, \ldots, d_{j_s}) \tag{103c}$$

for $i = 1, 2, \ldots m$ and $j_1, j_2, \ldots, j_s = 1, 2, \ldots, n$, and by analogy with (103b) we have the further notation

$$f_{j_1 j_2 \cdots j_s}^i(x) = \frac{\partial^s f^i(x^1, x^2, \ldots, x^n)}{\partial x^{j_1} \partial x^{j_2} \ldots \partial x^{j_s}}.$$

The matrix of partial derivatives $f_j^i(x)$, $i = 1, 2, \ldots m$, $j = 1, 2, \ldots, n$, is usually known as the *Jacobian* of f at x.

104 The characteristic and minimal polynomials.

For $M \in L(\mathbb{C}^n, \mathbb{C}^n)$, where we are now generalizing to complex vector spaces but specializing to square matrices, the following three statements are known to be equivalent:

$\det(M) = 0.$ (104a)
There exists a non-zero $v \in \mathbb{C}^n$ such that $Mv = 0$. (104b)
There exists $y \in \mathbb{C}^n$ for which there is no $x \in \mathbb{C}^n$ such that $Ax = y$. (104c)

When these statements are true, M is said to be *singular*.

For $A \in L(\mathbb{C}^n, \mathbb{C}^n)$, a number $\lambda \in \mathbb{C}$ is said to be an *eigenvalue* of A if $A - \lambda I$ is singular. If $M = A - \lambda I$ then v in (104b) is said to be a corresponding *eigenvector*. The set of all eigenvalues of A is called the spectrum of A and denoted by $\sigma(A)$. Substituting $M = A - \lambda I$ in (104a) and noting that $\det(A - \lambda I)$ is a polynomial in λ, we define this as the *characteristic polynomial* p. Thus we have

$$p(\lambda) = \det(A - \lambda I) \tag{104d}$$

and in terms of p the spectrum of A can be written

$$\sigma(A) = \{\lambda \in \mathbb{C} \ni p(\lambda) = 0\}.$$

The *spectral radius* $\rho(A)$ is defined as the greatest value of $|\lambda|$ for $\lambda \in \sigma(A)$.

If S is a non-singular matrix then $S^{-1}AS$ is said to be *similar* to A. It is easily seen that similarity is an equivalence relation and that the characteristic polynomial of $S^{-1}AS$ is the same as for A. If (λ, v) is an eigenvalue–eigenvector pair for A so that

$$Av = \lambda v$$

then

$$(S^{-1}AS)(S^{-1}v) = \lambda(S^{-1}v)$$

and $(\lambda, S^{-1}v)$ is an eigenvalue–eigenvector pair for $S^{-1}AS$.

If p has n distinct zeros then $\#\sigma(A)$, the number of members of $\sigma(A)$, exactly equals n. In this case, let (λ_1, v_1), (λ_2, v_2), ..., (λ_n, v_n) denote eigenvalue–eigenvector pairs, let V be the matrix whose successive columns are $v_1, v_2, \ldots, v_n$ and let $\Lambda = \text{diag}(\lambda_1, \lambda_2, \ldots, \lambda_n)$ be the diagonal matrix formed from the eigenvalues. It is then easy to verify that

$$AV = V\Lambda \tag{104e}$$

and that V is non-singular. It is now seen that $\Lambda = V^{-1}AV$ is similar to A and that the eigenvalue–eigenvector pairs of Λ are $(\lambda_1, e_1), (\lambda_2, e_2), \ldots, (\lambda_n, e_n)$ with the standard basis vectors appearing as eigenvectors.

If $B = S^{-1}AS$ then it also follows that $B^m = S^{-1}A^mS$ for any positive integer m and that a polynomial function in the matrix B is related in a corresponding way to the same polynomial in A. Thus using the characteristic polynomial of $A = V\Lambda V^{-1}$ we find

$$p(A) = Vp(\Lambda)V^{-1}.$$

However, $p(\Lambda)$ is the diagonal matrix $\text{diag}(p(\lambda_1), p(\lambda_2), \ldots, p(\lambda_n)) = 0$ because $p(x_1) = p(\lambda_2) = \cdots = p(\lambda_n) = 0$. Hence, in this special case in which p has distinct zeros, we have the Cayley–Hamilton theorem which states that the characteristic polynomial of any square matrix with the argument of the polynomial replaced by the matrix itself gives the zero matrix.

In cases where p has repeated zeros it may not be possible to choose a diagonal matrix similar to the given matrix. It is, however, always possible to choose a matrix S such that $S^{-1}AS$ has a *Jordan canonical form*. That is,

$$S^{-1}AS = \begin{bmatrix} J_1 & 0 & \cdots & 0 \\ 0 & J_2 & \cdots & 0 \\ \vdots & \vdots & & \vdots \\ 0 & 0 & \cdots & J_\nu \end{bmatrix},$$

where $J_1, J_2, \ldots, J_\nu$ are square matrices of the form

$$J_i = \begin{bmatrix} \lambda_i & \mu_i & 0 & \cdots & 0 \\ 0 & \lambda_i & \mu_i & \cdots & 0 \\ 0 & 0 & \lambda_i & \cdots & 0 \\ \vdots & \vdots & \vdots & & \vdots \\ 0 & 0 & 0 & \cdots & \lambda_i \end{bmatrix},$$

with λ_i an n_i-fold zero of p and J_i an $n_i \times n_i$ matrix in which $\mu_i \neq 0$ appears on the super-diagonal if $n_i > 1$. Since $A^{-1}AS$ is $n \times n$, we must have $\sum_{i=1}^{\nu} n_i = n$.

Because $(J_i - \lambda_i I)^{n_i} = 0$, the Cayley–Hamilton theorem can be proved from the existence of the Jordan canonical form. It may also happen that there is a polynomial P of lower degree than n for which $P(A) = 0$. Let g be a non-zero polynomial of lowest possible degree with this property. If there were two such polynomials g_1 and g_2, each with the same leading coefficient, then $(g_1 - g_2)(A) = 0$ and $g_1 - g_2$ would have an even lower degree. Hence g is unique to within a constant factor. This polynomial g is known as the *minimal polynomial* of A.

Let $p_i(z) = (\lambda_i - z)^{n_i}$ be the polynomial which is at the same time the characteristic polynomial and the minimal polynomial of J_i. We then have the following formulae for the characteristic and minimal polynomials of A:

$$p = \prod_{i=1}^{\nu} p_i$$

$$g = \bigvee_{i=1}^{\nu} p_i$$

where $\vee$ denotes the least common multiple operation.

105   Convergent and stable matrices.

In this subsection, A will denote an arbitrary member of $L(\mathbb{C}^n, \mathbb{C}^n)$. We will be concerned with the behaviour of A^m for large m. In the first definition, the particular norm is chosen merely for convenience and could equally well be replaced by another norm.

DEFINITION 105A.

The matrix A is *convergent* if $\| A^m \|_\infty \to 0$ as $m \to \infty$. We now give two alternative necessary and sufficient conditions for this property.

THEOREM 105B.

The following three statements are equivalent:
(a) A is convergent.
(b) $\rho(A) < 1$.
(c) There exists B similar to A such that $\| B \|_\infty < 1$.

Proof.

We will prove (a)$\Rightarrow$(b), (b)$\Rightarrow$(c) and (c)$\Rightarrow$(a). Let λ be an eigenvalue of greatest magnitude and let v be a corresponding eigenvector. It then follows that $\| A^m \|_\infty = \sup_{x \neq 0} \| A^m x \|_\infty / \| x \|_\infty \geq \| A^m v \|_\infty / \| v \|_\infty = |\lambda|^m$ so that $\| A^m \|_\infty \to 0$ implies that $|\lambda| < 1$. If (b) holds then choose B as the Jordan canonical form of A with every non-zero off-diagonal element equal to $\frac{1}{2}[1 - \rho(A)]$. Each row of B has at most two non-zero elements with the total sum of magnitudes bounded by $\rho(A) + \frac{1}{2}[1 - \rho(A)] = \frac{1}{2}[1 + \rho(A)] < 1$. Finally, if $\| B \|_\infty < 1$, with $B = V^{-1} A V$ then $A^m = V B^m V^{-1}$, implying that $\| A^m \|_\infty \leq \| V \|_\infty \| V^{-1} \|_\infty \| B \|_\infty^m$, which converges to zero. ∎

Note that in condition (c) an alternative norm could be used to give a sufficient condition for the convergence of A.

A weaker property than convergence, sometimes known as 'power-boundedness', is introduced in the following definition.

DEFINITION 105C.

The matrix A is *stable* if there exists a constant C such that $\| A^m \|_\infty \leq C$ for all $m = 1, 2, \ldots$.

We also introduce a relevant property of polynomials.

DEFINITION 105D.

Let P be a polynomial. Then P is said to *satisfy the root condition* if every zero of P lies in the closed unit disc and every repeated zero of P lies in the open unit disc.

We can now give necessary and sufficient conditions for A to be stable.

THEOREM 105E.

The following three statements are equivalent:
(a) A is stable.
(b) The minimal polynomial of A satisfies the root condition.
(c) There exists B similar to A such that $\| B \|_\infty \leq 1$.

Proof.

We will again prove (a)⇒(b), (b)⇒(c), (c)⇒(a). If λ is an eigenvalue of A and therefore a zero of the minimal polynomial we again $\| A^m \|_\infty \geq |\lambda|^m$ to prove that $|\lambda| \leq 1$. If λ is a repeated zero of the minimal polynomial then there exist non-zero vectors u, v such that $Au = \lambda u + v$ and $Av = \lambda v$. We now see by induction that $A^m u = \lambda^m u + m \lambda^{m-1} v$, implying that $\| A^m \|_\infty \geq |\lambda|^{m-1}$ $(m \| v \|_\infty / \| u \|_\infty - |\lambda|)$, which cannot be bounded as $m \to \infty$ unless $|\lambda| < 1$. If (b) is true then B can be chosen as the Jordan canonical form where the corresponding off-diagonals do not exist if an eigenvalue λ satisfies $|\lambda| = 1$

and where they can be chosen as $1 - |\lambda|$ if $|\lambda| < 1$. Finally, if $\| B \|_\infty \leq 1$ with $B = V^{-1}AV$ then $\| A^m \|_\infty \leq \| V \|_\infty \| V^{-1} \|_\infty$. ∎

106 Metric spaces.

A metric space is a set M together with a function $d : M \times M \to \mathbb{R}$, known as a metric, such that, for all $x, y, z \in M$,

$$d(x, y) = d(y, x), \tag{106a}$$

$$d(x, y) > 0, \quad \text{unless } x = y, \tag{106b}$$

$$d(x, x) = 0, \tag{105c}$$

$$d(x, z) \leq d(x, y) + d(y, z). \tag{106d}$$

An important example of a metric space is obtained by choosing M as an arbitrary subset of a vector space $\mathbb{R}^n$ and defining $d(x, y) = \| x - y \|$. This example enables the key result of this subsection, the 'contraction mapping theorem' (theorem 106E), to be used to study iterative solution methods for non-linear algebraic equations. In a second example, M is the set of all continuous functions on an interval $[a, b]$ to $\mathbb{R}^n$ such that for all $y \in M$, $y(a)$ equals a fixed value $y_0 \in \mathbb{R}^n$ and for all $x \in [a, b]$, $\| y(x) - y_0 \| \leq K$, where K is a positive constant. The metric d is defined by $d(y, z) = \sup_{x \in [a, b]} \| y(x) - z(x) \|$ for all $y, z \in M$.

Two fundamental ideas concerning sequences of real numbers carry over very simply to sequences in a metric space.

DEFINITION 106A.

A sequence $x^{(1)}, x^{(2)}, \ldots$ in a metric space (M, d) *converges* to ξ if $d(x^{(m)}, \xi) \to 0$ as $m \to \infty$.

DEFINITION 106B.

A sequence $x^{(1)}, x^{(2)}, \ldots$ in a metric space (M, d) is a *Cauchy sequence* if $d(x^{(l)}, x^{(m)}) \to 0$ as $\min(l, m) \to \infty$.

It is easy to verify that every convergent sequence is also a Cauchy sequence. Those situations where the reverse is true are designated as in the following definition.

DEFINITION 106C.

A metric space (M, d) is *complete* if every Cauchy sequence is convergent.

The first example given above is complete if and only if M is a compact subset of $\mathbb{R}^n$ (that is, M is closed and bounded). The second example we have described is necessarily complete. With complete metric spaces, convergence can be characterized without a knowledge of the actual limit ξ referred to in definition 106A. An important type of sequence is that generated by successively applying a particular function to each member to obtain the next.

Those functions from M to M covered by the following definition have a central place in the construction of convergent sequences.

DEFINITION 106D.

The function $\varphi : M \to M$, where (M, d) is a metric space, is a *contraction mapping* if there is a number $k < 1$ such that for all $x, y \in M$, $d(\varphi(x), \varphi(y)) \leq kd(x, y)$.

If $x^{(0)}$ is an arbitrary member of M and $x^{(m)} = \varphi(x^{(m-1)})$, for $m = 1, 2, \ldots$, then it is easy to prove that $x^{(0)}, x^{(1)}, x^{(2)}, \ldots$ is a Cauchy sequence. Furthermore, if this Cauchy sequency converges, it does so to a unique solution of the equation

$$x = \varphi(x), \tag{106e}$$

that is to a 'fixed point of φ'. We have the formal statement of this 'contraction mapping principle' which we state without proof.

THEOREM 106E.

If φ is a contraction mapping on a complete metric space (M, d), then (106e) has a unique solution which is the limit of a convergent sequence $x^{(0)}, x^{(1)}, x^{(2)}, \ldots$, where $x^{(0)} \in M$ is arbitrary and $x^{(1)} = \varphi(x^{(0)})$, $x^{(2)} = \varphi(x^{(1)}), \ldots$.

It is possible to widen this result slightly by using in place of contraction mappings the generalized contraction mappings in the definition which follows. In this definition φ^N denotes the composition of N copies of φ.

DEFINITION 106F.

The function $\varphi : M \to M$, is a *generalized contraction* if it is continuous and there is a positive integer N and a number $k < 1$ such that for all $x, y \in M$, $d(\varphi^N(x), \varphi^N(y)) \leq kd(x, y)$.

107 Uniform boundedness and equicontinuity.

We consider sequences of continuous functions on a closed bounded interval $I = [a, b]$ to a Euclidean space X. If the sequence f of such functions has members $f_1, f_2, \ldots$ then by a *subsequence* we mean a sequence made up from just some of $f_1, f_2, \ldots$ but with the order of the (infinite set of) subscripts preserved. Thus if $(n_1, n_2, \ldots)$ is a strictly increasing sequence of positive integers then $(f_{n_1}, f_{n_2}, \ldots)$ would be a subsequence of $(f_1, f_2, \ldots)$. We are interested in conditions which will guarantee that for a given sequence of functions there will exist a convergent subsequence. In preparation for this we introduce two definitions.

DEFINITION 107A.

The sequence $f = (f_1, f_2, \ldots)$ is *uniformly bounded* if the set $\{\, \| f_1 \|, \| f_2 \|, \ldots \}$ is bounded.

Note that in this definition, for $i = 1, 2, \ldots, \| f_i \|$ is defined as $\sup_{x \in I} \| f_i(x) \|$. Thus the definition means that there is a constant C such that for any $i = 1, 2, \ldots$ and any $x \in I$, $\| f_i(x) \| \le C$.

DEFINITION 107B.

The sequence $f = (f_1, f_2, \ldots)$ is *equicontinuous* on I if, for any $x \in I$ and any $\varepsilon > 0$, there is a δ such that for any $y \in I$ such that $|x - y| < \delta$, and for any $i = 1, 2, \ldots, \| f_i(y) - f_i(x) \| < \varepsilon$.

Note that, for a given choice of ε, the same choice of δ can be made for all members of the sequence.

We will show that the properties expressed in these definitions are appropriate conditions for the existence of convergent subsequences but first two preliminary lemmas are needed.

LEMMA 107C.

Given a bounded set S of a finite-dimensional vector space and a positive real number ε, there exists a finite collection of bounded sets $\{S_1, S_2, \ldots, S_m\}$ such that their union contains S and such that for any two points $x, y \in S_i$, for $i = 1, 2, \ldots, m$, $\| x - y \| < \varepsilon$.

Proof.

Without loss of generality (because we may replace S by a larger set) we may assume that

$$S = \{(x_1, x_2, \ldots, x_N) \in \mathbb{R}^N : |x_1| \le H, |x_2| \le H, \ldots, |x_N| \le H\}$$

We may also assume, because of (100e), that $\|\cdot\| = \|\cdot\|_\infty$.

Choose an integer $M > 2H/\varepsilon$ and $m = M^N$. Let $i_1, i_2, \ldots, i_N$ be the base M digits in $i - 1$ so that $i = 1 + i_1 + i_2 M + i_3 M^2 + \cdots + i_N M^{N-1}$ with $i_1, i_2, \ldots, i_N$ integers in the set $\{0, 1, \ldots, M - 1\}$ and choose S_i as the set of $(x_1, x_2, \ldots, x_N)$

values such that, for $j = 1, 2, \ldots, N$,

$$H\left(\frac{-M + 2i_j}{M}\right) \leq x_j \leq H\left(\frac{-M + 2(i_j + 1)}{M}\right). \quad \blacksquare$$

LEMMA 107D.

If $f = (f_1, f_2, \ldots)$ is equicontinuous on $I = [a, b]$ and ε is a positive real number then there exists a finite set of points $x_0, x_1, \ldots, y_k$ such that $a = x_0 < x_1 < \cdots < x_k = b$ and such that for any $j = 1, 2, \ldots, k$ and any two $x, y \in [x_{j-1}, x_j]$, $\| f_i(x) - f_i(y) \| < \varepsilon$ for all $i = 1, 2, \ldots$.

Proof.

Given ε, let T be the set of real numbers in I such that if $\bar{b} \in T$ then the statement of the lemma holds with the equality $x_k = b$ replaced by $x_k = \bar{b}$. Since T contains a, it is certainly not empty. Let b_0 be the supremum of T. We suppose that $b_0 < b$ and obtain a contradiction by applying definition 107B with x replaced by b_0. Hence $b_0 = b$. We suppose that although sup $T = b$, b is not a member of T and obtain a further contradiction by applying definition 107B with x replaced by b, finding a member of T greater than $b - \delta$ and adding a single extra member to the set $\{x_0, x_1, \ldots, x_k\}$. $\blacksquare$

We are now in a position to state the main result of this subsection.

THEOREM 107E.

If $f = (f_1, f_2, \ldots)$ is a uniformly bounded and equicontinuous sequence of functions on I to X, then there exists a subsequence which converges to a continuous function on I.

Proof.

We will obtain a Cauchy subsequence which necessarily converges to a function on I to X. That this limiting function is continuous and bounded easily follows from the uniform boundedness and equicontinuity and we will omit the details.

Define subsequences $f^0 = (f_1^0, f_2^0, \ldots)$, $f^1 = (f_1^1, f_2^1, \ldots)$, $f^2 = (f_1^2, f_2^2, \ldots)$, $\ldots)$, $\ldots$ in such a way that $f^0 = f$, for each $n = 1, 2, \ldots$, f^n is a subsequence of f^{n-1} and such that for any positive integers i and j, $\| f_i^{n-1} - f_j^{n-1} \| < 1/n$. To achieve this last requirement, select $x_0, x_1, \ldots, x_k$ in I as provided by lemma 107D with ε chosen as $1/(3n)$. Let S be the bounded part of X defined by $S = \{\xi \in X : \| \xi \| \leq C\}$ where C is as in definition 107A and form $S_1, S_2, \ldots, S_m$ as provided by lemma 107C with $\varepsilon = 1/(3n)$. Consider the m^{k+1} vectors of positive integers $[i_0, i_1, \ldots, i_k]$ where $i_0, i_1, \ldots, i_k \in \{1, 2, \ldots, m\}$

and associate with each such vector the set consisting of each integer j such that $f_j^{n-1}(x_0) \in S_{i_0}, f_j^{n-1}(x_1) \in S_{i_1}, \ldots, f_j^{n-1}(x_k) \in S_{i_k}$. At least one of this finite collection of sets of integers must be infinite since their union is infinite. This infinite set furnishes the required subsequence of f^{n-1} since for any x in $[x_{l-1}, x_l]$ and any $i, j \in \{1, 2, \ldots\}$ we have

$$\| f_i^{n-1}(x) - f_j^{n-1}(x) \| \leq \| f_i^{n-1}(x) - f_i^{n-1}(x_{l-1}) \|$$

$$+ \| f_i^{n-1}(x_{l-1}) - f_j^{n-1}(x_{l-1}) \| + \| f_j^{n-1}(x_{l-1}) - f_j^{n-1}(x) \|$$

$$< \frac{1}{3n} + \frac{1}{3n} + \frac{1}{3n} = \frac{1}{n}.$$

Having found the chain of subsequences f^1, f^2, ... we form the 'diagonal' subsequence $f_1^1, f_2^2, \ldots$ and note that it is the required Cauchy sequence since $\| f_m^m - f_n^n \| < 1/\min(m, n)$. ∎

11   ORDINARY DIFFERENTIAL EQUATIONS

110   Introduction to ordinary differential equations.

Let X denote an N-dimensional vector space and I an interval of $\mathbb{R}$. Usually, X will be a real vector space, but occasionally it will be convenient to take X as complex. If $f: I \times X \to X$ is given, then, by a first-order differential equation system, we will mean a relation of the form

$$y'(x) = f(x, y(x)). \tag{110a}$$

By a solution to (110a), we will mean a function $y \in C^1(I, X)$, the set of continuously differentiable functions on I to X, such that (110a) is satisfied for all $x \in I$ and such that certain subsidiary conditions are fulfilled. These subsidiary conditions, which we will briefly discuss now, are introduced because the solution to (110a) alone is not, in general, unique.

By an *initial value problem*, and it is with this type of problem that we are mainly concerned, we mean one in which the subsidiary condition is the value of y given at some particular point in I. By a *boundary value problem*, we mean one in which the subsidiary information is given in part at one point and in part at one or more different points. Another important class of problems is that in which the function f has one or more parameters built into it, and these are to be determined so that (110a) is satisifed as well as some additional boundary conditions.

In addition to first-order differential equations, we will pay some attention to equations of higher order. Thus, an nth-order system of equations is of the form

$$y^{(n)}(x) = f(x, y(x), y'(x), \ldots, y^{(n-1)}(x)), \tag{110b}$$

where $f: I \times X^n \to X$ is a given function and the solution is $y \in C^n(I, X)$ satisfying (110b) for all $x \in I$, together with subsidiary conditions. For example, an initial value problem for this type of equation is one in which the values of $y, y', \ldots, y^{(n-1)}$ are given at some particular point in I.

By an *autonomous* equation, we mean a first-order system (110a) or a higher-order equation (110b) ($n > 1$), in which the value of the function f is independent of the first variable. By a *special* equation of order n, we mean an equation of the form (110b) in which $f(x, u_0, u_1, \ldots, u_{n-1})$ is independent of $u_1, u_2, \ldots, u_{n-1}$. By a *linear* (first-order) system, we mean an equation of the form (110a), in which f is of the form

$$f(x, u) = A(x)u + \varphi(x), \qquad (110c)$$

where A is a matrix-valued function and φ is an arbitrary vector-valued function. If φ is the zero function, then (110a) is said to be *homogeneous*. We also use the phrase 'homogeneous part of' to describe that function derived from f in (110c) by replacing φ with zero.

We will often restrict ourselves to first-order systems, and usually we make the further assumption that the system is autonomous. To justify this position we note that, given (110b), we can immediately write down a first-order system on the space X^n:

$$z'(x) = g(x, z(x)),$$

where

$$g\left(x, \begin{bmatrix} u_0 \\ u_1 \\ \vdots \\ u_{n-1} \end{bmatrix}\right) = \begin{bmatrix} u_1 \\ u_2 \\ \vdots \\ f(x, u_0, u_1, \ldots, u_{n-1}) \end{bmatrix},$$

for which a solution z is related to a solution y of (110b) by

$$z(x) = \begin{bmatrix} y(x) \\ y'(x) \\ \vdots \\ y^{(n-1)}(x) \end{bmatrix}.$$

Furthermore, given a non-autonomous system (110a), we can immediately write down an autonomous system on the space $\mathbb{R} \times X$:

$$z'(x) = g(z(x)),$$

where

$$g\left(\begin{bmatrix} x \\ u \end{bmatrix}\right) = \begin{bmatrix} 1 \\ f(x, u) \end{bmatrix}$$

and for which a solution z is related to a solution y of (110a) by

$$z(x) = \begin{bmatrix} x \\ y(x) \end{bmatrix}.$$

111 Examples of differential equations.

The first example is of a one-dimensional autonomous linear system

$$y'(x) = qy(x), \tag{111a}$$

where q is a (possibly complex) number. It is easily verified that (111a) is satisfied by any function of the form

$$y(x) = c \exp(qx),$$

where c is a constant. To solve an initial value problem in which $y(x_0)$ is also specified, we solve for c to find that

$$y(x) = y(x_0) \exp(q(x - x_0)).$$

For simplicity, we now suppose that $y(x_0) = y(0) = 1$, so that $y(x) = \exp(qx)$.

If q is real and positive (respectively negative), this represents a monotonically increasing (respectively decreasing) solution and if $q = u + iv$ with $v \neq 0$, the solution is oscillatory with the sign of u determining whether $|y(x)|$ increases or decreases.

Our second example is a generalization of (111a) to the system

$$y'(x) = Ay(x), \tag{111b}$$

where A is a constant $n \times n$ matrix. If A is a diagonal matrix $A = \text{diag}(q_1, q_2, \ldots, q_n)$, then the n components of y satisfy separate equations

$$y_1'(x) = q_1 y_1(x),$$
$$y_2'(x) = q_2 y_2(x),$$
$$\vdots \qquad \vdots$$
$$y_n'(x) = q_n y_n(x)$$

and we can immediately write down the solution

$$y(x) = [y_1(x_0) \exp(q_1(x - x_0)), y_2(x_0) \exp(q_2(x - x_0)), \ldots,$$
$$y_n(x_0) \exp(q_n(x - x_0))]^{\mathsf{T}}.$$

This special case in which A is diagonal is remarkably typical in that if A has distinct eigenvalues, then by rewriting (111b) using eigenvectors as basis vectors, we can obtain an equivalent system in which A is diagonal. Let V denote the matrix formed from the eigenvectors as in (104e), where $\Lambda = \text{diag}(q_1, q_2, \ldots, q_n)$. Also let $z(x) = V^{-1}y(x)$; then (111b) written in terms of z becomes

$$z'(x) = \Lambda z(x).$$

Solving this equation and rewriting in terms of y gives the result

$$y(x) = V \text{diag}(\exp(q_1(x - x_0)), \exp(q_2(x - x_0)), \ldots,$$
$$\exp(q_n(x - x_0)))V^{-1}y(x_0).$$

If $q_1, q_2, \ldots, q_n$ have widely separated real parts, then, supposing that $\text{Re}(q_m)$ is one of the least of these real parts, as $x - x_0$ increases, the contribution of $\exp(q_m(x - x_0))$ to the solution will become increasingly less significant and it will become increasingly appropriate to approximate the solution by replacing $\exp(q_m(x - x_0))$ with zero.

Systems whose solutions contain rapidly decaying components, such as $\exp(q_m(x - x_0))$ in this example, are referred to as *stiff* differential equations. They are important in numerical analysis because they frequently arise in practical problems and because they are difficult to solve by traditional numerical methods.

Our third example is the special second-order system

$$y_1''(x) = \frac{-y_1(x)}{[y_1(x)^2 + y_2(x)^2 + y_3(x)^2]^{3/2}},$$

$$y_2''(x) = \frac{-y_2(x)}{[y_1(x)^2 + y_2(x)^2 + y_3(x)^2]^{3/2}},$$

$$y_3''(x) = \frac{-y_3(x)}{[y_1(x)^2 + y_2(x)^2 + y_3(x)^2]^{3/2}},$$

representing the equations of motion of a planet relative to a point sun, with units so normalized that the product of the mass of the sun and the gravitaitonal constant is unity. The solution is, of course, determined by the position and velocities at some given time, that is the vectors $[y_1(x_0), y_2(x_0), y_3(x_0)]^T$ and $[y_1'(x_0), y_2'(x_0), y_3'(x_0)]^T$ for some given time x_0, and it is well known that the path traced out by $[y_1(x), y_2(x), y_3(x)]^T$ is a plane conic with $[0, 0, 0]$ as focus. We content ourselves by giving the solution for just one special case, where $[y_1(0), y_2(0), y_3(0)]^T = [1, 0, 0]^T$ and $[y_1'(0), y_2'(0), y_3'(0)]^T = [0, 1, 0]^T$. In this case

$$y_1(x) = \cos(x), \qquad y_1'(x) = -\sin(x),$$

$$y_2(x) = \sin(x), \qquad y_2'(x) = \cos(x),$$

$$y_3(x) = 0, \qquad y_3'(x) = 0.$$

To recast this second-order system as a first-order system, we introduce three further variables $y_4 = y_1', y_5 = y_2', y_6 = y_3'$, so that the new set of equations becomes

$$y_1'(x) = y_4(x),$$

$$y_2'(x) = y_5(x),$$

$$y_3'(x) = y_6(x),$$

$$y_4'(x) = \frac{-y_1(x)}{[y_1(x)^2 + y_2(x)^2 + y_3(x)^2]^{3/2}},$$

$$y_5'(x) = \frac{-y_2(x)}{[y_1(x)^2 + y_2(x)^2 + y_3(x)^2]^{3/2}},$$

$$y_6'(x) = \frac{-y_3(x)}{[y_1(x)^2 + y_2(x)^2 + y_3(x)^2]^{3/2}}.$$

While this dynamical problem of a single particle moving under an inverse-square force can be solved in full, more general problems of the same type, in which each of a number of particles is moving in the combined gravitational fields of the others, is usually intractable and is regarded as a problem for numerical analysis.

As our fourth example we consider a case of the so-called predator–prey equations:

$$y_1'(x) = y_1(x)[3 - 2y_1(x) + y_2(x)], \tag{111c}$$

$$y_2'(x) = y_2(x)[3 - y_1(x) - y_2(x)], \tag{111d}$$

with initial values satisfying $y_1(0) > 0, y_2(0) > 0$. We note that if $y_1(0) = 2$, $y_2(0) = 1$, then $y_1(x), y_2(x)$ take on these same values for all x. This equation system is a model of the size of two populations at time x, $y_1(x)$ representing a predator and $y_2(x)$ a prey. Although we will not try to find formulae for the general solutions to this problem, it is interesting to study the behaviour of the solution near the point $[2, 1]^T$ (which represents an equilibrium state for the two populations).

If we introduce the function

$$V(x) = [y_1(x) - 2]^2 + [y_2(x) - 1]^2,$$

which measures the square of the distance between a point on the trajectory and the equilibrium position, we find by differentiation, making use of (111c) and (111d), that

$$V'(x) = 2[y_1(x) - 2]y_1(x)[3 - 2y_1(x) + y_2(x)]$$

$$+ 2[y_2(x) - 1]y_2(x)[3 - y_1(x) - y_2(x)],$$

which can be arranged in the form

$$V'(x) = -A[y_1(x) - 2]^2 - 2B[y_1(x) - 2][y_2(x) - 1] - C[y_2(x) - 1]^2$$

where $A = 4y_1(x), B = 1 - y_1(x), C = 2y_1(x) + 2y_2(x) - 4$. Since $A > 0$ and $AC > B^2$ if $y_1(x) > 0$ and $y_1(x)[7y_1(x) + 8y_2(x) - 14] - 1 > 0$, we see that if $y_1(x)$ and $y_2(x)$ satisfy these conditions then $V'(x) \leq 0$. It is easy to verify that these conditions are satisfied if $[y_1(x) - 2]^2 + [y_2(x) - 1]^2 \leq 0.6^2$, so that if the initial value is within a distance 0.6 of $[2,1]^T$, then along the trajectory $V(x)$ is a non-increasing function. In the sense that points close to the equilibrium position stay close to it, the equilibrium is said to be *stable*. The type of analysis of stability we have illustrated here is known as the Lyapunov direct method.

For our final differential equation example, we consider the system

$$y_1'(x) = -ky_1(x)y_2(x)$$
$$y_2'(x) = -ky_1(x)y_2(x),$$
$$y_3'(x) = ky_1(x)y_2(x).$$

This system arises in the study of a chemical reaction of the form $A + B \rightarrow C$ where, at time x, $y_1(x)$, $y_2(x)$ and $y_3(x)$ represent concentrations of the species A, B and C. The constant k is characteristic of the particular reaction and determines the rate at which it takes place.

For this very simple reaction, it is an easy matter to find a formula for the solution determined by given initial values by noting that $y_1(0) - y_1(x) = y_2(0) - y_2(x) = y_3(x) - y_3(0) = \varphi(x)$, say, and showing that

$$\varphi(x) = \frac{\exp(-kxy_1(0)) - \exp(-kxy_2(0))}{\exp(-kxy_1(0))/y_1(0) - \exp(-kxy_2(0))/y_2(0)}$$

if $y_1(0) \neq y_2(0)$, but that if $y_1(0) = y_2(0)$ then

$$\varphi(x) = \frac{kxy_1(0)^2}{1 + kxy_1(0)}.$$

By comparing this formula with experimental results, the constant k can be estimated.

As it happens, many important systems of equations associated with more complicated chemical reactions are quite intractable and therefore come under the purview of numerical analysis. The enormous variations in reaction rates that can actually occur in the same overall system do in fact cause serious difficulties in the numerical treatment of such problems and accordingly they are of major interest to numerical analysts.

112 Fundamental theory of differential equations.

We will deal only with initial value problems and consider such questions as the existence and uniqueness of solutions and the dependence of the solution on the initial values. As we saw in subsection 110, there is no essential loss in generality in dealing only with autonomous first-order systems and, since this leads to some simplifications in the presentation, we will do this. Thus, we consider the problem

$$y'(x) = f(y(x)), \qquad x \in I = [x_0, x_1], \tag{112a}$$

$$y(x_0) = y_0, \tag{112b}$$

where $f: X \rightarrow X$ and $y_0 \in X$ are given. Since many questions about differential equations are questions about f, we introduce some definitions concerning such functions.

DEFINITION 112A.

The function $f: X \to X$ is said to satisfy a *Lipschitz condition* if there exists a number L, known as the *Lipschitz constant* for f, such that for all $u, v \in X$,

$$\| f(u) - f(v) \| \le L \| u - v \|. \qquad (112c)$$

We recognize the Lipschitz condition as the global version of Lipschitz continuity introduced in subsection 102.

The norm $\| \cdot \|$ occurring in (112c) is, of course, quite arbitrary, but in the special case when it is the Euclidean norm ($\| u \|^2 = u^T u$) we can introduce a closely related but weaker condition.

DEFINITION 112B.

The function $f: X \to X$ is said to satisfy a *one-sided Lipschitz condition* if there exists a number λ, known as the *one-sided Lipschitz constant* for f, such that for all $u, v \in X$,

$$[f(u) - f(v)]^T (u - v) \le \lambda \| u - v \|^2. \qquad (112d)$$

If f is both continuous and satisfies the requirements of definition 112B, it will be called one-sided Lipschitz continuous.

We will return to Lipschitz and one-sided Lipschitz continuity when we come to consider uniqueness. However, continuity is enough to guarantee local existence of a solution as stated in the following result of Peano.

THEOREM 112C.

If $D = \{u \in X : \| u - y_0 \| \le R\}$ and for all $u \in D$, $\| f(u) \| \le M$ where the numbers R and M satisfy $M(x_1 - x_0) \le R$, then there exists a solution y to (112a) and (112b).

Proof.

For each real number α satisfying $0 < \alpha < x_1 - x_0$ let η_α be the solution to the 'delay differential equation'

$$\eta_\alpha'(x) = f(\eta_\alpha(x - \alpha)), \qquad x \in [x_0 + \alpha, x_1],$$

$$\eta_\alpha(x) = y_0, \qquad x \in [x_0, x_0 + \alpha].$$

In any subinterval $[x_0 + (n - 1)\alpha, \min(x_1, x_0 + n\alpha)]$ where $n = 2, 3, 4, \ldots, \eta_\alpha$ is easily seen to be defined uniquely as

$$\eta_\alpha(x) = y_0 + \int_{x_0}^{x - \alpha} f(\eta_\alpha(u)) \, du,$$

where the integral on the right is over previous subintervals only, so that η_α is given by induction on $n = 1, 2, 3, \ldots$.

We now remark that $\eta_\alpha(x)$ lies in D for all $x \in I$. This is again verified by induction on the subintervals already used, using the estimate

$$\| \eta_\alpha(x) - y_0 \| \leq \int_{x_0}^{x-\alpha} M\, du \leq (x_1 - x_0)M \leq R.$$

Generalizing this remark, we see that if $x_0 \leq \bar{x} < x \leq x_1$ then $\| \eta_\alpha(x) - \eta_\alpha(\bar{x}) \| \leq M(x - \bar{x})$ so that the family η_α for all α is equicontinuous (definition 107B) as well as being uniformly bounded (definition 107A). Hence, from theorem 107E, there is a sequence of α values, say $\alpha_1, \alpha_2, \ldots$, converging to zero such that if $y_n = \eta_{\alpha_n}, n = 1, 2, 3, \ldots$, then $y_1, y_2, \ldots$ converges uniformly to some function $y \in C(I, X)$.

It remains to show that, for all $x \in I$,

$$y(x) = y_0 + \int_{x_0}^{x} f(y(u))\, du,$$

implying that y satisfies (112a) and (112b). We have

$$\| y(x) - y_0 - \int_{x_0}^{x} f(y(u))\, du \| \leq \| y_n(x) - y_0 - \int_{x_0}^{x} f(y_n(u))\, du \|$$

$$+ \| y(x) - y_n(x) \| + \int_{x_0}^{x} \| f(y(u)) - f(y_n(u)) \|\, du,$$

and each of the terms on the right converges to zero as $n \to \infty$, the first because it is bounded by $\int_{x-\alpha_n}^{x} M\, du = \alpha_n M$, the second by the uniform convergence of y_n and the third because of this same fact and the continuity of f. ∎

Usually, at least for equation systems of interest to numerical analysts, f has stronger continuity properties than that supposed in this theorem and we will usually assume some variant of the Lipschitz property. We first emphasize, by a trivial example, that continuity of f is not sufficient for uniqueness of the solution. Specifically we consider the problem defined by $I = [0, 1]$, $X = \mathbb{R}$ and $f(u) = 2u^{1/2}$ (if $u > 0$) and $f(u) = 0$ (if $u \leq 0$). It is readily verified that for any $\xi \in [0, 1]$ there is a solution defined by $y(x) = 0$ (if $x \leq \xi$) and $y(x) = (x - \xi)^2$ (if $x > \xi$).

We now come to the main version of the Picard existence and uniqueness theorem.

THEOREM 112D.

If $f : X \to X$ satisfies a Lipschitz condition with constant L then there exists a unique solution y to the initial value problem (112a) and (112b).

Proof.

Consider the metric space (M, d) where $M = C(I, X)$ and $d(y, z) =$

$\sup_{x \in I} \| y(x) - z(x) \|$ for $y, z \in M$. Consider the mapping $\varphi : M \to M$ defined by

$$\varphi(y)(x) = y_0 + \int_{x_0}^{x} f(y(u)) \, du, \tag{112e}$$

so that any solution to the initial value problem is a fixed point of φ. We now show that φ is a generalized contraction (definition 106F) and thus, by the appropriately generalized form of theorem 106E, the conclusion of the present theorem will follow.

If $n = 0, 1, 2, 3, \ldots$ then $\| [\varphi^n(y) - \varphi^n(z)] (x) \| \le [L^n(x - x_0)^n/n!] \, d(y, z)$ because we have equality if $n = 0$ and for $n > 0$ we carry out the induction argument, using the Lipschitz condition

$$\| [\varphi^n(y) - \varphi^n(z)] (x) \| \le \int_{x_0}^{x} \| f(\varphi^{n-1}(y)(u)) - f(\varphi^{n-1}(z)(u)) \| \, du$$

$$\le L \int_{x_0}^{x} \frac{L^{n-1}(u - x_0)^{n-1}}{(n-1)!} \, du \, d(y, z)$$

$$= \frac{L^n(x - x_0)^n}{n!} \, d(y, z)$$

so that

$$\| \varphi^n(y) - \varphi^n(z) \| \le \frac{\sup_{x \in I} L^n(x - x_0)^n}{n!} \, d(y, z)$$

$$= \frac{L^n(x_1 - x_0)^n}{n!} \, d(y, z)$$

and, for n sufficiently great, $L^n(x_1 - x_0)^n/n! < 1$. ∎

We now seek to obtain a bound for $y(x) - y_0$ where y is the solution referred to in the last result. It will be assumed that $L \ne 0$ since, if $L = 0$, the solution is known exactly as $y(x) = y_0 + (x - x_0)f(y_0)$.

THEOREM 112E.

If, under the conditions of theorem 112D, the Lipschitz constant $L > 0$, then

$$\| y(x) - y_0 \| \le \| f(y_0) \| \frac{\exp(L(x - x_0)) - 1}{L}$$

Proof.

Let y_0 denote not only the initial value in X but also the function $I \to X$ with this constant value. If $y_1 = \varphi(y_0)$, $y_2 = \varphi(y_1)$, $\ldots$, where φ is the mapping in the proof of theorem 112D, then $\| y_1(x) - y_0 \| \le \| f(y_0) \| (x - x_0)$ and, by an induction argument,

$$\| y_n(x) - y_{n-1}(x) \| = \| [\varphi^{n-1}(y_1) - \varphi^{n-1}(y_0)](x) \|$$

$$\leq \| f(y_0) \| \frac{L^{n-1}(x - x_0)^n}{n!}$$

so that

$$\| y_n(x) - y_0 \| \leq \| (y_1 - y_0)(x) \| + \| [\varphi(y_1) - \varphi(y_0)](x) \|$$

$$+ \ldots + \| [\varphi^{n-1}(y_1) - \varphi^{n-1}(y_0)](x) \|$$

$$\leq \| f(y_0) \| \left((x - x_0) + \frac{L(x - x_0)^2}{2!} + \ldots + \frac{L^{n-1}(x - x_0)^n}{n!} \right)$$

$$\leq \| f(y_0) \| \frac{\exp(L(x - x_0)) - 1}{L}.$$

Thus each iterate, evaluated at the point x, lies in the closed disc with centre y_0 and radius $\| f(y_0) \| (\exp(L(x - x_0)) - 1)/L$ and accordingly the value of $y(x)$ lies in the same closed disc. ∎

Since the maximum radius of the disc occurring in the proof of theorem 112E occurs when $x = x_1$ and since the range of every iterate lies completely within the maximal disc, we can state without detailed proof a generalization of theorem 112D in which a local Lipschitz condition is used.

THEOREM 112F.

Let f be such that

$$\| f(u) - f(v) \| \leq L \| u - v \|$$

for all u, v in the disc $\{u \in X : \| u - y_0 \| \leq R\}$, where

$$R \geq \frac{\exp(L(x_1 - x_0)) - 1}{L}.$$

Then there exists a unique solution y to the initial value problem (112a), (112b).

We now generalize the existence and uniqueness theory to the case of one-sided Lipschitz continuous functions.

THEOREM 112G.

If f is continuous and satisfies a one-sided Lipschitz condition with constant λ then the initial value problem (112a), (112b) has a unique solution on any interval $[x_0, \bar{x}]$, where $\bar{x}$ is a real number greater than x_0.

Proof.

We first establish uniqueness. Suppose there were two solutions y, z on an interval $[x_0, \bar{x}]$. Subtracting the two differential equations, we find

$$y'(x) - z'(x) = f(y(x)) - f(z(x)). \tag{112f}$$

Let $v(x) = \| y(x) - z(x) \|^2 \exp(-2\lambda x)$, where we use the norm based on the inner product used in (112d); then we find

$$v'(x) = 2 \exp(-2\lambda x)[y'(x) - z'(x)]^T[y(x) - z(x)]$$
$$- 2\lambda \exp(-2\lambda x) \| y(x) - z(x) \|^2,$$

which, by (112d) and (112f), implies

$$v'(x) \le 0.$$

Hence, v is a non-increasing function and, accordingly,

$$\| y(x) - z(x) \| \le \exp(\lambda(x - x_0)) \| y(x_0) - z(x_0) \|, \tag{112g}$$

for all $x > x_0$. If $y(x_0) = z(x_0)$, this implies that $y(x) = z(x)$.

To establish existence on an arbitrary interval with left point x_0, let $\bar{x}$ be a number greater than x_0 for which a solution is guaranteed on $[x_0, \bar{x}]$ by theorem 112C. If $h = \bar{x} - x_0$ and we wish to establish existence on $[x_0, \bar{x}]$, then we write M as a bound on $\| f(u) \|$ for all $u \in X$ satisfying

$$\| u \| \le \| y_0 \| + \| y(\bar{x}) - y(x_0) \| \frac{1 - \exp(\lambda(\bar{x} - x_0))}{1 - \exp(\lambda(\bar{x} - x_0))} = R,$$

say (in which the last factor is replaced by $(\bar{x} - x_0)/(\bar{x} - x_0)$ if $\lambda = 0$), and we use theorem 112C to step along the interval $[x_0, \bar{x}]$ in steps as great as R/M. This is possible because the value of $\| y(x) \|$ never exceeds

$$\| y_0 \| + \| y(\bar{x}) - y(x_0)\| \frac{1 - \exp(\lambda(x - x_0))}{1 - \exp(\lambda(\bar{x} - x_0))} \tag{112h}$$

as we see by an induction argument on subintervals of length h. We suppose that $\| y(x - h) \|$ is bounded by (112h), with x replaced by $x - h$, and we use (112g) with $z(x) = y(x - h)$ and x_0 replaced by $x_0 + h$ to give

$$\| y(x) \| \le \| y(x - h) \| + \exp(\lambda(x - h - x_0)) \| y(\bar{x}) - y(x_0) \|.$$

The right-hand side, with $\| y(x + h) \|$ replaced by its bound, is equal to (112h). Since $\| y(x) \|$ is bounded by (112h), it is bounded by R and therefore $\| f(y(x)) \|$ is always bounded by M. ■

Although we have concentrated on autonomous systems, it is useful to consider briefly the non-autonomous system consisting of (110a) and (112b). To obtain a generalization to theorem 112D, it is only necessary to assume that f is continuous in its first variable and that it satisfies a Lipschitz condition in its second variable. That is, for all $x \in I$ and all $u, v \in X$,

$$\| f(x, u) - f(x, v) \| \le L \| u - v \|.$$

We state the generalized result without proof.

THEOREM 112H.

If f is continuous and satisfies a Lipschitz condition in its second variable then the initial value problem (110a), (112b) has a unique solution.

We now come to questions concerning the dependence of solutions on the initial value and we first consider a fundamental tool for such investigations, known as Gronwall's inequality.

THEOREM 112I.

If ϕ, K are continuous non-negative real-valued functions on $[x_0, x_1]$ such that

$$\phi(x) \le C + \int_{x_0}^{x} K(u)\phi(u)\,du,$$

for all $x \in [x_0, x_1]$ where C is a positive number, then, for all $x \in [x_0, x_1]$,

$$\phi(x) \le C \exp\left(\int_{x_0}^{x} K(u)\,du\right).$$

Proof.

Let

$$\Phi(x) = C \exp\left(\int_{x_0}^{x} K(u)\,du\right)$$

so that

$$\Phi(x) = C + \int_{x_0}^{x} K(u)\Phi(u)\,du.$$

Since ϕ and Φ are each continuous, there is a maximal interval $[x_0, \tilde{x}]$ with $x_0 \le \tilde{x} \le x_1$ such that $\phi(x) \le \Phi(x)$ for all x in this interval. If $\tilde{x} < x_1$ then $\phi(\tilde{x}) = \Phi(\tilde{x})$, and for all x in some interval $I = [\tilde{x}, \tilde{x} + \alpha]$, with $\alpha > 0$, $\phi(x) > \Phi(x)$. If α is chosen so small that $K(x)\alpha \le \theta$ for all $x \in I$ and for some $\theta < 1$, then, for all $x \in I$,

$$\phi(x) - \Phi(x) \le \int_{x_0}^{\tilde{x}} K(u)\,[\phi(u) - \Phi(u)]\,du$$

$$+ \int_{\tilde{x}}^{x} K(u)[\phi(u) - \Phi(u)]\,du$$

$$\le \theta \sup_{u \in I}\,[\phi(u) - \Phi(u)],$$

implying that $\sup_{x \in I}[\phi(x) - \Phi(x)] \leq 0$. This contradicts the previous conclusion that $\phi(x) - \Phi(x) > 0$ for $x \in I$. ∎

Using this result, we now obtain a relationship between two solutions to the same equation, but with different initial values.

THEOREM 112J.

If y, z each satisfy the differential equation

$$y'(x) = f(y(x))$$

on $I = [x_0, x_1]$, where f satisfies a Lipschitz condition with constant L, then for all $x \in I$,

$$\| y(x) - z(x) \| \leq \| y(x_0) - z(x_0) \| \exp(L(x - x_0)).$$

Proof.

Taking the norm of the difference of

$$y(x) = y(x_0) + \int_{x_0}^{x} f(y(u))\, du$$

and

$$z(x) = z(x_0) + \int_{x_0}^{x} f(z(u))\, du$$

and using the Lipschitz condition, we find

$$\| y(x) - z(x) \| \leq \| y(x_0) - z(x_0) \| + \int_{x_0}^{x} L \| y(u) - z(u) \|\, du$$

so that applying theorem 112I with $C = \| y(x_0) - z(x_0) \|$, $K(x) = L$ and $\phi(x) = \| y(x) - z(x) \|$ we obtain the required result. ∎

The following corollary follows immediately.

COROLLARY 112K.

Let $f: X \to X$ denote a Lipschitz continuous function and $F: X \to C^1(I, X)$ denote the mapping that takes $y_0 \in X$ to y which satisfies the system (112a), (112b). Then F is continuous.

Proof.

That F is continuous as a mapping into $C(I, X)$ is clear. For $C^1(I, X)$ we use the norm given by $\| y \| = \| y(x_0) \| + \sup_{x \in I} \| y'(x) \|$ and the result follows from the Lipschitz condition. ∎

Not only is the function F continuous but it is also differentiable if f satisfies the conditions of the following theorem.

THEOREM 112L.

Suppose that f is continuously differentiable and f' is bounded by a constant L in a closed set $S \subseteq X$ such that all points on $y(x)$, where y satisfies (112a), for $x \in I = [x_0, x_1]$, lie in the interior of S. Then the function F of corollary 112K is differentiable at $y(x_0)$; furthermore, its derivative is given by

$$F'(y(x_0))(x) = u(x),$$

where u is the solution to the matrix-valued differential equation system

$$u'(x) = f'(y(x))u(x), \tag{112i}$$

$$u(x_0) = I. \tag{112j}$$

Proof.

We first remark that the initial value problem given by (112i), (112j) has a unique solution because of theorem 112H, noting that the function mapping (x, v) to $f'(y(x))v$ is continuous in its first variable and satisfies a Lipschitz condition in its second variable and that $\| f'(y(x)) \| \leq L$. Using the same norm as in the proof of corollary 112K, and taking $\delta \in X$ sufficiently small to ensure that $F(y(x_0) + \delta)(x)$ always lies in S, we compute

$$\| F(y(x_0) + \delta) - F(y(x_0)) - u\delta \| = \| y(x_0) + \delta - y(x_0) - I\delta \|$$

$$+ \sup_{x \in I} \| z'(x) - y'(x) - u'(x)\delta \| \tag{112k}$$

where $z = F(y(x_0) + \delta)$. We now estimate

$$\| z'(x) - y'(x) - u'(x)\delta \| = \| f(z(x)) - f(y(x)) - f'(y(x))u(x)\delta \|$$

$$\leq \| f(z(x)) - f(y(x)) - f'(y(x))(z(x) - y(x)) \|$$

$$+ \| f'(y(x))[z(x) - y(x) - u(x)\delta] \|$$

$$= T_1 + T_2, \text{ say.}$$

Because of the boundedness of $\| z(x) - y(x) \| / \| \delta \|$, T_1 can be bounded by $\| \delta \| \varepsilon$ where $\varepsilon \to 0$ as $\| \delta \| \to 0$, whereas

$$T_2 \leq L \int_{x_0}^{x} \| z'(\xi) - y'(\xi) - u'(\xi)\delta \| \, d\xi.$$

Hence, using the Gronwall inequality, taking the supremum of the resulting bound on $\| z'(x) - y'(x) - u'(x)\delta \|$ and noting that the first term on the right-hand side of (112k) is zero, we find that

$$\| F(y(x_0) + \delta) - F(y(x_0)) - u\delta \| \leq \| \delta \| \varepsilon \exp(L(x_1 - x_0)),$$

implying that $F'(y(x_0)) = u$. ∎

113 Linear differential equations.

We take as our standard problem the system of first-order equations

$$y'(x) = A(x)y(x) + \varphi(x), \tag{113a}$$

where A is an $N \times N$ matrix-valued function on $\mathbb{R}$ and φ is an N-dimensional vector-valued function. We will assume throughout that $\| A(x) \|$ is bounded for all x and that A and φ are continuous. We will also consider the corresponding homogeneous equation

$$y'(x) = A(x)y(x). \tag{113b}$$

Linear problems also arise in the form

$$y^{(n)}(x) + c_1(x)y^{(n-1)}(x) + \cdots + c_n(x)y(x) = \psi(x)$$

but, as we have seen in subsection 110, this is equivalent to a first-order system. In fact, for this type of linear problem, the equivalent first-order system is just (113a) with $N = n$ and $A(x)$, $\varphi(x)$ given by

$$A(x) = \begin{bmatrix} 0 & 1 & 0 & \cdots & 0 \\ 0 & 0 & 1 & \cdots & 0 \\ 0 & 0 & 0 & \cdots & 0 \\ \vdots & \vdots & \vdots & & \vdots \\ -c_n(x) & -c_{n-1}(x) & -c_{n-2}(x) & \cdots & -c_1(x) \end{bmatrix}, \quad \varphi(x) = \begin{bmatrix} 0 \\ 0 \\ 0 \\ \vdots \\ \psi(x) \end{bmatrix}.$$

Let S denote the set of all solutions to (113b). If $y, z \in S$ and α, β are scalars, then it is easy to see that $\alpha y + \beta z$ is also in S. Thus, S is a linear space and its dimension is N because of the correspondence between S and values of its members at one particular value of x. Let $\{y_1, y_2, \cdots, y_N\}$ denote some basis for S and consider a matrix-valued function Y whose columns at x are $y_1(x), y_2(x), \ldots, y_N(x)$. Such a function is known as a *fundamental matrix* for the problem (113b).

Since multiplication by $Y(x)$ of a constant vector c is equivalent to the formation of the linear combination

$$c_1 y_1(x) + c_2 y_2(x) + \ldots + c_n y_N(x),$$

we can state, without detailed proofs, the two results.

THEOREM 113A.

If Y is a fundamental matrix for (113b) then y is a solution to (113b) if there is a vector c such that

$$y(x) = Y(x)c.$$

THEOREM 113B.

If Y is a fundamental matrix and M is a non-singular $N \times N$ matrix, then $\widetilde{Y}$, given by

$$\widetilde{Y}(x) = Y(x)M, \tag{113c}$$

is also a fundamental matrix.

Conversely, we also have the following theorem.

THEOREM 113C.

If Y, $\widetilde{Y}$ are fundamental matrices for (113b) then there is a non-singular matrix M such that (113c) holds.

Proof.

Choose the columns of M so that the columns of $\widetilde{Y}$ are the appropriate linear combinations of the columns of Y. The columns of M cannot be linearly dependent, since this would imply a linear dependence on the columns of $\widetilde{Y}$. ■

The columns of Y are not only independent as functions but they are also independent vectors at any point.

THEOREM 113D.

If Y is a fundamental matrix for (113b), then for any x, $Y(x)$ is non-singular.

Proof.

Consider the solution y given by $y(\xi) = Y(\xi)c$, for all ξ. If, for $c \neq 0$, $y(x) = 0$ for some fixed x, then

$$\| y(\xi) \| = \| \int_x^{\xi} A(u)y(u) \, du \|$$

$$\leq L \int_x^{\xi} \| y(u) \| \, du$$

(where L bounds $\| A(u) \|$) and, by theorem 112I, this implies $y(\xi) = 0$ and the linear dependence of the columns of Y. ■

For a given fundamental matrix Y, it is convenient to form the matrix $Y(x)Y(x_0)^{-1}$ where x_0, x are given points. We call this the *Green's matrix* and denote it by $Y(x, x_0)$. According to theorem 113B, $Y(x, x_0)$ regarded as a function of x, with x_0 fixed, is also a fundamental matrix. It also has the property of giving the value of any solution in the form $y(x) = Y(x, x_0)y(x_0)$ in terms of its value at a fixed point x_0. We also have the following result.

THEOREM 113E.

The Green's matrix $Y(x, x_0) = Y(x)Y(x_0)^{-1}$ is independent of the fundamental matrix Y used to define it.

Proof.

Use theorem 113C. ∎

So far we have concerned ourselves only with the homogeneous problem (113b). However, an important use of Green's matrix is in constructing solutions to the related non-homogeneous problem (113a) as in the following result.

THEOREM 113F.

Let Y denote Green's matrix for (113b). Then the solution to the initial value problem

$$y'(x) = A(x)y(x) + \varphi(x),$$
$$y(x_0) = y_0 \tag{113d}$$

is given by

$$y(x) = Y(x, x_0)y_0 + \int_{x_0}^{x} Y(x, z)\varphi(z) \, dz. \tag{113e}$$

Proof.

Clearly (113e) satisfies the initial condition. Since $Y(x, x_0)y_0$ satisfies the homogeneous equation, it is only necessary to show that u defined by

$$u(x) = \int_{x_0}^{x} Y(x, z)\varphi(z) \, dz$$

satisfies (113d). We have

$$u'(x) = \int_{x_0}^{x} \frac{\partial Y}{\partial x}(x, z)\varphi(z) \, dz + Y(x, x)\varphi(x)$$

$$= A(x) \int_{x_0}^{x} Y(x, z)\varphi(z) \, dz + Y(x, x)\varphi(x)$$

$$= A(x)u(x) + \varphi(x). \ \blacksquare$$

Finally, in our study of linear differential equations, we consider the behaviour of the solution for large values of the independent variable. What we will look for is bounds on $\| y(x) \|$ and we will mainly restrict ourselves to the homogeneous problem (113b). If $A(x)$ is independent of x (we will write its constant value simply as A), it is a simple matter to compute the solution in the form

$$y(x) = \exp((x - x_0)A)y(x_0)$$

and the behaviour for large x is determined by the eigenvalues of A with the greatest real part. Specifically, we have the following result.

THEOREM 113G.

If $\mu > \text{Re}(\lambda)$ for all $\lambda \in \sigma(A)$, then given any solution y to (113b), there is a number C such that

$$\| y(x) \| \leq C \exp(\mu(x - x_0)), \qquad \text{for all } x \geq x_0.$$

Proof.

Let $x - x_0 = n + \xi$, where n is an integer and $0 \leq \xi < 1$. Then, identically,

$$y(x) = \exp(\mu(x - x_0))(\exp(-\mu)\exp(A))^n(\exp(-\mu\xi)\exp(\xi A))y(x_0).$$

Since $\exp(-\mu)\exp(A)$ has all eigenvalues in the open unit disc, it is a stable matrix. Furthermore, $\| \exp(-\mu\xi)\exp(\xi A) \|$ is continuous and therefore bounded for $\xi \subseteq [0,1]$. Thus, the result holds with

$$C = \sup_n \| (\exp(-\mu)\exp(A))^n \| \sup_{\xi \in [0,1]} \| \exp(-\mu\xi)\exp(\xi A) \| \| y(x_0) \|. \quad \blacksquare$$

We now seek to generalize this result to problems in which $A(x)$ is not necessarily constant. It is not possible to simply restate theorem 113G with $\text{Re}(\lambda) < \mu$ for λ an eigenvalue of $A(x)$ for any x, since we have the two-dimensional example in which

$$A(x) = \begin{bmatrix} 0 & \alpha \\ \beta & 0 \end{bmatrix}$$

with $\alpha = \sin x$, $\beta = 0$ if $\sin x > 0$ and $\alpha = 0$, $\beta = -\sin x$ otherwise. On the other hand, we can choose μ as a bound on the norm of $A(x)$, but this can lead to unnecessarily crude estimates.

To get an improvement on such an analysis we introduce the so called 'logarithmic norm' of a matrix. Let M be a square matrix; then its *logarithmic norm*, denoted by $\mu(M)$, is defined by

$$\mu(M) = \lim_{\xi \to 0+} \frac{\| I + \xi M \| - 1}{\xi}.$$

Properties of $\mu(M)$ are studied in Dahlquist (1959) but we will note simply that $\mu(M)$ is bounded above by $\| M \|$ and below by the real part of each eigenvalue and that it can take on negative values. We also have the following result.

LEMMA 113H.

If c is a real number and M a square matrix, then

$$\mu(M + cI) = \mu(M) + c.$$

Proof.

We have

$$\frac{\| I + \xi(M + cI) \| - 1}{\xi} = \frac{\| I + \bar{\xi}M \| - 1}{\bar{\xi}} + c,$$

where $\bar{\xi} = \xi/(1 + c\xi)$, and the result follows by taking the limit as $\xi \to 0+$. ■

By the convexity and continuity of the norm as a real-valued function on a vector space, we can see that it possesses a directional (Gâteaux) derivative

$$\lim_{\xi \to 0+} (\| u + \xi v \| - \| u \|)/\xi$$

and by taking the composition of y with a norm, it follows that

$$\delta = \lim_{\xi \to 0+} [\| y(x + \xi) \| - \| y(x) \|]/\xi$$

exists if y is differentiable. On the assumption that $y'(x) = A(x)y(x)$ for A continuous, we can estimate δ as follows:

$$\| y(x + \xi) \| - \| y(x) \| = \| y(x) + \xi A(x)y(x) + \xi C(\xi) \| - \| y(x) \|,$$

where $\| C(\xi) \| \to 0$ as $\xi \to 0+$, and hence

$$\| y(x + \xi) \| - \| y(x) \| \le \| [I + \xi A(x)]y(x) \| - \| y(x) \| + \xi \| C(\xi) \|$$

$$\le [\| I + \xi A(x) \| - 1] \| y(x) \| + \xi \| C(\xi) \|$$

$$= \xi\mu(A(x))\| y(x) \| + \xi D(\xi) + \xi \| C(\xi) \|,$$

where $D(\xi) \to 0$ as $\xi \to 0+$. Hence, taking the limit, we find

$$\delta \le \mu(A(x)) \| y(x) \|.$$

If $\mu(A(x))$ is always non-positive, then this tells us that $\| y(x) \|$ cannot be an increasing function. Accordingly we can state the following lemma.

LEMMA 113I.

If A is continuous and $\mu[A(x)] \le 0$, then y satisfying (113b) is such that for all $x > x_0$,

$$\| y(x) \| \le \| y(x_0) \|.$$

We can generalize this bound as follows.

THEOREM 113J.

If A is continuous and $\mu(A(x)) \le m$ then y satisfying (113b) is such that, for all $x > x_0$,

$$\| y(x) \| \le \exp(m(x - x_0)) \| y(x_0) \|.$$

Proof.

Let $z(x) = y(x)\exp(-mx)$, so that $z'(x) = [A(x) - mI]z(x)$ and, by lemma 113H, $\mu(A(x) - mI) = \mu(A(x)) - m$. ∎

114 Stability of differential equations.

From the work of the previous subsection, we have a criterion, in the case of homogeneous linear systems, that solutions necessarily converge to the zero solution, or at least have bounded values. We now extend this to the study of the stability of non-linear autonomous systems. In the first place, we identify points in the solution space to which solutions have any possibility of converging.

DEFINITION 114A.

The point $\bar{y}$ is a *critical point* for the equation $y'(x) = f(y(x))$ if $f(\bar{y}) = 0$.

Clearly, if the initial value is $\bar{y}$, then the solution is given by $y(x) = \bar{y}$ for all x. What we will consider, however, is circumstances under which an initial value *near* $\bar{y}$ will yield a solution which remains near $\bar{y}$. We distinguish two levels of stability in the behaviour of a solution near a critical point.

DEFINITION 114B.

The critical point $\bar{y}$ for the equation $y'(x) = f(y(x))$ is said to be *stable* if, given any positive number ε, there is a number δ such that, for a solution y satisfying $\| y(x_0) - \bar{y} \| \le \delta$, it holds that $\| y(x) - \bar{y} \| \le \varepsilon$ for all $x > x_0$.

DEFINITION 114C.

The critical point $\bar{y}$ for the equation $y'(x) = f(y(x))$ is said to be *asymptotically stable* if there exists a number δ_0 such that, for a solution satisyfing $\| y(x_0) - \bar{y} \| \le \delta_0$, it holds that $y(x) \to \bar{y}$ as $x \to \infty$.

The method we will employ in analysing these concepts is known as Lyapunov's direct method and, when it is applicable, it provides sufficient conditions for a critical point to have one or the other of these properties. The basic idea is to devise a function, often referred to as a Lyapunov function, which behaves near a critical point like a 'distance' from the critical point and is such that its value cannot increase as the solution develops. The sort of function we want is partly specified in the following.

DEFINITION 114D.

Let D be an open set containing $\bar{y}$. Then $V: D \to \mathbb{R}$ is said to be *positive definite* for $\bar{y}$ and D if $V(\bar{y}) = 0$ and $V(z) > 0$ for all $z \in D \setminus \{\bar{y}\}$.

If V is also differentiable, we can form the function W defined by

$$W(z) = - V'(z)(f(z)), \tag{114a}$$

making use of the differential equation under scrutiny. With this machinery we are in a position to consider two of Lyapunov's theorems.

THEOREM 114E.

If there exists a mapping V such that V is positive definite in the neighbourhood D of a critical point $\bar{y}$ for the differential equation $y'(x) = f(y(x))$ and V is continuously differentiable with W defined by (114a) non-negative on D, then $\bar{y}$ is stable.

Proof.

Let S_0 denote a closed sphere with centre $\bar{y}$ lying entirely in D and with radius no greater than ε and let α be the infimum of $V(z)$ for z on the boundary of S_0. Let D_0 be the subset of S_0 containing a point z iff $V(z) \le \alpha$ and let S_1 be a closed sphere with centre $\bar{y}$ lying entirely within D_0. If δ is the radius of S_1, we will show that $\| y(x_0) - \bar{y} \| \le \delta$ implies $\| y(x) - \bar{y} \| \le \varepsilon$ for all $x > x_0$.

In fact, we have $V(y(x_0)) \le \alpha$ and $(V \circ y)'(x) = - W(y(x)) \le 0$ so that $V(y(x)) \le \alpha$ for all $x > x_0$ and $y(x)$ always lies within D_0 and hence S_0. ∎

THEOREM 114F.

If, in addition to the assumptions of theorem 114E, W is positive definite, then $\bar{y}$ is asymptotically stable.

Proof.

Choose δ_0 as described for the choice of δ for any $\varepsilon > 0$ in the proof of theorem 114E. With $\| y(x_0) - \bar{y} \| \le \delta_0$, we have seen that $V(y(x))$ cannot increase and we construct a sequence $x_0, x_1, x_2, \ldots$ such that $V(y(x_n)) \le \frac{1}{2} V(y(x_{n-1}))$. It will thus follow that $y(x_n) \to \bar{y}$. Let $\beta > 0$ be the infimum of $W(z)$ for all z such that $\frac{1}{2} V(y(x_{n-1})) \le V(z) \le V(y(x_{n-1}))$ and choose

$x_n = x_{n-1} + V(y(x_{n-1}))/2\beta$. By the mean value theorem,

$$V(y(x_{n-1})) - V(y(x_n)) > \beta(x_n - x_{n-1})$$
$$= \tfrac{1}{2}V(y(x_{n-1})),$$

implying that $V(y(x_n)) \le \tfrac{1}{2}V(y(x_{n-1}))$. ■

12 DIFFERENCE EQUATIONS

120 Introduction to difference equations.

In much the same way as a relationship between a function and its first, and possibly higher, derivatives is known as a differential equation, a relationship between a function and its first, and possibly higher, differences is known as a *difference equation*. Since we will always assume the existence of a natural mesh size h and define, for a given function y, such quantities as $\Delta y(x)$, $\Delta^2 y(x)$, ... evaluated with this spacing only at mesh points, it is appropriate to represent y as a sequence $(y_0, y_1, y_2, \ldots)$ where y_n denotes $y(x_0 + nh)$ for x_0 an initial tabular point. In this notation, we would write, for example, $Ey_n = y_{n+1}$, $\Delta y_n = y_{n+1} - y_n$.

If a difference equation is a relation between y_n, Δy_n, ..., $\Delta^k y_n$ (for some k), then it could equally well be regarded as a relation between y_n, $y_{n+1}, \ldots, y_{n+k}$, or, replacing n by $n - k$, as a relation between $y_{n-k}, y_{n-k+1}, \ldots, y_n$. In the equations we deal with, we will always assume that this last type of formulation can be rearranged in the form

$$y_n = \phi(n, y_{n-1}, \ldots, y_{n-k}) \tag{120a}$$

and we will take this as our standard problem. The integer k is known as the *order* of the difference equation. We refer to such an equation as a *kth order* difference equation.

Usually, we will restrict n to being non-negative and we will specify a particular solution by imposing values on $y_0, y_1, \ldots, y_{k-1}$. These *initial values* together with (120a) which holds for $n \ge k$ constitute the complete posing of a difference equation problem.

By induction on n, we can easily verify the following result which we state without explicit proof.

THEOREM 120A.

If $\phi : \mathbb{Z}^+ \times X^k \to X$ is given for any set X, and $y_0, y_1, \ldots, y_{k-1} \in X$ are also given, then there exists a unique solution $y : \mathbb{Z}^+ \to X$ satisfying (120a) for $n = k, k+1, \ldots$.

Usually X will be a vector space and, in particular, is often just $\mathbb{R}$. We now present some examples of difference equations.

In the first example, $X = \mathbb{R}$, $k = 2$, $y_0 = 0$, $y_1 = 1$, $\phi(n, y_{n-1}, y_{n-2}) = y_{n-1} + y_{n-2}$. This is the famous Fibonacci equation with solution

$$y_n = \frac{1}{\sqrt{5}} \left[\left(\frac{1 + \sqrt{5}}{2} \right)^n - \left(\frac{1 - \sqrt{5}}{2} \right)^n \right].$$

It is easy to see that y_n behaves asymptotically like a power of $(1 + \sqrt{5})/2$ in the sense that $\lim_{n \to \infty} (y_n)^{1/n} = (1 + \sqrt{5})/2$ and that $\lim_{n \to \infty} (y_n/y_{n-1}) = (1 + \sqrt{5})/2$.

In contrast to this second-order difference equation over $\mathbb{R}$, we now consider a first-order equation over $\mathbb{R}^2$. Denoting the solution sequence by z, the equation is

$$z_n = \begin{bmatrix} 0 & 1 \\ 1 & 1 \end{bmatrix} z_{n-1}, \tag{120b}$$

with initial vector $z_0 = [0, 1]^T$. It is interesting that this superficially different equation has a closely related solution to that of the first example and it is easy to verify that $z_n = [y_n, y_{n+1}]^T$ for all n, where $(y_0, y_1, \ldots)$ is the Fibonacci sequence. By repeated squaring of the matrix which occurs in (120b), distant members of the Fibonacci sequence can be calculated much more readily than by using the Fibonacci difference equation itself.

We next consider the difference equation $\varepsilon_n = \varepsilon_{n-1}\varepsilon_{n-2}$ which represents the approximate behaviour of errors, measured in units natural to the particular problem, of the sequence of approximations to the root of an algebraic equation as computed by the secant method (see for example, Householder, [1970]). If we take logarithms, $y_n = \ln |\varepsilon_n|$, then y_n obeys the Fibonacci difference equation. A consequence of this is for large n we have, approximately, $|\varepsilon_n| = |\varepsilon_{n-1}|^{(1 + \sqrt{5})/2}$, giving the order of convergence of $(1 + \sqrt{5})/2$ for the secant method.

The next example is of first order over $\mathbb{R}^2$. From an initial vector of positive numbers, an arbitrary member of the sequence is computed as $y_n = \phi(y_{n-1})$, where

$$\phi\left(\begin{bmatrix} u \\ v \end{bmatrix} \right) = \begin{bmatrix} (u + v)/2 \\ (uv)^{1/2} \end{bmatrix}.$$

The solution to this problem is the sequence of terms converging to the arithmetic–geometric mean of the components of the initial vector.

Our final example suposes that a real number h and a function $f : X \to X$ are given. If initial values y_0, y_1 are specified then we consider the sequence defined by

$$y_n = y_{n-2} + 2hf(y_{n-1}) \tag{120c}$$

for all $n = 2, 3, \ldots$. As we shall see later the sequence computed from (120c) can, in many cases, be used to obtain a sequence of approximations y_n to $y(x_0 + nh)$ satisfying a differential equation $y'(x) = f(y(x))$ where $y_0 = y(x_0)$ and $y_1 = y(x_0 + h)$ are known.

121 Linear difference equations with constant coefficients.

We will restrict ourselves to the situation where $X = \mathbb{R}$ and where the difference equation takes the form

$$c_0 y_n + c_1 y_{n-1} + \cdots + c_k y_{n-k} = a_n, \tag{121a}$$

with $c_0, c_1, \ldots, c_k$ constant numbers $(c_0 \neq 0)$ and $(a_k, a_{k+1}, \ldots)$ a given sequence. An equation of this type, in which $c_0, c_1, \ldots, c_k$ were allowed to depend on n would be called *linear*, but since they are constant, it is called *linear with constant coefficients*. If $a_k = a_{k+1} = \cdots = 0$, then the equation is said to be *homogeneous* and, moreover, the homogeneous equation

$$c_0 y_n + c_1 y_{n-1} + \cdots + c_k y_{n-k} = 0 \tag{121b}$$

is known as the *homogeneous part of* (121a).

THEOREM 121A.

If u is a solution to (121a) and $v^{(1)}, v^{(2)}, \ldots$ are solutions to (121b), then for any constants $\alpha_1, \alpha_2, \ldots,$

$$y = u + \alpha_1 v^{(1)} + \alpha_2 v^{(2)} + \cdots \tag{121c}$$

is a solution to (121a).

Proof.

If (121c) is substituted into (121a), the result is

$$(c_0 u_n + c_1 u_{n-1} + \cdots + c_k u_{n-k}) + \alpha_1 (c_0 v_n^{(1)} + c_1 v_{n-1}^{(1)} + \cdots + c_k v_{n-k}^{(1)})$$
$$+ \alpha_2 (c_0 v_n^{(2)} + c_1 v_{n-1}^{(2)} + \cdots + c_k v_{n-k}^{(2)})$$
$$+ \cdots$$

which equals a_n, contributed entirely from the first term. ∎

The way this result is used in practice is to construct the general solution making use of an appropriate collection $v^{(1)}, v^{(2)}, \ldots, v^{(m)}$ of solutions to (121b) and then to allocate the values to $\alpha_1, \alpha_2, \ldots, \alpha_m$ in such a way that $y_0, y_1, \ldots, y_{k-1}$ agree with the given initial values. Assuming that $v^{(1)}, v^{(2)}, \ldots, v^{(m)}$ are linearly independent we must in general choose $m = k$ since if $m < k$ then the vector space spanned by $\{[v_0^{(1)}, v_1^{(1)}, \ldots, v_{k-1}^{(1)}], [v_0^{(2)}, v_1^{(2)}, \ldots, v_{k-1}^{(2)}], \ldots, [v_0^{(m)}, v_1^{(m)}, \ldots, v_{k-1}^{(m)}]\}$ would not include all possible choices of $[y_0, y_1, \ldots, y_{k-1}]$. On the other hand, we cannot possibly have $m > k$ since there would then be constants $c_1, c_2, \ldots, c_m$ not all zero, such that $y_j = \sum_i c_i v_j^{(i)} = 0$ for $j = 0, 1, \ldots, k-1$ and hence, by induction, for $j = k, k+1, \ldots$ because y satisfies (121b). Thus y would be the zero sequence and the requirement that $v^{(1)}, v^{(2)}, \ldots, v^{(m)}$ be linearly independent would not be satisfied.

The approach we have described depends for its success on identifying the solution u which, of course, depends on the sequence $(a_k, a_{k+1}, a_{k+2}, \ldots)$. For the moment we avoid this aspect of the construction of y and turn instead to an investigation of the nature of $v^{(1)}, v^{(2)}, \ldots, v^{(k)}$. Associated with (121b), we have the polynomial p given by

$$p(z) = c_0 z^k + c_1 z^{k-1} + \cdots + c_k, \qquad (121\mathrm{d})$$

which can be factorized as

$$p(z) = c_0 (z - z_1)^{k_1} (z - z_2)^{k_2} \ldots (z - z_m)^{k_m}, \qquad (121\mathrm{e})$$

where $k_1 + k_2 + \cdots + k_m = k$ and $z_1, z_2, \ldots, z_m$ are the distinct zeros of p. We shall refer to p as the polynomial associated with the difference equation (121b) and to (121b) as the difference equation associated with p.

As a step towards understanding the relationship between the difference equation associated with p and the difference equations associated with the factors of p, we suppose that p is the product of two factors q and r. We then have the following two results.

LEMMA 121B.

If $p = qr$ and y is a solution to the difference equations associated with q, then y is a solution to the difference equation associated with p.

Proof.

If $r_m(z) = z^m$ for m a non-negative integer and y satisfies the difference equation associated with q, then y satisfies the difference equation associated with $q r_m$. The result of the lemma now follows by writing r as a linear combination of $r_0, r_1, \ldots$ and noting that if y satisfies the difference equation associated with each of a number of polynomials, then it satisfies the difference equation associated with a linear combination of these polynomials. ∎

LEMMA 121C.

If q and r are relatively prime polynomials and y is a solution to the difference equations associated with each of them, then y is the zero sequence.

Proof.

Let polynomials s, t be chosen so that $p = sq + tr$ is the polynomial such that $p(z) = 1$. Then y satisfies the difference equation associated with p and therefore $y_n = 0$ for all n. ∎

Combining the results of lemmas 121B and 121C we immediately find the following.

THEOREM 121D.

If $p = qr$ where the polynomials q and r are relatively prime, then any solution y to the difference equation associated with p can be written uniquely in the form $u + v$, where u satisfies the difference equation associated with q and v satisfies the difference equation associated with r.

We see from this result that in studying the solution of the difference equation associated with p given by (121e) it is only necessary to study the difference equations associated with its relatively prime factors. This study is, however, summed up by following theorem and its corollary.

THEOREM 121E.

If $p(z) = (z - z_1)^{k_1}$ then for $l = 0, 1, \ldots, k_1 - 1$, each of the sequences v given by

$$
v_n = \begin{cases}
0, & n < l, \\
1, & n = l, \\
\binom{n}{l} z_1^{n-l}, & n > l.
\end{cases}
\tag{121f}
$$

is a solution to the difference equation associated with p.

Proof.

We compute

$$
\sum_{i=0}^{k_1} \binom{k_1}{i} z_1^i (-1)^i v_{n-i} = \sum_{i=0}^{k_1} (-1)^i \binom{n-i}{l} \binom{k_1}{i} z_1^{n-l},
$$

which is seen to be equal to the lth derivative evaluted at z_1 of the polynomial q given by $q(z) = z^{n-k_1}(z - z_1)^{k_1}/l!$. Hence, since $q(z)$ contains a factor $(z - z_1)^{k_1}$, $q^{(l)}(z)$ contains a factor $(z - z_1)^{k_1-l}$ and hence $q^{(l)}(z_1) = 0$. ∎

COROLLARY 121F.

If $p(z) = (z - z_1)^{k_1}$, the general solution to the difference equation associated with p is given by

$$
y = \alpha_0 v^{(0)} + \alpha_1 v^{(1)} + \cdots + \alpha_{k_1-1} v^{(k_1-1)},
\tag{121g}
$$

where for $l = 0, 1, \ldots, k_1 - 1$, $v^{(l)}$ is given by v in (121f).

Proof.

Each of the terms in (121g) is a solution and $\alpha_0, \alpha_1, \ldots, \alpha_{k_1-1}$ can be chosen uniquely so that y_n takes on any specified set of initial values $y_0, y_1, \ldots, y_{k_1-1}$. ∎

Finally, in our study of homogeneous linear difference equations with constant coefficients we combine the results of theorem 121D and corollary 121F.

THEOREM 121G.

If p is given by (121e) where $z_1, z_2, \ldots, z_k$ are distinct, then the general solution of the difference equation associated with p is given by

$$y = \sum_{i=1}^{m} \sum_{j=0}^{k_i-1} \alpha_{ij} v^{(i,j)}, \tag{121h}$$

where $v^{(i,j)}$ is given by

$$v_n^{(i,j)} = \begin{cases} 0, & n < j, \\ 1, & n = j, \\ \binom{n}{j} z_i^{n-j}, & n > j. \end{cases}$$

Among the set of solutions given by theorem 121G, we can evidently select values for the α_{ij} such that the solution takes on the value 1 when $n = k - 1$ and 0 when $n = 0, 1, 2, \ldots, k - 2$. If this solution is $\tilde{y}$, then we define

$$Y_n = \begin{cases} 0, & n = -1, -2, -3, \ldots, \\ \tilde{y}_{n+k-1}, & n = 0, 1, 2, 3, \ldots, . \end{cases} \tag{121i}$$

Using this function we can construct a solution to a non-homogeneous linear difference equation with constant coefficients. In fact the following theorem, together with theorems 121A and 121G completes the task of finding the general solution to this sort of equation.

THEOREM 121H.

The difference equation (121a) has a particular solution given by y where

$$y_n = \begin{cases} 0, & n < k, \\ \dfrac{1}{c_0} \sum_{i=k}^{n} Y_{n-i} a_i, & n \geq k, \end{cases} \tag{121j}$$

where Y is given by (121i).

Proof.

We first note that $c_0 Y_m + c_1 Y_{m-1} + \cdots + c_k Y_{m-k}$ is equal to zero if m is positive (because of the definition of Y from the solution to the homogeneous difference equation) or negative (because $Y_{-1} = Y_{-2} = \cdots = 0$) but is equal to

c_0 if $m = 0$. Using this result, we compute $c_0 y_n + c_1 y_{n-1} + \cdots + c_k y_{n-k}$ where y is defined by (121j). The result is

$$\frac{1}{c_0} \sum_{i=0}^{k} c_i \sum_{j=k}^{n-i} Y_{n-i-j} a_j = \frac{1}{c_0} \sum_{i=0}^{k} c_i \sum_{j=k}^{n} Y_{n-i-j} a_j$$

$$= \frac{1}{c_0} \sum_{j=k}^{n} \left(\sum_{i=0}^{k} c_i Y_{n-i-j} \right) a_j$$

$$= \frac{1}{c_0} \sum_{j=k}^{n} c_0 \delta_{0, n-j} a_j$$

$$= a_n. \quad \blacksquare$$

122 Matrix difference equations.

We consider first-order difference equations of the form

$$y_n = A y_{n-1} + v_n, \qquad n = 1, 2, \ldots, \tag{122a}$$

where $y_0, y_1, \ldots$ is a sequence of vectors in $\mathbb{R}^N$ and A is a matrix. It is supposed that the vectors $y_0, v_1, v_2, \ldots$ are given.

THEOREM 122A.

The solution to (122a) is

$$y_n = A^n y_0 + A^{n-1} v_1 + A^{n-2} v_2 + \cdots + v_n. \tag{122b}$$

Proof.

Substitute y_n and y_{n-1} into (122a). $\quad \blacksquare$

The great simplicity of this result compared with the corresponding developments of subsection 121 leads us to widen its applicability. In fact we will see that all our conclusions concerning kth-order difference equations can be seen as consequences of theorem 122A.

Given a sequence of real numbers, $z_0, z_1, \ldots$, we can write down a related sequence $y_0, y_1, \ldots$ of points in $\mathbb{R}^k$ as follows:

$$y_0 = \begin{bmatrix} z_0 \\ z_1 \\ \vdots \\ z_{k-1} \end{bmatrix}, \qquad y_1 = \begin{bmatrix} z_1 \\ z_2 \\ \vdots \\ z_k \end{bmatrix}, \qquad \ldots. \tag{122c}$$

We have a relationship between difference equations satisfied by these two types of sequences.

THEOREM 122B.

If $z_0, z_1, \ldots$ satisfy the linear difference equation with constant coefficients

$$c_0 z_n + c_1 z_{n-1} + \cdots + c_k z_{n-k} = a_n, \tag{122d}$$

then $y_0, y_1, \ldots$ given by (122c) satisfy the equation

$$y_n = A y_{n-1} + v_n, \tag{122e}$$

where

$$A = \begin{bmatrix} 0 & 1 & 0 & \cdots & 0 & 0 \\ 0 & 0 & 1 & \cdots & 0 & 0 \\ 0 & 0 & 0 & \cdots & 0 & 0 \\ \vdots & \vdots & \vdots & & \vdots & \vdots \\ 0 & 0 & 0 & \cdots & 0 & 1 \\ -\dfrac{c_k}{c_0} & -\dfrac{c_{k-1}}{c_0} & -\dfrac{c_{k-2}}{c_0} & \cdots & -\dfrac{c_2}{c_0} & -\dfrac{c_1}{c_0} \end{bmatrix}, \quad v_n = \begin{bmatrix} 0 \\ 0 \\ 0 \\ \vdots \\ 0 \\ \dfrac{a_{n+k-1}}{c_0} \end{bmatrix}.$$

Proof.

Write out the components of (122e) making use of (122c). Each of the resulting k equations except the last is of the form $z_{n+i} = z_{n+i}$, $i = 0, 1, 2, \ldots, k - 2$, and the last is equivalent to (122d). ∎

We conclude this subsection by generalizing the solution to (122a) given in theorem 122A to the case where A depends on n.

THEOREM 122C.

Given matrices $A_1, A_2, \ldots$ and vectors $y_0, v_1, v_2, \ldots$, the solution to the equation

$$y_n = A_n y_{n-1} + v_n \tag{122f}$$

is

$$y_n = A_{n1} y_0 + A_{n2} v_1 + \cdots + A_{nn} v_{n-1} + v_n, \tag{122g}$$

where, for integers m, n satisfying $1 \le m \le n$,

$$A_{nm} = A_n A_{n-1} \cdots A_m.$$

Proof.

Substitute y_n, y_{n-1} from (122g) into (122f). ∎

123 Stability of solutions to difference equations.

We investigate the circumstances under which the solution to a linear difference equation with constant coefficients is guaranteed to be bounded and, in particular, when it converges to zero. We will consider equations only of the form

$$y_n = Ay_{n-1} + v_n, \tag{123a}$$

since we have seen that the kth order equations studied in subsection 121 can be reduced to this form.

DEFINITION 123A.

The difference equation (123a) is *convergent* if for any choice of y_0, $\| y_n \| \to 0$ as $n \to \infty$.

DEFINITION 123B.

The difference equation (123a) is *stable* if for any choice of y_0, $\| y_n \|$ is bounded as $n \to \infty$.

For homogeneous equations, criteria for these properties can be found from theorems 105B and 105E since we have the following result.

THEOREM 123C.

The linear difference equation

$$y_n = Ay_{n-1}$$

is convergent iff A is convergent and stable iff A is stable.

Proof.

(If). Since $y_n = A^n y_0$, $\| y_n \| \le \| A^n \| \, \| y_0 \|$ which converges to zero (respectively is bounded) if A is convergent (respectively is stable).
(Only if). If A^n is not convergent (respectively not stable) there must exist some i, j such that the (i, j) element of A^n is not convergent to zero (respectively not bounded). If y_0 is chosen as the basis vector e_j, it follows that y_n has its ith component not convergent to zero (respectively not bounded). ∎

The question of convergence is easily extended to non-homogeneous equations.

THEOREM 123D.

The linear difference equation

$$y_n = A y_{n-1} + v_n$$

is convergent iff A is convergent and the sequence v converges to zero.

Proof.

(Only if). By taking the difference of two solutions we immediately see the necessity of A being convergent. Furthermore, $\| v_n \| \le \| y_n \| + \| A \| \, \| y_{n-1} \|$ which converges to zero.

(If). From theorem 105B, we see that A is similar to a matrix $S^{-1}AS$ such that $1 > \| S^{-1}AS \| = k$, say. Hence, we can write (122b) in the form

$$y_n = S[(S^{-1}AS)^n(S^{-1}y_0) + (S^{-1}AS)^{n-1}(S^{-1}v_1) + \cdots + S^{-1}v_n]$$

so that

$$\| y_n \| \le \| S \| \, \| S^{-1} \| (k^n \| y_0 \| + k^{n-1} \| v_1 \| + \cdots + \| v_n \|)$$

Since $0 \le k < 1$, the infinite matrix

$$\frac{1}{1-k}\begin{bmatrix} 1 & 0 & 0 & \cdots & 0 & \cdots \\ k & 1 & 0 & \cdots & 0 & \cdots \\ k^2 & k & 1 & \cdots & 0 & \cdots \\ \vdots & \vdots & \vdots & \ddots & \vdots & \\ k^n & k^{n-1} & k^{n-2} & \cdots & 1 & \cdots \\ \vdots & \vdots & \vdots & & \vdots & \end{bmatrix}$$

is easily seen to satisfy the Toeplitz conditions (see, for example, Hardy [1949]) for transforming the convergent sequence

$$((1-k) \| S \| \, \| S^{-1} \| \, \| y_0 \|, (1-k) \| S \| \, \| S^{-1} \| \, \| v_1 \|, \ldots)$$

into another sequence, $(\| y_0 \|, \| y_1 \|, \ldots)$, with identical limit. ∎

The conditions for stability of the equation (123a) are rather more involved. We content ourselves by stating without its easy proof a necessary set of conditions and giving an example to show that these conditions are not sufficient.

THEOREM 123E.

The difference equation

$$y_n = A y_{n-1} + v_n$$

is stable only if A is stable and the sequence v is bounded.

That these conditions are not sufficient is easily seen by a one-dimensional example with $A = \lambda$ and $v_n = \mu^n$. If $|\lambda| = |\mu| = 1$ the conditions are satisfied but if, in addition, $\lambda = \mu$, we see that the solution $y_n = n\lambda^n$ is unstable.

13 NUMERICAL APPROXIMATION

130 Introduction to numerical approximation.

In this section, we will consider the question of approximating a function, or some quantity that depends on this function, in terms of various pieces of data given concerning it. For example, if the value of the function f is given at a number of distinct points, say $x_1, x_2, \ldots, x_n$, then we may wish to have some way of approximating $f(x)$ in terms of $f(x_1), f(x_2), \ldots, f(x_n)$. We may also wish to approximate such quantities as $f'(x)$ or $\int_a^b f(x)\, dx$ (for given numbers a and b) in terms of these same values.

The traditional tool for carrying out this work is polynomial interpolation. That is, we approximate f by the polynomial p of lowest possible degree (in general, $n - 1$) which agrees in value with f at $x_1, x_2, \ldots, x_n$, and then simply approximate the required property of f by the corresponding property of p. The rationale for this procedure is that the result will be exact whenever f happens to be a polynomial of sufficiently low degree (because p will then be the same as f) and, according to the Weierstrass approximation theorem, all continuous functions can be approximated, at least on a bounded closed interval, by polynomials to any required accuracy.

For many approximation problems, this approach is not satisfactory and alternative types of approximating functions are often used. For example, rational functions, spline approximations and exponential approximations are often found to be suitable alternatives.

131 The interpolational polynomial.

Given a number of distinct points $x_1, x_2, \ldots, x_n$ and a function f whose domain contains each of these points, we will use the name *interpolational polynomial* for that polynomial p of degree less than n such that $p(x_i) = f(x_i)$, $i = 1, 2, \ldots, n$. As a preliminary to proving the existence and uniqueness of this polynomial, we introduce two particular bases for the n-dimensional linear space of polynomials of degree less than n.

DEFINITION 131A.

Let $e_1, e_2, \ldots, e_n$ be defined by

$$e_i(x) = \prod_{j \neq i} \frac{x - x_j}{x_i - x_j}, \qquad i = 1, 2, \ldots, n.$$

Then $\{e_1, e_2, \ldots, e_n\}$ is known as the *Lagrange basis*.

DEFINITION 131B.

Let $e_0, e_1, \ldots, e_{n-1}$ be defined by

$$e_0(x) = 1, \ e_i(x) = (x - x_i)e_{i-1}(x), \qquad i = 1, 2, \ldots, n - 1.$$

Then $\{e_0, e_1, \ldots, e_{n-1}\}$ is known as the *Newton basis*.

THEOREM 131C.

Each of the Lagrange and Newton bases is a basis for the space of polynomials of degree less than n.

Proof.

In the case of the Lagrange basis, it is sufficient to prove that there do not exist constants $c_1, c_2, \ldots, c_n$, not all zero, such that $c_1 e_1 + c_2 e_2 + \cdots + c_n e_n = 0$. Evaluating this inequality at x_i and noting that $e_j(x_i) = \delta_{ij}$, we see that $c_i = 0$. Since this holds for each $i = 1, 2, \ldots, n$ we have a contradiction.

In the case of the Newton basis, if there were constants $c_0, c_1, \ldots, c_{n-1}$ such that $c_0 e_0 + c_1 e_1 + \cdots + c_{n-1} e_{n-1} = 0$ then by evaluating at $x_1, x_2, \ldots, x_n$ in turn, we see that $c_0, c_1, \ldots, c_{n-1}$ are all zero. ∎

Using these bases, we can prove the key result concerning polynomial interpolation.

THEOREM 131D.

There exists a unique polynomial p of degree less than n such that $p(x_i) = f(x_i)$, $i = 1, 2, \ldots, n$.

Proof.

We give two alternative proofs using each of the Lagrange and Newton bases. If $\{e_1, e_2, \ldots, e_n\}$ is the Lagrange basis, then the polynomial

$$f(x_1)e_1 + f(x_2)e_2 + \cdots + f(x_n)e_n$$

does satisfy the requirements of the interpolational polynomial. If there were a second such polynomial, the difference of these two polynomials would be of degree less than n and yet vanish at each of the n distinct points $x_1, x_2, \ldots, x_n$. Thus this difference would be the zero polynomial.

If $\{e_0, e_1, \ldots, e_{n-1}\}$ is the Newton basis, then we will show by induction on $N = 1, 2, \ldots, n$ that there is a unique interpolational polynomial for the set of points $\{x_1, x_2, \ldots, x_N\}$. In the case $N = 1$, there is clearly a unique polynomial of degree 0, given by $f(x_1)e_0$. Now assuming the result for the interpolational polynomial through the points $\{x_1, x_2, \ldots, x_{N-1}\}$, where $N > 1$, we write this polynomial as

$$p_{N-1} = c_0 e_0 + c_1 e_1 + \cdots + c_{N-2} e_{N-2}.$$

If we now consider the polynomial of degree less than N given by $p = p_{N-1} + c_{N-1} e_{N-1}$, we see that this also agrees with f at $x_1, x_2, \ldots, x_{N-1}$ and there is a unique choice of c_{N-1} such that $p(x_N) = f(x_N)$. ∎

From the construction of the interpolational polynomial in terms of the Newton basis, we see that the coefficients c_0 (respectively $c_1, c_2, \ldots$) depend on x_1 (respectively on x_1 and x_2, on $x_1, x_2, x_3, \ldots$) but not on later-numbered x values.

DEFINITION 131E.

The coefficient of e_i in the expansion of the interpolational polynomial in terms of the Newton basis is written as $f(x_1, x_2, \ldots, x_{i+1})$ and known as an ith-order *divided difference*.

THEOREM 131F.

The divided difference $f(x_1, x_2, \ldots, x_n)$ is a symmetric function in $x_1, x_2, \ldots, x_n$.

Proof.

The value of $f(x_1, x_2, \ldots, x_n)$ is the coefficient of x^{n-1} in the unique interpolational polynomial through $x_1, x_2, \ldots, x_n$. ∎

To compute divided differences in practice it is convenient to make use of the recursion given in the following result.

THEOREM 131G.

The nth−order divided difference is given by

$$f(x_1, x_2, \ldots, x_n, x_{n+1}) = \frac{f(x_2, \ldots, x_n, x_{n+1}) - f(x_1, x_2, \ldots, x_n)}{x_{n+1} - x_1} \qquad (131a)$$

Proof.

Let p_{n-2} (respectively p_n) be the unique interpolational polynomial for f at

$x_2, x_3, \ldots, x_n$ (respectively $x_1, x_2, \ldots, x_n, x_{n+1}$). We then have

$$p_n(x) = p_{n-2}(x) + f(x_2, \ldots, x_n, x_{n+1})(x - x_2) \cdots (x - x_n)$$
$$+ f(x_1, x_2, \ldots, x_n, x_{n+1})(x - x_2) \cdots (x - x_n)(x - x_{n+1}) \quad (131b)$$

and

$$p_n(x) = p_{n-2}(x) + f(x_1, x_2, \ldots, x_n)(x - x_2) \cdots (x - x_n)$$
$$+ f(x_1, x_2, \ldots, x_n, x_{n+1})(x - x_1)(x - x_2) \cdots (x - x_n), \quad (131c)$$

so that subtracting (131c) from (131b) and dividing by $(x - x_2) \cdots (x - x_n)$ gives the required result. ∎

By a divided-difference table we mean a configuration as follows:

$f(x_1)$

    $f(x_1, x_2)$

$f(x_2)$ $f(x_1, x_2, x_3)$

    $f(x_2, x_3)$ $f(x_1, x_2, x_3, x_4)$

$f(x_3)$ $f(x_2, x_3, x_4)$

    $f(x_3, x_4)$

                              $f(x_1, x_2, \ldots, x_n)$

    $f(x_{n-2}, x_{n-1})$ $f(x_{n-3}, x_{n-2}, x_{n-1}, x_n)$

$f(x_{n-1})$ $f(x_{n-2}, x_{n-1}, x_n)$

    $f(x_{n-1}, x_n)$

$f(x_n)$

in which the first column (the zeroth-order divided difference) is a table of f itself and numbers in later columns are found from the closest members of the immediately previous columns using (131a).

If the entries in the original table are computed from equally spaced arguments so that

$$x_2 - x_1 = x_3 - x_2 = x_4 - x_3 = \cdots = = h, \text{ say,}$$

then it is customary to form differences as on the right-hand side of (131a) without the corresponding division. Since the factor omitted in forming this nth-order divided difference is $1/(nh)$ these simple differences of order n are simply $n!h^n$ times the corresponding divided differences.

In this equal-interval case, it is customary to use one of several well-established notations in which the functions Δf, ∇f and δf are defined by

$$(\Delta f)(x) = f(x + h) - f(x),$$

$$(\nabla f)(x) = f(x) - f(x - h),$$

$$(\delta f)(x) = f\left(x + \frac{h}{2}\right) - f\left(x - \frac{h}{2}\right),$$

so that the (non-divided) difference table can be written in one of three terminologies:

$$f(x_1)$$
$$\Delta f(x_1)$$
$$f(x_2 \qquad\qquad \Delta^2 f(x_1)$$
$$\Delta f(x_2) \qquad\qquad .$$
$$f(x_3) \qquad\qquad . \qquad\qquad .$$
$$\vdots \qquad\qquad \vdots \qquad\qquad .$$
$$\vdots \qquad .$$

or

$$f(x_1)$$
$$\nabla f(x_2)$$
$$f(x_2) \qquad\qquad \nabla^2 f(x_3)$$
$$\nabla f(x_3) \qquad\qquad .$$
$$f(x_3) \qquad\qquad . \qquad\qquad .$$
$$\vdots \qquad\qquad \vdots \qquad\qquad .$$
$$\vdots$$

or, more symmetrically,

$$f(x_1)$$
$$\delta f(x_{3/2})$$
$$f(x_2) \qquad\qquad \delta^2 f(x_2)$$
$$\delta f(x_{5/2}) \qquad\qquad .$$
$$f(x_3) \qquad\qquad . \qquad\qquad .$$
$$\vdots \qquad\qquad \vdots \qquad\qquad .$$
$$\vdots$$

Note that we have elided the first pair of parentheses that properly belong in such expressions as $(\Delta^r f)(x)$. This follows established practice and does not lead to any ambiguity.

The operators Δ, ∇ and δ (which map functions to functions) are known as the *forward-difference, backward-difference* and *central-difference* operators respectively. It is not usually necessary to use a notation which shows their dependence on the value of the mesh size h since in particular applications of this notation, h is kept constant.

By an interpolation formula, we mean an approximation of the value of f at a particular point by the value of the interpolational polynomial at that point. From the results we have established here, we can write such a formula in various ways. In giving these formulae it is to be assumed that where finite-difference operators Δ, ∇ are used, the values of $x_2 - x_1$, $x_3 - x_2$, $\ldots$, $x_n - x_{n-1}$ are all the same and all equal to h.

Langrange interpolation formula:

$$f(x) \approx \sum_{i=1}^{n} f(x_i) \prod_{j \neq i} \frac{x - x_j}{x_i - x_j}.$$

Newton divided-difference interpolation formula:

$$f(x) \approx f(x_1) + (x - x_1) f(x_1, x_2) + \cdots$$
$$+ (x - x_1)(x - x_2) \cdots (x - x_{n-1}) f(x_1, x_2, \ldots, x_n). \qquad (131d)$$

Newton forward-difference interpolation formula:

$$f(x_1 + \theta h) \approx f(x_1) + \theta \, \Delta f(x_1) + \frac{\theta(\theta - 1)}{2!} \Delta^2 f(x_1) + \cdots$$
$$+ \frac{\theta(\theta - 1) \cdots (\theta - n + 2)}{(n - 1)!} \Delta^{n-1} f(x_1).$$

Reversing the order of $x_1, x_2, \ldots, x_n$ in interpreting the divided-difference formula in the case of equally spaced arguments, we have

Newton backward-difference interpolation formula:

$$f(x_n + \theta h) \approx f(x_n) + \theta \nabla f(x_n) + \frac{\theta(\theta + 1)}{2!} \nabla^2 f(x_n)$$
$$+ \cdots + \frac{\theta(\theta + 1) \cdots (\theta + n - 2)}{(n - 1)!} \nabla^{n-1} f(x_n).$$

Finally, let $f(x; x_1, x_2, \ldots, x_n)$ denote the interpolated value given by (131d) so that this equation, together with a similar one with x_n and x_1 interchanged, can be written in the forms

$$f(x; x_1, x_2, \ldots, x_n) = f(x; x_1, x_2, \ldots, x_{n-1}) + (x - x_1)A,$$
$$f(x; x_1, x_2, \ldots, x_n) = f(x; x_2, \ldots, x_n) + (x - x_n)A,$$

where $A = (x - x_2) \cdots (x - x_{n-1}) f(x_1, x_2, \ldots, x_n)$. Eliminating the term involving A from these two formulae we obtain an iterative version of the interpolation formula which we give together with a formula for $f(x; x_1)$:

Neville interpolation formula:

$$f(x; x_1) = f(x_1),$$

$$f(x; x_1, x_2, \ldots, x_n) = \frac{1}{x_n - x_1} [(x - x_1) f(x; x_2, \ldots, x_n)$$
$$- (x - x_n) f(x; x_1, x_2, \ldots, x_{n-1})].$$

132 Hermite interpolation.

We consider the problem of constructing a polynomial p of degree less than $m_1 + m_2 + \cdots + m_n$ which not only agrees with a sufficiently differentiable function f at the points $x_1, x_2, \ldots, x_n$ but is such that $p^{(k)}(x_i) = f^{(k)}(x_i)$

whenever $k < m_i$. For example, if $m_1 = m_2 = \cdots = m_n = 2$ then p' and f' agree at $x_1, x_2, \ldots, x_n$, as do p and f at these same points. In the case $m_1 = m_2 = \cdots = m_n = 1$ we simply have the interpolational polynomial of the last subsection, and when $n = 1$ we simply have the Taylor series expansion about x.

To carry out the construction of p, we can modify either the Lagrange or the Newton basis to suit the needs of this more general situation.

DEFINITION 132A.

Let e_{ik} be defined by

$$e_{ik}(x) = (x - x_i)^{k-1} \prod_{j \neq i} \frac{x - x_j}{x_i - x_j}.$$

Then $\{e_{ik} : i = 1, 2, \ldots, n; k = 1, 2, \ldots, m_i\}$ is known as the *generalized Lagrange basis*.

DEFINITION 132B.

Let e_{ik} be defined by

$$e_{ik}(x) = (x - x_i)^{k-1} \prod_{j < i} (x - x_j)^{m_j}.$$

Then $\{e_{ik} : i = 1, 2, \ldots, n; \ k = 1, 2, \ldots, m_i\}$ is known as the *generalized Newton basis*.

We state without proof the generalizations of theorems 131C and 131D. The proofs, which we omit, are similar to those given for the special cases.

THEOREM 132C.

Each of the generalized Lagrange and generalized Newton bases is a basis for the space of polynomials of degree less than $m_1 + m_2 + \cdots + m_n$.

THEOREM 132D.

There exists a unique polynomial p of degree less than $m_1 + m_2 + \cdots + m_n$ such that, for $i = 1, 2, \ldots, n$, $p^{(k)}(x_i) = f^{(k)}(x_i)$ for $k = 0, 1, 2, \ldots, m_i - 1$.

Although we will not give full details for expressing the more general (Hermite) interpolation formula in the styles of the previous subsection, we will introduce a convenient technique for obtaining Hermite interpolation formulae in particular cases. First, however, we introduce a preliminary lemma.

LEMMA 132E.

If p, q are polynomials such that $\deg(p) \le \deg(q) - 2$ and such that the open region in the complex plane bounded by a contour C contains all zeros of q, then

$$\frac{1}{2\pi i} \int_C \frac{p(z)}{q(z)} \, dz = 0. \tag{132a}$$

Proof.

From the Cauchy integral formula, it can be seen that the assertion of (132a) is equivalent to saying that this formula is true whenever C is a circular contour with centre 0 and sufficiently large radius R. Estimating $|p(z)/q(z)|$ for z on this contour and bounding the integral by the product of this estimate and the length of the circumference, we obtain a result which tends to zero as $R \to \infty$. ■

To express $p(x)$ in terms of $p^{(k)}(x_i)(i = 1, 2, \ldots, n; k = 0, 1, \ldots, m_i - 1)$, we choose q as the special polynomial given by

$$q(z) = (z - x)(z - x_1)^{m_1} \cdots (z - x_n)^{m_n}$$

and then write $1/q(z)$ in partial fractions prior to integrating. The terms in the resulting expression, which by lemma 132E add to zero, are of the form $p(x)$ or $p^{(k)}(x_i)$ with coefficients depending on $x, x_1, \ldots, x_n$ and on $m_1, m_2, \ldots, m_n$. Solving this for $p(x)$ gives the required Hermite interpolation formula. To use this to approximate $f(x)$ we simply replace p by f in every term.

As an example of this technique, consider the problem of finding a polynomial p, of degree 3, which agrees with f at 0 and 1 and such that p' agrees with f' at these same points. The partial fraction expansion of $1/q(z)$ is of the form

$$\frac{1}{(z - x)z^2(z - 1)^2} = \frac{A}{z - x} + \frac{B}{z} + \frac{C}{z^2} + \frac{D}{z - 1} + \frac{E}{(z - 1)^2}$$

and it is found that $A = 1/[x^2(x - 1)^2]$, $B = -(1 + 2x)/x^2$, $C = -1/x$, $D = (-3 + 2x)/(1 - x)^2$, $E = 1/(1 - x)$. Hence, since p has degree not exceeding 3, an application of lemma 132E gives the result

$$0 = \frac{A}{2\pi i} \int \frac{p(z)}{z - X} \, dz + \frac{B}{2\pi i} \int \frac{p(z)}{z} \, dz + \frac{C}{2\pi i} \int \frac{p(z)}{z^2} \, dz$$

$$+ \frac{D}{2\pi i} \int \frac{p(z)}{z - 1} \, dz + \frac{E}{2\pi i} \int \frac{p(z)}{(z - 1)^2} \, dz$$

which leads, using the Cauchy integral formula, to

$$p(x) = (1 - x)^2(1 + 2x)p(0) + x(1 - x)^2 p'(0) \\ + (3 - 2x)x^2 p(1) - x^2(1 - x)p'(1)$$

and hence to the approximation

$$f(x) \approx (1 - x)^2(1 + 2x)f(0) + x(1 - x)^2 f'(0) \\ + (3 - 2x)x^2 f(1) - x^2(1 - x)f'(1).$$

Finally, in this subsection, we derive the important case of Hermite interpolation in which $m_1 = m_2 = \cdots = m_n = 2$. Let $\phi(z) = (z - x_1)(z - x_2) \cdots (z - x_n)$ and consider the partial fraction expansion

$$\frac{1}{(z - x)(z - x_1)^2 \cdots (z - x_n)^2} = \frac{A}{z - x} + \sum_{j=1}^{n} \left[\frac{B_j}{z - x_j} + \frac{C_j}{(z - x_j)^2} \right], \quad (132b)$$

where we find that $A = 1/\phi(x)^2$, $C_j = 1/(x_j - x)\phi'(x_j)^2$ and

$$B_j = -C_j \left(\frac{1}{x_j - x} + 2 \sum_{k \neq j} \frac{1}{x_j - x_k} \right).$$

Applying the Cauchy integral formula to (132b) leads to the interpolation formula

$$f(x) \approx \sum_{j=1}^{n} \frac{1}{\phi'(x_j)^2} \left(\frac{\phi(x)^2}{(x - x_j)^2} - \frac{2\phi(x)^2}{x - x_j} \sum_{k \neq j} \frac{1}{x_j - x_k} \right) f(x_j) \\ + \sum_{j=1}^{m} \frac{1}{\phi'(x_j)^2} \frac{\phi(x)^2}{x - x_j} f'(x_j).$$

133 The calculus of finite differences.

Let X denote the linear space of all polynomials and let h denote a fixed non-zero real number. We introduce a basis $\{e_0, e_1, e_2, \ldots\}$ for X defined by $e_0(x) = 1$, $e_1(x) = x/h$, $e_2(x) = x^2/(2!h^2)$, $\ldots$, $e_n(x) = x^n/(n!h^n)$, $\ldots$. Any polynomial can then be represented as an infinite-dimensional vector whose components eventually become equal to zero. Thus,

$$\begin{bmatrix} c_0 \\ c_1 \\ c_2 \\ \vdots \end{bmatrix}$$

represents $c_0 e_0 + c_1 e_1 + c_2 e_2 + \cdots$. Consider the linear operator hD on X which carries out the action of differentiating a given polynomial and then multiplying the result by h. We easily see that $hD e_n = e_{n-1}$ so that hD is represented by the (infinite) matrix

$$\begin{bmatrix} 0 & 1 & 0 & \cdot & \cdot & \cdot \\ 0 & 0 & 1 & \cdot & \cdot & \cdot \\ 0 & 0 & 0 & \cdot & \cdot & \cdot \\ \vdots & \vdots & \vdots & & & \end{bmatrix}$$

Consider a linear operator E^θ on X which maps a polynomial p to a polynomial q, say, where $q(x) = p(x + \theta h)$. For any real numbers θ and φ, we see that

$$E^\theta \circ E^\varphi = E^{\theta + \varphi},$$

showing the notation E^θ to be a sensible one. We will, of course, write E for E^1. If we operate E^θ on a basis polynomial e_n, it is found that

$$(E^\theta e_n)(x) = \frac{(x + h\theta)^n}{n! h^n}$$

$$= \sum_{i=0}^{n} \binom{n}{i} \frac{(h\theta)^i x^{n-i}}{n! h^n}$$

$$= \sum_{i=0}^{n} \frac{\theta^i}{i!(n-i)!} \left(\frac{x}{h}\right)^{n-i}$$

$$= \sum_{i=0}^{n} \frac{\theta^i}{i!} e_{n-i}(x),$$

so that $E^\theta e_n = \sum_{i=0}^{n}(\theta^i/i!)e_{n-i}$ and E^θ can be represented by the infinite matrix

$$E^\theta = \begin{bmatrix} 1 & \theta & \dfrac{\theta^2}{2!} & \dfrac{\theta^3}{3!} & \cdot & \cdot & \cdot \\ 0 & 1 & \theta & \dfrac{\theta^2}{2!} & \cdot & \cdot & \cdot \\ 0 & 0 & 1 & \theta & \cdot & \cdot & \cdot \\ 0 & 0 & 0 & 1 & \cdot & \cdot & \cdot \\ \vdots & \vdots & \vdots & \vdots & & & \end{bmatrix}.$$

Interpreting h as the mesh size in an equally spaced table of a function, we can, using E^θ, obtain representations of the finite difference operators Δ, ∇, δ introduced in subsection 131, since for any polynomial p, $\Delta p = Ep - p$, $\nabla p = p - E^{-1}p$ and $\delta p = E^{\frac{1}{2}}p - E^{-\frac{1}{2}}p$:

$$\Delta = \begin{bmatrix} 0 & 1 & \dfrac{1}{2!} & \dfrac{1}{3!} & \dfrac{1}{4!} & \cdots \end{bmatrix},$$

$$\nabla = [0 \quad 1 \quad \frac{-1}{2!} \quad \frac{1}{3!} \quad \frac{-1}{4!} \cdots],$$

$$\delta = [0 \quad 1 \quad 0 \quad \frac{1}{2^2.3!} \quad 0 \quad \frac{1}{2^4.5!} \cdots],$$

where, for conciseness, we have shown only the first row of each of these matrices. As is the case also with $hD = [0\ 1\ 0\ 0 \cdots]$ and $E^\theta = [1\ \theta\ \theta^2/2!\ \theta^3/3! \cdots]$, these matrices have zeros below the main diagonal and all diagonals are of constant value.

Noting that $(hD)^n = [0\ 0 \cdots 0\ 1\ 0 \cdots]$, where n zeros precede the single '1', we can write

$$E^\theta = \sum_{n=0}^{\infty} \frac{\theta^n}{n!} (hD)^n,$$

$$\Delta = \sum_{n=1}^{\infty} \frac{1}{n!} (hD)^n,$$

$$\nabla = \sum_{n=1}^{\infty} \frac{(-1)^{n-1}}{n!} (hD)^n,$$

$$\delta = \sum_{n=0}^{\infty} \frac{1}{2^{2n}(2n+1)!} (hD)^{2n+1}.$$

Let φ denote a real-valued function of a real variable which possesses a Taylor expansion about zero with coefficients $c_0, c_1, \ldots$. Thus $\varphi(x) = \sum_{n=0}^{\infty} c_n x^n$. Let L denote a linear operator on X which is nilpotent in the sense that, given any polynomial p, there exists an integer k such that $L^k(p) = 0$. In this case we can define the operator $M = \sum_{n=0}^{\infty} c_n L^n$ by the formula $M(p) = \sum_{n=0}^{\infty} c_n L^n(p)$ in which the infinite sum is the sum of the finite set of non-vanishing terms. It is natural to write $M = \varphi(L)$ and we adopt this as a convention. Since hD is nilpotent, we can write, for example,

$$E^\theta = \exp(hD),$$

$$\Delta = \exp(hD) - 1,$$

$$\nabla = 1 - \exp(-hD),$$

$$\delta = 2 \sinh\left(\frac{hD}{2}\right),$$

where the appropriate function φ in each case is obtained by replacing hD by x in these formulae.

Let A denote the set of functions which possess all derivatives at zero and let A_0 denote the subset of A such that if $\varphi \in A_0$ then $\varphi(0) = 0$. Also let A_1 denote the subset of A_0 such that if $\varphi \in A_1$ then $\varphi'(0) \neq 0$. Since we will need to manipulate various expressions of the form $\varphi(L)$, where L is a nilpotent operator, we formally state a number of results that furnish the justification for what we wish to do. The proofs, which hinge on elementary properties of the coefficients in power series, are omitted.

THEOREM 133A.

If $\varphi, \psi \in A$, $a \in \mathbb{R}$ and L is nilpotent, then

$$(\varphi + \psi)(L) = \varphi(L) + \psi(L),$$

$$(\varphi - \psi)(L) = \varphi(L) - \psi(L),$$

$$(\varphi \cdot \psi)(L) = \varphi(L) \circ \psi(L),$$

$$(a\varphi)(L) = a\varphi(L).$$

If, furthermore, $\psi \in A_0$ then $\psi(L)$ is nilpotent, $\varphi \circ \psi \in A$ and

$$(\varphi \circ \psi)(L) = \varphi(\psi(L)),$$

and if $\psi \in A_1$ then ψ^{-1} exists and

$$L = \psi^{-1}(\psi(L)). \tag{133a}$$

Since the functions $x \mapsto e^x - 1$, $x \to 1 - e^{-x}$ and $x \to 2 \sinh(x/2)$ are all in A_1, the operators Δ, ∇ and δ are nilpotent and from (133a) it follows that

$$hD = \ln(1 + \Delta),$$

$$hD = -\ln(1 - \nabla),$$

and that

$$hD = 2 \sinh^{-1}\left(\frac{\delta}{2}\right),$$

so that, taking the first of these operating on a polynomial p, we find that

$$p'(x_1) = \frac{1}{h}\left[\Delta p(x_1) - \frac{1}{2}\Delta^2 p(x_1) + \frac{1}{3}\Delta^3 p(x_1) - \cdots\right], \tag{133b}$$

which is just the same result that could be obtained by differentiation of the Newton forward-difference interpolation formula. Corresponding to the identity (133b) we have for f, a non-polynomial function, the approximation

$$f'(x_1) \approx \frac{1}{h}\left(\Delta f(x_1) - \frac{1}{2}\Delta^2 f(x_1) + \frac{1}{3}\nabla^3 f(x_1) - \cdots\right).$$

To conclude this subsection, we use the methods introduced here to give an alternative proof of the Newton forward-difference interpolation formula. We have

$$E^\theta = \exp(\theta h D)$$

$$= \exp(\theta[\ln(1 + \Delta)])$$

$$= \exp(\ln((1 + \Delta)^\theta))$$

$$= (1 + \Delta)^\theta$$

$$= 1 + \theta\Delta + \binom{\theta}{2} \Delta^2 + \cdots,$$

so that we find the formula

$$f(x_1 + \theta h) \approx f(x_1) + \theta\Delta f(x_1) + \binom{\theta}{2} \Delta^2 f(x_1) + \cdots.$$

134 Numerical quadrature.

Consider the function φ defined by $\varphi(0) = 1$ and $\varphi(x) = [\exp(x) - 1]/x$ for $x \neq 0$. We will denote by J the operator $\varphi(hD)$. Since $hDJ = \Delta$ and $J(e_0) = e_0$, it is easy to see that

$$(Jp)(x) = \frac{1}{h} \int_x^{x+h} p(u) \, du,$$

for any polynomial p. We shall use the operator J as a tool in obtaining a number of different types of formula for approximating an integral. Since $hD = \ln(1 + \Delta)$, we can also write J as $\psi(\Delta)$ where $\psi(0) = 1$ and $\psi(x) = x/\ln(1 + x)$ for $x \neq 0$. Thus

$$\psi(x) = \frac{1}{1 - x/2 + x^2/3 - x^3/4 + \ldots}$$

$$= 1 + \frac{x}{2} - \frac{x^2}{12} + \frac{x^3}{24} - \frac{19x^4}{720} + \frac{3x^5}{160} - \frac{863x^6}{60\,480}$$

$$+ \frac{275x^7}{24\,192} - \frac{33\,953x^8}{3\,628\,800} + \frac{8183x^9}{1\,036\,800} - \frac{3\,250\,433x^{10}}{479\,001\,600} + \cdots, \tag{134a}$$

so that we have the approximation

$$\int_x^{x+h} f(x) \, dx \approx h[f(x) + \tfrac{1}{2}\Delta f(x) - \tfrac{1}{12}\Delta^2 f(x) + \tfrac{1}{24}\Delta^3 f(x) - \cdots].$$

If we truncate this at the term $\Delta f(x) = f(x + h) - f(x)$, then we obtain the so-called trapezoidal rule

$$\int_x^{x+h} f(x) \, dx \approx h[\tfrac{1}{2} f(x) + \tfrac{1}{2} f(x + h)],$$

which is evidently exact for f a polynomial of degree 1. Many formulae for numerical quadrature are generalizations of this approximation.

The first generalization we consider is that of the (closed) Newton–Cotes formulae which express an integral $\int_x^{x+nh} f(x) \, dx$ in terms of $f(x)$, $f(x + h), \ldots, f(x + nh)$. In the case when f is a polynomial, we can find this integral by expanding the operator $J(1 + E + \cdots + E^{n-1})$ in powers of Δ and truncating at the term in Δ^n. Since $1 + E + \cdots + E^{n-1} = \varphi(\Delta)$ where $\varphi(0) = n$ and $\varphi(n) = [(1 + x)^n - 1]/x$, we find

$$J(1 + E + \cdots + E^{n-1}) = (\varphi\psi)(\Delta),$$

where $(\varphi\psi)(x) = [(1 + x)^n - 1]/\ln(1 + x)$ for $x \neq 0$ and $(\varphi\psi)(0) = n$. Carrying out the programme of multiplying the series for φ by the series given by (134a), truncating at the appropriate term and then rewriting Δ as $E - 1$, we obtain the (closed) Newton–Cotes formulae in the form

$$\int_x^{x+nh} f(x)\,dx \approx \frac{h}{D_n} \left[c_{n,0}f(x) + c_{n,1}f(x+h) + \cdots + c_{n,n}f(x+nh) \right],$$

where the numbers $D_n, c_{n,0}, c_{n,1}, \ldots, c_{n,n}$ are given in the cases $n = 1, 2, 3, \ldots, 6$ in table 134.

Table 134   Denominators and numerators for closed Newton–Cotes coefficients

n	D_n	$c_{n,0}$	$c_{n,1}$	$c_{n,2}$	$c_{n,3}$	$c_{n,4}$	$c_{n,5}$	$c_{n,6}$
1	2	1	1					
2	3	1	4	1				
3	8	3	9	9	3			
4	45	14	64	24	64	14		
5	288	95	375	250	250	375	95	
6	140	41	216	27	272	27	216	41

In a similar way we can derive the (open) Newton–Cotes formulae, in which $\int_x^{x+nh} f(x)\,dx$ is approximated by a linear combination of $f(x+h)$, $f(x+2h)$ $,\ldots, f(x+(n-1)h)$ (with $f(x)$ and $f(x+nh)$ excluded). In this case $J(1 + E + \cdots + E^{n-1})$ is written as E composed with a series in Δ which is truncated at the term in Δ^{n-2}. It is found that

$$J(1 + E + \cdots + E^{n-1}) = E\chi(\Delta),$$

where $\chi(x) = [(1 + x)^n - 1]/(1 + x)\ln(1 + x)$ for $x \neq 0$ with $\chi(0) = n$. Corresponding to table 134, we present table 134' which gives the open

Table 134'   Denominators and numerators for open Newton–Cotes coefficients

n	D'_n	$c'_{n,1}$	$c'_{n,2}$	$c'_{n,3}$	$c'_{n,4}$	$c'_{n,5}$	$c'_{n,6}$
2	1	2					
3	2	3	3				
4	3	8	-4	8			
5	24	55	5	5	55		
6	10	33	-42	78	-42	33	
7	1440	4277	-3171	3934	3934	-3171	4277

Newton–Cotes formulae for $n = 2,3,4,5,6,7$ in the form

$$\int_x^{x+nh} f(x)\,dx \approx \frac{h}{D_n'}\,[c_{n,1}'f(x+h) + c_{n,2}'f(x+2h) + \cdots$$

$$+ c_{n,\,n-1}'f(x+(n-1)h)].$$

Returning now to the operator J and subtracting from it $\frac{1}{2}(1+R)$ (which is just the trapezoidal rule approximation) gives an operator which we attempt to write in the form $ER - R$. This is easily seen to be possible if the operator R is $\varphi(hD)$, where $\varphi(x) = \{[\exp(x)-1]/x - \frac{1}{2}[1+\exp(x)]\}/[\exp(x)-1]$ for $x \neq 0$, and $\varphi(0) = 0$. Expanding R in powers of hD gives the Euler–MacLaurin formula, in powers of δ (using the relation $hD = 2\sinh^{-1}(\delta/2)$ but first taking out a factor $\frac{1}{2}(E^{\frac12} + E^{-\frac12})$) gives the Gauss-Encke formula and in powers of Δ or ∇ (the ER in $ER - R$) gives the Newton–Gregory formula. Furthermore, since $(1 + E + \cdots + E^{n-1})(E-1) = E^n - 1$, it is possible to sum each of these formulae over n successive intervals to get the following forms:

$$\int_x^{x+nh} f(x)\,dx \approx h[\tfrac{1}{2}f(x) + f(x+h) + \cdots + f(x+(n-1)h) + \tfrac{1}{2}f(x+nh)] + C,$$

where C is given in various cases by the following formulae:

Euler–MacLaurin formula:

$$C = -\frac{h^2}{12}\,[f'(x+nh) - f'(x)] + \frac{h^4}{720}\,[f^{(3)}(x+nh) - f^{(3)}(x)]$$

$$-\frac{h^6}{30\,240}\,[f^{(5)}(x+nh) - f^{(5)}(x)] + \cdots.$$

Gauss-Encke formula:

$$C = -\frac{h}{12}\,[\mu\delta f(x+nh) - \mu\delta f(x)] + \frac{11h}{720}\,[\mu\delta^3 f(x+nh) - \mu\delta^3 f(x)]$$

$$-\frac{191h}{60\,480}\,[\mu\delta^5 f(x+nh) - \mu\delta^5 f(x)] + \cdots,$$

where μ is the operator $\frac{1}{2}(E^{\frac12} + E^{-\frac12})$.

Newton–Gregory formula:

$$C = -\frac{h}{12}\,[\nabla f(x+nh) - \Delta f(x)] - \frac{h}{24}\,[\nabla^2 f(x+nh) + \Delta^2 f(x)]$$

$$-\frac{19h}{720}\,[\nabla^3 f(x+nh) - \Delta^3 f(x)] - \frac{3h}{160}\,[\nabla^4 f(x+nh) + \Delta^4 f(x)] - \cdots.$$

The various quadrature formulae we have so far considered all express the value of an integral, say $\int_a^b f(x)\,dx$, as a linear combination of $f(x_1)$, $f(x_2),\dots,f(x_n)$, where $x_1,x_2,\dots,x_n$ are preassigned numbers. For an arbitrary choice of $x_1, x_2, \dots, x_n$ the result can be made to be exact for f any polynomial of degree less than n, but in general not for higher degrees. By a Gaussian quadrature formula we will mean a formula which expresses the integral in this way, but with $x_1, x_2, \dots, x_n$ specifically chosen to make the result exact for f as high a degree as possible.

Let ϕ be a polynomial of degree n, orthogonal on $[a, b]$ to all polynomials of lower degree. It is well known that the zeros of ϕ are distinct and all lie in (a, b) and that if $a = -1, b = +1$ then ϕ may be taken as P_n, the Legendre polynomial of degree n (see, for example, Abramowitz and Stegun [1965]). For any polynomial p, let q, r be the quotient and remainder when p is divided by ϕ. Thus $p = \phi q + r$ and r has degree less than n.

THEOREM 134A.

The is a unique quadrature formula

$$\int_a^b f(x)\,dx \approx \sum_{i=1}^n w_i f(x_i) \tag{134b}$$

which is precise when f is a polynomial of degree less than $2n$. The numbers $x_1, x_2, \dots, x_n$ are the zeros of ϕ and $w_1, w_2, \dots, w_n$ are positive numbers given by

$$w_i = \frac{1}{\phi'(x_i)} \int_a^b \frac{\phi(x)}{x - x_i}\,dx, \qquad i = 1, 2, \dots, n. \tag{134c}$$

Proof.

If f is replaced by $p = \phi q + r$, where q and r each has degree less than n, then the fact that (134b) is satisfied exactly is equivalent to

$$\int_a^b r(x)\,dx = \sum_{i=1}^n w_i r(x_i), \tag{134d}$$

since $\int_a^b \phi(x)q(x)\,dx = 0$ by orthogonality and $\sum_{i=1}^n w_i \phi(x_i)q(x_i) = 0$ by choice of $x_1, x_2, \dots, x_n$. To guarantee that (134d) holds for any polynomial r of degree less than n, choose as a basis for such polynomials $\{e_1, e_2, \dots, e_n\}$, where

$$e_i(x) = \frac{\phi(x)}{x - x_i} \quad (x \neq x_i), \qquad e_i(x_i) = \phi'(x_i), \qquad i = 1, 2, \dots, n.$$

The condition with r replaced by e_i is exactly (134e). To prove uniqueness, suppose that there existed a different set of points $\{\bar{x}_1, \bar{x}_2, \dots, \bar{x}_n\}$ for which

(134b) could be satisfied exactly for f any polynomial of degree less than $2n$. If $f = \bar\phi q$ where q is any polynomial of degree less than n and $\bar\phi(x) = (x - \bar x_1)$ $(x - \bar x_2) \cdots (x - \bar x_n)$, then we see that

$$\int_a^b \bar\phi(x)q(x)\,dx = 0,$$

so that $\bar\phi$ is, apart from a constant factor, the same as ϕ. To show that $w_1, w_2, \ldots, w_n$ are all positive, substitute $r = e_i^2$ into (134d) to give, for $i = 1, 2, \ldots, n$,

$$w_i e_i^2(x_i) = \int_a^b e_i^2(x)\,dx. \quad \blacksquare$$

135 Errors in numerical approximation.

The type of approximation so far considered expresses a quantity depending on a function in terms of other information about that function. For example, the Lagrange interpolation formula expresses $f(x)$ in terms of, say, $f(x_1)$, $f(x_2), \ldots, f(x_n)$. The error, that is the amount by which the approximate value falls short of the quantity it is approximating, can be expressed in a variety of different forms and we will consider three of these. In the first of these, applicable to all interpolation formulae of the general Hermite form, the error is written as a mean value, that is as a computable number multiplied by some high-order derivative of the function f evaluated at an unspecified point.

The second is applicable to a wide variety of applications including various types of differentation and integration formulae. It is, in effect, an exact error formula and expresses the error as the integral of a high-order derivative of f multiplied by some known function. Finally, we will consider situations in which the error can be expressed in some sort of asymptotic form.

We consider first of all the general problem of Hermite interpolation, in which a polynomial p is found with degree less than $N = m_1 + m_2 + \cdots + m_n$ such that for $i = 1, 2, \ldots, n$ and $j = 0, 1, 2, \ldots, m_i - 1$,

$$p^{(j)}(x_i) = f^{(j)}(x_i)$$

and we wish to obtain a formula for $f(x) - p(x)$.

THEOREM 135A.

If $f \in C^N[a, b]$, $x, x_1, x_2, \ldots, x_n \in [a, b]$, then

$$f(x) - p(x) = \prod_{i=1}^n \frac{(x - x_i)^{m_i} f^{(N)}(\xi)}{N!,}$$

where ξ is some number in (a, b).

Proof.

The result of the theorem is obviously true when $x \in \{x_1, x_2, \ldots, x_n\}$, so we may assume this is not the case. Let R be a number such that $f(x) - p(x) = \prod_{i=1}^{n}(x - x_i)^{m_i} R$ so that the function ϕ defined by

$$\phi(t) = f(t) - p(t) - \prod_{i=1}^{n} (t - x_i)^{m_i} R$$

vanishes $N + 1$ times (counting multiple zeros) in the interval $[a, b]$. Differentiating this N times gives a function $f^{(N)} - N!R$, which vanishes at least once in the interior of this interval. If the point where this vanishes is ξ then we solve for R and substitute into the formula assumed for $f(x) - p(x)$ to obtain the required result. ∎

Naturally, in particular applications of this result, we take $[a, b]$ to be the smallest interval containing each of $x, x_1, x_2, \ldots, x_n$. It is interesting to observe that the error formula given by theorem 135A covers a wide range of situations from Lagrange interpolation to Taylor series.

In the Taylor series case, however, it is well known that the error committed by truncating after the term in $f^{(n-1)}$ has an error in the form of an integral

$$f(x) - p(x) = \frac{1}{(n-1)!} \int_{x_1}^{x} (x - t)^{n-1} f^{(n)}(t) \, dt,$$

where p is the polynomial given by

$$p(t) = f(x_1) + tf'(x_1) + \cdots + \frac{t^{n-1}}{(n-1)!} f^{(n-1)}(x_1).$$

We now consider how errors for other types of approximation formulae can be written in the form $\int K(t) f^{(n)}(t) \, dt$. Although the method we will describe is applicable in more general situations, we will confine ourselves to approximations that take the form

$$\sum_{i=1}^{m} (a_i F(x_i) + b_i f(x_i) + c_i f'(x_i)) \approx 0, \qquad (135a)$$

where F is the indefinite integral of f and it is assumed that $\sum_{i=1}^{m} a_i = 0$ and that the approximation is exact when f is a polynomial of degree less than n.

By setting each of the a_i and all but one of the c_i equal to zero, we obtain from (135a) the error in a numerical differentiation formula. We could also consider the situation when all but two of the a_i are zero, and in this case we have an error formula for a large class of numerical quadrature approximations.

In the statement of the theorem which follows, the notation $(\ )_+$ is used to denote the value within the parentheses if this is positive and a value of zero otherwise.

THEOREM 135B.

If (135a) is exact whenever f is replaced by a polynomial of degree less than n, and if $f \in C^{(n)} I$, where I is a closed interval containing each of $x_1, x_2, \ldots, x_m$, then

$$\sum_{i=1}^{m} [a_i F(x_i) + b_i f(x_i) + c_i f'(x_i)] = \int_I K(t) f^{(n)}(t)\, dt,$$

with

$$K(t) = \sum_{i=1}^{m} \left[a_i \frac{(x_i - t)_+^n}{n!} + b_i \frac{(x_i - t)_+^{n-1}}{(n-1)!} + c_i \frac{(x_i - t)_+^{n-2}}{(n-2)!} \right]. \qquad (135b)$$

Proof.

Let d denote the left-hand end of the interval I and write each of $F(x_i)$, $f(x_i)$ and $f'(x_i)$ as a Taylor series about d up to the term in $f^{(n-1)}(d)$ and with error in one of the forms

$$\frac{1}{n!} \int_d^{x_i} (x_i - t)^n f^{(n)}(t)\, dt,$$

$$\frac{1}{(n-1)!} \int_d^{x_i} (x_i - t)^{n-1} f^{(n)}(t)\, dt,$$

$$\frac{1}{(n-2)!} \int_d^{x_i} (x_i - t)^{n-2} f^{(n)}(t)\, dt$$

respectively. Multiplying by the constants a_i, b_i, c_i and summing we find that

$$\sum_{i=1}^{m} [a_i F(x_i) + b_i f(x_i) + c_i f'(x_i)] = (a_1 + a_2 + \cdots + a_m) F(d) + C_0 f(d) +$$

$$C_1 f'(d) + \cdots + C_{n-1} f^{(n-1)}(d) + \int_I K(t) f^{(n)}(t)\, dt, \qquad (135c)$$

where $C_0, C_1, \ldots, C_{n-1}$ are constants and K is given by (135b). Since $\Sigma a_i = 0$ and the left-hand side of (135c) vanishes whenever f is a polynomial of degree less than n, $C_0 = C_1 = \cdots = C_{n-1} = 0$ and the result follows. ∎

Since the interval I may be replaced by a larger interval, it is clear that $K(t)$ vanishes outside I so that we may, if we wish, formally replace the integral expression for the error by $\int_{-\infty}^{\infty} K(t) f^{(n)}(t)\, dt$.

It is interesting to consider the case when K does not change sign in I, since in this case we obtain a generalization of theorem 135A.

COROLLARY 135C.

If, under the conditions of theorem 135B, K has constant sign, then

$$\int_I K(t) f^{(n)}(t) \, dt = f^{(n)}(\xi) \int_I K(t) \, dt,$$

for some $\xi \in I$.

Note that K does not depend on the function f and is a piecewise polynomial whose integral is easily evaluated in any particular situation. Even if K does not have constant sign, we may bound the magnitude of the error by

$$\sup_{t \in I} |f^{(n)}(t)| \int |K(t)| \, dt.$$

Finally, in this subsection, we consider the behaviour of a family of approximations in which each is of the form

$$\sum_{i=1}^m \left[\frac{1}{h} a_i F(x_0 + h\xi_i) + b_i f(x_0 + h\xi_i) + h c_i f'(x_0 + h\xi_i) \right] \approx 0, \quad (135d)$$

in which f is a given function, $x_0, \xi_1, \xi_2, \ldots, \xi_m$ are real numbers and h is a variable parameter. Of particular interest is the behaviour of this approximation as $h \to 0$. Note that the coefficients $1/h$ and h are incorporated into the term involving F and f' respectively, so that the conditions on a_i, b_i, c_i $(i = 1, 2, \ldots . m)$ for (135d) to be exact for all polynomials of a given degree are independent of h.

If, in fact, (135d) is exact for all polynomials of degree less than n, with the numbers $a_i, b_i, c_i (i = 1, 2, \ldots, m)$ constant, then the kernel K in the error term occurring in theorem 135B takes the form

$$\sum_{i=1}^m \left[\frac{1}{h} a_i \frac{(h\xi_i + x_0 - t)_+^n}{n!} + b_i \frac{(h\xi_i + x_0 - t)_+^{n-1}}{(n-1)!} + h c_i \frac{(h\xi_i + x_0 - t)_+^{n-2}}{(n-2)!} \right]$$

$$= h^{n-1} \sum_{i=1}^m \left(a_i \frac{(\xi_i - u)_+^n}{n!} + b_i \frac{(\xi_i - u)_+^{n-1}}{(n-1)!} + c_i \frac{(\xi_i - u)_+^{n-2}}{(n-2)!} \right)$$

$$= h^{n-1} \bar{K}(u), \text{ say,}$$

where $u = (t - x_0)/h$. If $\bar{I}$ is an interval containing each of $\xi_1, \xi_2, \ldots, \xi_m$, the error becomes $h^n \int_{\bar{I}} \bar{K}(u) f^{(n)}(x_0 + hu) \, du$.

If $f \in C^{n+1} I$ for some interval I containing x_0 as an interior point then for h small enough, $x_0 + hu \in I$, whenever $u \in \bar{I}$. In this case, $|f^{(n)}(x_0 + hu) - f^{(n)}(x_0)|/h$ is bounded as $h \to 0$, so that we may write the error term in the asymptotic form

$$Ch^n f^{(n)}(x_0) + O(h^{n+1}),$$

where

$$C = \int_{\bar{I}} \bar{K}(u) \, du$$

and where we have adopted the widely used convention of writing $O(g(h))$ for an unspecified function which is bounded by a constant multiplied by $g(h)$.

136 Rational interpolation.

In this subsection, we will consider the problem of interpolating through a set of points $x_1, x_2, \ldots, x_n$ not with a polynomial, but with a rational function with degrees l (numerator) and m (denominator), such that $l + m < n$. Mainly we will be concerned with the case that $l \leq [n/2]$ and that $m \leq [n/2]$, where $[x]$ denotes the integer part of x. First, however, we will establish a basic uniqueness result.

THEOREM 136A

If R, $\bar{R}$ are rational functions each with degrees not exceeding l (numerator) and not exceeding m (denominator), such that $l + m < n$ and if R and $\bar{R}$ each satisfy the requirement that $R(x) = \bar{R}(x) = f(x)$ whenever $x \in \{x_1, x_2, \ldots, x_n\}$, a set of n distinct points, and f is a given function, then $R = \bar{R}$.

Proof.

If $R = N/D$ and $\bar{R} = \bar{N}/\bar{D}$, where N and $\bar{N}$ are polynomials with degrees not exceeding l, and D and $\bar{D}$ are polynomials with degrees not exceeding m, then for all $x \in \{x_1, x_2, \ldots x_n\}$, $p(x) = 0$ where p is the polynomial $p = N\bar{D} - \bar{N}D$ of degree less than n. Thus p is the zero polynomial and therefore $N/D = \bar{N}/\bar{D}$. ∎

Although there is no corresponding general existence result, we will, in the cases of interest to us, give formulae for the construction of rational interpolates. The status of these formulae is that they are valid unless in a particular instance the evaluation of them leads to the occurrence of a zero division.

Confining ourselves to the case where the numerator has degree $l \leq [n/2]$ and the denominator has degree $m \leq [(n-1)/2]$, we write

$$f(x; x_1, x_2, \ldots, x_n) = N(x; x_1, x_2, \ldots, x_n)/D(x; x_1, x_2, \ldots, x_n)$$

for this rational interpolation function, known as Thiele's interpolation formula, which agrees with $f(x)$ when $x = x_1, x_2, \ldots, x_n$.

THEOREM 136B.

There exist numbers $c_0, c_1, c_2, \ldots, c_n$ such that

$$f(x; x_1, x_2, \ldots, x_n) = c_0 + \frac{x - x_1}{c_1 +} \frac{x - x_2}{c_2 +} \cdots \frac{x - x_{n-1}}{c_{n-1}}. \tag{136a}$$

Proof.

We prove the result by induction on the number of interpolated points. If (136a) is known to hold for $n - 1$ such points, then we choose $c_0 = f(x_1)$ and we then consider the problem of finding $g(x; x_2, \ldots x_n)$ where

$$g(x) = \frac{x - x_1}{f(x) - f(x_1)}, \qquad (x \neq x_1),$$

$$g(x_1) = \frac{1}{f'(x_1)}.$$

If, as the induction hypothesis assures us is possible,

$$g(x; x_2, \ldots, x_n) = c_1 + \frac{x - x_2}{c_2 +} \cdots \frac{x - x_{n-1}}{c_{n-1}},$$

for some numbers $c_1, \ldots c_{n-1}$, then (136a) follows. ∎

It is convenient to introduce further numbers $d_0, d_1, \ldots$ defined by

$$d_0 = c_0, \qquad\qquad d_1 = c_1,$$

$$d_2 = c_0 + c_2, \qquad\qquad d_3 = c_1 + c_3,$$

$$d_4 = c_0 + c_2 + c_4, \qquad d_5 = c_1 + c_3 + c_5,$$

$$\vdots \qquad\qquad\qquad\qquad \vdots$$

so that conventionally writing $d_{-1} = d_{-2} = 0$, we see that $c_i = d_i - d_{i-2}$ for $i = 0, 1, 2, \ldots$. It will turn out that $d_0, d_1, d_2, \ldots$ play an analogous role to the divided differences of subsection 131, and clearly d_i depends only on the values of $x_1, x_2, \ldots, x_{i+1}$ and not on any of $x_{i+2}, \ldots, x_n$.

DEFINITION 136C.

For $i = 0, 1, 2, \ldots$, the quantity d_i is known as an ith-order *reciprocal difference* and written as $f(x_1, x_2, \ldots, x_{i+1})$.

Although we have used the same notation here as for divided differences, we will never use these two concepts in the same context and ambiguity will thus be avoided.

Before considering an analogue of theorem 131G we introduce two preliminary lemmas.

LEMMA 136D.

If $N_{-1}, N_0, N_1, \ldots, D_{-1}, D_0, D_1, \ldots$ are defined by

$$N_{-1} = 1, \qquad N_0 = c_0,$$

$$D_{-1} = 0, \qquad D_0 = 1,$$

and, for $i = 1, 2, \ldots, n-1$,

$$\frac{N_i}{D_i} = c_i \frac{N_{i-1}}{D_{i-1}} + (x - x_i) \frac{N_{i-2}}{D_{i-2}}, \tag{136b}$$

then for $i = 0, 1, 2, \ldots, n-1$, $f(x; x_1, x_2, \ldots, x_{i+1}) = N_i/D_i$.

Proof.

The result follows from the standard recursion formula for the successive convergents to a continued fraction. ∎

From now on we will write N_i and D_i as $N(x; x_1, x_2, \ldots, x_{i+1})$ and $D(x; x_1, x_2, \ldots, x_{i+1})$ and standardize the numerator and denominator of $f(x; x_1, x_2, \ldots, x_{i+1})$ in this way.

LEMMA 136E.

For $i = 0, 2, 4, 6, \ldots$, $N(x; x_1, x_2, \ldots, x_{i+1})$ is a polynomial of degree $i/2$ with highest-degree coefficient $f(x_1, x_2, \ldots, x_{i+1})$ and $D(x; x_1, x_2, \ldots, x_{i+1})$ is a polynomial of degree $i/2$ with highest-degree coefficient 1. For $i = 1, 3, 5, 7, \ldots$, $N(x; x_1, x_2, \ldots, x_{i+1})$ is a polynomial of degree $(i+1)/2$ with highest-degree coefficient 1 and $D(x; x_1, x_2, \ldots, x_{i+1})$ is a polynomial of degree $(i-1)/2$ with highest-degree coefficient $f(x_1, x_2, \ldots, x_{i+1})$.

Proof.

The ratio of the highest-degree coefficients in the numerator and denominator of $f(x; x_1, x_2, \ldots, x_n)$ is either $f(x_1, x_2, \ldots, x_n)$ or $1/f(x_1, x_2, \ldots, x_n)$ according to whether n is odd or even. ∎

When dealing with divided differences, we found a simple recurrence for expressing a difference of one order in terms of those of lower order. In the case of reciprocal differences, we see that $f(x_1, x_2) = (x_1 - x_2)/[f(x_1) - f(x_2)]$, the reciprocal of the ordinary first-order divided difference. We now state the generalization to higher-order reciprocal differences.

THEOREM 136F.

The $(n-1)$th-order reciprocal difference is given for $n = 2, 3, \ldots$ by

$$f(x_1, x_2, \ldots, x_n) = \frac{x_n - x_1}{f(x_2, \ldots x_n) - f(x_1, \ldots, x_{n-1})} + f(x_2, \ldots, x_{n-1}), \tag{136c}$$

where the last term is omitted when $n = 2$.

Proof.

Assuming that $n > 2$, we have, from lemma 136D,

$$N(x; x_1, \ldots, x_n) = [f(x_1, \ldots, x_n) - f(x_2, \ldots, x_{n-1})] N(x; x_1, \ldots, x_{n-1})$$
$$+ (x - x_1)N(x; x_2, \ldots, x_{n-1}) \tag{136d}$$

and, interchanging the roles of x_1 and x_n,

$$N(x; x_1, \ldots, x_n) = [f(x_1, \ldots, x_n) - f(x_2, \ldots, x_{n-1})] N(x; x_2, \ldots, x_n)$$
$$+ (x - x_n)N(x; x_2, \ldots, x_{n-1}) \tag{136e}$$

so that, subtracting (136e) from (136d), we find

$$[f(x_1, \ldots, x_n) - f(x_2, \ldots, x_{n-1})] [N(x; x_1, \ldots, x_{n-1}) - N(x; x_2, \ldots, x_n)]$$
$$= (x_1 - x_n)N(x; x_2, \ldots, x_{n-1}). \tag{136f}$$

Again using lemma 136D we have the two formulae:

$$N(x; x_1, \ldots, x_{n-1}) = [f(x_1, \ldots, x_{n-1}) - f(x_2, \ldots, x_{n-2})] N(x; x_2, \ldots, x_{n-1})$$
$$+ (x - x_{n-1})N(x; x_2, \ldots, x_{n-2}),$$

$$N(x; x_2, \ldots, x_n) = [f(x_2, \ldots, x_n) - f(x_2, \ldots, x_{n-2})] N(x; x_2, \ldots, x_{n-1})$$
$$+ (x - x_{n-1})N(x; x_2, \ldots, x_{n-2}),$$

whose difference gives

$$N(x; x_1, \ldots, x_{n-1}) - N(x; x_2, \ldots, x_n)$$
$$= [f(x_1, \ldots, x_{n-1}) - f(x_2, \ldots, x_n)] N(x; x_2, \ldots, x_{n-1}). \tag{136g}$$

We now divide (136f) by (136g) to give the result (136c). ∎

Using theorem 136F we can construct reciprocal-difference tables analogous to divided-difference tables:

$f(x_1)$					
0	$f(x_1, x_2)$				
$f(x_2)$		$f(x_1, x_2, x_3)$			
0	$f(x_2, x_3)$		$f(x_1, x_2, x_3, x_4)$		
$f(x_3)$		$f(x_2, x_3, x_4)$		·	·
0	·	$f(x_3, x_4)$	·	·	$f(x_1, x_2, \ldots, x_n)$
	·	·	·	·	·
	·	·	·	$f(x_{n-3}, x_{n-2}, x_{n-1}, x_n)$	·
$f(x_{n-1})$	·	$f(x_{n-2}, x_{n-1}, x_n)$			
0	$f(x_{n-1}, x_n)$				
$f(x_n)$					

in which we have inserted a column of zeros so that any reciprocal difference of order at least unity can be computed in essentially the same way from its two closest neighbours in the previous column and from the remaining corner of the rhombus that includes these three table entries:

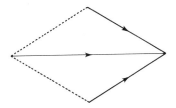

We now consider how to generalize the Neville interpolation formula to the rational case. For convenience, we will write

$$\alpha A = N(x; x_1, \ldots, x_n),$$

$$\alpha = D(x; x_1, \ldots, x_n),$$

$$\beta B = [f(x_1, \ldots, x_n) - f(x_2, \ldots, x_{n-1})] N(x; x_1, \ldots, x_{n-1}),$$

$$\beta = [f(x_1, \ldots, x_n) - f(x_2, \ldots, x_{n-1})] D(x; x_1, \ldots, x_{n-1}),$$

$$\gamma C = [f(x_1, \ldots, x_n) - f(x_2, \ldots, x_{n-1})] N(x; x_2, \ldots, x_n),$$

$$\gamma = [f(x_1, \ldots, x_n) - f(x_2, \ldots, x_{n-1})] D(x; x_2, \ldots, x_n),$$

$$\delta D = N(x; x_2, \ldots, x_{n-1}),$$

$$\delta = D(x; x_2, \ldots, x_{n-1}),$$

so that $A = f(x; x_1, \ldots, x_n)$, $B = f(x; x_1, \ldots, x_{n-1})$, $C = f(x; x_2, \ldots, x_n)$ and $D = f(x; x_2, \ldots, x_{n-1})$ are in the rhombus arrangement:

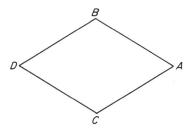

It can be seen that (136d) and (136e) together with the two similar equations with N replaced by D throughout can be written in the forms

$$\alpha A = \beta B + (x - x_1)\delta D, \tag{136h}$$

$$\alpha = \beta + (x - x_1)\delta, \tag{136i}$$

$$\alpha A = \gamma C + (x - x_n)\delta D, \tag{136j}$$

$$\alpha = \gamma + (x - x_n)\delta. \tag{136k}$$

Subtracting B times (136i) from (136h), subtracting C times (136k) from (136j) and dividing the resulting equations gives

$$\frac{A - B}{A - C} = \frac{x - x_1}{x - x_n} \frac{D - B}{D - C}.$$

If we write $\theta = (x - x_n)/(x - x_1)$ and rearrange we obtain

$$A = B + \frac{C - B}{1 - \theta(C - D)/(B - D)}, \tag{136l}$$

which is precisely the same as the Neville interpolation formula except for the factor $(C - D)/(B - D)$ in the denominator of the last term.

Finally, we look briefly at the case of rational interpolation in which the denominator has degree $[n/2]$ and the numerator only $[(n - 1)/2]$. Although the complete theory we have presented could be recast in this case, we confine ourselves to the rational form of the Neville interpolation formula. If the function g is the reciprocal of the function f it is clear that to obtain the interpolation formula we want, it is sufficient to interpolate g and take the reciprocal of this interpolation formula. We find that the first column of the approximation table is just of the function f itself, whereas the second column is computed from the formula

$$A = B + \frac{C - B}{1 - \theta C/B}.$$

From the third column onwards, we employ the formula (136l) just as for the case we considered in detail.

137 Padé approximations.

Just as there is a general type of Hermite interpolation that includes the Lagrange formula and Taylor's series as extreme cases, so there is a corresponding range of rational approximations with the methods of the last subsection at one end of the scale and, at other, a type of approximation of a given function by a rational function such that the given function and its approximation agree in value and in the value of its first, second, ... derivatives at one particular point. This last type of approximation is associated with the name Padé and, for convenience and standardization, we will take the point of agreement as being zero.

Thus, given a function f and two non-negative integers l and m, we consider the problem of finding a rational function $R = N/D$ such that $\deg(N) \leq l$, $\deg(D) \leq m$ and such that $R(0) = f(0)$, $R'(0) = f'(0)$, ... $R^{(l+m)}(0) = f^{(l+m)}(0)$. The full array of approximations of this form for $l, m = 0, 1, 2, \ldots$ is known as the Padé table for f. We will sometimes attach subscripts $R_{l,m} = N_{l,m}/D_{l,m}$ to indicate the degrees of a particular approximation, and we arrange the table in the pattern:

$$
\begin{array}{cccccc}
R_{00} & R_{10} & R_{20} & \cdot & \cdot & \cdot \\
R_{01} & R_{11} & R_{21} & \cdot & \cdot & \cdot \\
R_{02} & R_{12} & R_{22} & \cdot & \cdot & \cdot \\
\\
\cdot & \cdot & \cdot & \cdot & \cdot & \cdot \\
\cdot & \cdot & \cdot & \cdot & \cdot & \cdot
\end{array}
$$

given that all the entries do exist. Note that the top row is the sequence of Taylor series approximations and the first column gives the sequence of reciprocals of Taylor series for $1/f$.

By the diagonal of the Padé table, we will mean the sequence of approximations $R_{00}, R_{11}, R_{22}, \ldots$ and we will also call the sequences $R_{01}, R_{12}, R_{23}, \ldots,$ $R_{02}, R_{13}, R_{24}, \ldots$ the first, second, $\ldots$ subdiagonals of the Padé table.

Although for a given function not all members of the table necessarily exist, we at least have a uniqueness result.

THEOREM 137A.

If $R_{lm} = N/D$ and $\bar{R}_{lm} = \bar{N}/\bar{D}$ are each (l, m) Padé approximations to f, then $R_{lm} = \bar{R}_{lm}$.

Proof.

If $P = N\bar{D} - \bar{N}D$ then $R_{lm} - \bar{R}_{lm} = P/D\bar{D}$, where P is a polynomial of degree not exceeding $l + m$. Since $|P(x)/D(x)\bar{D}(x)| = O(|x|^{l+m+1})$ as $x \to 0$, it follows that P is the zero polynomial. ∎

We will, from now on, suppose that the formal Taylor series of f is

$$
f(x) = c_0 + c_1 x + c_2 x^2 + \cdots
$$

and, assuming that $R_{lm} = N/D$ exists, we will write

$$
N(x) = a_0 + a_1 x + a_2 x^2 + \cdots + a_l x^l,
$$

$$
D(x) = b_0 + b_1 x + b_2 x^2 + \cdots + b_m x^m.
$$

We will also conventionally write

$$
a_{l+1} = a_{l+2} = \cdots = b_{m+1} = b_{m+2} = \cdots = c_{-1} = c_{-2} = \cdots = 0.
$$

With this notation, we can immediately see that N/D is in fact the (l, m) Padé approximation if $N(x) = f(x)D(x) + O(x^{l+m+1})$, or if

$$
a_i = \sum_{j=0}^{i} b_j c_{i-j}, \qquad i = 0, 1, 2, \ldots, l + m, \tag{137a}
$$

so that $b_0, b_1, \ldots, b_m$ satisfy the homogeneous linear system found by writing

$i = l + 1, l + 2, \ldots, l + m$ in $(137a)$:

$$\begin{bmatrix} c_{l+1} & c_l & \cdots & c_{l-m+1} \\ c_{l+2} & c_{l+1} & \cdots & c_{l-m+2} \\ \vdots & \vdots & \vdots & \vdots \\ c_{l+m} & c_{l+m-1} & \cdots & c_l \end{bmatrix} \cdot \begin{bmatrix} b_0 \\ b_1 \\ \vdots \\ b_m \end{bmatrix} = 0. \qquad (137b)$$

Given that this is satisfied, $a_0, a_1, \ldots, a_l$ are given by writing $i = 0, 1, \ldots, l$ in $(137a)$.

To avoid generality, for which we have no real need, we will consistently confine ourselves to functions covered by the following definition.

DEFINITION 137B.

A function f with the formal Taylor series

$$f(x) = c_0 + c_1 x + c_2 x^2 + \cdots$$

is *normal* if, for every pair (λ, μ) of non-negative integers,

$$\det \begin{bmatrix} c_\lambda & c_{\lambda-1} & \cdots & c_{\lambda-\mu} \\ c_{\lambda+1} & c_\lambda & \cdots & c_{\lambda-\mu+1} \\ \vdots & \vdots & & \vdots \\ c_{\lambda+\mu} & c_{\lambda+\mu-1} & \cdots & c_\lambda \end{bmatrix} \neq 0. \qquad (137c)$$

We have the characteristic property of such functions expressed in the following theorem.

THEOREM 137C.

If f is a normal function and l, m are non-negative integers then the Padé approximation $R_{lm} = N/D$ exists, the degrees of N and D are given by $\deg(N) = l$ and $\deg(D) = m$, and if $| f(x) - R_{lm}(x)| = O(|x|^{n+1})$ as $x \to 0$ then n does not exceed $l + m$.

Proof.

If f is normal then $(137b)$ can be solved with $b_0 \neq 0$ (writing $\lambda = l$ and $\mu = m - 1$ in $(137c)$), and with $b_m \neq 0$ (writing $\lambda = l + 1$ and $\mu = m - 1$ in $(137c)$), so the existence of $R_{lm} = N/D$ with $\deg(D) = m$ follows. If $\deg(N) < l$ then R_{lm} would also be an $(l - 1, m + 1)$ Padé approximation but with $\deg(D) < m + 1$, and this we have just shown to be impossible. If $| f(x) - R_{lm}(x)| = O(|x|^{n+1})$ with $n > l + m$ then we would have an $(l, m + 1)$ Padé approximation with $\deg(D) < m + 1$, again impossible. ∎

As an example, consider the exponential function

$$\exp(x) = 1 + x + \frac{x^2}{2!} + \frac{x^3}{3!} + \cdots,$$

so that, for this function, $c_n = 1/n!$.

THEOREM 137D.

The exponential function is normal.

Proof.

If, on the contrary, there existed a null vector $[v_0, v_1, \ldots, v_n]^T$ for the matrix occurring in (137c), then

$$\sum_{i=0}^{\mu} \frac{v_i}{(j-i)!} = 0, \qquad j = \lambda, \lambda + 1, \ldots, \lambda + \mu. \tag{137d}$$

If we define a polynomial p by

$$p(x) = \sum_{i=0}^{\mu} \frac{v_i x^{\lambda + \mu - i}}{(\lambda + \mu - i)!},$$

then (137d) is equivalent to

$$p^{(\mu)}(1) = p^{(\mu-1)}(1) = \cdots = p'(1) = p(1) = 0,$$

implying that p has a $(\mu + 1)$-fold zero at unity. Adding to this the λ-fold zero at zero gives more zeros then the degree of p, which is therefore the zero polynomial and hence $v_0 = v_1 = \cdots = v_\mu = 0$. ∎

Assured that the full Padé table exists for this function, we now proceed to establish the form of a member of the table.

THEOREM 137E.

The (l, m) member of the Padé table for the exponential function is given by N/D, where

$$N(x) = \sum_{i=0}^{l} \frac{l!(l+m-i)!}{i!(l-i)!} x^i,$$

$$D(x) = \sum_{i=0}^{m} \frac{m!(l+m-i)!}{i!(m-i)!} (-x)^i.$$

Proof.

We will multiply the series for $\exp(x)$ by the polynomial given for $D(x)$ and obtain the polynomial given for $N(x)$ up to terms in x^{l+m}. The coefficient of

x^i in $D(x)\exp(x)$ is

$$\sum_{j=0}^{i} \frac{m!(l+m-j)!}{j!(m-j)!}(-1)^j \frac{1}{(i-j)!} = \frac{l!m!}{i!} \sum_{j=0}^{i} \binom{l+m-j}{m-j}(-1)^j \binom{i}{j}$$

$$= \frac{l!m!}{i!} \sum_{j=0}^{i} \binom{-l-1}{m-j}(-1)^m \binom{i}{j}$$

$$= \frac{l!m!}{i!}(-1)^m C, \tag{137e}$$

where C is the coefficient of x^m in $(1+x)^{-l-1}(1+x)^i = (1+x)^{i-l-1}$.
Thus

$$C = \binom{i-l-1}{m} = (-1)^m \binom{l+m-i}{m}$$

so that, substituting into (137e) we obtain the result

$$\frac{l!m!}{i!}\binom{l+m-i}{m} = \frac{l!(l+m-i)!}{i!(l-i)!}. \quad \blacksquare$$

In table 137 we present expressions for $R_{lm}(x)$ for $l \le 3$ and $m \le 6$. These expressions are those given by theorem 137E but with a common factor $[\min(l,m)]!$ divided out from both N and D. To obtain corresponding expressions with l and m interchanged, so as to extend this table slightly, we note that $R_{lm}(x) = 1/R_{ml}(-x)$.

Returning now to an arbitrary normal function, a detailed study of the Padé

Table 137   Padé approximations to exp(x)

1	$\dfrac{1+x}{1}$
$\dfrac{1}{1-x}$	$\dfrac{2+x}{2-x}$
$\dfrac{2}{2-2x+x^2}$	$\dfrac{6+2x}{6-4x+x^2}$
$\dfrac{6}{6-6x+3x^2-x^3}$	$\dfrac{24+6x}{24-18x+6x^2-x^3}$
$\dfrac{24}{24-24x+12x^2-4x^3+x^4}$	$\dfrac{120+24x}{120-96x+36x^2-8x^3+x^4}$
$\dfrac{120}{120-120x+60x^2-20x^3+5x^4-x^5}$	$\dfrac{720+120x}{720-600x+240x^2-60x^3+10x^4-x^5}$
$\dfrac{720}{720-720x+360x^2-120x^3+30x^4-6x^5+x^6}$	$\dfrac{5040+720x}{5040-4320x+1800x^2-480x^3+90x^4-12x^5+x}$

table reveals a number of quite remarkable relations between members of the table and some of its neighbours. We will examine here one small family of such relationships as an introduction to this body of work and as a tool for our immediate use in section 35.

If $R_{lm} = N/D$, we will write X_{lm} for the two-dimensional vector of polynomials given by

$$X_{lm} = \begin{bmatrix} N \\ D \end{bmatrix}.$$

We then have the following relationship between three nearby vectors.

THEOREM 137F.

If f is a normal function, $n = 0, 1$ or 2 and l, m are integers greater than unity, then there exist numbers α, β, γ such that

$$X_{l-n,m}(x) = (\alpha x + \beta)X_{l-1,m-1}(x) + \gamma x^2 X_{l-2,m-2}(x). \tag{137f}$$

Proof.

Consider the family of rational functions

$$\frac{(\alpha x + \beta)N_{l-1,m-1}(x) + \gamma x^2 N_{l-2,m-2}(x)}{(\alpha x + \beta)D_{l-1,m-1}(x) + \gamma x^2 D_{l-2,m-2}(x)}$$

with $\beta \neq 0$ but α, β, γ otherwise arbitrary. If

$$N_{l-1,m-1}(x) - D_{l-1,m-1}(x)f(x) = tx^{l+m-1} + ux^{l+m} + O(x^{l+m+1})$$

$$N_{l-2,m-2}(x) - D_{l-2,m-2}(x)f(x) = vx^{l+m-3} + wx^{l+m-2} + O(x^{l+m-1}),$$

$\dfrac{2 + 2x + x^2}{2}$	$\dfrac{6 + 6x + 3x^2 + x^3}{6}$
$\dfrac{6 + 4x + x^2}{6 - 2x}$	$\dfrac{24 + 18x + 6x^2 + x^3}{24 - 6x}$
$\dfrac{12 + 6x + x^2}{12 - 6x + x^2}$	$\dfrac{60 + 36x + 9x^2 + x^3}{60 - 24x + 3x^2}$
$\dfrac{60 + 24x + 3x^2}{60 - 36x + 9x^2 - x^3}$	$\dfrac{120 + 60x + 12x^2 + x^3}{120 - 60x + 12x^2 - x^3}$
$\dfrac{360 + 120x + 12x^2}{360 - 240x + 72x^2 - 12x^3 + x^4}$	$\dfrac{840 + 360x + 60x^2 + 4x^3}{840 - 480x + 120x^2 - 16x^3 + x^4}$
$\dfrac{2520 + 720x + 60x^2}{2520 - 1800x + 600x^2 - 120x^3 + 15x^4 - x^5}$	$\dfrac{6720 + 2520x + 360x^2 + 20x^3}{6720 - 4200x + 1200x^2 - 200x^3 + 20x^4 - x^5}$
$\dfrac{20\,160 + 5040x + 360x^2}{20\,160 - 15\,120x + 5400x^2 - 1200x^3 + 180x^4 - 18x^5 + x^6}$	$\dfrac{60\,480 + 20\,160x + 2520x^2 + 120x^3}{60\,480 - 40\,320x + 12\,600x^2 - 2400x^3 + 300x^4 - 24x^5 + x^6}$

where, by normality, $t \neq 0$, $v \neq 0$, then we can compute

$$(\alpha x + \beta)N_{l-1,m-1}(x) + \gamma x^2 N_{l-2,m-2}(x)$$
$$- f(x)[(\alpha x + \beta)D_{l-1,m-1}(x) + \gamma x^2 D_{l-2,m-2}(x)]$$
$$= (\beta t + \gamma v)x^{l+m-1} + (\alpha t + \beta u + \gamma w)x^{l+m} + O(x^{l+m+1}),$$

which is $O(x^{l+m})$ iff

$$\beta t + \gamma v = 0 \tag{137g}$$

and is $O(x^{l+m+1})$ iff (137g) holds and also

$$\alpha t + \beta u + \gamma w = 0. \tag{137h}$$

Suppose that the highest-degree terms of $N_{l-1,m-1}(x)$ and $N_{l-2,m-2}(x)$ are respectively $ax^{l-2} + bx^{l-1}$ and $cx^{l-3} + dx^{l-2}$ so that $N_{l-1,m-1}(x) - ax^{l-2} - bx^{l-1}$ is a polynomial in x of degree less than $l - 2$ and $N_{l-2,m-2}(x) - cx^{l-3} - dx^{l-2}$ is a polynomial of degree less than $l - 3$. The higher-degree terms in $(\alpha x + \beta)N_{l-1,m-1}(x) + \gamma x^2 N_{l-2,m-2}(x)$ are

$$(\alpha a + \beta b + \gamma c)x^{l-1} + (\alpha b + \gamma d)x^l,$$

so that this polynomial is of degree less than $l + 1$ iff

$$\alpha b + \gamma d = 0 \tag{137i}$$

and of degree less than l if (137i) holds and also

$$\alpha a + \beta b + \gamma c = 0. \tag{137j}$$

We now establish the result of the theorem in its three cases. If $n = 0$, select α, β, γ to satisfy (137g) and (137h) and to satisfy whatever normalization is appropriate for $X_{l,m}$. If $n = 1$, select α, β, γ suitably normalized to satisfy (137 g) and (137 i). Finally, if $n = 2$, select α, β, γ suitably normalized to satisfy (137i) and (137j). ∎

As an example of the use of this theorem, we consider the exponential function, for which the components of X_{lm} are given by theorem 137E. It is found that, for $n = 0$,

$$\alpha = \frac{l + m - 1}{l + m - 2}(m - l),$$

$$\beta = (l + m)(l + m - 1),$$

$$\gamma = \frac{(l - 1)(m - 1)(l + m)}{l + m - 2},$$

for $n = 1$,

$$\alpha = \frac{-l + 1}{l + m - 2}, \qquad \beta = l + m - 1, \qquad \gamma = \frac{(l - 1)(m - 1)}{l + m - 2},$$

and, for $n = 2$,

$$\alpha = -\frac{1}{l + m - 2}, \qquad \beta = 1, \qquad \gamma = \frac{m - 1}{l + m - 2}.$$

The justification for our assertion of these values of α, b, γ in the three cases is, of course, trivial, since it is only necessary to verify that for $f = \exp$, the coefficients of $1, x, x^2$ in the first components agree in (137f).

14 GRAPHS AND COMBINATORICS

140 Graphs.

In this section we will introduce the ideas of graphs and directed graphs and of rooted trees in particular. Although our choice of subject matter is biased towards questions applicable in later parts of this book, this is done in a general context. For further reading for graph theory, see, for example, Berge [1962].

By a graph, we will mean a finite set V together with a subset E of the set of pairs of distinct members of V. The members of V are known as vertices and the members of E are known as edges. We will represent a graph by using points in a diagram to denote the vertices and lines joining points to represent the edges. Thus, the graph (V, E) where $V = \{x, y, z\}$ and $E = \{\{x, y\}, \{y, z\}\}$ can be represented by the following diagram:

DEFINITION 140A.

A graph $\Gamma' = (V', E')$ is *isomorphic* to a graph $\Gamma = (V, E)$ if there is a bijection $\varphi: V \to V'$ such that for all $x, y \in V$, $\{x, y\} \in E$ iff $\{\varphi(x), \varphi(y)\} \in E'$.

The sense of this definition is that two graphs are isomorphic if the diagrams for them are the same, except for any labelling added to show the names of vertices or edges. We shall use the convention of using the word 'graph' to mean either a particular pair (V, E) or the isomorphic class corresponding to a particular diagram. If the more abstract meaning is used in a particular context, we will feel free to introduce vertex or edge names with which to label the graph.

DEFINITION 140B.

A graph $\Gamma' = (V', E')$ is a *subgraph* of the graph $\Gamma = (V, E)$ if $V' \subseteq V$ and $E' \subseteq E \cap (V' \times V')$.

A common type of subgraph (and the name 'subgraph' is often reserved for this case) is when V' is a subset of V and $E' = E \cap (V' \times V')$. This is known as the subgraph induced by V'. In terms of diagrams, the subgraph induced by V' is formed by simply removing all vertices that are not in V' and removing all edges whose endpoints are not both in V'.

DEFINITION 140C.

A *chain* is a sequence of distinct edges $e_1, e_2, \ldots, e_n$ such that $e_i = \{v_{i-1}, v_i\}$ for $i = 1, 2, \ldots, n$, where $v_0, v_1, \ldots, v_n \in V$. Such a chain is said to be a *chain from v_0 to v_n*. A *circuit* is a chain from a vertex to itself.

DEFINITION 140D.

A pair of vertices in a graph is *connected* if there is a chain from one to the other. A graph is *connected* if every pair of vertices is connected. A *connected component* of a graph (V, E) is a subset V' of V such that the subgraph induced by V' is connected but V' is not properly contained in a subset V'' of V for which the subgraph induced by V'' is connected.

DEFINITION 140E.

A *non-rooted tree* is a connected graph with no circuits.

Note that our terminology is not standard in that we reserve the term 'tree' for what is usually known as a rooted tree rather than the entities defined in definition 140E. The following graphs are not non-rooted trees since they are disconnected or else have circuits:

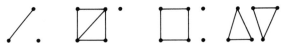

By contrast, we list below all the non-rooted trees with less than six vertices:

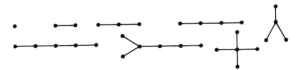

If one particular vertex of V is distinguished from the rest and called the 'root' of the tree we have what is called a 'rooted tree'. Formally, a rooted tree is a triple (V, E, r), where (V, E) is a non-rooted tree and $r \in V$ is the root. Two rooted trees (V', E', r') and (V, E, r) would only be isomorphic if there exists a bijection of the type specified in definition 140A which maps r to r'. The rooted trees with less than six vertices are shown below with the root distinguished by the style of point used to represent this vertex:

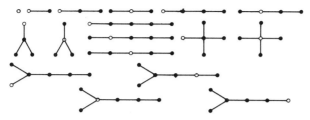

DEFINITION 140F.

The *symmetry group* of a (non-rooted or rooted) tree is the group of iso-morphisms of the tree with itself. The *symmetry* of a tree is the order of its symmetry group.

For example, consider the tree

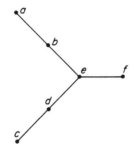

for which $V = \{a, b, c, d, e, f\}$, $E = \{\{a, b\}, \{b, e\}, \{c, d\}, \{d, e\}, \{e, f\}\}$. The only isomorphisms are the identity mapping on V and the mapping that inter-changes a and c and also interchanges b and d. Thus the symmetry is 2 and the symmetry group is the unique group of this order. Note that the rooted tree formed by making e or f the root has this same symmetry, whereas if one of a, b, c or d is the root, the symmetry is 1.

141 A tree enumeration problem.

We consider the problem of counting all non-rooted trees (V, E) in which V is a fixed set of cardinality n, say, but E is allowed to vary. In other words, we wish to know how many fully labelled non-rooted trees there are with n vertices. Since the result depends only on n, we will, for convenience, take V to be a subset of some totally ordered infinite set such as the set of all integers.

THEOREM 141A.

The number of completely labelled non-rooted trees with n vertices is n^{n-2}.

Proof.

The result is easily verified in the trivial cases $n = 1$ and $n = 2$. For $n = 3, 4, \ldots$ we will establish a bijective relation between the set of all non-rooted trees with vertex set V (where $\# V = n$) and the set V^{n-2} of $(n - 2)$-dimensional vectors with components in V. This will be done by induction on n. For $n = 3$, the bijection relates the tree with centre vertex $v \in V$ to the vector $[v]$, and the result is obvious in this case.

Suppose now that the result has been proved for cardinalities less than n where $n > 3$. If t is some tree labelled with the elements of V, let v denote the greatest vertex such that it lies in a single edge $\{v, w\}$. (That such 'terminal vertices' as v exists follows easily from the non-existence of circuits in a non-rooted tree.) If $V' = V \backslash \{v\}$ so that $\# V' = n - 1$, then the subgraph, t' say, induced by V' is also a non-rooted tree and, by the induction hypothesis, it will have a corresponding vector in $(V')^{n-3}$ in the bijection associated with the cardinality $n - 1$. If this vector is $[x_1, x_2, \ldots, x_{n-3}]$ then we define the vector in V^{n-2} corresponding to t to be $[w, x_1, x_2, \ldots, x_{n-3}]$. Clearly, every such vector arises just once and we have the required bijection.

The number of completely labelled non-rooted trees is thus the number of members of V^{n-2} which is just n^{n-2}. ∎

We now extend this result to rooted trees.

COROLLARY 141B.

The number of rooted trees with n vertices such that every vertex except the root is labelled from a fixed set of $n - 1$ labels is n^{n-2}. The number of fully labelled rooted trees is n^{n-1}.

Proof.

The first part of the result follows by regarding the designation 'root' as an additional label and using theorem 141A. The second part follows by noting that for each fully labelled unrooted tree with n vertices there are n distinct positions where a root may be placed. ∎

142 Directed graphs.

In contrast to the concept of graph introduced in subsection 140, we consider here the idea of a directed graph consisting of a set V of vertices and a set R of *ordered* pairs of members of V. Thus R is a relation on V and we shall call members of R the arcs of the graph (V, R). Just as for graphs, we will exclude the possibility that an arc begins and ends at the same vertex. Thus R is an irreflexive relation. If (x, y) is an arc, y will be called the *successor* of x and x will be called the *predecessor* of y.

DEFINITION 142A.

A directed graph (V', R') is *isomorphic* to a directed graph (V, R) if there is a bijection $\varphi\colon V \to V'$, such that for all $x, y \in V$, $(x, y) \in R$ iff $(\varphi(x), \varphi(y)) \in R'$.

As with graphs, we will use the words 'directed graph' to refer to a particular pair (V, R) or to an isomorphic class corresponding to a particular diagram. In diagrammatic representations, an arc (x, y) will be indicated as

DEFINITION 142B.

A directed graph (V, R) is said to be *injective* if each vertex has no more than one predecessor.

DEFINITION 142C.

A *path* is a sequence of distinct arcs $r_1, r_2, \ldots, r_n$ such that $r_i = (v_{i-1}, v_i)$ for $i = 1, 2, \ldots, n$, where $v_0, v_1, \ldots, v_n \in V$. Such a path is said to be a *path from* v_0 *to* v_n. A *cycle* is a path from a vertex to itself.

DEFINITION 142D.

Two vertices v_1, v_2 are *connected from below* if there is a path from v_1 to v_2, a path from v_2 to v_1 or a path from v_0 to v_1 and from v_0 to v_2, where v_0 is some other vertex. A directed graph is said to be *connected from below* if every two vertices are connected from below.

DEFINITION 142E.

A directed graph (V, R) is *ordered* if there exists a total ordering on V such that if (v_1, v_2) is an arc then $v_1 < v_2$.

THEOREM 142F.

A directed graph is *ordered* iff it has no cycles.

Proof.

If an ordered directed graph had a cycle $[(v_0, v_1), (v_1, v_2), \ldots, (v_{n-1}, v_0)]$ then $v_0 < v_1 < v_2 < \cdots < v_0$, which is impossible. On the other hand, let (V, R) be a directed graph without cycles and let R' be a maximal subset of R such that (V, R') is ordered. Let $r = (v, w)$ be a member of R but not of R' and let V' be the set of vertices to which there is a path from w. Since (V, R) has no cycles, $v \notin V'$. Define a new ordering on V such that $x < y$ if x, $y \in V'$ have this relationship in the old ordering, that $x < y$ if $x, y \in V \backslash V'$ have this relationship in the old ordering and that $x < y$ whenever $x \in V \backslash V'$, $y \in V'$. That this is still a total ordering implies that $(V, R' \cup \{r\})$ is an ordered directed graph, contradicting the assumption that R' is maximal. ∎

DEFINITION 142G.

A *directed tree* is an injective ordered directed graph which is connected from below. The *root* of a directed tree is the infimum of the set of all vertices.

Note that in this definition it is tacitly assumed that the root is independent of the choice of ordering. This, however, is clear since if there were two roots they would not be connected from below.

We now relate the concept of directed tree with that of rooted tree.

THEOREM 142H.

Let (V, R) be a directed tree and r its root. Let E, a set of vertex pairs, be defined by $E = \{\{v, w\}: (v, w) \in R\}$; then (V, E, r) is a rooted tree.

Proof.

It is clear that (V, E) is connected. To show that it has no circuits, suppose, on the contrary, that $e_1, e_2, \ldots, e_n$ is a circuit where $e_1 = \{v_0, v_1\}, \ldots, e_n = \{v_{n-1}, v_0\}$. Without loss of generality we may suppose that $(v_0, v_1) \in R$ (if this were not the case, $(v_1, v_0) \in R$ and we could reverse the order of members of the circuit). Since (V, R) is injective, it follows in turn that $(v_1, v_2) \in R$, $(v_2, v_3) \in R, \ldots, (v_{n-1}, v_0) \in R$, so that we have a cycle, contradicting the assumption that (V, R) is a directed tree. ∎

We now consider a converse to this result and thereby establish the essential equivalence of the concepts of directed trees and rooted trees.

THEOREM 142I.

Let (V, E, r) be a rooted tree and let $R \subseteq V \times V$ be defined so that $(v, w) \in R$ iff there is a chain from r to w whose last member is $\{v, w\}$. Then (V, R) is a directed tree with root r.

Proof.

We first show that (V, R) is injective. Let w be a vertex different from r. If there were chains $e_1, e_2, \ldots, e_m$ and $f_1, f_2, \ldots, f_n$ from r to w where $e_m = (u, w)$, $f_n = (v, w)$ and $u \neq v$, the sequence $e_m, e_{m-1}, \ldots, e_1, f_1, f_2, \ldots,$

f_n is a circuit or else there is an edge of the first chain equal to one from the second chain. In the latter case, let p be the greatest member of $\{1, 2, \ldots, m\}$, such that $e_p = f_q$ for some $q \in \{1, 2, \ldots, n\}$. Clearly, $p = m$, $q = n$ cannot both hold and hence $e_m, e_{m-1}, \ldots, e_{p+1}, f_{q+1}, \ldots, f_n$ is a circuit. This contradicts the assumption that (V, E) is a tree. It is now easy to see that (V, R) is connected from below since, if the vertices v, w are each different from r, then there are paths from r, to v and from r to w. ∎

Throughout this book, we will call a directed or a rooted tree simply a tree. Usually we will use the directed graph interpretation of trees.

143 Notations for trees.

We will generally write T for the set of all trees and τ for the trivial tree with exactly one vertex. Let $t_1 = (V_1, R_1)$, $t_2 = (V_2, R_2)$, ..., $t_s = (V_s, R_s)$ be trees where we can suppose that $V_1, V_2, \ldots, V_s$ are disjoint. Let $r_1, r_2, \ldots, r_s$ be the roots of these trees and let $t = (V, R)$ where $V = \{r\} \cup V_1 \cup V_2 \cup \cdots \cup V_s$ with $r \notin V_1 \cup V_2 \cup \cdots \cup V_s$ and $R = \{(r, r_1), (r, r_2), \ldots, (r, r_s)\} \cup R_1 \cup R_2 \cup \cdots \cup R_s$. It is easy to see that t is a tree with root r. We write this as

$$t = [t_1 t_2 \cdots t_s]. \tag{143a}$$

It is also clear that if we are given a tree t with more than one vertex, then we can find a decomposition of the form of (143a) by simply removing the root and all arcs from it and identifying $t_1, t_2, \ldots, t_s$ with the connected components that remain.

We will introduce two aids to conciseness in using this notation. In the first place, if there are repeated trees among $t_1, t_2, \ldots, t_s$ then they can be combined together using a power notation. Thus, for example, $[t_1 t_1 t_1 t_2 t_2]$ can also be written $[t_1^3 t_2^2]$. The second aid to conciseness is to place subscripts on [and on] to indicate (matching) repetitions of these symbols. Thus, we will write $[[[\tau]\tau]]$ also as $[_2[\tau]\tau]_2$ (but *not* as $[_3\tau]\tau]_2$).

Using the notation introduced here, it is a simple matter to write any particular tree as a sequence of the symbols $\tau, [,]$, with the optional use of numerical superscripts and subscripts where this gives a more concise representation. As an example, consider the tree:

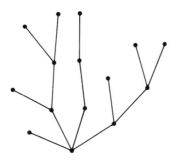

86

Table 143   Notations for trees

•	τ	τ
	$[\tau]$	$\tau\tau$
	$[\tau^2]$	$\tau\tau \cdot \tau$
	$[_2\tau]_2$	$\tau \cdot \tau\tau$
	$[\tau^3]$	$(\tau\tau \cdot \tau)\tau$
	$[\tau[\tau]]$	$\tau\tau \cdot \tau\tau$
	$[_2\tau^2]_2$	$\tau(\tau\tau \cdot \tau)$
	$[_3\tau]_3$	$\tau(\tau \cdot \tau\tau)$
	$[\tau^4]$	$(\tau\tau \cdot \tau)\tau \cdot \tau$
	$[\tau^2[\tau]]$	$(\tau\tau \cdot \tau) \cdot \tau\tau$
	$[\tau[\tau^2]]$	$\tau\tau \cdot (\tau\tau \cdot \tau)$

Continued

Table 143 Continued

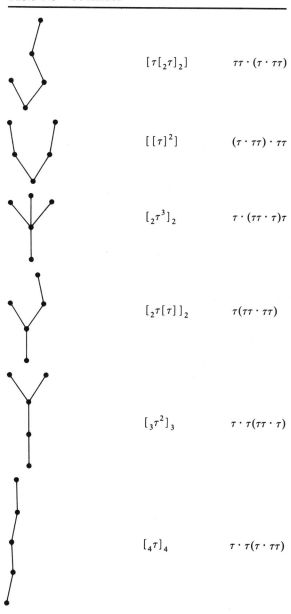

$[\tau[_2\tau]_2]$	$\tau\tau \cdot (\tau \cdot \tau\tau)$
$[[\tau]^2]$	$(\tau \cdot \tau\tau) \cdot \tau\tau$
$[_2\tau^3]_2$	$\tau \cdot (\tau\tau \cdot \tau)\tau$
$[_2\tau[\tau]]_2$	$\tau(\tau\tau \cdot \tau\tau)$
$[_3\tau^2]_3$	$\tau \cdot \tau(\tau\tau \cdot \tau)$
$[_4\tau]_4$	$\tau \cdot \tau(\tau \cdot \tau\tau)$

which can be written as

$$[\tau[\tau[\tau^2]]^2[_2\tau]_2].$$

A second notation which we will sometimes find convenient to use is based on a binary operation on T. Define the operation $\cdot : T \times T \to T$ by the formulae

$$\tau \cdot u = [u]$$
$$[t_1 \, t_2 \cdots t_s] \cdot u = [t_1 \, t_2 \cdots t_s u].$$

In using this notation, we will frequently elide the symbol $\cdot$ where this cannot cause confusion. If juxtaposition and $\cdot$ occur at the same level in an expression involving trees, we will, conventionally, perform the juxtapositions first. Thus, for example, $\tau \cdot \tau\tau$ means $\tau \cdot (\tau \cdot \tau)$.

it is clearly possible to write any tree (excepting τ) as the product of two trees with fewer vertices since $t = [t_1 \, t_2 \cdots t_s] = u \cdot v$, where $u = [t_1 \, t_2 \cdots t_{s-1}]$, $v = t_s$. Thus, we can write any tree in terms of τ using sufficient $\cdot$ operations. For example, the tree $[\tau[\tau[\tau^2]]^2[_2\tau]_2]$ which we have considered previously can be written as $(\tau\tau \cdot (\tau\tau \cdot (\tau\tau \cdot \tau)))(\tau \cdot \tau\tau) \cdot (\tau\tau \cdot (\tau\tau \cdot \tau))$.

It should be noted that the following identity holds for all $t, u, v \in T$:

$$tu \cdot v = tv \cdot u.$$

In table 143, all trees with less than six vertices are shown with corresponding expressions in terms of the two notations introduced here.

144 Some functions defined on trees.

Let $r(t)$ denote the *order* or number of vertices in a tree t. Clearly, r satisfies each of two recursions with initial value $r(\tau) = 1$, as follows:

$$r([t_1 t_2 \cdots t_s]) = 1 + r(t_1) + r(t_2) + \cdots + r(t_s),$$

$$r(tu) = r(t) + r(u).$$

The *height* H of a tree is defined as 1 plus the length of the longest possible path in the tree. Thus,

$$H(\tau) = 1,$$

$$H([t_1 t_2 \cdots t_s]) = 1 + \max_i H(t_i),$$

$$H(tu) = \max[H(t), 1 + H(u)].$$

In defining the *width* of a tree as the number of terminal vertices, that is as the number of vertices with no successors, it is sometimes convenient to define the width of τ as 0, rather than 1. We define $w(t)$ as the width of t using the first convention and $\bar{w}(t)$ using the second convention. Thus $w(t) = \bar{w}(t)$ if $t \neq \tau$ and $w(\tau) = 0$, $\bar{w}(\tau) = 1$. We have the recursions

$$w([t_1 t_2 \cdots t_s]) = \bar{w}([t_1 t_2 \cdots t_s]) = \bar{w}(t_1) + \bar{w}(t_2) + \cdots + \bar{w}(t_s),$$

$$w(tu) = \bar{w}(tu) = w(t) + \bar{w}(u).$$

The *density* γ of a tree is defined to satisfy the recursion

$$\gamma(\tau) = 1,$$

$$\gamma([t_1 t_2 \cdots t_s]) = r([t_1 t_2 \cdots t_s])\gamma(t_1)\gamma(t_2) \cdots \gamma(t_s). \tag{144a}$$

Although the density does not have such an obvious significance as the other functions we have introduced, it can be thought of as a measure of the non-bushiness of a tree. In two extreme types of trees with $H(t) = 2$ and $H(t) = r(t)$ we have $\gamma(t) = r(t)$ and $\gamma(t) = r(t)!$ respectively. The function σ, the *symmetry* introduced in definition 140F, also satisfies a recursion as expressed in the following theorem.

THEOREM 144A.

If $t = [t_1^{n_1} t_2^{n_2} \ldots t_s^{n_s}]$, where $t_1, t_2, \ldots, t_s$ are all distinct, then

$$\sigma(\tau) = 1, \tag{144b}$$

$$\sigma(t) = n_1! n_2! \ldots n_s! \sigma(t_1)^{n_1} \sigma(t_2)^{n_2} \cdots \sigma(t_s)^{n_s}. \tag{144c}$$

Proof.

Equation (144b) is trivial. In the case where t has more than one vertex, let (V, R) be a labelling of t such that the root is v. For each $i = 1, 2, \ldots, s$, let the n_i copies of t_i have vertex sets V_{ij} ($j = 1, 2, \ldots, n_i$). Let G_{ij} denote the symmetry group on the copy number j of t_i so that the order of G_{ij} is $\sigma(t_i)$. Let G_i denote a group of order $n_i!$ which permutes the sets V_{ij} ($j = 1, 2, \ldots, n_i$) amongst themselves in such a way as to preserve isomorphism of the subtrees with vertex sets V_{ij}. A member of the symmetry group of t can be written uniquely as a product

$$g_1 g_2 \cdots g_s g_{11} g_{12} \cdots g_{1n_1} g_{21} g_{22} \cdots g_{sn_s}$$

where

$$g_i \in G_i, \ g_{ij} \in G_{ij} \qquad (i = 1, 2, \ldots, s, \ j = 1, 2, \ldots, n_i),$$

and, accordingly, the order of this group is the product of the orders of the groups $G_1, G_2, \ldots, G_{11}, G_{12}, \ldots, G_{sn_s}$. This fact is equivalent to (144c). ∎

145 Some enumeration problems.

We first obtain a count of all rooted trees with a given number of vertices. Let a_i denote the number of distinct trees with i vertices.

THEOREM 145A.

The numbers $a_1, a_2, \ldots$ satisfy the relation

$$a_1 + a_2 x + a_3 x^2 + \cdots = \prod_{n=1}^{\infty} (1 - x^n)^{-a_n}. \tag{145a}$$

Note that this means that the formal expansion of the infinite product on the right-hand side agrees term by term with the left-hand side.

Proof.

Let U be a finite subset of T and let U' denote the set of all trees of the form $[t_1 t_2 \cdots t_s]$ where $t_1, t_2, \ldots, t_s \in U$. Conventionally, we suppose that $\tau \in U'$. Let $\phi_V(x) = \Sigma_{t \in V} x^{r(t)-1}$ for V any subset of T. Note that $\phi_V(x)$ denotes a formal series.

We first prove by induction on the cardinality of U that

$$\phi_{U'}(x) = \prod_{t \in U} (1 - x^{r(t)})^{-1}. \tag{145b}$$

This is certainly true for U empty. Let $U = U_0 \cup \{u\}$ where $u \notin U_0$ and suppose (145b) has been established with U replaced by U_0. Every tree in U' is of the form $[t_1 t_2 \ldots t_s u^n]$ where $t_1, t_2, \ldots, t_s \in U_0$ and $n = 0, 1, \ldots$. Hence, $\phi_{U'}(x) = \phi^0(x) + \phi^1(x) + \phi^2(x) + \cdots$ where $\phi^0, \phi^1, \ldots$ denote the contributions in which $n = 0, 1, \ldots$ respectively. However, $\phi^0(x) = \phi_{U_0'}(x)$, $\phi^1(x) = x^{r(u)} \phi_{U_0'}(x)$ (because the trees of the form $[t_1 t_2 \cdots t_s u]$ are the same in number as those in U' except that their orders are higher by $r(u)$), $\phi^2(x) = x^{2r(u)} \phi_{U_0'}(x), \ldots$; Hence,

$$\phi_{U'}(x) = (1 + x^{r(u)} + x^{2r(u)} + \cdots) \phi_{U_0'}(x)$$

$$= (1 - x^{r(u)})^{-1} \prod_{t \in U_0} (1 - x^{r(t)})^{-1}$$

so that (145b) follows.

If U is the set of all trees with less than n vertices, then U' certainly contains all with less than $n + 1$ vertices. Hence, a_n is the coefficient of x^{n-1} in

$$\phi_{U'}(x) = \prod_{t \in U} (1 - x^{r(t)})^{-1}$$

$$= \prod_{i=1}^{n-1} (1 - x^i)^{-a_i}.$$

Multiplying the right-hand side by $\prod_{i \geq n} (1 - x_i)^{-a_i}$ does not effect the coefficient of x^{n-1}, so that a_n is the coefficient of x^{n-1} in $\prod_{i=1}^{\infty} (1 - x_i)^{-a_i}$ and (145a) follows. ∎

From theorem 145A, $a_1, a_2, \ldots$ can be computed recursively. By comparing coefficients of x^0 we find $a_1 = 1$; by substituting on the right-hand side of (145a) and then comparing coefficients of x^1 we find $a_2 = 1$. Continuing in this way enables us to find a_n for any required n. We show these values up to $n = 10$, together with values of $\Sigma_{i=1}^n a_i$, in table 145.

We now consider a particular rooted tree and ask how many ways it can be labelled under various rules.

Let $\beta(t)$ denote the number of ways of labelling a tree t with $r(t) - 1$ distinct labels on condition that the root is not labelled but every other vertex is labelled. Also, let $\bar{\beta}(t)$ denote the number of ways of carrying out this labelling with $r(t)$ labels so that every vertex is labelled.

Table 145   Numbers of trees of various orders

n	a_n	$\sum_{i=1}^{n} a_i$	n	a_n	$\sum_{i=1}^{n} a_i$
1	1	1	6	20	37
2	1	2	7	48	85
3	2	4	8	115	200
4	4	8	9	286	486
5	9	17	10	719	1205

THEOREM 145B.

For any $t \in T$,

$$\beta(t) = \frac{[r(t) - 1]!}{\sigma(t)},$$

$$\bar{\beta}(t) = \frac{r(t)!}{\sigma(t)}.$$

Proof.

We regard two labellings of all vertices (in the case of $\bar{\beta}$) or all except the root (in the case of β) as equivalent if there is a member of the symmetry group mapping from one to the other. Since the order of the symmetry group is $\sigma(t)$, we obtain the required result by dividing the order of the symmetric group on the set of all vertices that are to be labelled by $\sigma(t)$. ∎

The functions β and $\bar{\beta}$ will have an important application in chapter 3. In that same chapter, we will also need to consider functions $\beta_n, \bar{\beta}_n$ (for n a non-negative integer) defined so that, for a given tree t, $\beta_n(t)$ denotes the number of ways of labelling n of the vertices of t with n distinct labels on condition that (a) every terminal is labelled and (b) the root is not labelled.

Similarly, the function $\bar{\beta}_n(t)$ is defined in just the same way but with the omission of the restriction (b) on the labelling of the root. Conventionally, we will regard the single vertex of τ as being a terminal for the definition of $\bar{\beta}$ but as not being a terminal for the definition of β. Thus $\beta_n(\tau)$ is unity if $n = 0$ but zero otherwise and $\bar{\beta}_n(\tau)$ is unity if $n = 1$ but zero otherwise.

THEOREM 145C.

For all $t \in T$ and all $n = 0, 1, 2, \ldots$,

$$\beta_n(t) = \frac{n! [r(t) - 1 - w(t)]!}{\sigma(t)[n - w(t)]! [r(t) - 1 - n]!},$$

$$\bar{\beta}_n(t) = \frac{n! [r(t) - \bar{w}(t)]!}{\sigma(t)[n - \bar{w}(t)]! [r(t) - n]!}.$$

Proof.

Allocating $w(t)$ (respectively $\bar{w}(t)$) of the n labels to the terminals can be done in $n!/[n - w(t)]!$ (respectively $n!/[n - \bar{w}(t)]!$) ways while the allocation of the remaining labels to the remaining vertices can be done in $[r(t) - 1 - w(t)]!/[r(t) - 1 - n]!$ (respectively $[r(t) - \bar{w}(t)]!/[r(t) - n]!$) ways. Multiplying these factors and dividing by $\sigma(t)$ to allow for isomorphic labellings gives the required results. ∎

COROLLARY 145D.

For all $t \in T$,

$$\beta(t) = \beta_{r(t) - 1}(t), \tag{145c}$$

$$\bar{\beta}(t) = \beta_{r(t)}(t), \tag{145d}$$

and for all $t \in T$ and $n = 1, 2, 3, \ldots$,

$$\bar{\beta}_n(t) = \beta_n(t) + n\beta_{n-1}(t). \tag{145e}$$

Proof.

Each of (145c) and (145d) follows immediately from theorems 145B and 145C whereas (145e) follows after a little manipulation from theorem 145C. More directly, (145e) can be seen by counting the labellings that contribute to $\bar{\beta}_n(t)$ in two subsets. The first are the $\beta_n(t)$ labellings for which the root does not receive a label and the remainder are made up from n ways of labelling the root times $\beta_{n-1}(t)$ ways of labelling the other vertices with the remaining labels. ∎

The final enumeration question we will consider concerns a function α on T, defined so that $\alpha(t)$ is the number of ways of labelling t with a given totally ordered set V (with $\# V = r(t)$) in such a way that if (m, n) is an arc then $m < n$.

THEOREM 145E.

For all $t \in T$,

$$\alpha(t) = \frac{r(t)!}{\gamma(t)\sigma(t)}$$

Proof.

Both sides equal unity for $t = \tau$. We complete a proof by induction for the tree $t = [t_1^{n_1} t_2^{n_2} \cdots t_s^{n_s}]$ where $t_1, t_2, \ldots, t_s$ are all distinct. There is a unique choice (the minimum of V) for the label attached to the root, and the number of ways

Table 145'   Various functions on trees

t	$r(t)$	$\sigma(t)$	$\gamma(t)$	$w(t), \bar{w}(t)$	$H(t)$	$\alpha(t)$	$\beta(t)$	$\bar{\beta}(t)$
τ	1	1	1	0, 1	1	1	1	1
$[\tau]$	2	1	2	1	2	1	1	2
$[\tau^2]$	3	2	3	2	2	1	1	3
$[_2\tau]_2$	3	1	6	1	3	1	2	6
$[\tau^3]$	4	6	4	3	2	1	1	4
$[\tau[\tau]]$	4	1	8	2	3	3	6	24
$[_2\tau^2]_2$	4	2	12	2	3	1	3	12
$[_3\tau]_3$	4	1	24	1	4	1	6	24
$[\tau^4]$	5	24	5	4	2	1	1	5
$[\tau^2[\tau]]$	5	2	10	3	3	6	12	60
$[\tau[\tau^2]]$	5	2	15	3	3	4	12	60

continued

Table 145' Continued

t	$r(t)$	$\sigma(t)$	$\gamma(t)$	$w(t), \overline{w}(t)$	$H(t)$	$\alpha(t)$	$\beta(t)$	$\overline{\beta}(t)$
$[\tau[_2\tau]_2]$	5	1	30	2	4	4	24	120
$[[\tau]^2]$	5	2	20	2	3	3	12	60
$[_2\tau^3]_2$	5	6	20	3	3	1	4	20
$[_2\tau[\tau]]_2$	5	1	40	2	4	3	24	120
$[_3\tau^2]_3$	5	2	60	2	4	1	12	60
$[_4\tau]_4$	5	1	120	1	5	1	24	120

of allocating the remaining labels among the $n_1 + n_2 + \cdots + n_s$ subtrees is

$$\frac{[r(t) - 1]!}{r(t_1)!^{n_1} n_1! r(t_2)!^{n_2} n_2! \cdots r(t_s)!^{n_s} n_s!}.$$

Multiplying this by

$$\prod_{i=1}^{s} \left[\frac{r(t_i)!}{\gamma(t_i)\sigma(t_i)} \right]^{n_i}$$

to allow for the actual labelling of the subtrees and making use of (144a) and (144c) gives the required result. ∎

In table 145′, values of $r(t)$, $\sigma(t)$, $\gamma(t)$, $w(t)$ (respectively $\bar{w}(t)$), $H(t)$, $\alpha(t)$, $\beta(t)$ and $\bar{\beta}(t)$ are given for each $t \in T$ such that $r(t) \leq 5$. It will be observed that $\Sigma_{r(t) = n}\alpha(t) = (n - 1)!$, $\Sigma_{r(t) = n}\beta(t) = n^{n-2}$, $\Sigma_{r(t) = n}\bar{\beta}(t) = n^{n-1}$, as provided by corollary 141B.

146 An algebraic structure related to trees.

We will introduce some algebraic structures appropriate for analysing some of the numerical methods introduced in chapters 3 and 4.

Let $T^{\#}$ denote the set of all (rooted) trees together with the empty tree which we will denote by $\emptyset$. We will write $\mathbb{R}^T$ and $\mathbb{R}^{T^{\#}}$ respectively for the set of all real-valued functions on T or on $T^{\#}$.

If a particular tree t is labelled as (V, R) and $r \notin V$, we consider the set of all partitions of $V \cup \{r\}$ into subsets V_1 and V_2 such that the root of V is in V_1 but r is in V_2 and such that if $(x, y) \in R$ then either x and y are both in V_1 or both in V_2 or x is in V_1 and y is in V_2. We define corresponding arc sets R_1 and R_2 such that if $(x, y) \in R$ with both x and y in V_1 then $(x, y) \in R_1$, such that if $(x, y) \in R$ with both x and y in V_2 then $(x, y) \in R_2$ and such that if $(x, y) \in R$ with $x \in V_1$ and $y \in V_2$ then $(r, y) \in R_2$. Such a partitioning of V, together with the construction of R_1 and R_2, represents a kind of pruning and grafting operation. Starting with t and a single stump (the isolated root r), we prune various parts from t to leave the tree (V_1, R_1) and then graft the cuttings to r to form a second tree (V_2, R_2). If $t_1 = (V_1, R_1)$ and $t_2 = (V_2, R_2)$ then we denote by $P(t)$ the set of all such pairs as (t_1, t_2) where differently labelled copies of the same pairs are counted separately.

We now define a binary operation $*$ on $\mathbb{R}^T$ by

$$(\alpha * \beta)(t) = \sum_{(t_1, t_2) \in P(t)} \alpha(t_2)\beta(t_1), \tag{146a}$$

for all α, $\beta \in \mathbb{R}^T$. We now show that this operation is associative.

LEMMA 146A.

If α, β, $\gamma \in \mathbb{R}^T$ then

$$(\alpha * \beta) * \gamma = \alpha * (\beta * \gamma).$$

Proof.

Let $t = (V, R)$ be an arbitrary tree, then we must prove that

$$[(\alpha * \beta) * \gamma](t) = \sum_{(t_1, u) \in P(t)} \sum_{(t_2, t_3) \in P(u)} \alpha(t_3)\beta(t_2)\gamma(t_1) \qquad (146b)$$

and

$$[\alpha * (\beta * \gamma)](t) = \sum_{(v, t_3) \in P(t)} \sum_{(t_1, t_2) \in P(v)} \alpha(t_3)\beta(t_2)\gamma(t_1) \qquad (145c)$$

are equal. If, in the two summations, r_2 and r_3 are the roots of t_2 and t_3 respectively and $t_i = (V_i, R_i)$ for $i = 1, 2, 3$ then the restrictions on possible choices of V_1, V_2, V_3 are the same in each of (146b) and (146c); that is if $(x, y) \in R$ and $x \in V_m$, $y \in V_n$ for $m, n \in \{1, 2, 3\}$ then each restriction is that $m \le n$. ∎

Since the right-hand side of (146a) is a bilinear functional on the vector of values of α and β, we will introduce a basis for such functionals. We write $b(t_2, t_1)$ for that mapping on $\mathbb{R}^T \times \mathbb{R}^T \to \mathbb{R}$ such that for all $\alpha, \beta \in \mathbb{R}^T$,

$$b(t_2, t_1)(\alpha, \beta) = \alpha(t_2)\beta(t_1).$$

Hence, if we define a function $L : T \to M$ where M is the set of bilinear functionals on $\mathbb{R}^T$, by the formula

$$L(t) = \sum_{(t_1, t_2) \in P(t)} b(t_2, t_1),$$

then (146a) can be written

$$(\alpha * \beta)(t) = L(t)(\alpha, \beta).$$

We will wish to have a convenient way of computing $L(t)$ for the set of $t \in T$ so we will seek a formula for expressing $L(tu)$ in terms of $L(t)$ and $L(u)$. Note that tu represents the product introduced in subsection 143. In addition to using this and other notations introduced there we will write $t \circ u$ as that tree formed from t and u by identifying their roots. Thus, if $t = [t_1 t_2 \cdots t_m]$, $u = [u_1 u_2 \cdots u_n]$ then $t \circ u = [t_1 t_2 \cdots t_m u_1 u_2 \cdots u_n]$.

If $a = \Sigma_i a_i b(s_i, t_i)$ and $b = \Sigma_j b_j b(u_j, v_j)$ are given members of M, where each of the summations is over some finite set of integers, we define their product by the formula

$$ab = \sum_i \sum_j a_i b_j b(s_i \circ u_j, t_i v_j).$$

Furthermore, we define an operation on $M \times T$ to M by

$$at = \sum_i a_i b(s_i t, t_i).$$

Using these two operations, we give a recursion for L.

LEMMA 146B.

The function $L : T \to M$ satisfies

$$L(\tau) = b(\tau, \tau), \tag{146d}$$

$$L(uv) = L(u)L(v) + L(u)v, \tag{146e}$$

for all $u, v \in T$.

Proof.

The truth of (146d) follows immediately from (146a) since $P(\tau) = \{(\tau, \tau)\}$. If $t = uv$, then $P(t)$ consists of the set of all pairs $(u_1 v_1, u_2 \circ v_2)$ for $(u_1, u_2) \in P(u)$, $(v_1, v_2) \in P(v)$ together with the set of all pairs $(u_1, u_2 v)$ for $(u_1, u_2) \in P(u)$. Hence, we establish (146e) as follows:

$$L(t) = \sum_{(u_1, u_2) \in P(u)} \sum_{(v_1, v_2) \in P(v)} b(u_2 \circ v_2, u_1 v_1) + \sum_{(u_1, u_2) \in P(u)} b(u_2 v, u_1)$$

$$= \sum_{(u_1, u_2) \in P(u)} \sum_{(v_1, v_2) \in P(v)} b(u_2, u_1) b(v_2, v_1) + \sum_{(u_1, u_2) \in P(u)} b(u_2, u_1) v$$

$$= L(u)L(v) + L(u)v. \quad \blacksquare$$

Using this result, we can tabulate L for low values of t and hence write down expressions for $(\alpha * \beta)(t)$.

For example,

$$L(\tau) = b(\tau, \tau),$$

$$L(\tau\tau) = b(\tau, \tau\tau) + b(\tau\tau, \tau),$$

$$L(\tau\tau \cdot \tau) = b(\tau, \tau\tau \cdot \tau) + 2b(\tau\tau, \tau\tau) + b(\tau\tau \cdot \tau, \tau),$$

$$L(\tau \cdot \tau\tau) = b(\tau, \tau \cdot \tau\tau) + b(\tau\tau, \tau\tau) + b(\tau \cdot \tau\tau, \tau),$$

so that

$$(\alpha * \beta)(\tau) = \alpha(\tau)\beta(\tau),$$

$$(\alpha * \beta)(\tau\tau) = \alpha(\tau)\beta(\tau\tau) + \alpha(\tau\tau)\beta(\tau),$$

$$(\alpha * \beta)(\tau\tau \cdot \tau) = \alpha(\tau)\beta(\tau\tau \cdot \tau) + 2\alpha(\tau\tau)\beta(\tau\tau) + \alpha(\tau\tau \cdot \tau)\beta(\tau),$$

$$(\alpha * \beta)(\tau \cdot \tau\tau) = \alpha(\tau)\beta(\tau \cdot \tau\tau) + \alpha(\tau\tau)\beta(\tau\tau) + \alpha(\tau \cdot \tau\tau)\beta(\tau).$$

We now introduce a unary relation $(\ )'$ on $\mathbb{R}^{T^*}$ to $\mathbb{R}^T$ by defining

$$\alpha'([t_1 t_2 \ldots t_s]) = \alpha(\emptyset)\alpha(t_1)\alpha(t_2) \ldots \alpha(t_s),$$

for all $t_1, t_2, \ldots, t_s \in T$ with $\alpha'(\tau) = \alpha(\emptyset)$.

LEMMA 146C.

If $\alpha, \beta \in \mathbb{R}^{T^*}$ with $\alpha(\emptyset) = 1$, then there is a $\gamma \in \mathbb{R}^{T^*}$ such that $\alpha' * \beta' = \gamma'$.

Proof.

It is easy to see that, for all u, $v \in T$, $\alpha'(u \circ v) = \alpha'(u)\alpha'(v)$, $\alpha'(uv) = \alpha'(u)\alpha'([v])$, $\beta'(uv) = \beta'(u)\beta'([v])$ and that the result of the lemma is equivalent to the assertion that for all u, $v \in T$, $(\alpha' * \beta')(uv) = (\alpha' * \beta')(u)(\alpha' * \beta')([v])$. We verify this last statement, where

$$L(u) = \sum_i a_i b(\bar{u}_i, u_i)$$

$$L(v) = \sum_j b_j b(\bar{v}_j, v_j),$$

so that

$$(\alpha' * \beta')(uv) = [L(u)L(v) + L(u)v](\alpha', \beta')$$

$$= \sum_{i,j} a_i b_j \alpha'(\bar{u}_i \circ \bar{v}_j)\beta'(u_i v_j) + \sum_i a_i \alpha'(\bar{u}_i v)\beta'(u_i)$$

$$= \sum_i a_i \alpha'(\bar{u}_i)\beta'(u_i)\left[\sum_j b_j \alpha'(\bar{v}_j)\beta'(\bar{v}_j) + \alpha'(v)\right]$$

$$= (\alpha' * \beta')(u)(\alpha' * \beta')([v]) \quad \blacksquare$$

We are now in a position to define an algebraic structure on $\mathbb{R}^{T^\#}$. Let G_1 denote the set of those real-valued functions on $T^\#$ such that if $\alpha \in G$, then $\alpha(\emptyset) = 1$. We also denote by G the set of all members of $\mathbb{R}^{T^\#}$ when equipped with the usual vector space structure and with the product $(\ ) \cdot (\ ): G_1 \times G \to G$ defined as follows.

DEFINITION 146D.

If $\alpha \in G_1$ and $\beta \in G$ then $\alpha \cdot \beta \in G$ (also written $\alpha\beta$) is defined as γ where $\alpha' * \beta' = \gamma'$.

Note that the existence of the γ in this definition is assured by lemma 146C. Its uniqueness is trivial.

The main properties of the structure consisting of G, its subset G_1, the vector space operations of addition and scalar multiplication and the operation of definition 146D, are summed up in the following theorem.

THEOREM 146E.

Let $e \in G_1$ be defined by $e(\emptyset) = 1$, $e(t) = 0$ $(t \in T)$. If $\alpha, \beta \in G_1, \gamma, \delta \in G$ and $c \in \mathbb{R}$, then there exists $\alpha^{-1} \in G_1$ such that

$$\alpha\beta \in G_1, \tag{146f}$$

$$\alpha e = e\alpha = \alpha, \tag{146g}$$

$$\alpha\alpha^{-1} = \alpha^{-1}\alpha = e, \tag{146h}$$

$$(\alpha\beta)\gamma = \alpha(\beta\gamma), \tag{146i}$$

$$\alpha(c\gamma) = c(\alpha\gamma), \tag{146j}$$

$$\alpha(\gamma + \delta) = \alpha\gamma + \alpha\delta. \tag{146k}$$

Proof.

To prove (146f) we must show that $(\alpha\beta)(\emptyset) = 1$. If $\alpha\beta = \gamma$ then $\alpha' * \beta' = \gamma'$ so that

$$\gamma(\emptyset) = \gamma'(t) = (\alpha' * \beta')(\tau) = \alpha'(\tau)\beta'(\tau) = \alpha(\emptyset)\beta(\emptyset) = 1.$$

The second result, (146g), is equivalent to $\alpha' * e' = e' * \alpha' = \alpha'$ and this follows immediately from (146a) because $e'(\emptyset) = 0$, $e'(\tau) = 1$ and $e'(t) = 0$ for $t \neq \tau$. To prove (146h), we note that $(\alpha\beta)(t) = \alpha(t) + \beta(t) + c(\alpha, \beta, t)$ where $c(\alpha, \beta, t)$ involves values of α and β for trees with lower numbers of vertices than t. Thus, both left and right inverses exist and their equality follows from the associativity (which makes $(G_1, \cdot)$ a group) expressed as (146i) which we see follows in turn from lemma 146A. Finally, we must prove (146j) and (146k) which express the fact that multiplication on the left by an element of G_1 is linear on the vector space G. If $\alpha\gamma = \delta$ then $\alpha' * \gamma' = \delta'$ so that evaluating this at $[t]$ where $t = \emptyset$ or $t \in T$ we find $(\alpha' * \gamma')([t]) = \delta(t)$. In the expansion of $(\alpha' * \gamma')([t])$ as a bilinear function, the dependence on γ' is only through factors of the form $\gamma'([u])$ where $u = \emptyset$ or $u \in T$. However, $\gamma'([u]) = \gamma(u)$ and thus $\delta(t)$ depends linearly on terms of the form $\gamma(u)$. ∎

We now wish to study the detailed nature of the multiplication operation on $G_1 \times G$. If t is any member of T, define $\hat{t}$ (also written as $t\hat{\ }$) as that linear functional on G such that $\hat{t}(\alpha) = \alpha(t)$ for all $\alpha \in G$. Also let G^* denote the set of all linear functionals on G of the form $\Sigma_t c(t)\hat{t}$, where the summation is over some finite subset of members of T and c is a real-valued function on this subset. If $\Sigma_t c(t)\hat{t}$ and $\Sigma_u d(u)\hat{u}$ are two such functionals, then we define their product as $\Sigma_t \Sigma_u c(t)d(u)\hat{t}\hat{u}$, where $\hat{t}\hat{u} = (tu)\hat{\ }$.

THEOREM 146F.

If $\alpha \in G_1$ and $t \in T^\#$ then there is a member $\lambda(\alpha, t)$ of G^* such that, for all $\beta \in G$,

$$(\alpha\beta)(t) = \lambda(\alpha, t)(\beta) + \alpha(t)\beta(\emptyset) \tag{146l}$$

and, furthermore, $\lambda: G_1 \times T^\# \to G^*$ satisfies the conditions for any $\alpha \in G_1$:

$$\lambda(\alpha, \emptyset) = 0, \tag{146m}$$

$$\lambda(\alpha, \tau) = \hat{\tau}, \tag{146n}$$

$$\lambda(\alpha, uv) = \lambda(\alpha, u)\lambda(\alpha, v) + \alpha(v)\lambda(\alpha, u) \quad \text{for all } u, v \in T. \tag{146o}$$

Proof.

We first verify that $(\alpha\beta)(t)$ is, as claimed by (146l), linear in β with $\alpha(t)$ the coefficient of $\beta(\emptyset)$. We have, using lemma 146B,

$$(\alpha\beta)(t) = (\alpha' * \beta')([t])$$

$$= L(\tau t)(\alpha', \beta')$$

$$= L(\tau)L(t)(\alpha', \beta') + [L(\tau)t](\alpha', \beta').$$

If $L(t) = \Sigma_i a_i b(u_i, v_i)$, then this has value

$$(\alpha\beta)(t) = \sum_i a_i b(\tau, \tau) b(u_i, v_i)(\alpha', \beta') + [b(\tau, \tau)t](\alpha', \beta')$$

$$= \sum_i a_i b(u_i, \tau v_i)(\alpha', \beta') + b(\tau t, \tau)(\alpha', \beta')$$

$$= \sum_i a_i \alpha'(u_i)\beta(v_i) + \alpha(t)\beta(\emptyset).$$

It remains to show that $\lambda(\alpha, t)$ defined by $\Sigma_i a_i \alpha'(u_i)\hat{v}_i$, where $L(t) = \Sigma_i a_i b(u_i, v_i)$, satisfies (146n) and (146o). Since we already know that $(\alpha\beta)(\emptyset) = \alpha(\emptyset)\beta(\emptyset)$. In the case $t = \tau$, we have $L(\tau) = b(\tau, \tau)$ so that $\lambda(\alpha, t) = \alpha'(\tau)\hat{\tau} = \hat{\tau}$ since $\alpha(\emptyset) = 1$. To prove (146o), let $L(u) = \Sigma_i a_i b(s_i, t_i)$, $L(v) = \Sigma_j b_j b(u_i, v_j)$ so that

$$L(uv) = L(u)L(v) + L(u)v$$

$$= \sum_{i,j} a_i b_j b(s_i \circ u_j, t_i v_j) + \sum_i a_i b(s_i v, t_i)$$

and, accordingly,

$$\lambda(\alpha, uv)(\beta) = \sum_{i,j} a_i b_j \alpha'(s_i \circ u_j)\beta(t_i v_j) + \sum_i a_i \alpha'(s_i v)\beta(t_i)$$

$$= \sum_{i,j} a_i \alpha'(s_i)b_j \alpha'(u_j)\beta(t_i v_j) + \sum_i a_i \alpha'(s_i)\alpha(v)\beta(t_i)$$

$$= \lambda(\alpha, u)\lambda(\alpha, v)(\beta) + \alpha(v)\lambda(\alpha, u)(\beta). \quad \blacksquare$$

Even though G and G_1 have been introduced in a rather abstract way, the importance of these algebraic structures in later chapters makes it appropriate to illustrate some of their properties in a pictorially evocative manner.

Since the members of G are mapping to real numbers of the set $T^\#$ of trees, including the empty tree, it is convenient to represent them as infinite vectors where the components are the images of the elements of $T^\#$ listed in some conventional order. We will adopt an order in which $\emptyset$ appears as tree number 0, τ as tree number 1, and so on, using the order as in tables 143 and 145′ and continuing on from these in a similar manner.

Besides explaining pictorially how products are formed, we will attempt to make the function λ more palatable by relating it to the same description. For $\alpha \in G$, let α_i, $i = 0, 1, \ldots$, denote $\alpha(t_i)$ where t_i is tree number i. If $\alpha \in G_1$, we must have $\alpha_0 = 1$.

To see how $(\alpha\beta)_i = (\alpha\beta)(t_i)$ is formed we draw the tree t_i with labels attached to its vertices and form subtrees which retain the original root in all possible ways. In addition, we include in this collection of possible subtrees the empty tree. For each subtree u we form a term in the formula for $(\alpha\beta)_i$ by writing down a factor $\beta(u)$ and multiply this by $\alpha(v)$ for v each of the connected components in the graph that remains when u is removed from t_i. Similarly to forming $(\alpha\beta)_i$ by adding these terms, we form $\lambda(\alpha, t_i)$ by omitting the empty tree from the choices of u and otherwise adding terms of the form $\hat{u}$ multiplied by products of all $\alpha(v)$.

For example, with the first five trees:

$$t_0 = \emptyset, \; t_1 = \bullet \;, \; t_2 = \Big\lvert \;, \; t_3 = \bigvee \;, \; t_4 = \Big\}\;,$$

we have

$$(\alpha\beta)_0 = \beta_0,$$

$$(\alpha\beta)_1 = \alpha_1\beta_0 + \beta_1,$$

$$(\alpha\beta)_2 = \alpha_2\beta_0 + \alpha_1\beta_1 + \beta_2,$$

$$(\alpha\beta)_3 = \alpha_3\beta_0 + \alpha_1^2\beta_1 + 2\alpha_1\beta_2 + \beta_3,$$

$$(\alpha\beta)_4 = \alpha_4\beta_0 + \alpha_2\beta_1 + \alpha_1\beta_2 + \beta_4,$$

and

$$\lambda(\alpha, t_0) = 0,$$

$$\lambda(\alpha, t_1) = \hat{t}_1,$$

$$\lambda(\alpha, t_2) = \alpha_1\hat{t}_1 + \hat{t}_2,$$

$$\lambda(\alpha, t_3) = \alpha_1^2\hat{t}_1 + 2\alpha_1\hat{t}_2 + \hat{t}_3,$$

$$\lambda(\alpha, t_4) = \alpha_2\hat{t}_1 + \alpha_1\hat{t}_2 + \hat{t}_4.$$

Note that in the formulae for $(\alpha\beta)_3$ and $\lambda(\alpha, t_3)$ the terms $\alpha_1\beta_2$ or $\alpha_1\hat{t}_2$ occur twice because t_2 occurs as a labelled subtree of t_3 in two different ways. The trees t_1, t_2, t_3, t_4 are related by $t_2 = t_1t_1$, $t_3 = t_2t_1$, $t_4 = t_1t_2$ and it is easy to verify that (146o) in theorem 146F is satisfied in the case of these products. For

Table 146   The function λ used in multiplication of $G_1 \times G$

i	t_i		$\lambda(\alpha, t_i)$
0	$\emptyset$		0
1	τ	$=$ •	$\hat{t}_1$
2	$[\tau]$	$=$	$\alpha_1\hat{t}_1 + \hat{t}_2$
3	$[\tau^2]$	$=$	$\alpha_1^2\hat{t}_1 + 2\alpha_1\hat{t}_2 + \hat{t}_3$
4	$[{}_2\tau]_2$	$=$	$\alpha_2\hat{t}_1 + \alpha_1\hat{t}_2 + \hat{t}_4$
5	$[\tau^3]$	$=$	$\alpha_1^3\hat{t}_1 + 3\alpha_1^2\hat{t}_2 + 3\alpha_1\hat{t}_3 + \hat{t}_5$
6	$[\tau[\tau]]$	$=$	$\alpha_1\alpha_2\hat{t}_1 + (\alpha_1^2 + \alpha_2)\hat{t}_2 + \alpha_1\hat{t}_3 + \alpha_1\hat{t}_4 + \hat{t}_6$
7	$[{}_2\tau^2]_2$	$=$	$\alpha_3\hat{t}_1 + \alpha_1^2\hat{t}_2 + 2\alpha_1\hat{t}_4 + \hat{t}_7$
8	$[{}_3\tau]_3$	$=$	$\alpha_4\hat{t}_1 + \alpha_2\hat{t}_2 + \alpha_1\hat{t}_4 + \hat{t}_8$
9	$[\tau^4]$	$=$	$\alpha_1^4\hat{t}_1 + 4\alpha_1^3\hat{t}_2 + 6\alpha_1^2\hat{t}_3 + 4\alpha_1\hat{t}_5 + \hat{t}_9$
10	$[\tau^2[\tau]]$	$=$	$\alpha_1^2\alpha_2\hat{t}_1 + (\alpha_1^3 + 2\alpha_1\alpha_2)\hat{t}_2 + (2\alpha_1^2 + \alpha_2)\hat{t}_3 + \alpha_1^2\hat{t}_4 + \alpha_1\hat{t}_5 + 2\alpha_1\hat{t}_6 + \hat{t}_{10}$

Continued

Table 146 Continued

i	t_i	$\lambda(\alpha, t_i)$

11    $[\tau[\tau^2]] =$

$$\alpha_1\alpha_3\hat{t}_1 + (\alpha_1^3 + \alpha_3)\hat{t}_2 + \alpha_1^2\hat{t}_3 + 2\alpha_1^2\hat{t}_4 + 2\alpha_1\hat{t}_6 + \alpha_1\hat{t}_7 + \hat{t}_{11}$$

12    $[\tau[_2\tau]_2] =$

$$\alpha_1\alpha_4\hat{t}_1 + (\alpha_1\alpha_2 + \alpha_4)\hat{t}_2 + \alpha_2\hat{t}_3 + \alpha_1^2\hat{t}_4 + \alpha_1\hat{t}_6 + \alpha_1\hat{t}_8 + \hat{t}_{12}$$

13    $[[\tau]^2] =$

$$\alpha_2^2\hat{t}_1 + 2\alpha_1\alpha_2\hat{t}_2 + \alpha_1^2\hat{t}_3 + 2\alpha_2\hat{t}_4 + 2\alpha_1\hat{t}_6 + \hat{t}_{13}$$

14    $[_2\tau^3]_2 =$

$$\alpha_5\hat{t}_1 + \alpha_1^3\hat{t}_2 + 3\alpha_1^2\hat{t}_4 + 3\alpha_1\hat{t}_7 + \hat{t}_{14}$$

15    $[_2\tau[\tau]]_2 =$

$$\alpha_6\hat{t}_1 + \alpha_1\alpha_2\hat{t}_2 + (\alpha_1^2 + \alpha_2)\hat{t}_4 + \alpha_1\hat{t}_7 + \alpha_1\hat{t}_8 + \hat{t}_{15}$$

16    $[_3\tau^2]_3 =$

$$\alpha_7\hat{t}_1 + \alpha_3\hat{t}_2 + \alpha_1^2\hat{t}_4 + 2\alpha_1\hat{t}_8 + \hat{t}_{16}$$

17    $[_4\tau]_4 =$

$$\alpha_8\hat{t}_1 + \alpha_4\hat{t}_2 + \alpha_2\hat{t}_4 + \alpha_1\hat{t}_8 + \hat{t}_{17}$$

example,

$$\lambda(\alpha, t_4) = \lambda(\alpha, t_1 t_2) = \lambda(\alpha, t_1)\lambda(\alpha, t_2) + \alpha(t_2)\lambda(\alpha, t_1)$$
$$= \hat{t}_1(\alpha_1\hat{t}_1 + \hat{t}_2) + \alpha(t_2)\hat{t}_1$$
$$= \alpha_2\hat{t}_1 + \alpha_1\hat{t}_2 + \hat{t}_4.$$

In table 146, the formulae for $\lambda(\alpha, t_i)$ are taken as far as trees with five vertices. As we have seen, formulae for $(\alpha\beta)(t_i)$ can be found immediately by replacing $\hat{t}_i$ by β_j for $j = 1, 2, \ldots, 17$ and adding the term $\alpha_i\beta_0$.

Further properties of the algebraic system introduced here, particularly concerning the group G_1, are given in Butcher (1972).

2

The Euler method and its generalizations

20   THE EULER METHOD

200   Introduction.

In introducing the famous method of Euler which has been republished in his collected works (Euler, 1913) a variety of model problems could be used. Our choice is to use a first-order autonomous equation on a normed vector space X. This equation we will write as

$$y'(x) = f(y(x)), \qquad \text{(200a)}$$

and we will write the intial condition as $y(x_0) = y_0 \in X$. As we saw in subsection 110, this standard problem is general enough to include equations equivalent to non-autonomous systems of equations of first or higher orders.

The Euler method for solving (200a) numerically makes use of a sequence of values of the independent variable starting from x_0. Suppose this sequence is $(x_0, x_1, x_2, \ldots)$ and that the so-called step sizes $x_1 - x_0, x_2 - x_1, \ldots$ are denoted by $h_1, h_2, \ldots$. In addition to y_0, which equals the solution value $y(x_0)$, a sequence of approximations $y_1 \approx y(x_1)$, $y_2 \approx y(x_2)$, $\ldots$ is defined by the formula

$$y_n = y_{n-1} + h_n f(y_{n-1}), \qquad n = 1, 2, \ldots .$$

Thus, for each value of n, the approximation to $y(x_n)$ is computed as though y_{n-1} were exactly equal to $y(x_{n-1})$ and that the slope of the y function were constant throughout the interval $[x_{n-1}, x_n]$. Clearly the quality of this sequence of approximations will depend on the magnitudes of $h_1, h_2, \ldots$ and on how rapidly $f(u)$ varies as u varies.

In addition to yielding a value at the step points $x_0, x_1, \ldots$, we will widen the scope of the method slightly by approximating the solution for $x \in (x_{n-1}, x_n]$ as

$$y_{n-1} + (x - x_{n-1})f(y_{n-1}). \qquad \text{(200b)}$$

We denote by Y the function whose value is the approximate solution. That is, $Y(x_0) = y_0$ and if $x \in (x_{n-1}, x_n]$, then $Y(x)$ is given as the value of (200b).

To measure how rapidly $f(u)$ varies with u, we assume that f satisfies a Lipschitz condition and use the Lipschitz constant L as the appropriate measure. As we saw in subsection 112, this condition guarantees the existence and uniqueness of the solution function y on any bounded interval which includes the initial value x_0. We will assume, without explicitly stating it on every occasion, that this condition holds throughout the present section. We will use a standard interval $[x_0, \bar{x}]$ and suppose, for notational convenience, that $\bar{x} > x_0$ and also that when the points $x_1, x_2, \ldots$ are selected, the last of these is identical with $\bar{x}$ and that $x_0 < x_1 < x_2 \cdots$. The greatest of $x_1 - x_0, x_2 - x_1, \ldots$ will be denoted by H.

In the next three subsections, we will develop results to show that the computed solution converges uniformly to the solution of the differential equation as $H \to 0$.

201   Local truncation error.

If $x \in (x_{n-1}, x]$ then we are interested in the value of

$$y(x) - y(x_{n-1}) - (x - x_{n-1})f(y(x_{n-1})) = y(x) - y(x_{n-1}) - (x - x_{n-1})y'(x_{n-1})$$

$$(201a)$$

This quantity would be zero if y were replaced by Y(but with $Y'(x_{n-1})$ replaced by the right-hand derivative of Y at x_{n-1}) and thus is a measure of the difference in behaviour of Y from that of y. If we needed to compare the behaviour of Y with that of some other approximation to the the solution, say z, then we could consider the quantity $z(x) - z(x_{n-1}) - (x - x_{n-1})f(z(x_{n-1}))$.

DEFINITION 201A.

The *local truncation error* in computing $Y(x)$ by the Euler method is

$$l(x) = y(x) - y(x_{n-1}) - (x - x_{n-1})y'(x_{n-1}).$$

More generally, the local truncation error *relative to z* is

$$l(z, x) = z(x) - z(x_{n-1}) - (x - x_{n-1})f(z(x_{n-1})).$$

At first glance, the quantity $l(x)$ seems to have little relationship to the computed solution, but it can be interpreted in a way that makes its significance clear. As we carry the solution over many steps we would expect the values of $Y(x)$ and $y(x)$ to diverge further and further apart. The *local* truncation error is concerned only with the divergence produced within the present step so that it is appropriate to reinitialize $Y(x_{n-1})$ to the value of $y(x_{n-1})$ in studying this source of error. At the beginning of the step we then have $Y(x_{n-1}) = y(x_{n-1})$ so that $y(x) - Y(x)$ has grown at the point

$x \in (x_{n-1}, x_n]$ to

$$y(x) - Y(x) = [y(x) - (x - x_{n-1})f(y(x_{n-1}))] - [Y(x) - (x - x_{n-1})f(Y(x_{n-1}))]$$
$$= l(x) - 0$$
$$= l(x).$$

In subsection 202, we will study the behaviour of the total error in a series of steps of the Euler method but for the present we will seek bounds on $\| l(x) \|$.

LEMMA 201B.

If f satisfies a Lipschitz condition with constant L and $p \in X^*$ then

$$p(y(x) - y(x_{n-1})) = (x - x_{n-1})p(f(y(\xi))), \tag{201b}$$

for some $\xi \in (x_{n-1}, x)$, and

$$\| y(x) - y(x_{n-1}) \| \leq m(x - x_{n-1}), \tag{201c}$$

where

$$m = \sup_{u \in [x_0, \bar{x}]} \| f(y(u)) \|.$$

Proof.

We make use of the mean value theorem

$$\phi(b) - \phi(a) = (b - a)\phi'(\xi),$$

for some $\xi \in (a, b)$ where $\phi \in C^1[a, b]$.

Choose $a = x_{n-1}$, $b = x$ and $\phi = p \circ y$. We then have

$$p(y(x) - y(x_{n-1})) = (x - x_{n-1})p(y'(\xi)),$$

where $\xi \in (x_{n-1}, x)$. Hence, making use of (200a) we obtain (201b). Taking the modulus of each side of (201b) and bounding the right-hand side we have

$$| p(y(x) - y(x_{n-1})) | \leq m(x - x_{n-1}) \| p \|,$$

so that

$$\| y(x) - y(x_{n-1}) \| = \sup_{p \in X^*, p \neq 0} \frac{| p(y(x) - y(x_{n-1})) |}{\| p \|}$$
$$\leq m(x - x_{n-1}). \quad \blacksquare$$

THEOREM 201C.

Under the same assumptions as for lemma 201B,

$$\| l(x) \| \leq mL(x - x_{n-1})^2. \tag{201d}$$

Proof.

For any $p \in X^*$ we have (201b) so that

$$p(l(x)) = p(y(x) - y(x_{n-1}) - (x - x_{n-1})f(y(x_{n-1})))$$
$$= (x - x_{n-1})p(f(y(\xi)) - f(y(x_{n-1}))),$$

where $\xi \in (x_{n-1}, x)$. Hence

$$|p(l(x))| \leq (x - x_{n-1})L \| p \| \, \| y(\xi) - y(x_{n-1}) \|. \tag{201e}$$

Since $\xi - x_{n-1} \leq x - x_{n-1}$ we can use (201c) with x replaced by ξ to give

$$\| y(\xi) - y(x_{n-1}) \| \leq m(\xi - x_{n-1})$$
$$\leq m(x - x_{n-1}). \tag{201f}$$

From (201e) and (201f) we find

$$\| l(x) \| = \sup_{p \in X^*, p \neq 0} \frac{|p(l(x))|}{\| p \|}$$
$$\leq mL(x - x_{n-1})^2. \quad \blacksquare$$

The bound on $\| l(x) \|$ given in theorem 201C is made under minimal assumptions on the function f. In practice, however, it is often known that f is sufficiently smooth as to guarantee corresponding smoothness for y. If, for example, it is known that f is continuously differentiable then the identity $y''(x) = f'(y(x))(y'(x)) = f'(y(x))(f(y(x)))$ implies the existence and continuity of y''. This type of situation is covered by the following result.

THEOREM 201D.

If for all $p \in X^*$, $p \circ y \in C^2[x_0, \bar{x}]$ and $M = \max_{u \in [x_0, \bar{x}]} \| y''(u) \|$, then for $x \in (x_{n-1}, x_n]$,

$$\| l(x) \| \leq \tfrac{1}{2} M(x - x_{n-1})^2. \tag{201g}$$

Proof.

Write Taylor's theorem in the form

$$\phi(b) - \phi(a) - (b - a)\phi'(a) = \tfrac{1}{2}(b - a)^2 \phi''(\xi),$$

for some $\xi \in (a, b)$, where $\phi \in C^2[a, b]$. Choose $a = x_{n-1}$, $b = x$ and $\phi = p \circ y$, where $p \in X^*$, to obtain

$$p(l(x)) = \tfrac{1}{2}(x - x_{n-1})^2 p(y''(\xi)),$$

where ξ depends on p. The result of the theorem follows. $\quad \blacksquare$

202 Global truncation error.

In this subsection we will make a study of the difference between the solution y and the computed solution given by the piecewise linear function Y. As an alternative means of expressing the work of subsections 200 and 201 we will use a mapping $s: [x_0, \bar{x}] \rightarrow \{x_0, x_1, \ldots\}$ defined by $s(x_0) = x_0$ and $s(x) = x_{n-1}$ if $x \in (x_{n-1}, x_n]$. With this notation, the equation giving $Y(x)$ as the expression (200b) can be rewritten as

$$Y(x) = Y(s(x)) + [x - s(x)] f(Y(s(x))), \qquad (202a)$$

which, along with the value of $Y(x_0)$, determines Y. In this subsection we will not assume that $Y(x_0) = y(x_0)$. Recalling the definition of H from subsection 200, we see that for all $x \in [x_0, \bar{x}]$, $x - s(x) \le H$.

To study the global error we write

$$\alpha(x) = y(x) - Y(x),$$

$$\beta(x) = f(y(x)) - f(Y(x)).$$

The first of these is the global truncation error and, by the Lipschitz condition, the second is related to it by the inequality

$$\| \beta(x) \| \le L \| \alpha(x) \|. \qquad (202b)$$

From the bounds we have derived on $l(x)$, it is convenient to write this quantity in the form $[x - s(x)]^2 E(x)$, where $E(x)$ is bounded.

LEMMA 202A.

If $l(x) = [x - s(x)]^2 E(x)$ and $\| E(x) \| \le C$ then, if $L \ne 0$,

$$\| \alpha(x) \| \le \| \alpha(x_0) \| \exp(L(x - x_0)) + \frac{HC}{L} [\exp(L(x - x_0)) - 1], \qquad (202c)$$

and if $L = 0$,

$$\| \alpha(x) \| \le \| \alpha(x_0) \| + HC(x - x_0). \qquad (202d)$$

Proof.

By the definition of local truncation error (definition 201A) we have

$$y(x) - y(s(x)) - [x - s(x)] f(y(s(x))) = [x - s(x)]^2 E(x),$$

which, together with (202a), gives

$$\alpha(x) = \alpha(s(x)) + [x - s(x)] \beta(s(x)) + [x - s(x)]^2 E(x),$$

so that taking norms, using (202b) and the bounds on $\| E(x) \|$ and $[x - s(x)]$, we find that

$$\| \alpha(x) \| \le \{1 + L[x - s(x)]\} \| \alpha(s(x)) \| + H[x - s(x)]C. \qquad (202e)$$

In the case $L \neq 0$, add HC/L to each side of (202e) and arrange in the form

$$\| \alpha(x) \| + \frac{HC}{L} \leq \{1 + L[x - s(x)]\}\left[\| \alpha(s(x)) \| + \frac{HC}{L}\right]$$

so that, using the inequality

$$1 + L[x - s(x)] \leq \exp(L[x - s(x)]) = \exp(L(x - x_0))\exp(-L[s(x) - x_0]),$$

we find that

$$\left[\| \alpha(x) \| + \frac{HC}{L}\right]\exp(-L(x - x_0)) \leq \left[\| \alpha(s(x)) \| + \frac{HC}{L}\right]\exp(-L[s(x) - x_0]).$$

$$(202f)$$

Let $\varphi(x)$ denote the left-hand side of (202f) so that this inequality can be written as $\varphi(x) \leq \varphi(s(x))$. If this result is used recursively we find $\varphi(x) \leq \varphi(s^n(x))$ for all $n = 1, 2, \ldots$, where s^n denotes the composition of n copies of s. But for any x there is an n such that $s^n(x) = x_0$. Hence,

$$\| \alpha(x) \| + \frac{HC}{L} \leq \exp(L(x - x_0))\left[\| \alpha(x_0) \| + \frac{HC}{L}\right],$$

and (202c) follows.

In the case $L = 0$, (202e) is equivalent to

$$\| \alpha(x) \| - HC(x - x_0) \leq \| \alpha(s(x)) \| - HC[s(x) - x_0],$$

so that, again using a recursion argument, we find that

$$\| \alpha(x) \| - HC(x - x_0) \leq \| \alpha(x_0) \|,$$

and (202d) follows. ∎.

THEOREM 202B.

$$\| Y(x) - y(x) \| \leq \| Y(x_0) - y(x_0) \| \exp(L(x - x_0))$$
$$+ H \sup_{u \in [x_0, \bar{x}]} \| f(y(u)) \| [\exp(L(x - x_0)) - 1].$$

Proof.

Substitute $C = mL$ (from theorem 201C) into (202c) and (202d). Note that there is no need to use a separate form of this result when $L = 0$ since, in this case, $C = 0$. ∎.

This error estimate can be replaced by a sharper result under a further smoothness condition.

THEOREM 202C.

If $L \neq 0$ and, for all $p \in X^*$, $p \circ y \in C^2[x_0, \bar{x}]$, then

$$\| Y(x) - y(x) \| \leq \| Y(x_0) - y(x_0) \| \exp(L(x - x_0))$$

$$+ \tfrac{1}{2} H \sup_{u \in [x_0, \bar{x}]} \| y''(u) \| \frac{\exp(L(x - x_0)) - 1}{L}.$$

Proof.

Use theorem 201D and lemma 202A. ∎.

203 Convergence of the Euler method.

Let $(Y_1, Y_2, \ldots)$ be a sequence of solutions computed using the Euler method and let $(H_1, H_2, \ldots)$ be the corresponding sequence of maximum step sizes (that is the quantity H introduced in subsection 200).

THEOREM 203A.

If $H_n \to 0$ as $n \to \infty$ and $Y_n(x_0) \to y(x_0)$ as $n \to \infty$ then Y_n converges uniformly to y as $n \to \infty$.

Proof.

For a given real number $\varepsilon > 0$, choose N such that, for all $n \geq N$, the inequality

$$\| Y_n(x_0) - y(x_0) \| \leq \tfrac{1}{2} \varepsilon \exp(- L(\bar{x} - x_0))$$

holds, as does the inequality

$$H_n \sup_{u \in [x_0, \bar{x}]} \| f(y(u)) \| \leq \begin{cases} \tfrac{1}{2}\varepsilon L \left[\exp(L(\bar{x} - x_0)) - 1 \right]^{-1}, & L \neq 0, \\ \tfrac{1}{2}\varepsilon (\bar{x} - x_0)^{-1}, & L = 0. \end{cases}$$

Hence, if $x \in [x_0, \bar{x}]$ and $n \geq N$, it follows from theorem 202B that $\| Y_n(x) - y(x) \| \leq \varepsilon$. ∎.

204 Order of convergence.

In this subsection, we will study the *rate* at which the computed solution, found from Euler's method, converges to the exact solution.

As a simplification, we will suppose that $x_1, x_2, \ldots$ are always chosen so that $x_1 - x_0 = x_2 - x_1 = \cdots = h$, say. In this case $H = h$. We will consider the solution only at one fixed point $\bar{x}$ and h will always be chosen so that $(\bar{x} - x_0)/h$ is an integer. We will also suppose for simplicity that the computed solution is exact at x_0.

From theorem 203A it immediately follows that $\| Y(\bar{x}) - y(\bar{x}) \| /h$ is bounded as $h \to 0$. If it were the case that, for some positive integer p, $\| Y(\bar{x}) - y(\bar{x}) \| /h^p$ is bounded, then we would say that 'the numerical solution is of order p' and if p is the highest such value we would say that 'the order of the numerical solution is p'. Thus, the order of the Euler method is at least one. That it cannot be higher (except perhaps for some special problems) is easily seen by an example.

Let $X = \mathbb{R}^2$ and let $f : X \to X$ be defined by $f([u, v]) = [v, 1]$. Select $x_0 = 0$ and $\bar{x} = 1$ and the initial value $y(0) = [0, 0]$. The exact solution to this initial value problem is $y(x) = [\frac{1}{2}x^2, x]$ so that $y(1) = [\frac{1}{2}, 1]$. On the other hand, if $h = 1/N$ for N a positive integer, the solution computed by the Euler method is found to be $[\frac{1}{2}(1 - h), 1]$, leading to the global error at one equal to $\| Y(1) - y(1) \|_\infty = \frac{1}{2}h$. Thus the method cannot have an order higher than one.

In general, it is an advantage to use methods with a higher order than one since a decrease in the value of h (requiring an increase in the computational effort) results in a greater gain in accuracy. Later in this chapter we will survey some devices that can be used to obtain higher orders.

205 Asymptotic error formula.

We saw in theorem 201D that the error committed in a step, that is the local truncation error, satisfies (201g). We now derive a closely related result which gives an asymptotic formula for the global error. Using the function s introduced in subsection 202 we have the following theorem.

THEOREM 205A.

If for each $p \in X^*$, $p \circ y \in C^3[x_0, \bar{x}]$ then there is a constant C such that for all $x \in [x_0, \bar{x}]$,

$$\| l(x) - \tfrac{1}{2}[x - s(x)]^2 y''(s(x)) \| \le C[x - s(x)]^3. \tag{205a}$$

Proof.

The result follows from the Taylor expansion of $y(x) = y(s(x) + [x - s(x)])$ where we note that y''' is bounded. ∎.

Since we have a bound on the global truncation error, we can obtain a region R in which solution values are certain to lie for all stepsize sequences for which $H \le h_0$, say. We will assume for the rest of this subsection that f'' exists throughout R and that $\| f''(u) \|$ is bounded for all $u \in R$. Since $f' \circ y$ is necessarily continuous on $[x_0, \bar{x}]$ it makes sense to consider the differential equation system

$$e'(x) = f'(y(x))e(x) + [x - s(x)] y''(s(x)),$$
$$e(x_0) = 0, \tag{205b}$$

whose solution is uniquely defined on $[x_0, \bar{x}]$ and where it is easily verified that $\| e(x) \|/H$ is bounded.

Using these notations and assumptions and the function Y, introduced in subsection 202, we have the following result.

THEOREM 205B.

The exact and computed solutions are related in such a way that there exist constants h_0 and D such that, for $H \le h_0$.

$$\| y(x) - Y(x) - e(x) \| \le H^2 D. \tag{205c}$$

Proof.

Let $z(x) = y(x) - e(x)$. We will compute the local truncation error for the method as though it were being used to approximate z rather than y. We find

$$l(z, x) = y(x) - e(x) - y(s(x)) + e(s(x)) - [x - s(x)] f(y(s(x)) - e(s(x)))$$
$$= T_1 + T_2 + T_3 + T_4 + T_5,$$

where

$$T_1 = y(x) - y(s(x)) - [x - s(x)] f(y(s(x))) - \tfrac{1}{2}[x - s(x)]^2 y''(s(x)),$$

$$T_2 = [x - s(x)] [f(y(s(x))) - f(y(s(x)) - e(s(x))) - f'(y(s(x)))e(s(x))],$$

$$T_3 = -\int_{s(x)}^{x} \{e'(t) - f'(y(t))e(t) - [t - s(x)] y''(s(x))\} \, dt,$$

$$T_4 = -\int_{s(x)}^{x} [f'(y(t)) - f'(y(s(x)))] e(t) \, dt,$$

$$T_5 = -\int_{s(x)}^{x} f'(y(s(x)))[e(t) - e(s(x))] \, dt.$$

Because of theorem 205A, $\| T_1 \|$ is bounded by a multiple of $[x - s(x)]^3$. Because of Taylor's theorem and the known bounds on $\| e(s(x)) \|$ and on $x - s(x)$, $\| T_2 \|$ is bounded by a multiple of $[x - s(x)]H^2$. Because the integrand vanishes, $T_3 = 0$. It remains to estimate T_4 and T_5. We have

$$\| T_4 \| \le [x - s(x)] \sup_{t \in [s(x), x]} \| e(t) \| \sup_{t \in [s(x), x]} \| f'(y(t)) - f'(y(s(x))) \|$$

$$\le [x - s(x)]^2 \sup_{t \in [s(x), x]} \| e(t) \| \sup_{t \in [s(x), x]} \| f''(y(t)) \|,$$

which is bounded by a multiple of $[x - s(x)]^2 H$. Also,

$$\| T_5 \| \le [x - s(x)] \| f'(y(s(x))) \| \sup_{t \in [s(x), x]} \| e(t) - e(s(x)) \|$$

$$\le [x - s(x)]^2 \| f'(y(s(x))) \| \sup_{t \in [s(x), x]} \| e'(t) \|$$

$$\leq [x - s(x)]^2 \| f'(y(s(x))) \| M,$$

where

$$M = \sup_{t \in [s(x), x]} \{ \| f'(y(s(x))) \| \| e(t) \| + [x - s(x)] \| y''(s(x)) \| \}.$$

Since $[x - s(x)] \leq H$, $\| T_5 \|$, as well as $\| T_1 \|$, $\| T_2 \|$, $\| T_4 \|$, is bounded by a multiple of $[x - s(x)]H^2$. Hence, for some D_0, we have

$$\| l(z, x) \| \leq [x - s(x)]H^2 D_0.$$

Use lemma 202A, with $C = HD_0$, and the result follows. ∎

Theorem 205B can be interpreted as an asymptotic formula for the error in approximating $y(x)$ by $Y(x)$ since (205c) can be written as

$$y(x) = Y(x) + e(x) + O(H^2).$$

Even though it is rarely practicable to evaluate the function e, its very existence is useful in interpreting computed results. This is especially the case in the use of what Stetter (1973) calls 'coherent' stepsize sequences. Let $\theta : [x_0, \bar{x}] \rightarrow (0, 1]$ denote a 'relative stepsize' function and H an overall bound on step sizes. For a stepsize sequence characterized by the pair (θ, H), the steps at $x_1, x_2, \ldots$ are defined by

$$\int_{x_0}^{x_n} \frac{dx}{\theta(x)} = Hn.$$

If two sequences defined by (θ, H_1) and (θ, H_2) are chosen then the functions e_1 and e_2 corresponding to the global truncation errors are approximately related by

$$H_1^{-1} e_1 \approx H_2^{-1} e_2$$

and a comparison of computed results for different step sizes makes possible reasonably realistic estimates of the truncation errors.

206 Stability characteristics.

Besides knowing that the Euler method converges and that it has a certain error behaviour when integrating over a fixed interval, it is interesting to study the behaviour of numerical solutions when developed over an extended interval. To limit the scope of such an enquiry as this, we consider only the constant coefficient linear system given by

$$y'(x) = My(x),$$

where M is a constant matrix. Using fixed stepsize h, the Euler method gives as the approximate solution at $x_n = x_0 + nh$, the result

$$y_n = (I + hM)y_{n-1},$$

so that

$$y_n = (I + hM)^n y_0. \qquad (206a)$$

The exact solution is, of course, $y(nh) = \exp(nhM)y(x_0)$ and we wish to examine some features of the approximation of this by y_n given by (206a). By making a change of basis so that $y(x) = Sz(x)$, where S is a (constant) non-singular matrix, the differential equation system becomes

$$z'(x) = S^{-1}MSz(x),$$

with solution

$$z(nh) = \exp(nhS^{-1}MS)z(x_0)$$

and Euler approximation

$$z_n = (I + hS^{-1}MS)^n z_0.$$

Thus, if we work in a complex vector space, there is no loss of generality in supposing that M is a Jordan canonical form since, if it is a general matrix, S can be chosen so that $S^{-1}MS$ is in canonical form. As a further simplification, we will assume that the zeros of the characteristic polynomial of M are distinct, so that this Jordan form is diagonal. If $M = \text{diag}\{\lambda_1, \lambda_2, \ldots, \lambda_m\}$ and $y(x)$ has components $y_1(x), y_2(x), \ldots, y_m(x)$, then the differential equation system becomes

$$y_i'(x) = \lambda_i y_i(x), \qquad i = 1, 2, \ldots, m,$$

with solution

$$y_i(x_0 + nh) = \exp(nh\lambda_i)y_i(x_0)$$

and numerical approximation

$$(y_i)_n = (1 + h\lambda_i)^n (y_i)_0.$$

Since there is no interaction between the various components in either the numerical or the exact solution, we are, in effect, considering a set of scalar equations of the form

$$y'(x) = qy(x),$$

where q is a (complex) number. If h is the stepsize then we will write $z = hq$. In this case the computed solution after n steps is $(1 + z)^n y_0$, whereas the exact solution is $\exp(zn)y_0$. Rather than study the precise quantitative behaviour of the approximation, we shall, in this subsection, look at one qualitative aspect, that is the boundedness or otherwise of the two solutions as $n \to \infty$.

For the exact solution, it is easy to see that $|\exp(nz)|$ is bounded iff $|\exp(z)| \le 1$. If $z = x + iy$, then $|\exp(z)| = \exp(x)$, which is less than or equal to unity iff x is negative or zero. Thus the set of values for which the exact solution is bounded is precisely the non-positive half-plane $\{z \in \mathbb{C} : \text{Re}(z) \le 0\}$. On the other hand, in the case of the numerical solution, boundedness is achieved iff $|1 + z| \le 1$. The set of points with this property is the closed disc in the complex plane with centre -1 and radius 1.

The stronger property of converging to zero as $n \to \infty$ requires that $|\exp(z)|$ and $|1 + z|$ respectively are strictly less than unity. It is easy to see that for the exact solution this is the negative half-plane $\{x \in \mathbb{C} : \mathrm{Re}(z) < 0\}$ and in the case of the computed solution, it is the *open* disc with centre -1 and radius 1.

The regions $|1 + z| \le 1$ (for boundedness) and $|1 + z| < 1$ (for convergence to zero) are important characteristics of the Euler method. By comparing with the exact solution, we see that for z a complex number with non-positive real part which is not in $|1 + z| \le 1$, the numerical solution is a poor approximation in that it produces a sequence with increasing magnitude to represent a solution which is bounded in magnitude.

To overcome this difficulty in practice, we must bring $z = hq$ into this disc. Since q depends on the problem, we have no recourse but to reduce h to achieve this. As it happens, many problems arise in practice which are, at least approximately, of the linear–constant coefficient type, where at least some of the eigenvalues possess extremely negative real parts.

It is a serious limitation of the Euler method as applied to these so-called 'stiff problems' that the very negative eigenvalues, which correspond to fast decaying components whose contribution to the exact solution is negligible, impose a limitation on h which makes the method so inefficient as to be unusable.

The closed disc referred to above will be called the *stability region* of the

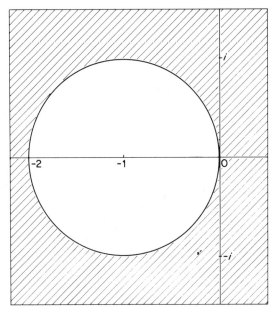

Figure 206 Stability region for the Euler method

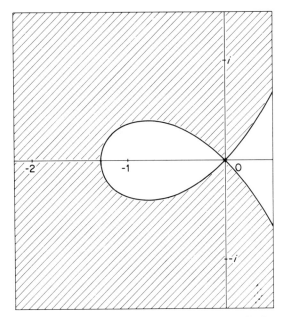

Figure 206′   Order star for the Euler method

Euler method and its interior, where the numerical result converges to zero, will be called the *strict stability region*. They are illustrated in figure 206 as the unshaded part of the complex plane represented there. The boundary is not included in the case of the strict stability region.

Since the function $1 + z$ representing the growth of the numerical solution is considered alongside $\exp(z)$, representing the growth of the exact solution, it is also interesting to consider the rate at which the solution grows *relative* to the exact solution. Thus we might ask when $|1 + z| \le |\exp(z)|$ (or perhaps when $|1 + z| < |\exp(z)|$.) These would naturally be called the *relative stability region* and the *strict relative stability region* respectively, but we will follow the terminology of Wanner, Hairer and Nørsett (1978) and call the set where $|1 + z| > |\exp(z)|$ the *order star* for the Euler method. This is illustrated in figure 206′. Notice that the behaviour near $z = 0$ is determined by the fact that $(1 + z)\exp(-z) = 1 - \frac{1}{2}z^2 + O(z^3)$ so that $|(1 + z)\exp(-z)| = 1 - \frac{1}{2}r^2 \cos(2\theta) + O(r^3)$ where $r \exp(i\theta) = x + iy$, the behaviour for $|\operatorname{Re}(z)|$ large is determined by the dominant effect of the exponential function, the behaviour for $\operatorname{Re}(z) = 0$ is the same as for the absolute stability region and the behaviour near $z = -1$ is determined by the pole there.

207 Rounding error.

The analysis of the Euler method that has been presented has ignored the presence of rounding errors in computer arithmetic. Suppose in the computation of y_n that a rounding error r_n is committed so that

$$y_n = y_{n-1} + hf(y_{n-1}) + r_n, \tag{207a}$$

where $h = x_n - x_{n-1}$, which we will now assume to be constant. Combining (207a) with the formula

$$y(x_n) = y(x_{n-1}) + hf(y(x_{n-1})) + l(x_n) \tag{207b}$$

and writing α_n, β_n for $\alpha(x_n)$, $\beta(x_n)$ as introduced in subsection 202, we find

$$\alpha_n = \alpha_{n-1} + h\beta_{n-1} + l(x_n) - r_n.$$

If some bound on r_n can be provided, then the analysis of subsection 202 can be generalized to give an estimate of the total error in the integration by the Euler method.

Although we will not attempt such an analysis, it can perhaps be remarked that a major source of rounding error occurs in the adding of $hf(y_{n-1})$ to y_{n-1} and, since the components of y_{n-1} are normally greater in magnitude than those of $hf(y_{n-1})$, this error is bounded by a vector whose components are proportional to the magnitudes of the corresponding components of y_{n-1}. A consequence of this remark is that there is not likely to be a marked dependence of the magnitude of the rounding errors on the value of h. Thus a bound on the total local error for step n would be of the form $A(x_n) + h^2 B(x_n)$ and, accordingly, a bound on the total global error would be of the form $h^{-1}(C + h^2 D)$, where C represents the contribution of the rounding error and D the contribution of the truncation error.

It is not an easy matter to obtain usable values of C and D but we can at least see that C/h becomes the dominant term for small h. Thus, the convergence result in theorem 203A becomes meaningless in the presence of rounding error.

As a means of minimizing the contribution of rounding error, it is possible to program the performance of a single Euler step so that the error produced in adding $hf(y_{n-1})$ to y_{n-1} is kept in a separate array and added into the increment of the next step. This technique is used in algorithm 207' below and, for comparison, the straightforward algorithm 207 is also given. In each case, N and h denote the dimension of the differential equation system and the step size respectively and y denotes an array which holds the current value of the solution vector. In the case of algorithm 207', z denotes an array (initially set to zero) which holds the rounding error estimated in a step and intended to be added into a subsequent step. The procedure *diffeqn*, when called as *diffeqn*(y, dy), causes the array dy to be replaced by $f(y)$.

Algorithm 207: A single step of the Euler method

diffeqn(y, dy);
for $i := 1$ **to** N **do** $y[i] := y[i] + h * dy[i]$

Algorithm 207': A single Euler step with improved accuracy

diffeqn(y, dy);
for $i := 1$ **to** N **do**
begin
  $u := z[i] + h*dy[i]$;
  $v := y[i] + u$;
  $z[i] := u - (v - y[i])$;
  $y[i] := v$
end

Various proposals for keeping rounding error at a manageable level have been made by Kahan [1965] and by Vitásek (1969). The technique embodied in algorithm 207' is due to Møller (1965, 1965a) although the essential idea can be traced back to Gill (1951). It is sometimes known as the Gill–Møller algorithm.

To illustrate the use of these procedures, we present the solution of a differential equation solved using each of them. In this example, $N = 3$ and the

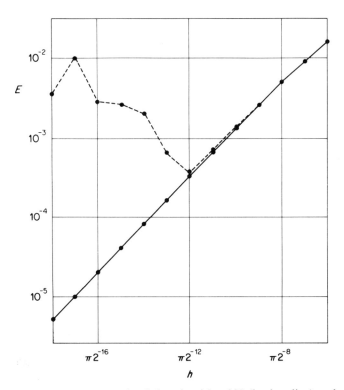

Figure 207 Error E versus step size h for algorithm 207 (broken line) and algorithm 207' (full line)

equations are

$$y_1'(x) = y_2(x) + y_3(x)[1 - y_3(x)], \qquad y_1(0) = 0,$$
$$y_2'(x) = -y_1(x), \qquad\qquad\qquad\quad y_2(0) = 0,$$
$$y_3'(x) = 1, \qquad\qquad\qquad\qquad\quad y_3(0) = 0,$$

with exact solution

$$y_1(x) = 1 - 2x + 2\sin(x) - \cos(x),$$
$$y_2(x) = x^2 - x - 2 + 2\cos(x) + \sin(x),$$
$$y_3(x) = x,$$

and special value

$$y(\pi) = [2 - 2\pi, \pi^2 - \pi - 4, \pi]^{\mathrm{T}}.$$

In an environment in which numbers are carried to six or seven decimal digits, each of the two procedures was used to compute $y(\pi)$ with a variety of constant step sizes $h = \pi 2^{-m}$ with $m = 6$ to 20. For illustrative purposes, we present results in figure 207 as far as $m = 18$ and for the first component only.

For moderately small step sizes (say $\pi 2^{-5}$ down to $\pi 2^{-10}$), the two sets of results are almost identical and they closely agree with the error behaviour that would be expected if rounding errors were negligible. With the algorithm 207′ results, this behaviour continues down at least to $h = \pi 2^{-20}$, whereas for such step sizes the results found using algorithm 207 are completely destroyed by rounding error.

21 GENERALIZATIONS OF THE EULER METHOD

210 Introduction.

In our study of the Euler method in section 20, we saw that it had a number of desirable properties which were offset by a number of limitations. In the present section, we consider generalizations of the Euler method that will yield improved numerical behaviour but will retain, as much as possible, its characteristic quality of simplicity.

An important aim will be to obtain methods for which the asymptotic errors behave like high powers of the step size h. For such a method, the gain in accuracy resulting from a given reduction in step size would be better than for the Euler method where, as we have seen, the error behaves asymptotically like the first power of h. We will also wish to examine the stability characteristics of various methods, as we did for the Euler method in subsection 206, with the intention of overcoming the limitation from which that method suffered when used to solve stiff differential equations. The two major aims, improved accuracy and better stability, have to be balanced against the need to avoid expending unnecessary computational effort.

In the next few subsections, we explore techniques for constructing methods that achieve some of these aims.

211   More computations in a step.

Instead of computing f once in a step, as in the Euler method we might look for methods in which f is computed (with different arguments, of course) two or more times. For example, if y_n is computed as

$$y_n = y_{n-1} + \frac{h}{2} [f(y_{n-1}) + f(y_{n-1} + hf(y_{n-1}))], \qquad (211a)$$

then we have a method of second order. To see why this is the case, replace y_{n-1} and y_n by $y(x_{n-1})$ and $y(x_n)$ respectively and estimate the difference of the two sides of (211a) on the assumption that f is sufficiently smooth to ensure that each component of y has continuous third derivatives.

We find that

$$\| y(x_n) - y(x_{n-1}) - \frac{h}{2} [f(y(x_{n-1})) + f(y(x_{n-1}) + hf(y(x_{n-1})))] \|$$

$$\leq \ \| y(x_n) - y(x_{n-1}) - hf(y(x_{n-1})) - \frac{h^2}{2} y''(x_{n-1}) \|$$

$$+ \frac{h}{2} \| f(y(x_n)) - f(y(x_{n-1})) - hy''(x_{n-1}) \|$$

$$+ \frac{h}{2} \| f(y(x_n)) - f(y(x_{n-1}) + hy'(x_{n-1})) \|$$

$$\leq \ \| y(x_n) - y(x_{n-1}) - hy'(x_{n-1}) - \frac{h^2}{2} y''(x_{n-1}) \|$$

$$+ \frac{h}{2} \| y'(x_n) - y'(x_{n-1}) - hy''(x_{n-1}) \|$$

$$+ \frac{hL}{2} \| y(x_n) - y(x_{n-1}) - hy'(x_{n-1}) \|$$

$$= \ O(h^3),$$

where we have made use of the fact that $f(y(x)) = y'(x)$ and the Lipschitz condition, and where we have estimated the remainder term in each of three Taylor series.

The method defined by (211a) is an example of an important class of one-step methods (that is to say, methods in which the formula for y_n depends only on y_{n-1} and not on any of $y_{n-2}, y_{n-3}, \ldots$) known as Runge-Kutta methods. We will briefly review their properties in section 22.

212 Dependence on more previous values.

Rather than computing y_n in a complicated manner from the value of y_{n-1}, we can consider methods in which only one f value is computed in each step but where y_n depends not only on y_{n-1} and $f(y_{n-1})$ but also on y_{n-2}, $f(y_{n-2}), \ldots, y_{n-k}, f(y_{n-k})$. Such a method is known as a k-step method. An example of a 2-step method is

$$y_n = y_{n-2} + 2hf(y_{n-1}). \tag{212a}$$

Like the Runge-Kutta method introduced in (211a), (212a) is also a second-order method. We see this by replacing y_{n-2}, y_{n-1}, y_n respectively by $y(x_{n-2})$, $y(x_{n-1})$, $y(x_n)$ and estimating the difference of the two sides of (212a), on the assumption that each component of y has continuous third derivatives on $[x_{n-2}, x_n]$. We have

$$\| y(x_n) - y(x_{n-2}) - 2hf(y(x_{n-1})) \|$$

$$\leq \| y(x_n) - y(x_{n-1}) - hy'(x_{n-1}) - \frac{h^2}{2} y''(x_{n-1}) \|$$

$$+ \| y(x_{n-2}) - y(x_{n-1}) + hy'(x_{n-1}) - \frac{h^2}{2} y''(x_{n-1}) \|$$

$$= O(h^3),$$

where we have again used the remainder term in Taylor series.

A brief survey of the properties of this sort of *linear* (because y_n depends linearly on $y_{n-1}, f(y_{n-1}), \ldots$) *multistep method* will be presented in section 23.

213 Use of higher derivatives.

Since the Euler method is related to the use of the truncated Taylor series

$$y(x_{n-1} + h) = y(x_{n-1}) + hy'(x_{n-1}) + O(h^2),$$

it is natural to consider the generalization to more accurate Taylor series approximations. For example, the formula

$$y(x_{n-1} + h) = y(x_{n-1}) + hy'(x_{n-1}) + \frac{h^2}{2} y''(x_{n-1}) + O(h^3)$$

yields the method

$$y_n = y_{n-1} + hf(y_{n-1}) + \frac{h^2}{2} f'(y_{n-1})(f(y_{n-1})). \tag{213a}$$

We note that if y_{n-1} is replaced by $y(x_{n-1})$ in (213a) then the last term is precisely $(h^2/2)y''(x_{n-1})$ because

$$y''(x) = \frac{d}{dx} y'(x) = \frac{d}{dx} f(y(x))$$

$$= f'(y(x))(y'(x)) = f'(y(x))(f(y(x))),$$

using the chain rule and the fact that $y'(x) = f(y(x))$. Thus, (213a) is a method of order 2.

Obviously, this idea can be extended to higher derivatives. Further consideration will be given to Taylor series methods in section 24.

Besides methods which compute higher derivatives of y, there exist methods which make use of f' (and possibly f'', etc.) in their own right. In particular, the so-called *Rosenbrock methods* fall into this category.

214 Multistep—multistage—multiderivative methods.

The generalizations introduced in the last three subsections can be combined in various ways. For example, multistep methods which have more f evaluations per step than is the case for *linear* multistep methods have been proposed by a number of authors and are sometimes called *hybrid methods*. The name *pseudo-Runge–Kutta* is also used for certain closely related methods.

A general class of methods of this type is the object of study of chapter 4, whereas some of the important special cases will be considered in the sequel to this volume.

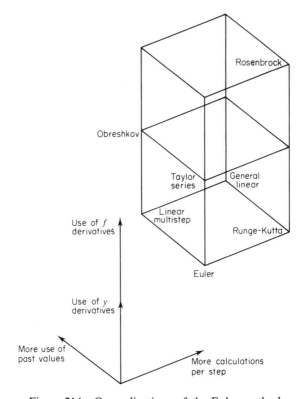

Figure 214   Generalizations of the Euler method

Methods that are basically of the Runge–Kutta type but which make use of derivatives of f (this includes the Rosenbrock methods mentioned in the last subsection) also exist, as do methods of the linear multistep type combined with higher y derivatives (Obreshkov methods).

Finally, there exist methods, sometimes called *multistep–multistage–multiderivative methods*, which combine each of the three types of generalization of the Euler method already mentioned. As a means of classifying these methods and the simplier types already mentioned, figure 214 is presented. This represents a lattice of method types based on the ordering 'is a generalization of' and also uses three coordinate directions, shown at the lower left of the figure, which represent the different types of generalization. Not all points in the lattice are labelled but, where they are, a caption is given indicating that the class of methods in question has been studied in some detail.

215 Implicit methods

Methods so far considered have been explicit in the sense that the algorithm for determining an approximation at the end of a step consists of a definite sequence of derivative calculations and rational operations. In contrast to this, method can be constructed in which the solution at every step involves the solution of a functional equation. For example, the method which defines y_n (an approximation to $y(x_n)$) in terms of y_{n-1} by the equation

$$y_n - hf(y_n) = y_{n-1} \tag{215a}$$

has an obvious relationship to the Euler method and is known as the *implicit Euler method*.

To evaluate y_n, (215a) requires the solution for z of the equation

$$z - hf(z) = y_{n-1}, \tag{215b}$$

and this can be carried out, for example, by using an iteration scheme

$$z^{(m)} = y_{n-1} + hf(z^{(m-1)}), \tag{215c}$$

for $m = 1, 2, 3, \ldots$ with $z^{(0)} = y_{n-1}$, say. If f satisfies a Lipschitz condition with constant L and h satisfies

$$hL < 1, \tag{215d}$$

then (by the contraction mapping theorem) this scheme converges to the unique solution of (215b).

As it happens, the main application of the implicit Euler method (and of implicit methods in general) is to stiff problems for which L is an extremely large number. This particular method possesses an important theoretical property (A-stability) which makes it ideal for stiff problems.

To illustrate the nature of this concept we present in figure 215 the stability region for the implicit Euler method corresponding to figure 206 for the Euler

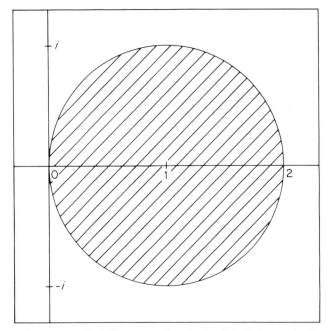

Figure 215   Stability region for the implicit Euler method

method itself. We see from figure 215 that the stability region includes the entire left half-plane so that the implicit Euler method gives bounded results for $y' = qy$ whenever the same is true for the exact solution.

Since the restriction on h for (215c) to be convergent, that is the inequality (215d), is much too severe to be practical for stiff problems, (215b) would normally be solved using a Newton–Raphson process or a modification of this process.

Implicit methods of Runge–Kutta and multistep types are well-established methods for treating stiff problems and are considered in later chapters of this book.

216   Local error estimates.

Along with an approximation to the solution at the end of a step, it is desirable to have information on the accuracy of this approximation so that a sensible strategy for adjustment of step size, so as to control the error, can be used. For example, the Euler method can be used to compute an approximation at the end of a step and the formula given by (211a), which gives a higher-order approximation, can be used as a result to which the Euler result can be compared to obtain an approximation to the local error.

If this computation is arranged in a suitable way, then no extra derivative calculations need actually be performed in addition to those required for the Euler method itself. One way of programming this is as shown in algorithm

216. The various parameters have the same meanings as in algorithm 207 with the additional parameters dy (denoting an array with dimension $[1 .. N]$ used to hold the derivative at the end of a step and used in the succeeding step) and l (an array of the same dimension denoting an approximation to the local truncation error generated within the step).

Algorithm 216: A single Euler step with error estimate

```
for i := 1 to N do
begin
    y[i] := y[i] + h*dy[i];
    temp[i] := dy[i]
end;
diffeqn (y, dy);
for i := 1 to N do l[i] := h*0.5*(temp[i] - dy[i])
```

The way this algorithm might be used is that, after a step is completed, the norm of l would be computed and compared with a preassigned constant. If $\| l \|$ exceeds this constant, then the step is deemed to be sufficiently inaccurate to warrant repeating the step with a smaller h value. If h_1, h_2 are the old and new h values and l_1, l_2 are the corresponding l vectors, then, because of theorem 201D, $\| l_1 \| h_1^{-2}$ will approximately equal $\| l_2 \| h_2^{-2}$. Hence, if $\| l_2 \|$ is required to have a particular value, then h_2 should be set equal to $h_1 (\| l_2 \| / \| l_1 \|)^{1/2}$.

Similarly, if $\| l \|$ is unnecessarily small, then the value of h can be increased for subsequent steps on the assumption that $\| l \| h^{-2}$ is approximately constant over adjacent steps.

Methods for estimating local errors are available for a large variety of methods, especially linear multistep methods. These will be considered in detail later.

217   Composite cyclic methods.

Situations exist in which two or more members of a particular family of methods (say linear multistep), when used in a cyclic fashion, exhibit some desirable properties which the individual members do not possess.

For example, the order of the Euler method and of the implicit Euler method is 1. However, if these two methods are used alternatively over successive steps, then the overall result behaves as though an order 2 method were used. Furthermore, this combined method has a stability region which includes the negative half-plane (a property possessed by the implicit Euler method but not by the classical Euler method).

This particular combination of methods can be regarded as a single implicit Runge–Kutta method over a double step. While a similar remark applies to other types of composite methods, there is nevertheless an interest in studying these methods in their own right.

218 Extrapolation methods.

We saw in subsection 205 that the Euler method had an asymptotic error of the form $\frac{1}{2} he(x)$. This means that if two different step-sizes h_1, h_2 are used and if the computed results at $\bar{x}$ are η_1 and η_2 respectively then

$$y(\bar{x}) = \eta_1 + \tfrac{1}{2} h_1 e(\bar{x}) + o(h_1), \tag{218a}$$

$$y(\bar{x}) = \eta_2 + \tfrac{1}{2} h_2 e(\bar{x}) + o(h_2), \tag{218b}$$

where the function e was introduced in subsection 205. Multiplying (218a) by h_2 and (218b) by h_1, subtracting and dividing by $h_2 - h_1$, we find

$$y(\bar{x}) = \frac{h_1 \eta_2 - h_2 \eta_1}{h_1 - h_2} + \frac{h_1 o(h_2) - h_2 o(h_1)}{h_1 - h_2}. \tag{218c}$$

If $h_1 = c_1 h$, $h_2 = c_2 h$ where c_1, c_2 are fixed constants, then (218c) implies

$$y(\bar{x}) = \frac{c_1 \eta_2 - c_2 \eta_1}{c_1 - c_2} + o(h), \tag{218d}$$

as $h \to 0$, as long as $c_1 \neq c_2$. Thus, approximating $y(\bar{x})$ by $(c_1 \eta_2 - c_2 \eta_1)/(c_1 - c_2)$ gives an error that tends to zero as $h \to 0$ more quickly than the error in either (218a) or (218b).

In fact, if the exact solution y satisfies certain smoothness assumptions in addition to those assumed in subsection 205, then the $o(h)$ term may be replaced by $O(h^2)$, and this term itself has an expansion in the form $e_2(x) h^2 + O(h^3)$. Thus, the procedure for generating the result (218d) may be extended to further steps and we can arrive at a sequence of approximations of ever-increasing speed of convergence.

Let us apply this extrapolation idea to the particular case of the problem $y'(x) = y(x)$, $y(0) = 1$, with $c_1 = 1$, $c_2 = \frac{1}{2}$. The result given by the Euler method at $\bar{x} = 1$ is

or

$$(1 + h)^{1/h} \qquad \text{(using step size } h\text{)}$$

$$(1 + h/2)^{2/h} \qquad \text{(using step size } h/2\text{)}.$$

On the other hand, the result given by (218d) is

$$2(1 + h/2)^{2/h} - (1 + h)^{1/h}$$

The work required to compute this can be taken as the number of derivative calculations, that is to say, $3/h$. To do the same amount of work using a straightforward Euler method a step of size $h/3$ would be required and the computed result would then be $(1 + h/3)^{3/h}$.

It can be shown that

$$\exp(1) > 2(1 + h/2)^{2/h} - (1 + h)^{1/h} > (1 + h/3)^{3/h},$$

which implies that the extrapolated result is more accurate than the result with step size $h/3$.

Extrapolation can be applied to a wide variety of different methods and is particularly successful in the case of certain linear multistep methods and certain Runge–Kutta methods. We will refer again to extrapolation in the case of Runga–Kutta methods in subsection 368.

22   RUNGE–KUTTA METHODS

220   Historical introduction.

The idea of extending the Euler method by allowing for a multiplicity of evaluations of the function f within each step was originally proposed by Runge (1895). Further contributions were made by Heun (1900) and by Kutta (1901). The latter completely characterized the set of Runge–Kutta methods of order 4 and proposed the first methods of order 5. Special methods for second-order differential equations were proposed by Nyström (1925) who also contributed to the development of methods for first-order equations. It was not until the work of Huťa (1956, 1957) that sixth-order methods were introduced.

Since the advent of digital computers, fresh interest has been focused on Runge–Kutta methods, and a large number of research workers have contributed to recent extensions to the theory and the development of particular methods. Although early studies were devoted entirely to explicit Runge–Kutta methods, interest has now extended to implicit methods which are now recognized as appropriate for stiff differential equations.

A number of different approaches have been used in the analysis of Runge–Kutta methods but the one used in this section, and in the more detailed analysis in Chapter 3, is the one developed by the present author, Butcher (1963) following on from the work of Gill (1951) and Merson (1957).

221   General form of explicit Runge–Kutta methods.

Given the vector y_{n-1} as an approximation to $y(x_{n-1})$, where y satisfies the differential equation system

$$y'(x) = f(y(x)), \tag{221a}$$

the approximation y_n to $y(x_n)$ is computed by evaluating, for $i = 1, 2, \ldots, s$,

$$F_i = f(Y_i), \tag{221b}$$

where $Y_1, Y_2, \ldots, Y_s$ are given by

$$Y_i = y_{n-1} + h \sum_{j < i} a_{ij} F_j, \tag{221c}$$

and then evaluating

$$y_n = y_{n-1} + h \sum_{j=1}^{s} b_j F_j. \tag{221d}$$

The quantities $Y_1, Y_2, \ldots, Y_s$, as given by (221c), are approximations to solution values $y(x)$ for x ranging through various values near x_{n-1}. Also, from (221b), we see that $F_1, F_2, \ldots, F_s$ are approximations to $y'(x)$ at the same x values. In particular, because the sum in (221c) is empty when $i = 1$, $Y_1 = y_{n-1}$ is an approximation to $y(x_{n-1})$ and F_1 is an approximation to $y'(x_{n-1})$. The integer s is the number of stages of the method and measures its complexity since the number of evaluations of f per step equals s. The set of numbers $a_{21}, a_{31}, a_{32}, \ldots, a_{s, s-1}$, $b_1, b_2, \ldots, b_s$ are constants which characterize a particular method of this type.

Algorithm 221 will now be presented for carrying out a single step of a Runge–Kutta method. It is assumed that the a_{ij} and b_j are stored in arrays $a[1 .. s, 1 .. s - 1]$ and $b[1 .. s]$ and that their values have been given outside the algorithm. It is further assumed that h is the step size to be used and that N is the number of differential equations in the system to be solved. The array y holds the vector y_{n-1} when the algorithm is entered but on return this is replaced by y_n. The function f is represented by a user-supplied procedure $diffeqn$. When a call $diffeqn(u, dy)$ is made on this procedure it is intended that the array dy be overwritten by the value of $f(u)$.

Algorithm 221: General Runge–Kutta step

```
for i := to s do
begin
      for k := 1 to N do
      begin
            u[k] := y[k];
            for j := 1 to i - 1 do
                  u[k] := u[k] + h * a[i, j] * F[j, k]
      end;
      diffeqn(u, dy);
      for k := 1 to N do F[i, k] := dy[k]
end;
for k := 1 to N do
for j := 1 to s do y[j] := y[j] + h * b[j] * F[j, k]
```

If, instead of dealing with the autonomous equation (221a), we consider the non-autonomous system

$$z'(x) = g(x, z(x)), \tag{221e}$$

then, as we saw in subsection 150, this can be reduced to the form of (221a) by writing

$$y(x) = \begin{bmatrix} x \\ z(x) \end{bmatrix}$$

and

$$f\left(\begin{bmatrix} u \\ v \end{bmatrix}\right) = \begin{bmatrix} 1 \\ g(u, v) \end{bmatrix},$$

where the initial values are chosen as $y_0 = \begin{bmatrix} x_0 \\ z_0 \end{bmatrix}$.

We now examine the effect of applying the algorithm to the more general form. Let

$$Y_i = \begin{bmatrix} X_i \\ Z_i \end{bmatrix}, \qquad i = 1, 2, \ldots, s,$$

$$y_{n-1} = \begin{bmatrix} x_{n-1} \\ z_{n-1} \end{bmatrix}.$$

Then the formulae for the computation of $X_1, X_2, \ldots, X_s, Z_1, Z_2, \ldots, Z_s$ are

$$X_i = x_{n-1} + h \sum_{j<i} a_{ij}, \tag{221f}$$

$$Z_i = z_{n-1} + h \sum_{j<i} a_{ij} g(X_j, Z_j).$$

The values of the components of y_n are found to be

$$x_n = x_{n-1} + h \sum_j b_j,$$

$$z_n = z_{n-1} + h \sum_j b_j g(X_j, Z_j),$$

so that, since $x_n = x_{n-1} + h$ by definition, we must have $\sum_j b_j = 1$. The coefficient of h in (221f) will be denoted by c_i. We may thus write the algorithm for the non-autonomous system (221e) as

$$\left. \begin{array}{l} Z_i = z_{n-1} + h \displaystyle\sum_{j<i} a_{ij} G_j, \\[2mm] G_i = g(x_{n-1} + hc_i, Z_i), \\[2mm] z_n = z_{n-1} + h \displaystyle\sum_j b_j G_j. \end{array} \right\} \qquad i = 1, 2, \ldots, s,$$

222 Consistency and convergence.

The condition

$$\sum_{j=1}^{s} b_j = 1, \tag{222a}$$

which we saw in the last subsection to be necessary for the method to at least compute the independent variable correctly, is known as the *consistency condition*. We shall see that this condition is also necessary and sufficient for the local truncation error of the method to have asymptotic behaviour $O(h^2)$. Since the Euler method also has the local truncation error $O(h^2)$, we can think of (222a) as being the condition that the Runge–Kutta method is no worse than the Euler method from this point of view.

Before proving this result we establish a preliminary lemma.

LEMMA 222A.

If $F_1, F_2, \ldots, F_s$ are given by (221b), where f satisfies a Lipschitz condition with the constant L and the value of y_{n-1} equals $y(x_{n-1})$, then there exist positive numbers h_0 and k such that, for all $h \in (0, h_0)$,

$$\| F_i - f(y(x_{n-1})) \| \le kh, \qquad i = 1, 2, \ldots, s. \tag{222b}$$

Proof.

Let h_0 be chosen so that

$$d = 1 - h_0 L \max_i \left(\sum_{j < i} | a_{ij} | \right) > 0$$

and let $k = \max_i(| c_i |)L \| f(y(x_{n-1})) \|/d$. We now prove (222b) by induction on i. It is true for $i = 1$ since the left-hand side is zero. If it has been proved for $j < i$, then we have

$$\| F_i - f(y(x_{n-1})) \| = \left\| f\left(y(x_{n-1}) + h \sum_{j < i} a_{ij}F_j \right) - f(y(x_{n-1})) \right\|$$

$$\le hL \left\| \sum_{j < i} a_{ij}F_j \right\|$$

$$\le hL \left\| \sum_{j < i} a_{ij}[F_j - f(y(x_{n-1}))] \right\| + hL \left\| \sum_{j < i} a_{ij}f(y(x_{n-1})) \right\|$$

$$\le h^2 Lk \sum_{j < i} | a_{ij} | + hL | c_i | \| f(y(x_{n-1})) \|$$

$$\le hk(1 - d) + hkd$$

$$= hk. \quad \blacksquare$$

THEOREM 222B.

The local truncation error of a consistent Runge–Kutta method is $O(h^2)$ as $h \to 0$.

Proof.

We estimate

$$\| y(x_n) - y(x_{n-1}) - h\sum_i b_iF_i \| \leq \| y(x_n) - y(x_{n-1}) - hf(y(x_{n-1})) \|$$

$$+ h \sum_i | b_i | \| F_i - f(y(x_{n-1})) \|.$$

Each term is $O(h^2)$, the first because of the corresponding property of Euler's method and the terms in the summation because of the result of lemma 222A.
∎

An analysis of the global truncation error can be made in a similar way to that for the Euler method, but we do not present it in this introductory section.

223   Order conditions.

The local truncation error $y(x_n) - y(x_{n-1}) - h\Sigma_i b_iF_i$, where $F_i = f(Y_i)$ and $Y_i = y(x_{n-1}) + h\Sigma_{j < i} a_{ij}F_j$, clearly depends on the values of the numbers $a_{21}, \ldots, b_1, \ldots, b_s$. Although we will postpone a detailed analysis of the dependence of the local truncation error on these numbers until chapter 3, we review briefly the general form of the conditions for a given order of accuracy.

Let T denote the set of rooted trees introduced in subsections 140 and 142. With any particular $t \in T$ we will associate a certain polynomial in $a_{21}, \ldots, b_1, \ldots, b_s$ denoted by $\Phi(t)$ and a certain numerical constant denoted by $\gamma(t)$. As we will see in section 30, necessary and sufficient conditions that the order of a method should be m are that $\Phi(t) = 1/\gamma(t)$ for all t with no more than m vertices.

The form of the $\Phi(t)$ polynomials can be written down from the structure of the particular t. We simply attach labels $i, j, k, \ldots$ to the various vertices of the tree, with i the label associated with the root, and then form $b_i\Pi a_{jk}$ where the product is over all pairs of vertices (j, k) such that there is an arc of t from j to k. Having formed this product, we sum over $i, j, k, \ldots$ from 1 to s with the proviso that a_{jk} is set to zero if $j \leq k$.

To form the value of $\gamma(t)$ for a particular t, associate a value 1 with all terminal vertices and a value $1 + i$ with all vertices for which i is the sum of the numbers attached to vertices which branch outwards from this one. The value of $\gamma(t)$ is then simply the product of the integers associated with each vertex. It is easy to see that $\gamma(t)$ is the density of t introduced in subsection 144.

For example, if t is the tree given below, we attach labels and integers to form $\Phi(t)$ and $\gamma(t)$ as shown:

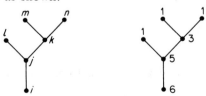

$$\Phi(t) = \sum b_i a_{ij} a_{jl} a_{jk} a_{km} a_{kn},$$

$$\gamma(t) = 90.$$

Making use of the numbers $c_1 = 0, c_2, \ldots, c_s$ introduced in subsection 221, we can write the formula for $\Phi(t)$ more compactly. In this case it is equal to $\sum b_i a_{ij} c_j a_{jk} c_k^2$.

In table 223, we show the trees with up to four vertices and the corresponding equations $\Phi(t) = 1/\gamma(t)$.

Table 223

t	$\Phi(t)$	$= 1/\gamma(t)$
	$\sum b_i$	$= 1$
	$\sum b_i c_i$	$= \frac{1}{2}$
	$\sum b_i c_i^2$	$= \frac{1}{3}$
	$\sum b_i a_{ij} c_j$	$= \frac{1}{6}$
	$\sum b_i c_i^3$	$= \frac{1}{4}$
	$\sum b_i c_i a_{ij} c_j$	$= \frac{1}{8}$
	$\sum b_i a_{ij} c_j^2$	$= \frac{1}{12}$
	$\sum b_i a_{ij} a_{jk} c_k = \frac{1}{24}$

224 Examples of methods.

To illustrate the use of the equations in table 223, we present here methods of various orders from one to four in which the number of stages is actually equal to the order. The first of these is, of course, the Euler method.

One stage, order 1:

$$b_1 = 1.$$

Two stages, order 2:

(a) $a_{21} = \frac{1}{2}$, $b_1 = 0$, $b_2 = 1$,
(b) $a_{21} = 1$, $b_1 = b_2 = \frac{1}{2}$.

Three stages, order 3:

(a) $a_{21} = \frac{2}{3}$, $a_{31} = a_{32} = \frac{1}{3}$, $b_1 = \frac{1}{4}$, $b_2 = 0$, $b_3 = \frac{3}{4}$
(b) $a_{21} = \frac{1}{2}$, $a_{31} = -1$, $a_{32} = 2$, $b_1 = b_3 = \frac{1}{6}$, $b_2 = \frac{2}{3}$.

Four stages, order 4:

(a) $a_{21} = a_{32} = \frac{1}{2}$, $a_{31} = a_{41} = a_{42} = 0$, $a_{43} = 1$,
    $b_1 = b_4 = \frac{1}{6}$, $b_2 = b_3 = \frac{1}{3}$,
(b) $a_{21} = \frac{1}{4}$, $a_{31} = 0$, $a_{32} = \frac{1}{2}$, $a_{41} = 1$, $-a_{42} = a_{43} = 2$,
    $b_1 = b_4 = \frac{1}{6}$, $b_2 = 0$, $b_3 = \frac{2}{3}$.

225 Numerical examples.

To illustrate the behaviour as $h \to 0$ for the seven methods quoted in the last subsection, one very simple problem has been examined. This is the one-dimensional initial value problem

$$y'(x) = \frac{y(x)(1 - y(x))}{2y(x) - 1},$$

$$y(0) = \frac{5}{6},$$

for which the solution is $y(x) = \frac{1}{2} + [(\frac{1}{4} - \frac{5}{36} \exp(-x)]^{1/2}$. Using a step size $h = 1/n$ for various $n = 2, 3, 4, 6, 8, 12, \ldots, 2048$ (where we alternatively increase the value of n by factors $\frac{3}{2}$ and $\frac{4}{3}$) and each of these methods over n steps, the value of $y(1)$ was computed in seven different ways. The magnitude of the error E for different h for each of these methods is shown in figure 225. The Euler method is marked 1, the second-order methods are marked 21, 22, the third-order methods 31, 32 and the fourth-order methods 41, 42. Note that the exact value of $y(1)$, to 12 decimals, is 0.945988377843. It will be seen that, on the logarithmic scale which has been used for these graphs, the error for each method is represented very closely by a straight line whose slope equals the order of the method.

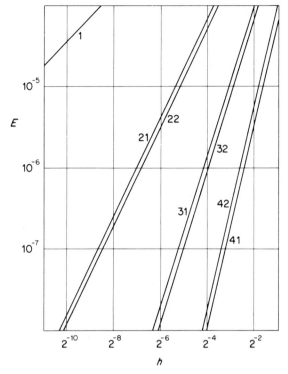

Figure 225   Error versus step size for various Runge–Kutta methods

23 LINEAR MULTISTEP METHODS

230 Historical introduction.

The idea of extending the Euler method by allowing the approximate solution at a point to depend on the solution values and derivative values before the immediately previous point was proposed in a special case by Bashforth and Adams (1883). This special type of method, the Adams–Bashforth method, was introduced and a further method, later to be known as the Adams–Moulton method, was alluded to. The latter implicit type of method was developed further by Moulton (1926). Other special types of linear multistep methods were proposed by Nyström (1925) and Milne (1926, 1953). The idea of predictor–corrector methods is associated with the name of Milne, especially because of his advocacy of a simple type of error estimate available with such methods.

The modern theory of linear multistep methods was developed by Dahlquist (1956) and has become widely known through the exposition by Henrici (1962, 1963).

231   General form of linear multistep methods.

As we saw in section 21, linear multistep methods compute y_n, an approximation to $y(x_n)$, as a linear combination of $y_{n-1}, y_{n-2}, \ldots, y_{n-k}$ and of the derivatives computed at $y_n, y_{n-1}, \ldots, y_{n-k}$. Thus

$$
\begin{aligned}
y_n = \alpha_1 y_{n-1} + \alpha_2 y_{n-2} + \cdots + \alpha_k y_{n-k} \\
+ h[\beta_0 f(y_n) + \beta_1 f(y_{n-1}) + \cdots + \beta_k f(y_{n-k})],
\end{aligned} \tag{231a}
$$

for some fixed numbers $\alpha_1, \alpha_2, \ldots, \alpha_k, \beta_0, \beta_1, \ldots, \beta_k$.

It may happen, of course, that α_k or β_k is zero, but we assume that this is not the case for both of them, since if it were, the value of k could be reduced. Under the assumption that k cannot be so reduced, the method given by (231a) is known as a linear k-step method.

It is rather crucial whether or not the value of β_0 is zero. If it is, the method is said to be explicit; otherwise it is implicit. For an explicit method y_n can be computed directly from (231a), whereas in the implicit case, we must solve an equation of the form

$$
y_n - h\beta_0 f(y_n) = v, \tag{231b}
$$

where the vector v is independent of y_n and is given by

$$
v = \sum_{i=1}^{k} [\alpha_i y_{n-i} + h\beta_i f(y_{n-i})].
$$

If f has a small Lipschitz constant, then (231b) can be solved by fixed-point iteration as the limit of the sequence $z^{(0)}, z^{(1)}, \ldots$, where, for $j \geq 1$, $z^{(j)}$ is given by

$$
z^{(j)} = v + h\beta_0 f(z^{(j-1)}).
$$

In the case of stiff problems, where the Lipschitz constant is necessarily large, this iterative method is not convergent and it is necessary to use some variant of the Newton method. A further possible approach to the use of implicit methods is in the context of predictor–corrector pairs. This will be discussed briefly in subsection 234.

When discussing Runge–Kutta methods in section 22, we considered the modifications to the formulae that become necessary if a non-autonomous system is used. We will now conduct a similar analysis in the case of a linear multistep method.

Consider the problem

$$
z'(x) = g(x, z(x)), \tag{231c}
$$

which can be written

$$
y'(x) = f(y(x)), \tag{231d}
$$

where

$$
f\left(\begin{bmatrix} u \\ v \end{bmatrix} \right) = \begin{bmatrix} 1 \\ g(u, v) \end{bmatrix}
$$

and

$$y(x) = \begin{bmatrix} x \\ z(x) \end{bmatrix}.$$

If exact values of the first component are given at $x_{n-1}, x_{n-2}, \ldots, x_{n-k}$, then applying (231a) to (231d) and writing the result in terms of its components, we have

$$x_n = \alpha_1 x_{n-1} + \alpha_2 x_{n-2} + \cdots + \alpha_k x_{n-k} + h(\beta_0 + \beta_1 + \cdots + \beta_k), \quad (231e)$$

$$z_n = \alpha_1 z_{n-1} + \alpha_2 z_{n-2} + \cdots + \alpha_k z_{n-k}$$

$$+ h[\beta_0 g(x_n, z_n) + \beta_1 g(x_{n-1}, z_{n-1}) + \cdots + \beta_k g(x_{n-k}, z_{n-k})],$$
$$(231f)$$

where we assume that the first component is computed exactly as $x_n = x_{n-1} + h$. Substituting $x_{n-1} = x_n - h$, $x_{n-2} = x_n - 2h, \cdots, x_{n-k} = x_n - kh$ into (231e), we find

$$x_n = (\alpha_1 + \alpha_2 + \cdots + \alpha_k)x_n + h(\beta_0 + \beta_1 + \cdots + \beta_k - \alpha_1 - 2\alpha_2 - \cdots - k\alpha_k).$$

For this to be satisfied independently of the values of x_n and h, we must have

$$\alpha_1 + \alpha_2 + \cdots + \alpha_k = 1, \quad (231g)$$

$$\alpha_1 + 2\alpha_2 + \cdots + k\alpha_k = \beta_0 + \beta_1 + \cdots + \beta_k, \quad (231h)$$

where (231g) and (231h) are known as the *consistency conditions*.

232 Consistency, stability and convergence.

Just as for Runge–Kutta methods, the consistency conditions are necessary and sufficient that the local truncation error behaves like $O(h^2)$, as is the case for the Euler method. However, unlike single-step methods, a complication arises in transferring this result to a global setting. In fact, an additional condition on the coefficients, known as the *stability condition*, is necessary and sufficient for consistent methods to converge. Although the technical details of the relationship between stability and convergence will be postponed until the sequel to this volume, we define the stability concept at this point.

DEFINITION 232A.

The method given by (231a) is *stable* if all solutions to the difference equation

$$y_n = \alpha_1 y_{n-1} + \alpha_2 y_{n-2} + \cdots + \alpha_k y_{n-k}$$

are bounded as $n \to \infty$.

Let the polynomial p be defined by

$$p(z) = (z - z_1)(z - z_2) \cdots (z - z_k)$$

$$= z_k - \alpha_1 z^{k-1} - \cdots - \alpha_k.$$

We now define a condition on $\alpha_1, \alpha_2, \ldots, \alpha_k$ expressed in terms of $z_1, z_2, \ldots, z_k$.

DEFINITION 232B.

The method given by (231a) is said to *satisfy the root condition* if $|z_1| \leq 1, |z_2| \leq 1, \ldots, |z_k| \leq 1$ and if, for $i \neq j, z_i = z_j$ then $|z_i| < 1$.

Using the theory of difference equations from section 12, we see that a method is stable iff it satisfies the root condition. This will be proved as the consequence of a more general result in chapter 4.

The convergence concept is obviously more complicated than for one-step methods, since we must specify k starting values to begin the numerical integration, and how we specify them will, in general, depend on the value of h. Let ϕ_n denote the functional relationship between y_n, these initial values and the value of h, so that

$$y_n = \phi_n(y_0, y_1, \ldots, y_{k-1}, h).$$

Since the exact solution is continuous and $y(x_i) \to y(x_0)$ as $h \to 0$ for $i = 0, 1, \ldots, k - 1$, it is appropriate to consider rules for generating starting values $y_0, y_1, \ldots, y_{k-1}$ so that each of these has the limit $y(x_0)$ as $h \to 0$.

DEFINITION 232C.

The method (231a) is said to be *convergent* if for any standard problem on the interval $[x_0, \bar{x}]$,

$$\phi_n(y_0, y_1, \ldots, y_{k-1}, (\bar{x} - x_0)/n) \to y(\bar{x})$$

as $y_0, y_1, \ldots, y_{k-1} \to y(x_0)$ and $n \to \infty$.

As we have remarked, consistency and stability are together necessary and sufficient for convergence. This will follow from the more general considerations of section 42.

233 Order conditions.

Just as for Runge–Kutta methods, we define the local truncation error of the method given by (231a) by replacing $y_n, y_{n-1}, \ldots, y_{n-k}$ by $y(x_n)$, $y(x_{n-1}), \ldots, y(x_{n-k})$ respectively in this equation and taking the difference of the two sides. Since $f(y(x)) = y'(x)$ we obtain the following expression for the local truncation error in the nth step:

$$y(x_n) - \alpha_1 y(x_{n-1}) - \cdots - \alpha_k y(x_{n-k})$$
$$- h[\beta_0 y'(x_n) + \beta_1 y'(x_{n-1}) + \cdots + \beta_k y'(x_{n-k})]. \tag{233a}$$

Unlike Runge–Kutta methods, this expression for the local truncation error for linear multistep methods takes the simple form of being linear in y.

DEFINITION 233A.

The method given by (231a) is of order m if (233a) vanishes whenever y is a polynomial of degree m.

THEOREM 233B.

The method given by (231a) is of order m iff

$$\alpha_1 + \alpha_2 + \cdots + \alpha_k = 1, \tag{233b}$$

$$\alpha_1 + 2\alpha_2 + \cdots + k\alpha_k = \beta_0 + \beta_1 + \cdots + \beta_k, \tag{233c}$$

$$\alpha_1 + 2^j\alpha_2 + \cdots + k^j\alpha_k = j(\beta_1 + 2^{j-1}\beta_2 + \cdots + k^{j-1}\beta_k),$$
$$j = 2, 3, \ldots, m. \tag{233d}$$

Proof.

Since the polynomials $p_0, p_1, \ldots, p_m$ defined by $p_0(x) = 1$, $p_1(x) = (x - x_n)$, $p_j(x) = (x - x_n)^j (j = 2, 3, \ldots, m)$ are a basis for the linear space of polynomials of degree m, it is necessary and sufficient, for the method to be of order m, that (233a) vanishes for $y = p_0$, $y = p_1$, $y = p_j$, $j = 2, 3, \ldots, m$. Evaluating the local truncation error in these cases gives (233b), (233c) and (233d). ∎

Note that (233b) and (233c) are identical to the consistency conditions, so that we have the following results without further proof.

COROLLARY 233C.

A linear multistep method is consistent if and only if it is of order 1.

Although we will not present a detailed error analysis of linear multistep methods in this volume, we will at this point establish a relationship between the order of methods and the asymptotic behaviour of the local truncation error for small $|h|$, when the method is applied to problems with sufficiently smooth solutions.

THEOREM 233D.

If $y \in C^{m+1}[x_{n-k}, x_n]$ and the method given by (231a) is of order m, then the local truncation error is $O(h^{m+1})$ as $h \to 0$.

Proof.

Expand $y(x_{n-1})$, $y'(x_{n-1})$, $\ldots$, $y(x_{n-k})$, $y'(x_{n-k})$ in Taylor series about x_n. We have

$$y(x) = y(x_n)p_0(x) + \frac{y'(x_n)p_1(x)}{1!} + \cdots + \frac{y^{(m)}(x_n)p_m(x)}{m!}$$

$$+ \frac{y^{(m+1)}(\xi(x))p_{m+1}(x)}{(m+1)!}, \tag{233e}$$

$$y'(x) = \frac{y'(x_n)p_1'(x)}{1!} + \frac{y''(x_n)p_2'(x)}{2!} + \cdots + \frac{y^{(m)}(x_n)p_m'(x)}{m!}$$

$$+ \frac{y^{(m+1)}(\eta(x))p_{m+1}'(x)}{(m+1)!}, \tag{233f}$$

where $p_0, p_1, \ldots$ were as introduced into the proof of theorem 233B and $\xi(x)$ and $\eta(x)$ denote certain mean values between x_n and x. If Ly denotes the expression (233a) then, by the order conditions, $Lp_0 = Lp_1 = \cdots = Lp_m = 0$. Hence using (233e) and (233f), we find

$$Ly = - \sum_{i=1}^{k} \alpha_i \frac{y^{(m+1)}(\xi(x_i))p_{m+1}(x_{n-i})}{(m+1)!}$$

$$- h \sum_{i=1}^{k} \beta_i \frac{y^{(m+1)}(\eta(x_i))p_{m+1}'(x_{n-i})}{(m+1)!},$$

so that

$$\| Ly \| \leq \sup_{t \in [x_{n-k}, x_n]} \| y^{(m+1)}(t) \| Mh^{m+1},$$

where

$$M = \frac{1}{(m+1)!} \sum_{i=1}^{k} |\alpha_i| i^{m+1} + \frac{1}{m!} \sum_{i=1}^{k} |\beta_i| i^m.$$

Hence,

$$\| Ly \| = O(h^{m+1}) \quad \text{as} \quad h \to 0. \quad \blacksquare$$

To actually obtain methods of a particular order m not greater than $k + 1$ we can first choose $\alpha_1, \alpha_2, \ldots, \alpha_k$ satisfying (233b) and then solve for $\beta_0, \beta_1, \ldots, \beta_k$ to satisfy the linear system (233c) and (233d) ($j = 2, 3, \ldots, m$). It may happen that for some choices of $\alpha_1, \alpha_2, \ldots, \alpha_k$ an order higher than $k + 1$ can be obtained. In fact with $m = 2k$, the linear system in $\alpha_1, \alpha_2, \ldots, \alpha_k, \beta_0, \beta_1, \ldots, \beta_k$ consisting of (233b), (233c) and (233d) ($j = 2, 3, \ldots, 2k$) is non-singular and we can actually find methods with orders this great. However, for $k > 2$ all such methods are unstable. It has been shown by Dahlquist (1956), that $k + 1$ is always the highest order that can be attained for stable methods unless k is even in which case $k + 2$ can be attained.

234 Predictor–corrector methods

As we remarked in subsection 231, implicit methods (those in which $\beta_0 \neq 0$) are usually implemented iteratively. However, if we have a pair of methods

$$y_n = \bar{\alpha}_1 y_{n-1} + \bar{\alpha}_2 y_{n-2} + \cdots + \bar{\alpha}_k y_{n-k}$$
$$+ h[\bar{\beta}_1 f(y_{n-1}) + \bar{\beta}_2 f(y_{n-2}) + \cdots + \bar{\beta}_k f(y_{n-k})], \qquad (234a)$$

$$y_n = \alpha_1 y_{n-1} + \alpha_2 y_{n-2} + \cdots + \alpha_k y_{n-k}$$
$$+ h[\beta_0 f(y_n) + \beta_1 f(y_{n-1}) + \cdots + \beta_k f(y_{n-k})], \qquad (234b)$$

of which the first (known as a *predictor*) is explicit and the second (known as a *corrector*) is implicit, then these can be used together as a so-called predictor–corrector pair. For example, (234a) can be used to compute a quantity $y_n^{(0)}$ (the *predicted* value) and (234b) can be modified by replacing the term $h\beta_0 f(y_n)$ on the right-hand side by $h\beta_0 f(y_n^{(0)})$. The value y_n computed from this modified version of (234b) (the *corrected* value) is used in succeeding steps. Used in this way, the pair of methods (234a), (234b) is said to be in PECE mode where this name refers to four stages in the computation of y_n and $f(y_n)$. These are (1) the computation of $y_n^{(0)}$ as the predicted value (P), (2) the evaluation (E) of $f(y_n^{(0)})$, (3) the computation of y_n as the corrected value (C) and (4) the evaluation (E) of $f(y_n)$ for use in succeeding steps.

By constrast the predictor–corrector method can be used in PEC mode in which step (4) is omitted. In this case the solution is found from the equations

$$y_n^{(0)} = \bar{\alpha}_1 y_{n-1} + \bar{\alpha}_2 y_{n-2} + \cdots + \bar{\alpha}_k y_{n-k}$$
$$+ h[\bar{\beta}_1 f(y_{n-1}^{(0)}) + \bar{\beta}_2 f(y_{n-2}^{(0)}) + \cdots + \bar{\beta}_k f(y_{n-k}^{(0)})],$$

$$y_n = \alpha_1 y_{n-1} + \alpha_2 y_{n-2} + \cdots + \alpha_k y_{n-k}$$
$$+ h[\beta_0 f(y_n^{(0)}) + \beta_1 f(y_{n-1}^{(0)}) + \cdots + \beta_k f(y_{n-k}^{(0)})].$$

Also we have PECEC and PECECE modes in which a second application of the corrector formula is made, using in the term $h\beta_0 f(y_n)$ the value of y_n found in the first use of the corrector.

There is an extensive literature on predictor–corrector methods, including comparisons of efficiency for different modes. At this point we make only a single observation concerning the local accuracy of such methods. Let $y_n^{(1)}, y_n^{(2)}, \ldots, y_n^{(N)}$ denote the sequence of N corrected values in a $P(EC)^N E$ or a $P(EC)^N$ method (note that $(EC)^N$ denotes a sequence of EC pairs). If $f_n^{(j)}$ denotes $f(y_n^{(j)})$ $(j = 1, 2, \ldots, N)$ then in a $P(EC)^N E$ method we have

$$y_n^{(0)} = \sum_{i=1}^{k} \bar{\alpha}_i y_{n-i}^{(N)} + h \sum_{i=1}^{k} \bar{\beta}_i f_{n-i}^{(N)},$$

and, for $j = 1, 2, \ldots, N$,

$$y_n^{(j)} = \sum_{i=1}^{k} \alpha_i y_{n-i}^{(N)} + h\left[\beta_0 f_n^{(j-1)} + \sum_{i=1}^{k} \beta_i f_{n-i}^{(N)}\right],$$

whereas in a P(EC)^N method,

$$y_n^{(0)} = \sum_{i=1}^{k} \bar{\alpha}_i y_{n-i}^{(N)} + h \sum_{i=1}^{k} \bar{\beta}_i f_{n-i}^{(N-1)},$$

and, for $j = 1, 2, \ldots, N$,

$$y_n^{(j)} = \sum_{i=1}^{k} \alpha_i y_{n-i}^{(N)} + h \left[\beta_0 f_n^{(j-1)} + \sum_{i=1}^{k} \beta_i f_{n-i}^{(N-1)} \right].$$

It is appropriate to estimate $\| y_n^{(N)} - y(x_n) \|$ for each of these schemes and $\| hf_n^{(N)} - hf(y(x_n)) \|$ or $\| hf_n^{(N-1)} - hf(y(x_n)) \|$ for the P(EC)^{N}E and P(EC)^N methods respectively, since these will represent contributions to errors that carry on to later steps.

THEOREM 234A.

If (234a) is of order $\bar{m}$ and (234b) is of order m and if $y_{n-i} = y(x_{n-i})$, $f_{n-i} = f(y(x_{n-i}))$ for $i = 1, 2, \ldots, k$ and if f satisfies a Lipschitz condition, then for a P(EC)^{N}E or a P(EC)^N method

$$\| y_n^{(N)} - y(x_n) \| = O(h^{\tilde{m}+1}),$$

$$\| hf_n^{(N-1)} - hf(y(x_n)) \| = = O(h^{\tilde{m}+1}),$$

$$\| hf_n^{(N)} - hf(y(x_n)) \| = O(h^{\tilde{m}+2}),$$

where $\tilde{m} = \min(\bar{m} + N, m)$.

Proof.

Since a method of any order is also of lower order, we can assume without loss of generality that $\tilde{m} = \bar{m} + N = m$. Because of the order property of the predictor, we have

$$\| y_n^{(0)} - y(x_n) \| = O(h^{\bar{m}+1}). \tag{234c}$$

We will show by induction on $j = 1, 2, \ldots, N$ that

$$\| y_n^{(j)} - y(x_n) \| = O(h^{\bar{m}+j+1}), \tag{234d}$$

and for $j = 0, 1, 2, \ldots, N$ that

$$\| hf_n^{(j)} - hf(y(x_n)) \| = O(h^{\bar{m}+j+2}). \tag{234e}$$

Since f satisfies a Lipschitz condition, the truth of (234e) follows immediately from the corresponding case of (234c) or (234d). We now prove (234d) assuming its truth for lower values of j. Let

$$v = \sum_{i=1}^{k} \alpha_i y(x_{n-i}) + h \sum_{i=1}^{k} \beta_i f(y(x_{n-i})),$$

so that, by the order condition for (234b),

$$\| y(x_n) - h\beta_0 f(y(x_n)) - v \| = O(h^{m+1}).$$

Hence,

$$\| y_n^{(j)} - y(x_n) \| \leqq \| y_n^{(j)} - h\beta_0 f(y_n^{(j-1)}) - v \|$$
$$+ | \beta_0 | \ \| hf(y_n^{(j-1)}) - hf(y(x_n)) \|$$
$$+ \| y(x_n) - h\beta_0 f(y(x_n)) - v \|$$
$$= 0 + O(h^{\bar{m}+j+1}) + O(h^{m+1}) = O(h^{\bar{m}+j+1}). \ \blacksquare$$

235 Examples of methods

We consider here special methods in which one of the numbers $\alpha_1, \alpha_2, \ldots, \alpha_k$, say α_l, is equal to unity and the rest are zero. Methods of this type are automatically stable since $k - l$ of the numbers $z_1, z_2, \ldots, z_k$ occurring in definition 232B are zero and the remaining l are the distinct numbers with magnitude 1 given by $\exp(2\pi i j/l)$ ($j = 0, 1, 2, \ldots, l - 1$). We will present formulae for both explicit and implicit methods with various values of k and l. For some methods it will be assumed that $\beta_k = 0$. In each case, however, free values of $\beta_0, \beta_1, \beta_2, \ldots, \beta_k$ will be so chosen as to make the order as high as possible.

Methods of the type considered here with $l = 1$ are known as Adams–Bashforth (explicit case) or Adams–Moulton (implicit case) methods, while explicit methods with $l = 2$ are known as Nyström methods. The implicit method with $k = l = 2$ was proposed by Milne (1926) and two generalizations are possible. Following the terminology of Henrici (1962), we will refer to all methods with $l = 2$ as Milne–Simpson methods, whereas for all methods of implicit type with $k = l$ we will use the name Cotes methods (because of their relationship to Newton–Cotes quadrature formulae). The corresponding explicit methods, where also $\beta_k = 0$, will be called open Cotes methods.

EXPLICIT METHODS

$k = 1$, $l = 1$, order = 1:

$$y_n = y_{n-1} + hf(y_{n-1}). \tag{235a}$$

$k = 2$, $l = 1$, order = 2:

$$y_n = y_{n-1} + \frac{h}{2} \left[3f(y_{n-1}) - f(y_{n-2}) \right]. \tag{235b}$$

$k = 2$, $l = 2$, order = 2:

$$y_n = y_{n-2} + 2hf(y_{n-1}). \tag{235c}$$

$k = 3$, $l = 3$, order = 2:

$$y_n = y_{n-3} + \frac{3h}{2} \left[f(y_{n-1}) + f(y_{n-2}) \right].$$ (235d)

$k = 3$, $l = 1$, order = 3:

$$y_n = y_{n-1} + \frac{h}{12} \left[23 f(y_{n-1}) - 16 f(y_{n-2}) + 5 f(y_{n-3}) \right].$$ (235e)

$k = 3$, $l = 2$, order = 3:

$$y_n = y_{n-2} + \frac{h}{3} \left[7 f(y_{n-1}) - 2 f(y_{n-2}) + f(y_{n-3}) \right].$$ (235f)

IMPLICIT METHODS

$k = 1$, $l = 1$, order = 1:

$$y_n = y_{n-1} + h f(y_n).$$ (235g)

$k = 1$, $l = 1$, order = 2:

$$y_n = y_{n-1} + \frac{h}{2} \left[f(y_n) + f(y_{n-1}) \right].$$ (235h)

$k = 2$, $l = 1$, order = 3:

$$y_n = y_{n-1} + \frac{h}{12} \left[5 f(y_n) + 8 f(y_{n-1}) - f(y_{n-2}) \right].$$ (235i)

$k = 2$, $l = 2$, order = 4:

$$y_n = y_{n-2} + \frac{h}{3} \left[f(y_n) + 4 f(y_{n-1}) + f(y_{n-2}) \right].$$ (235j)

$k = 3$, $l = 1$, order = 4:

$$y_n = y_{n-1} + \frac{h}{24} \left[9 f(y_n) + 19 f(y_{n-1}) - 5 f(y_{n-2}) + f(y_{n-3}) \right].$$ (235k)

$k = 3$, $l = 3$, order = 4:

$$y_n = y_{n-3} + \frac{3h}{8} \left[f(y_n) + 3 f(y_{n-1}) + 3 f(y_{n-2}) + f(y_{n-3}) \right].$$ (235l)

236 Numerical examples.

To illustrate the behaviour of some of the linear multistep methods quoted in the last subsection, we present some numerical results using the same problem and initial value and a similar sequence of step sizes as in subsection 225. The characteristic difficulty of multistep methods, that is the evaluation of

$y_1, y_2, \ldots, y_{k-1}$ to start off the integration, is avoided through the use of exact values at $y(h), y(2h), \ldots, y((k-1)h)$. Although this does not, of course, correspond to standard computational practice, it is felt here to be preferable for illustrative purposes than the use of, say, Runge–Kutta computed values at these points, since this would produce a final error compounded from local errors in two quite different types of method. Since this starting complication makes a step size $h = 1/n$ useless as a basis for comparison when n is small, the sequence begins at $n = 8$.

The six methods used in this experiment are:

Method 21: The second-order Adams–Bashforth method given by (235b).

Method 22: The second-order PECE method with the predictor as in method 21 and the corrector as in the Adams–Moulton method given by (235h).

Method 31: The third-order Nyström method given by (235f).

Method 32: The third-order PEC method with the Adams–Bashforth predictor (235e) and Adams–Moulton corrector (235i).

Method 41: The fourth-order PECEC method with the predictor as in method 31 and the corrector as in the Milne–Simpson method given by (235j).

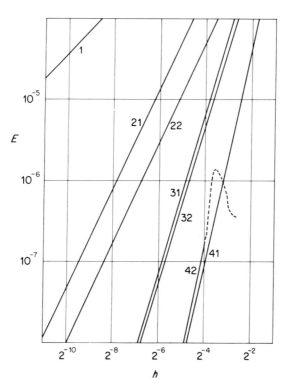

Figure 236   Error versus step size for various linear multistep methods

Method 42: The implicit fourth-order method consisting of the Cotes method
(2351).

The error E as a function of step size h for the approximations to $y(1)$ are given in figure 236. The names of the methods 21, 22, 31, 32, 41 and 42 are attached to the various graphs and, for comparison, the Euler method marked 1 is also shown. As for the Runge–Kutta methods, the slopes of these graphs (on logarithmic scales) are close to constants and are approximately equal to the orders. In the case of method 42 the haphazard behaviour for large h is represented by the use of a broken line.

24   TAYLOR SERIES METHODS

240   Introduction to Taylor series methods.

A differential equation $y'(x) = f(y(x))$, characterized by the function f, is presented to a computer as a procedure for evaluating $f(v)$ for an arbitrary choice of the vector v. The program for numerically solving the equation uses this procedure in a way exactly corresponding to the occurrence of the function f in the mathematical formulation of the method. In this brief introduction we contemplate the use of procedures which compute, for a given value of $y(x)$, the values not only of $y'(x)$ but also of $y''(x)$ and perhaps of $y'''(x)$ and further high derivatives. With such facilities available there is a wide range of possible methods, but the natural and straightforward choice of the Taylor series is the line almost always followed. Let $f_2(y(x))$, $f_3(y(x)), \ldots, f_m(y(x))$ be functions which give values respectively for $y''(x)$, $y'''(x), \ldots, y^{(m)}(x)$, so that the mth-order Taylor series for computing $y_n(\approx y(x_n) = y(x_{n-1} + h))$ becomes

$$y_n = y_{n-1} + hf(y_{n-1}) + \frac{h^2}{2!} f_2(y_{n-1}) + \cdots + \frac{h^m}{m!} f_m(y_{n-1}). \quad (240a)$$

Most serious investigations of this method have been concerned with the automatic generation of the procedures for computing $f_2, \ldots, f_m$ from the given procedure for computing f. While this aspect of Taylor series is more within the scope of algebraic manipulation than of numerical analysis, there are other important aspects which also arise with other methods, such as error estimation, order selection and stepsize control.

Although many individuals and teams have made important contributions to the use of Taylor series methods, we mention three in particular. The program of Gibbons (1960) using a computer with the limited memory available at that time, used a recursive technique to generate the Taylor coefficients automatically. A similar approach using greater sophistication and more powerful computational tools was used by Barton, Willers and Zahar (1971). The work of Moore (1964) is especially interesting in that it uses interval arithmetic and supplies rigorous error bounds for the computed solution.

241 Manipulation of power series.

It is convenient to scale the terms occurring in (240a) so that the $1/i!$ factor are absorbed into the $f_i(y_{n-1})$ coefficient. Thus each component of y_n as an approximation to the corresponding component of $y(x_{n-1} + h)$ takes the form $a_0 + a_1h + a_2h^2 + \cdots + a_mh^m$. If a second such expansion $b_0 + b_1h + b_2h^2 + \cdots + b_mh^m$ is added or subtracted we simply add or subtract corresponding coefficients. Multiplying series but truncating at the h^m term gives

$$(a_0 + a_1h + a_2h^2 + \cdots + a_mh^m)(b_0 + b_1h + b_2h^2 + \cdots + b_mh^m)$$
$$= c_0 + c_1h + c_2h^2 + \cdots + c_mh^m,$$

where

$$c_i = \sum_{j=0}^{i} a_{i-j}b_j, \tag{241a}$$

and a quotient

$$(c_0 + c_1h + c_2h^2 + \cdots + c_mh^m)(b_0 + b_1h + b_2h^2 + \cdots + b_mh^m)^{-1}$$
$$= a_0 + a_1h + a_2h^2 + \cdots + a_mh^m$$

is found by reinterpreting the relationship between a_i, b_i, c_i to give

$$a_i = \begin{cases} \dfrac{c_0}{b_0}, & i = 0, \\[3mm] \dfrac{c_i - \sum_{j=1}^{i} a_{i-j}b_j}{b_0}, & i > 0. \end{cases} \tag{241b}$$

Given a system of differential equations with dependent variables $y^1, y^2, \ldots, y^N$, we write the truncated power series expansion for $y_k(x_{n-1} + h)$ in the form $y_0^k + hy_1^k + h^2y_2^k + \cdots + h^my_m^k$ for $k = 1, 2, \ldots, N$. A procedure for evaluating y_i^k for $i = 0, 1, \ldots, m$ uses the known approximations at x_{n-1} to give y_0^k. For $i = 0, 1, \ldots, m - 1$ in turn, the ith coefficient for the expansions of each function occurring in the right-hand side of the differential equations is formed from known values of y_j^k, $k = 1, 2, \ldots, N$, $j \leq i$. Let f_i^k denote the ith coefficient in the formula for $\mathrm{d}y^k/\mathrm{d}x$. The relationship

$$\frac{\mathrm{d}}{\mathrm{d}h}(y_0^k + hy_1^k + h^2y_2^k + \cdots + h^my_m^k) = y_1^k + 2hy_2^k + \cdots + mh^{m-1}y_m^k$$

enables us to determine y_{i+1}^k as $f_i^k/(i+1)$.

When we have reached $i = m - 1$, all the required Taylor coefficients are known at x_{n-1} and a summation of the Taylor series for each component gives the approximation at $x_n = x_{n-1} + h$.

This approach will be illustrated in the next subsection.

242 An example of a Taylor series solution.

We consider the same differential equation used in the numerical examples of subsections 225 and 236. That is,

$$y'(x) = y(x)\frac{1 - y(x)}{2y(x) - 1} \tag{242a}$$

Let $a_0, a_1, \ldots, a_m$ denote Taylor coefficients for $y(x_{n-1} + h)$. Let $b_0, b_1, \ldots, b_m$ denote corresponding coefficients for $1 - y(x)$ so that $b_0 = 1 - a_0$, $b_i = -a_i$ $(i > 0)$. Let $c_0, c_1, \ldots, c_m$ denote the Taylor coefficients for $y(x_{n-1} + h)[1 - y(x_{n-1} + h)]$ so that these can be computed by (241a) once a_i and b_i are known. Coefficients $d_0, d_1, \ldots, d_m$ for $2y(x_{n-1} + h) - 1$ can be computed from $a_0, a_1, \ldots, a_m$ coefficients; $e_0, e_1, \ldots, e_m$ for $y(x_{n-1} + h)[1 - y(x_{n-1} + h)]/[2y(x_{n-1} + h) - 1]$ are then found from (241b) with a_j replaced by e_j and b_j by d_j. The recursive formation of $a_0, a_1, \ldots, a_m$ is completed by the relation $a_{i+1} = e_i/(i + 1)$.

To show how these steps are put together and how a step forward from $y_{n-1} \approx y(x_{n-1})$ to $y_n \approx y(x_{n-1} + h)$ is taken, we present algorithm 242, for this particular differential equation only. The variable y contains y_{n-1} initially and y_n on exit. The arrays a, b, c, d, e with dimensions $[0..m]$ are used to hold $a_0, a_1, \ldots, a_m, \ldots, e_0, e_1, \ldots, e_m$.

Algorithm 242: Taylor method for particular problem

```
a[0] := y;
for i := 0 to m − 1 do
begin
    if i = 0 then
    begin
        b[0] := 1 − a[0]; c[0] := a[0] * b[0];
        d[0] := 2 * a[0] − 1; e[0] := c[0]/d[0]
    end else
    begin
        b[i] := − a[i]; d[i] := 2 * a[i]; sum := 0;
        for j := 0 to i do sum := sum + a[i − j] * b[j];
        c[i] := sum;
        for j := 1 to i do sum := sum − e[i − j] * d[j];
        e[i] := sum/d[0]
    end;
    a[i + 1] := e[i]/(i + 1)
end;
y := a[m];
for i := m − 1 downto 0 do y := a[i] + h * y
```

243 Numerical results.

We present below, in figure 243, Taylor series results for the example problem

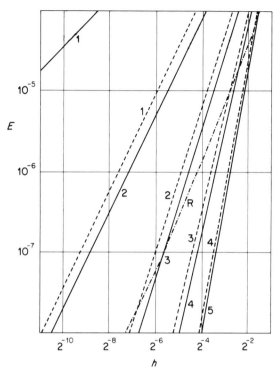

Figure 243   Error versus step size for Taylor series methods (——), MacLaurin series
methods (- -), and for a special Rosenbrock method R(· — ·)

of the last subsection with initial value $y(0) = \frac{5}{6}$, just as for the examples in
subsections 225 and 236. The value of $y(1)$ is found as in those computations
and the same constant step sizes are used. The orders $1, 2, 3, 4, 5$ are attached
to the unbroken curves which show the error behaviour for these methods. The
broken curves will be explained in the next subsection.

We have used one higher order than in the Runge–Kutta and linear
multistep examples. Just as for those cases, the slope of the graph on a
logarithmic scale agrees closely with the order of the method.

244   Other methods using higher derivatives.

To illustrate the range of methods made possible by allowing other
combinations of higher derivatives than those of the usual Taylor series, we
consider the combination of a Taylor series predictor with a corrector based
on the Euler–MacLaurin series which we can write in the form

$$y(x + h) = y(x) + \frac{h}{2} [y'(x) + y'(x + h)] + \frac{h^2}{12} [y''(x) - y''(x + h)]$$

$$- \frac{h^4}{720} [y^{(4)}(x) - y^{(4)}(x + h)] + \cdots \tag{244a}$$

If y_n^* denotes the predicted result given by (240a), then we could base a corrector on (244a) by defining the final value of y_n as

$$y_n = y_{n-1} + \frac{h}{2}\,[f(y_{n-1}) + f(y_n^*)] + \frac{h^2}{12}\,[f_2(y_{n-1}) - f_2(y_n^*)]$$

$$- \frac{h^4}{720}\,[f_4(y_{n-1}) - f_4(y_n^*)] + \cdots, \tag{244b}$$

where, in the final term, h would have degree $2[m/2]$. In contrast to this PECE formulation, it is also possible to use the PEC version of this method where, for $n > 1$, $f(y_{n-1}), f_2(y_{n-1}), \ldots$ are replaced in (244b) by $f(y_{n-1}^*), f_2(y_{n-1}^*), \ldots$.

To show how this PEC formulation works in practice, the same problem solved in subsection 243 was solved by this method also, and the results are shown by the broken curves in figure 243. We see that the errors for this method are considerably lower and exhibit an apparent enhancement up to the next higher order. Because the bulk of the computational effort is expended in the evaluation of Taylor coefficients at each new point and since this is done once per step both for the Taylor series and for the PEC method, the costs of the two methods are more or less equal so that a comparison of their efficiencies would place the predictor–corrector method in the favoured position.

Finally we mention the existence of another class of methods that has been studied systematically. These are the Obreshkov methods and can be regarded as the multistep generalizations of Taylor series. This class includes, for example, the Euler–MacLaurin method that we have just touched on. A general form for such methods is

$$y_n = \sum_{j=1}^{k} \alpha y_{n-j} + h \sum_{j=0}^{k} \beta_j f(y_{n-j}) + \sum_{i=2}^{m} h^i \sum_{j=0}^{k} \beta_{ij} f_i(y_{n-j}),$$

where $f_2, f_3, \ldots, f_m$ are the functions introduced in subsection 240 for yielding values of the second, third, $\ldots$, mth derivatives. The original work on these methods is due to Obreshkov (1940, 1942).

245 The use of f derivatives.

In the implementation of implicit methods by the Newton iteration method, the solution of systems of linear equations of the form

$$[I - h\beta f'(y)]\,v = u,$$

or sometimes of a more complicated form, is required in each iteration. The constant β is characteristic of the method, y is a point on the solution trajectory, and it is required to calculate v making use of a computed value of the vector u. This type of process is very costly for large problems because of the expense of evaluating $f'(y)$, of factorizing $I - h\beta f'(y)$ and of carrying out the back-substitutions in computing v. It is natural to consider methods which cut down at least part of these costs and this is the essence of the proposal of Rosenbrock (1963) that the function f' be used directly in the design of the method rather than as a tool for the solution of a system of non-linear equations.

For illustrative purposes we consider a method of only second-order taken from Rosenbrock's paper, in which y_n is computed making use of intermediate quantities F_1 and F_2 using the three equations

$$[I - h(1 - \tfrac{1}{2}\sqrt{2})f'(y_{n-1})] F_1 = f(y_{n-1}),$$

$$[I - h(1 - \tfrac{1}{2}\sqrt{2})f'(y_{n-1})] F_2 = f(y_{n-1} + \tfrac{1}{2}h(\sqrt{2} - 1)F_1), \qquad (245a)$$

$$y_n = y_{n-1} + hF_2.$$

Methods of various orders have been derived by Rosenbrock and others and they are known collectively either as *Rosenbrock methods* or by the more ambiguous name of *implicit Runge–Kutta methods*. They have been studied intensively in recent years and will be reviewed more fully in the sequel to this volume.

As far as the present author is aware, no one has seriously proposed the direct use of the bilinear operator $f''(y)$, for given y, as a possible generalization of Rosenbrock methods.

We conclude this section by pointing out that for comparison purposes, the method given by equations (245a) has been added, with the label R, to the numerical results displayed in figure 243.

3

Analysis of Runge–Kutta methods

30   THE ORDER CONDITIONS

300   The order of two-stage methods.

For simplicity, we will consider only explicit methods with exactly two stages. Suppose that, for such a method, y_n is the solution value computed at the end of n steps of calculation and that it is given by the formula

$$y_n = y_{n-1} + h[b_1 f(y_{n-1}) + b_2 f(y_{n-1} + ha_{21} f(y_{n-1}))]. \tag{300a}$$

Note that y_n is supposed to approximate $y(x_n)$, where $x_n = x_{n-1} + h$ and where y is the exact solution to the differential equation

$$y'(x) = f(y(x)). \tag{300b}$$

The numerical constants b_1, b_2, a_{21} are free to be chosen in any way, and we will consider how they should be chosen so that the order of the method is as high as possible. That is to say, if x_{n-1} is kept fixed, y_{n-1} is fixed with value $y(x_{n-1})$ and h is allowed to vary, we wish the order of approximation of y_n to y to be as high as possible.

To express this more precisely, let

$$Y(x) = y(x_{n-1}) + (x - x_{n-1})[b_1 f(y(x_{n-1}))$$
$$+ b_2 f(y(x_{n-1}) + (x - x_{n-1}) a_{21} f(y(x_{n-1})))], \tag{300c}$$

so that the condition for order N is

$$Y^{(m)}(x_{n-1}) = y^{(m)}(x_{n-1}), \qquad m = 0, 1, 2, \dots, N, \tag{300d}$$

and we wish (300d) to hold for as high a value of N as possible.

It is assumed, of course, that the function f is differentiable sufficiently often in a neighbourhood of $y(x_{n-1})$ for (300d) to have a meaning. From this point onwards we restrict ourselves, without loss of generality, to the case $n = 1$.

In the hope of obtaining orders higher than was possible for Euler's method we now find expressions for the first three derivatives of y and of Y.

Throughout these manipulations we use the notations for the derivatives of f of various orders introduced in subsection 102. That is, we will interpret $f'(y(x))$ as a linear operator, $f''(y(x))$ as a bilinear operator, and so on. We will make frequent use, for example in (300f), of the chain rule for the differentiation of the composition of two functions. We have

$$y'(x) = f(y(x)), \tag{300e}$$

$$y''(x) = (f \circ y)'(x) = f'(y(x))(y'(x)) = f'(y(x))(f(y(x))), \tag{300f}$$

$$y'''(x) = f''(y(x))(f(y(x)), y'(x)) + f'(y(x))(f'(y(x))(y'(x)))$$
$$= f''(y(x))(f(y(x)), f(y(x))) + f'(y(x))(f'(y(x))(f(y(x)))), \tag{300g}$$

and, by a more tedious computation,

$$Y'(x) = b_1 f(y(x_0)) + b_2 f(y(x_0) + (x - x_0)a_{21}f(y(x_0)))$$
$$+ (x - x_0)b_2 a_{21}f'(y(x_0) + (x - x_0)a_{21}f(y(x_0)))(f(y(x_0))), \tag{300h}$$

$$Y''(x) = 2b_2 a_{21}f'(y(x_0) + (x - x_0)a_{21}f(y(x_0)))(f(y(x_0)))$$
$$+ (x - x_0)b_2 a_{21}^2 f''(y(x_0) + (x - x_0)a_{21}f(y(x_0)))\ (f(y(x_0)), f(y(x_0))), \tag{300i}$$

$$Y'''(x) = 3b_2 a_{21}^2 f''(y(x_0) + (x - x_0)a_{21}f(y(x_0)))(f(y(x_0)), f(y(x_0)))$$
$$+ (x - x_0)b_2 a_{21}^3 f'''(y(x_0) + (x - x_0)a_{21}f(y(x_0)))$$
$$(f(y(x_0)), f(y(x_0)), f(y(x_0))). \tag{300j}$$

Evaluating (300e) to (300j) at $x = x_0$ and writing $y(x_0) = y_0$ we obtain as the conditions for order 3, the following equations:

$$(b_1 + b_2)f(y_0) = f(y_0), \tag{300k}$$

$$2b_2 a_{21}f'(y_0)(f(y_0)) = f'(y_0)(f(y_0)), \tag{300l}$$

$$3b_2 a_{21}^2 f''(y_0)(f(y_0), f(y_0)) = f''(y_0)(f(y_0), f(y_0)) + f'(y_0)(f'(y_0)(f(y_0))). \tag{300m}$$

For second order, only (300k) and (300l) need hold and for first order only (300k) need hold. Since in general $f(y_0) \neq 0$ and $f'(y_0)(f(y_0)) \neq 0$, it is clear that the necessary and sufficient conditions for first and second orders are:

First order: $\qquad b_1 + b_2 = 1$,

Second order: $\qquad b_1 + b_2 = 1$, $\qquad b_2 a_{21} = \frac{1}{2}$.

By considering, for example, the case when $f(u) = u$, we see that it is not possible to satisfy (300m). Hence, third order cannot be achieved with explicit two-stage methods.

These remarks enable us to assert the following theorem, which will eventually be superseded by one in which the restriction on N is removed.

THEOREM 300A

If an s-stage explicit Runge–Kutta method has order N where $N \leq 3$, then $s \geq N$.

301 Elementary differentials.

In the previous subsection, we computed $y'(x)$, $y''(x)$, $y'''(x)$ and found that they depended on the following four vectors: $\mathbf{f}, \mathbf{f}'(\mathbf{f}), \mathbf{f}''(\mathbf{f}, \mathbf{f}), \mathbf{f}'(\mathbf{f}'(\mathbf{f}))$, where $\mathbf{f} = f(y(x)), \mathbf{f}' = f'(y(x)), \mathbf{f}'' = f''(y(x))$. We shall extend this collection of vectors, which we shall call 'elementary differentials', so that we are able to express derivatives of y of any order.

To motivate the approach we will use, consider table 301. In this table we have shown the four rooted trees with less than four vertices and we have indicated a relation between these and the four elementary differentials listed in the second column by means of the operation diagrams in the third column. In these diagrams, $\mathbf{f}^{(n)}$ is an n-ary operator and is always attached to a vertex which has n outward branching arcs. The n operands on which $\mathbf{f}^{(n)}$ is to act in a particular diagram are found from the n subdiagrams rooted to these n outward branching arcs. In the case $n = 0$, corresponding to terminal vertices, $\mathbf{f}^{(0)} = \mathbf{f}$.

Table 301

Tree	Elementary differential	Operation diagram
•	$\mathbf{f}$	$\mathbf{f}$ •
	$\mathbf{f}'(\mathbf{f})$	$\mathbf{f}$; $\mathbf{f}'$
	$\mathbf{f}''(\mathbf{f}, \mathbf{f})$	$\mathbf{f}$ $\mathbf{f}$; $\mathbf{f}''$
	$\mathbf{f}'(\mathbf{f}'(\mathbf{f}))$	$\mathbf{f}$; $\mathbf{f}'$; $\mathbf{f}'$

It is clear that for any rooted tree we can form such an operation diagram and thus a corresponding elementary differential. For example, the tree

leads, via the operation diagram

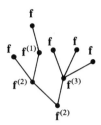

to $\mathbf{f}^{(2)}(\mathbf{f}^{(2)}(\mathbf{f}, \mathbf{f}^{(1)}(\mathbf{f})), \mathbf{f}^{(3)}(\mathbf{f}, \mathbf{f}, \mathbf{f}))$ which, as it happens, is a term occurring in the formula for $y^{(9)}(x)$. Note that we use the notation $\mathbf{f}', \mathbf{f}'', \ldots$ interchangeably with $\mathbf{f}^{(1)}, \mathbf{f}^{(2)}, \ldots$.

It is interesting to write these vectors making use of the partial derivative notation. Let $\mathbf{f}^i_{j_1 j_2 \cdots j_n} = f^i_{j_1 j_2 \cdots j_n}(y(x))$ denote an nth order partial derivative of component number i of $\mathbf{f}$. We then have, for example,

$$(\mathbf{f})^i = \mathbf{f}^i, \tag{301a}$$

$$(\mathbf{f}'(\mathbf{f}))^i = \mathbf{f}^i_j \mathbf{f}^j, \tag{301b}$$

$$(\mathbf{f}''(\mathbf{f}, \mathbf{f}))^i = \mathbf{f}^i_{jk} \mathbf{f}^j \mathbf{f}^k, \tag{301c}$$

$$(\mathbf{f}'(\mathbf{f}'(\mathbf{f})))^i = \mathbf{f}^i_j \mathbf{f}^j_k \mathbf{f}^k, \tag{301d}$$

where the summation convention applies (that is, there is an implicit summation over every superscript $j, k, \ldots$ which appears also as a subscript).

We note a simple rule for associating such expressions as occur in (301a) to (301d) with the corresponding rooted trees: to the root of the tree attach a label i and to the other vertices attach other labels, $j, k, l, \ldots$. For each vertex write down $\mathbf{f}$ and attach to it a superscript equal to its label and subscripts equal to the labels of each outwardly connected vertex. The (summed) product of these factors is the expression for the ith component of the required vector. For the example previously given, we obtain the labelled tree

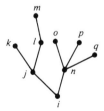

and the vector component

$$\mathbf{f}^i_{jn} \mathbf{f}^j_{kl} \mathbf{f}^k \mathbf{f}^l_m \mathbf{f}^m \mathbf{f}^n_{opq} \mathbf{f}^o \mathbf{f}^p \mathbf{f}^q.$$

We are now in a position to give a formal definition of the set of elementary differentials. We will assume that f is differentiable sufficiently often for the definition to be meaningful for any specified choice of t.

DEFINITION 301A

For a function $f: X \to X$, the *elementary differential* $F(t): X \to X$, corresponding to $t \in T$ (T denotes the set of rooted trees), is defined by

$$F(\tau)(y) = f(y),$$

where τ is the tree with a single vertex and by

$$F(t)(y) = f^{(s)}(y)(F(t_1)(y), F(t_2)(y), \ldots, F(t_s)(y)),$$

where $t = [t_1, t_2 \cdots t_s]$, for all $y \in X$.

It will sometimes become necessary (for example in subsection 305) to indicate the dependence of F on f. When this is so, we will write $F(f, t)$ in place of $F(t)$.

302   The Taylor series for the differential equation solution.

In this subsection we use the elementary differentials introduced in the previous subsection to express the formulae for $y^{(n)}(x)$, $n = 1, 2, \ldots$, where y satisfies the differential equation

$$y'(x) = f(y(x)). \tag{302a}$$

We first prove a version of the chain rule for high-order differentiation. For S an arbitrary finite set, $\#S$ will denote the number of members in S and $P(S)$ will denote the set of partitions of S. Thus if $p \in P(S)$ then $p = \{p_1, p_2, \ldots\}$ where $p_1, p_2, \ldots$ are non-empty pairwise-disjoint sets whose union is S. The number of components in a partition p is $\#p$.

LEMMA 302A

If y is $\#S$ times differentiable at x and f is $\#S$ times differentiable at $y(x)$ then $f \circ y$ is $\#S$ times differentiable at x and

$$(f \circ y)^{(\#S)}(x) = \sum_{p \in P(S)} f^{(\#p)}(y(x))(y^{(\#p_1)}(x), y^{(\#p_2)}(x), \ldots). \tag{302b}$$

Proof.

If $\#S = 1$ then $P(S)$ has only one member $\{S\}$ and (302b) is simply the chain rule. If $S = \{s\} \cup S_0$, where $s \notin S_0$, then

$$(f \circ y)^{(\#S)}(x) = \sum_{p \in P(S_0)} f^{(\#p+1)}(y(x))(y^{(1)}(x), y^{(\#p_1)}(x), y^{(\#p_2)}(x), \ldots)$$

$$+ \sum_{p \in P(S_0)} f^{(\#p)}(y(x))(y^{(\#p_1+1)}(x), y^{(\#p_2)}(x), \ldots)$$

$$+ \sum_{p \in P(S_0)} f^{(\#p)}(y(x))(y^{(\#p_1)}(x), y^{(\#p_2+1)}(x), \ldots)$$

$$+ \ldots \tag{302c}$$

The first sum on the right of (302c) corresponds to all those partitions of S containing $\{s\}$ as a separate element in the partition, while the remainder correspond to those partitions in which s appears in a larger set. Thus all terms on the right-hand side of (302c) correspond uniquely to terms on the right-hand side of (302b). Hence we have verified the result for $\#S = \#S_0 + 1$ so that the general result follows by induction. ∎

For each $t \in T$, we recall that $\alpha(t)$, introduced in subsection 145, denotes the number of ways of labelling t with a set of $r(t)$ ordered symbols such that along each outwardly directed arc the labels increase. For convenience, such a set of labels can be chosen from the integers. Thus for S a finite set of integers we consider the set of rooted trees with exactly $\#S$ vertices, each labelled with an element of S, such that along each outwardly directed arc the labels increase. Denote this set by T_S. For $u \in T_S$ let $|u|$ denote the corresponding member of T formed by removing the labels from u. Thus we have

$$\alpha(t) = \#\{u : |u| = t\}. \tag{302d}$$

LEMMA 302B.

If $f: X \to X$ is differentiable $\#S - 1$ times at $y(x)$ and $y' = f \circ y$ then y is $\#S$ times differentiable at x and

$$y^{(\#S)}(x) = \sum_{u \in T_S} F(|u|)(y(x)). \tag{302e}$$

Proof.

When $\#S = 1$, (302e) is equivalent to (302a). To complete the proof by induction on $\#S$, let $S = \{s\} \cup S_0$, where s is less than each member of S_0. We have

$$y^{(\#S)}(x) = y^{(\#S_0 + 1)}(x) = (f \circ y)^{(\#S_0)}(x)$$

$$= \sum_{p \in P(S_0)} f^{(\#p)}(y(x))\left(\sum_{u_1 \in T_{p_1}} F(|u_1|)(y(x)), \sum_{u_2 \in T_{p_2}} F(|u_2|)(y(x)), \ldots\right)$$

$$= \sum_{p \in P(S_0)} \left(\sum_{u_1 \in T_{p_1}} \sum_{u_2 \in T_{p_2}} \cdots\right) F([|u_1|, |u_2|, \ldots])(y(x)) \tag{302f}$$

$$= \sum_{u \in T_S} F(|u|)(y(x)), \tag{302g}$$

where $p, u_1, u_2, \ldots$ in (302f) are related to u in (302g) by the rule that when the root (labelled by s) in u is removed, the labelled trees that remain are $u_1, u_2, \ldots$ with labels from $p_1, p_2, \ldots$. ∎

Noting that (302d) holds, we can now state without further proof the following corollary to lemma 302B.

THEOREM 302C.

If $f: X \to X$ is differentiable $n - 1$ times at $y(x)$ and $y' = f \circ y$, then y is n times differentiable at x and

$$y^{(n)}(x) = \sum_{r(t) = n} \alpha(t) F(t)(y(x)). \tag{302h}$$

Note that the summation in this equation is over all trees for which $r(t)$, the number of vertices in t, is equal to n. The equation (302h) enables us to find the Taylor coefficients for y in a routine manner. We have the Taylor expansion given by

THEOREM 302D.

If $f: X \to X$ is continuously differentiable n times in an open set containing $y(\xi)$ for all $\xi \in [x_0, x]$ then

$$y(x) = y(x_0) + \sum_{r(t) \leq n} \frac{(x - x_0)^{r(t)} \alpha(t) F(t)(y(x_0))}{r(t)!}$$

$$+ \int_{x_0}^{x} \frac{(x - \xi)^n}{n!} \sum_{r(t) = n+1} \alpha(t) F(t)(y(\xi)) \, \mathrm{d}\xi. \tag{302i}$$

303 The Taylor series for the implicit Euler method.

The implicit Euler method, introduced briefly in subsection 215, computes the value of y_n in terms of y_{n-1} as the solution of the equation

$$y_n = y_{n-1} + h f(y_n).$$

In this subsection we will write $x_{n-1} = x_0$ and for a given $x = x_0 + h$ define $y(x)$ as the value of y_n satisfying this equation with $y(x_0) = y_0$. Thus the function y defined in this way from the form of the implicit Euler method satisfies the functional equation

$$y(x) = y(x_0) + (x - x_0) f(y(x)). \tag{303a}$$

Apart from its role as one of many methods appropriate for solving stiff ordinary differential equations, the implicit Euler method formulated in this way has an importance all of its own. Based on the Taylor series for the function y satisfying (303a), we will derive in subsection 305 the Taylor series for any Runge–Kutta method.

In studying the equation (303a), where $x_0 \in \mathbb{R}$, $y(x_0) \in X$ are given, we will assume that $f: X \to X$ is differentiable sufficiently often for $y^{(n)}$ to be defined in a neighbourhood of x_0 for various $n = 1, 2, \ldots$.

For S a finite set we introduce various sets of labelled trees in which each member of S occurs as a label once and only once. The set U_S of trees is labelled with the members of S in such a way that every terminal vertex receives a label but the root does not. Similarly V_S will denote the set of trees labelled with the members of S in such a way that every terminal vertex receives a label (and the root or any other non-terminal vertex may or may not). Also $W_S = V_S \setminus U_S$ is the set of trees labelled with S such that all terminal vertices as well as the root receive a label.

If $u \in V_S$, let $|u|$ denote the corresponding tree in T. If $\#S = n$, we recall that $\beta_n(t)$, introduced in subsection 145, denotes the number of members u of U_S such that $|u| = t$ and $\bar{\beta}_n(t)$ denotes the number of u in V_S such that $|u| = t$. Conventionally, when S is empty, we define $U_S = V_S$ as the set whose only member is the unlabelled tree with only one vertex. Thus, $\beta_0(t) = \bar{\beta}_0(t) = 1$ (if $t = \tau$) or 0 (if $t \neq \tau$). As in subsection 145, we will also write $\beta(t) = \beta_{r(t)-1}(t)$, $\bar{\beta}(t) = \bar{\beta}_{r(t)}(t)$.

We define a mapping $\varphi: V_S \to V_S$ such that for all $u \in V_S$, $|\varphi(u)|$ has exactly one more vertex than u, the root of $\varphi(u)$ is not labelled, only one arc leaves the root of $\varphi(u)$ and if the root of $\varphi(u)$ is removed, what remains is just u. The relation between u and $\varphi(u)$ is illustrated in the case of four examples, in table 303.

Table 303   Examples of φ.

S	$u \in V_S$	$\varphi(u)$
$\{1\}$		
$\{1,2\}$		
$\{1,2,3\}$		
$\{1,2,3,4\}$		

We now define $V_S^{(0)}$ as the set of those members of V_S which are not equal to $\varphi(u)$ for any $u \in V_S$. Also $V_S^{(n)}$, for $n = 1, 2, \ldots$, is defined by

$$V_S^{(n)} = \{\varphi(u) : u \in V_S^{(n-1)}\},$$

so that $V_S^{(0)}, V_S^{(1)}, \ldots$ are disjoint and

$$V_S^{(0)} \cup V_S^{(1)} \cup \cdots = V_S.$$

It is easy to see that $V_S^{(0)}$ is the union of W_S and the set of those members of U_S such that underlying trees have more than one arc branching from the root.

THEOREM 303A.

Suppose f is continuously differentiable n times in the set

$$X_0 = \left\{ z \in X : \| z - y(x_0) \| \le \frac{\theta \| f(y(x_0)) \|}{1 - \theta L} \right\},$$

where $\theta > 0$, $L > 0$ and $\theta L < 1$ and $\| f^{(1)}(z) \| \le L$ for all $z \in X_0$. Then a unique y satisfying (303a) exists and is n times continuously differentiable in $I_0 = [x_0 - \theta, x_0 + \theta]$. Furthermore, if $\#S \le n$, then for all $x \in I_0$,

$$(f \circ y)^{(\# S)}(x) = \sum_{u \in U_S} (x - x_0)^{r(|u|) - \# S - 1} F(|u|)(y(x)), \qquad (303b)$$

$$y^{(\# S)}(x) = \sum_{u \in V_S} (x - x_0)^{r(|u|) - \# S} F(|u|)(y(x)). \qquad (303c)$$

Proof.

For each $x \in I_0$ and $\xi \in X_0$, $g(\xi) = y(x_0) + (x - x_0)f(\xi)$ is in X_0 because

$$\| [y(x_0) + (x - x_0)f(\xi)] - y(x_0) \| \le \theta \| f(y(x_0)) + [f(\xi) - f(y(x_0))] \|$$

$$\le \theta \| f(y(x_0)) \| + \theta L \| \xi - y(x_0) \|$$

$$\le \theta \| f(y(x_0)) \| + \frac{\theta L \theta \| f(y(x_0)) \|}{1 - \theta L}$$

$$\le \frac{\theta \| f(y(x_0)) \|}{1 - \theta L}.$$

Hence, the function g is into X_0 and, because of the Lipschitz condition on f, is a contraction mapping. Thus, for each $x \in I_0$, $y(x)$ satisfying (303a) is the unique fixed point of g. The differentiability of y follows from the implicit function theorem (see, for example, Lang [1968]). We now establish the equivalence of (303b) and (303c) on the assumption that (303b) holds for lower-order derivatives than $\#S$. Differentiate (303a) $\#S$ times using the

Leibniz theorem, to give

$$y^{(\# S)}(x) - (x - x_0)(f \circ y)^{(\# S)}(x) = (\# S)(f \circ y)^{(\# S-1)}(x)$$

$$= \sum_{s \in S} (f \circ y)^{(\# (S \setminus \{s\}))}(x)$$

$$= \sum_{s \in S} \sum_{u \in U_{S \setminus \{s\}}} (x - x_0)^{r(|u|) - \# S} F(|u|)(y(x))$$

$$= \sum_{u \in W_S} (x - x_0)^{r(|u|) - \# S} F(|u|)(y(x))$$

$$= \sum_{u \in V_S} (x - x_0^{r(|u|) - \# S} F(|u|)(y(x))$$

$$- \sum_{u \in U_S} (x - x_0)^{r(|u|) - \# S} F(|u|)(y(x)),$$

$$(303d)$$

assuming that one of the two series on the last line of (303d) is convergent. To prove (303c) we use (303d) and lemma 302A to obtain

$$y^{(\# S)}(x) = \sum_{u \in W_S} (x - x_0)^{r(|u|) - \# S} F(|u|)(y(x))$$

$$+ (x - x_0) f'(y(x))(y^{(\# S)}(x))$$

$$+ (x - x_0) \sum_{p \in P_0(S)} f^{(\# p)}(y(x))(y^{(\# p_1)}(x), y^{(\# p_2)}(x), \ldots), \quad (303e)$$

where $P_0(S)$ is the set $P(S)$ with the single member $\{S\}$ omitted. Substituting the formulae for $y^{(\# p_1)}, y^{(\# p_2)}, \ldots$ into (303e), as we may do by the induction hypothesis since $\# p_1, \# p_2, \ldots$ are all less than $\# S$, we find

$$y^{(\# S)}(x) - (x - x_0) f'(y(x)) y^{(\# S)}(x) = \sum_{u \in V_S^{(0)}} (x - x_0)^{r(|u|) - \# S} F(|u|)(y(x))$$

$$(303f)$$

and noting that the linear operator $1 - (x - x_0) f'(y(x))$ has the inverse

$$1 + (x - x_0) f'(y(x)) + (x - x_0)^2 f'(y(x))^2 + \cdots,$$

since $\| (x - x_0) f'(y(x)) \| < 1$, we now have

$$y^{(\# S)}(x) = \sum_{n=0}^{\infty} \sum_{u \in V_S^{(n)}} (x - x_0)^{r(|u|) - \# S} F(|u|)(y(x))$$

$$= \sum_{u \in V_S} (x - x_0)^{r(|u|) - \# S} F(|u|)(y(x)). \quad \blacksquare \quad (303g)$$

Making use of the numbers $\beta_n(t)$, $\bar{\beta}_n(t)$ already introduced, we can express the results of theorem 303A in a more conventional form.

THEOREM 303B.

Under the conditions of theorem 303A,

$$(f \circ y)^{(n)}(x) = \sum_{t \in T} \beta_n(t)(x - x_0)^{r(t) - n - 1} F(t)(y(x)), \qquad (303\text{h})$$

$$y^{(n)}(x) = \sum_{t \in T} \bar{\beta}_n(t)(x - x_0)^{r(t) - n} F(t)(y(x)). \qquad (303\text{i})$$

Finally, we give various Taylor series results.

THEOREM 303C.

Under the conditions of theorem 303A (with the additional condition that f is $n + 1$ times continuously differentiable in the case of equation (303k)),

$$f(y(x)) = \sum_{r(t) \leq n} \frac{\beta(t)(x - x_0)^{r(t) - 1} F(t)(y(x_0))}{[r(t) - 1]!}$$

$$+ \int_{x_0}^{x} \frac{(x - \xi)^{n-1}}{(n - 1)!} \left[\sum_{r(t) \geq n + 1} \beta_n(t)(\xi - x_0)^{r(t) - n - 1} F(t)(y(\xi)) \right] d\xi \qquad (303\text{j})$$

$$y(x) = y(x_0) + \sum_{r(t) \leq n} \frac{\bar{\beta}(t)(x - x_0)^{r(t)} F(t)(y(x_0))}{r(t)!}$$

$$+ \int_{x_0}^{x} \frac{(x - \xi)^{n}}{n!} \left[\sum_{r(t) \geq n + 1} \bar{\beta}_{n+1}(t)(\xi - x_0)^{r(t) - n - 1} F(t)(y(\xi)) \right] d\xi, \qquad (303\text{k})$$

$$y(x) = y(x_0) + \sum_{r(t) \leq n} \frac{\beta(t)(x - x_0)^{r(t)} F(t)(y(x_0))}{[r(t) - 1]!}$$

$$+ (x - x_0) \int_{x_0}^{x} \frac{(x - \xi)^{n-1}}{(n - 1)!} \left[\sum_{r(t) \geq n + 1} \beta_n(t)(\xi - x_0)^{r(t) - n - 1} F(t)(y(\xi)) \right] d\xi. \qquad (303\text{l})$$

Proof.

Using (303h) we evaluate $(f \circ y)^{(i)}(x_0)$ for $i = 0, 1, \ldots, n - 1$. Because of the factors of $x - x_0$ in (303h), the only terms contributing to $(f \circ y)^{(i)}(x_0)$ are from those $t \in T$ such that $r(t) = i + 1$. We thus find that

$$\sum_{i=0}^{n-1} \frac{(f \circ y)^{(i)}(x_0)(x - x_0)^i}{i!} = \sum_{i=0}^{n-1} \sum_{r(t) = i + 1} \frac{\beta_i(t) F(t)(y(x_0))(x - x_0)^i}{i!}$$

$$= \sum_{r(t) \leq n} \frac{\beta(t)(x - x_0)^{r(t) - 1} F(t)(y(x_0))}{[r(t) - 1]!} \qquad (303\text{m})$$

since $\beta_{r(t)-1}(t) = \beta(t)$. The addition to (303m) of

$$\int_{x_0}^{x} \frac{(x-\xi)^{n-1}}{(n-1)!}(f \circ y)^{(n)}(\xi)\, d\xi,$$

where $(f \circ y)^{(n)}$ is given by (303h), yields the Taylor expansion (303j). In the same way, (303i) is used to compute the Taylor series for y given as (303k). Finally, (303l) follows from (303a) and (303j). ∎

It should be noted that $\bar{\beta}(t) = r(t)\beta(t)$ so that the only difference between (303k) and (303l) is in the expressions given for the remainder terms.

304 Elementary weights.

In our study of explicit two-stage Runge–Kutta methods (subsection 300) it was found that the conditions for order 2 involved the polynomials $b_1 + b_2$ and $b_2 a_{21}$. In this subsection we will see that these are just the first two examples of a set of polynomials corresponding to the set of rooted trees. To make the work sufficiently general for later considerations, we do not assume that we are dealing only with explicit Runge–Kutta methods. Thus, given the value of $y(x_0)$, we take as our standard Runge–Kutta method for computing $Y(x)$ as an approximation to $y(x)$, the method given by

$$Y_i(x) = y(x_0) + (x - x_0) \sum_{j=1}^{s} a_{ij} f(Y_j(x)), \qquad i = 1, 2, \ldots, s, \qquad (304a)$$

$$Y(x) = y(x_0) + (x - x_0) \sum_{j=1}^{s} b_j f(Y_j(x)), \qquad (304b)$$

where $Y_1(x), Y_2(x), \ldots, Y_s(x)$ are intermediate results.
    This method will be represented by the array of numbers

$$
\begin{array}{c|cccc}
c_1 & a_{11} & a_{12} & \cdots & a_{1s} \\
c_2 & a_{21} & a_{22} & \cdots & a_{2s} \\
\vdots & \vdots & \vdots & & \vdots \\
c_s & a_{s1} & a_{s2} & \cdots & a_{ss} \\
\hline
 & b_1 & b_2 & \cdots & b_s
\end{array}
$$

that characterizes the method. The numbers $c_1, c_2, \ldots, c_s$ are defined by $c_i = \sum_{j=1}^{s} a_{ij}$ for $i = 1, 2, \ldots, s$.
    We will sometimes, for convenience, write $b_j = a_{s+1,j}$, $c_{s+1} = \sum_{j=1}^{s} b_j$ and $Y = Y_{s+1}$, so that (304b) becomes an example of (304a) with $i = s + 1$.
    The *elementary weight* $\Phi_i(t)$ for stage i ($i = 1, 2, \ldots, s + 1$) and rooted tree

t is now defined recursively by the formulae

$$\Phi_i(\tau) = c_i, \tag{304c}$$

$$\Phi_i([t_1 t_2 \cdots t_m]) = \sum_{j=1}^{s} a_{ij}\Phi_j(t_1)\Phi_j(t_2) \cdots \Phi_j(t_m). \tag{304d}$$

We also write $\Phi(t) = \Phi_{s+1}(t)$ and call this *the* elementary weight for the rooted tree t.

In much the same way as elementary differentials have a simple geometrical relationship to the corresponding rooted trees, there is a rule for writing down the formula for $\Phi(t)$ directly from the diagram for the tree. To the root of the tree attach the label i and to the other vertices attach other labels, say $j, k, l, \ldots$. For each arc write down a factor a_{uv} where u, v are the labels at the ends of the arc. Insert a further factor b_i and then sum each index $i, j, k, l, \ldots$ through the numbers $1, 2, \ldots, s$.

For example, the tree t used as an example in subsection 300 leads, via the diagram

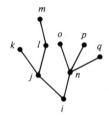

to the formula

$$\Phi(t) = \sum_{i,j,k,l,m,n,o,p,q=1}^{s} b_i a_{ij} a_{jk} a_{jl} a_{lm} a_{in} a_{no} a_{np} a_{nq}.$$

To simplify this, we can use the numbers $c_1, c_2, \ldots c_s$ to give

$$\Phi(t) = \sum_{i,j,l,n=1}^{s} b_i a_{ij} c_j a_{jl} c_l a_{in} c_n^3,$$

by summing over k, m, o, p and q.

305   The Taylor series for the Runge–Kutta solution.

In this subsection we will make use of the work in subsections 303 and 304 to find the Taylor coefficients for the function Y given by (304b). Let $\bar{y}$ be defined by

$$\bar{y}(x) = [Y_1(x), Y_2(x), \ldots, Y_{s+1}(x)]^T$$

so that

$$\bar{y}(x) = \bar{y}(x_0) + (x - x_0)\tilde{f}(\bar{y}(x)),$$

where

$$\bar{y}(x_0) = [y(x_0), y(x_0), \ldots, y(x_0)]^T$$

and $\tilde{f}: X^{s+1} \to X^{s+1}$ is defined by

$$\tilde{f}([z_1, z_2, \ldots, z_{s+1}]^T) = [\sum_j a_{1j}f(z_j), \sum_j a_{2j}f(z_j), \ldots, \sum_j a_{s+1,j}f(z_j)]^T.$$

If we define the linear mapping $P_i: X^{s+1} \to X$ by

$$P_i([z_1, z_2, \ldots, z_{s+1}]^T) = z_i,$$

then $\tilde{f}$ is related to f by

$$P_i(\tilde{f}(z)) = \sum_j a_{ij}f(P_j(z)), \tag{305a}$$

for $i = 1, 2, \ldots, s+1$.

It is clear from theorem 303C that the Taylor coefficients for $\bar{y}$ and thus for $Y = Y_{s+1}$ will depend on the elementary differentials $F(\tilde{f}, t)(y(x_0))$ for various $t \in T$. We now obtain formulae for these quantities.

LEMMA 305A.

If f is n times differentiable at each of $P_1z, P_2z, \ldots, P_{s+1}z$ for $z \in X^{s+1}$, then $\tilde{f}$ defined by (305a) is n times differentiable at z and for all $k_1, k_2, \ldots, k_n \in X$ and for $i = 1, 2, \ldots, s+1$,

$$P_i(\tilde{f}^{(n)}(z)(k_1, k_2, \ldots, k_n)) = \sum_{j=1}^{s} a_{ij}f^{(n)}(P_jz)(P_jk_1, P_jk_2, \ldots, P_jk_n). \tag{305b}$$

Proof.

The case when $n = 0$ reduces to (305a). We now proceed by induction and assume the result for orders of derivative less than $n > 0$. For fixed $k_1, k_2, \ldots, k_{n-1} \in X$ and $u \neq 0$ we compute

$$\frac{1}{\|u\|} \| P_i\tilde{f}^{(n-1)}(z+u)(k_1, k_2, \ldots, k_{n-1}) - P_i\tilde{f}^{(n-1)}(z)(k_1, k_2, \ldots, k_{n-1})$$

$$- \sum_{j=1}^{s} a_{ij}f^{(n)}(P_jz)(P_jk_1, P_jk_2, \ldots, P_jk_{n-1}, P_ju) \|$$

$$= \frac{1}{\|u\|} \| \sum_{j=1}^{s} a_{ij}[f^{(n-1)}(P_j(z+u))(P_jk_1, P_jk_2, \ldots, P_jk_{n-1})$$

$$- f^{(n-1)}(P_jz)(P_jk_1, \ldots, P_jk_{n-1}) - f^{(n)}(P_jz)(P_jk_1, P_jk_2, \ldots, P_jk_{n-1}, P_ju)] \|,$$

which tends to zero as $\|u\| \to 0$ because f is n times differentiable. Thus $\tilde{f}$ is n times differentiable at z and the derivative is given by (305b). ∎

THEOREM 305B.

If for some $t \in T$, $F(f, t)(z)$ exists, and if $z \in X^{s+1}$ is such that $P_i z = y_0$ for $i = 1, 2, \ldots, s + 1$, then $F(\tilde{f}, t)(z)$ exists and

$$P_i F(\tilde{f}, t)(z) = \Phi_i(t) F(f, t)(y_0), \quad i = 1, 2, \ldots, s + 1. \tag{305c}$$

Proof.

In the case when $t = \tau$, (305c) reduces to (305a). If $t = [t_1 t_2 \cdots t_m]$ we have, using (304d) and (305b), and the assumption that (305c) holds with t replaced by each of $t_1, t_2, \ldots, t_m$,

$$P_i F(\tilde{f}, t)(z) = P_i \tilde{f}^{(m)}(z)(F(\tilde{f}, t_1)(z), F(\tilde{f}, t_2)(z), \ldots, F(\tilde{f}, t_m)(z))$$

$$= \sum_{j=1}^{s} a_{ij} f^{(m)}(y_0)(P_j F(\tilde{f}, t_1)(z), P_j F(\tilde{f}, t_2)(z), \ldots, P_j F(\tilde{f}, t_m)(z))$$

$$= \sum_{j=1}^{s} a_{ij} f^{(m)}(y_0)(\Phi_j(t_1) F(f, t_1)(y_0), \Phi_j(t_2) F(f, t_2)(y_0), \ldots,$$

$$\Phi_j(t_m) F(f, t_m)(y_0))$$

$$= \sum_{j=1}^{s} a_{ij} \Phi_j(t_1) \Phi_j(t_2) \cdots \Phi_j(t_m) f^{(m)}(y_0)(F(f, t_1)(y_0), F(f, t_2)(y_0),$$

$$\ldots F(f, t_m)(y_0))$$

$$= \Phi_i(t) F(f, t)(y_0). \quad \blacksquare$$

Using this result, we can immediately write down the Taylor series for Y given by (304a) and (304b). For the present, we confine ourselves to computing the Taylor coefficients at x_0.

THEOREM 305C.

If f is $n - 1$ times differentiable at $y(x_0)$ then

$$Y^{(n)}(x_0) = \sum_{r(t)=n} \bar{\beta}(t) \Phi(t) F(f, t)(y(x_0)) \tag{305d}$$

and

$$Y^{(n)}(x_0) = n \sum_{r(t)=n} \beta(t) \Phi(t) F(f, t)(y(x_0)). \tag{305e}$$

Proof.

It has already been noted that $\bar{\beta}(t) = r(t) \beta(t)$ so that (305d) and (305e) are equivalent. To derive (305d), replace y by $\bar{y}$ in (303i) and set $x = x_0$. We have

$$\bar{y}^{(n)}(x_0) = \sum_{r(t)=n} \bar{\beta}_n(t)F(\tilde{f}, t)(\bar{y}(x_0)),$$

so that, by theorem 305B,

$$Y^{(n)}(x_0) = P_{s+1}\bar{y}^{(n)}(x_0)$$

$$= \sum_{r(t)=n} \bar{\beta}_n(t)P_{s+1}F(\tilde{f}, t)(\bar{y}(x_0))$$

$$= \sum_{r(t)=n} \bar{\beta}(t)\Phi_{s+1}(t)F(f, t)(y(x_0))$$

$$= \sum_{r(t)=n} \bar{\beta}(t)\Phi(t)F(f, t)(y(x_0)). \quad \blacksquare$$

306 Values of elementary weights and differentials.

Let T_0 be some finite subset of T. In this subsection we show that Runge–Kutta methods exist for which the elementary weights have any given values on T_0. In fact, the result will be proved under the assumption that the Runge–Kutta methods are explicit. A consequence of this result is that Runge–Kutta methods exist with all orders.

Similarly, we will show that there exists a finite-dimensional vector space X, an infinitely differentiable function $f: X \to X$, a point $y_0 \in X$ and a linear functional $p^* \in X^*$ such that the $p^*F(t)(y_0)$ takes on any given values for $t \in T_0$.

THEOREM 306A

If T_0 is a finite subset of T and $\theta: T_0 \to \mathbb{R}$ is given, then there exists an explicit Runge–Kutta method such that $\Phi(t) = \theta(t)$ for all $t \in T_0$.

Proof.

Let n denote the greatest of the orders of the members of T_0. The result of the theorem is obvious if $n \leq 1$ and we prove it by induction for $n > 1$. If T_1 is defined as the smallest subset of T which contains $t_1, t_2, \ldots, t_m$ if $[t_1 t_2 \cdots t_m] \in T_0$, then the greatest order of members of T_1 is evidently less than n, so that, by the induction hypothesis, we may assume the result of the theorem applied to T_1.

Let S denote the set of all possible mappings taking t to $\Phi(t)$ for $t \in T_0$ among the whole family of explicit Runge–Kutta methods. We will show that S is the set of all functions on T_0 to $\mathbb{R}$ by showing that it has four properties which are sufficient for this to be the case. These properties are (a) S is a linear space; (b) S is an algebra under pointwise multiplication, that is if α and β are in S then $\alpha\beta$ defined by $(\alpha\beta)(t) = \alpha(t)\beta(t)$ for all $t \in T_0$ is also in S; (c) S contains the identity element in this algebra, that is S contains e such that $e(t) = 1$ for all $t \in T_0$; and (d) S distinguishes points, that is given t, t', distinct members of T_0, there is a member α of S such that $\alpha(t) \neq \alpha(t')$.

To prove (a), consider two Runge–Kutta methods giving elementary weights

$$\Phi(t) = \sum_{i=1}^{s} b_i \prod_{k=1}^{m} \Phi_i(t_k),$$

and

$$\bar{\Phi}(t) = \sum_{j=1}^{\bar{s}} \bar{b}_j \sum_{k=1}^{m} \bar{\Phi}_j(t_k),$$

for $t = [t_1 t_2 \cdots t_m] \in T_0$. By considering a method with $s + \bar{s}$ stages made up from the combined stages of the original methods but with coefficients $Cb_1, Cb_2, \ldots, Cb_s, \bar{C}\bar{b}_1, \bar{C}\bar{b}_2, \ldots, \bar{C}\bar{b}_s$, we obtain an elementary weight

$$C\Phi(t) + \bar{C}\bar{\Phi}(t)$$

for C, $\bar{C}$ any real numbers.

To prove (b), form the product

$$\Phi(t)\bar{\Phi}(t) = \sum_{(i,j) \in I} b_i \bar{b}_j \prod_{k=1}^{m} \Phi_i(t_k)\bar{\Phi}_j(t_k)$$

where I is the product set $\{1, 2, \ldots, s\} \times \{1, 2, \ldots, \bar{s}\}$ and we can, by adding further stages if necessary, find a method for which the (i, j) stage gives an elementary weight equal to $\Phi_i(t_k)\bar{\Phi}_j(t_k)$ where $t_k \in T_1$.

To prove (c), consider the s-stage method for which $b_i = 0$ $(i < s)$, $b_s - 1$ and $\Phi_s(t) = 1$ for all $t \in T_1$.

To prove (d), let $t = [t_1^{k_1} t_2^{k_2} \cdots t_m^{k_m}]$, $t' = [t_1^{k_1'} t_2^{k_2'} \cdots t_m^{k_m'}]$, where $t_1, t_2, \ldots, t_m$ are distinct members of T_1 and one at least of (k_1, k_1'), $(k_2, k_2'), \ldots, (k_m, k_m')$ is a distinct pair. If it is (k_1, k_1'), choose an s-stage method such that $b_i = 0$ $(i < s)$, $b_s = 1$, $\Phi_s(t_1) = 2$ and $\Phi_s(t_i) = 1 (i > 1)$.

Having established that S has the required four properties, we see that the conditions of the Stone–Weierstrass theorem are satisfied in the trivial case of T_0 with the discrete topology (for an exposition of these topological notions, see, for example, Dugundji [1966]). Thus S is the set of all vectors on T_0. ∎

In proving the second result of this subsection, we will make use of a function $\varphi: \mathbb{R} \to \mathbb{R}$ such that derivatives of φ of all orders exist and are continuous and bounded and such that $\varphi(0) = \varphi'(0) = \cdots = \varphi^{(n-1)}(0) = 1$ for n the greatest order of members of a finite set T_0 of trees. Such a function may be constructed, for example, by the formula $\varphi(x) = P(x)\exp(-x^2)$, where P is an $(n-1)$th-degree polynomial with coefficients so chosen as to give the required values of $\varphi(0), \varphi'(0), \ldots, \varphi^{(n-1)}(0)$.

THEOREM 306B.

If T_0 is a finite subset of T and $\theta: T_0 \to \mathbb{R}$ is given, then there exists a finite-dimensional vector space X, a function $f: X \to X$ with bounded, continuous derivatives of all orders, a point $y_0 \in X$ and a linear functional $p^* \in X^*$ such that $p^* F(t)(y_0) = \theta(t)$ for all $t \in T_0$.

Proof.

Using the function φ introduced above and a Runge–Kutta method with coefficients

$$
\begin{array}{c|cccc}
c_1 & a_{11} & a_{12} & \cdots & a_{1s} \\
c_2 & a_{21} & a_{22} & \cdots & a_{2s} \\
\vdots & \vdots & \vdots & & \vdots \\
c_s & a_{s1} & a_{s2} & \cdots & a_{ss} \\
\hline
& b_1 & b_2 & \cdots & b_s
\end{array}
$$

such that $\Phi(t) = \theta(t)$ for all $t \in T_0$, we define X as the $(s+1)$-dimensional space of real-valued functions on $\{1, 2, \ldots, s+1\}$, f by the formula $f(y) = M(\varphi \circ y)$ for all $y \in X$, where M is the matrix

$$
M = \begin{bmatrix}
a_{11} & a_{12} & \cdots & a_{1s} & 0 \\
a_{21} & a_{22} & \cdots & a_{2s} & 0 \\
\vdots & \vdots & & \vdots & \vdots \\
a_{s1} & a_{s2} & \cdots & a_{ss} & 0 \\
b_1 & b_2 & \cdots & b_s & 0
\end{bmatrix},
$$

y is the zero vector in X and $p^* = [0\,0 \cdots 0\,1]$. It is easy to see that $f^{(m)}$ is defined by

$$
f^{(m)}(y)(k_1, k_2, \ldots, k_m) = M((\varphi^{(m)} \circ y) \cdot k_1 \cdot k_2 \cdots \cdot k_m)
$$

where '$\cdot$' denotes pointwise multiplication. Hence, for $m < n$,

$$
f^{(m)}(y_0)(k_1, k_2, \cdots, k_m) = M(k_1 \cdot k_2 \cdots \cdot k_m)
$$

and it follows that

$$
p^* F(t)(y_0) = \Phi(t) = \theta(t),
$$

for all $t \in T_0$. ∎

307 The order conditions.

We now generalize the work in subsection 300 so that we can find the order

Table 307   *Order conditions related to trees.*

$r(t)$	t	$\Phi(t) = 1/\gamma(t)$
1	τ	$\sum b_i = 1$
2	$[\tau]$	$\sum b_i c_i = \frac{1}{2}$
3	$[\tau^2]$	$\sum b_i c_i^2 = \frac{1}{3}$
3	$[_2\tau]_2$	$\sum b_i a_{ij} c_j = \frac{1}{6}$
4	$[\tau^3]$	$\sum b_i c_i^3 = \frac{1}{4}$
4	$[\tau[\tau]]$	$\sum b_i c_i a_{ij} c_j = \frac{1}{8}$
4	$[_2\tau^2]_2$	$\sum b_i a_{ij} c_j^2 = \frac{1}{12}$
4	$[_3\tau]_3$	$\sum b_i a_{ij} a_{jk} c_k = \frac{1}{24}$
5	$[\tau^4]$	$\sum b_i c_i^4 = \frac{1}{5}$
5	$[\tau^2[\tau]]$	$\sum b_i c_i^2 a_{ij} c_j = \frac{1}{10}$

Table 307 Continued

$r(t)$		t	$\Phi(t) = 1/\gamma(t)$
5		$[\tau[\tau^2]]$	$b_i c_i a_{ij} c_j^2 = \frac{1}{15}$
5		$[\tau[_2\tau]_2]$	$\sum b_i c_i a_{ij} a_{jk} c_k = \frac{1}{30}$
5		$[[\tau]^2]$	$\sum b_i a_{ij} c_j a_{ik} c_k = \frac{1}{20}$
5		$[_2\tau^3]_2$	$\sum b_i a_{ij} c_j^3 = \frac{1}{20}$
5		$[_2\tau[\tau]]_2$	$\sum b_i a_{ij} c_j c_j a_{jk} c_k = \frac{1}{40}$
5		$[_3\tau^2]_3$	$\sum b_i a_{ij} a_{jk} c_k^2 = \frac{1}{60}$
5		$[_4\tau]_4$	$\sum b_i a_{ij} a_{jk} a_{kl} c_l = \frac{1}{120}$

conditions for an arbitrary Runge–Kutta method. This method will be taken to be

$$
\begin{array}{c|cccc}
c_1 & a_{11} & a_{12} \cdots & a_{1s} \\
c_2 & a_{21} & a_{22} \cdots & a_{2s} \\
\vdots & \vdots & \vdots & \vdots \\
c_s & a_{s1} & a_{s2} \cdots & a_{ss} \\
\hline
 & b_1 & b_2 \cdots & b_s
\end{array}
$$

and the coefficients used to form $\Phi_i(t)$ for $t \in T$ and $i = 1, 2, \ldots, s + 1$ will come from this method, which we will refer to as the standard Runge–Kutta method. We will assume that $f: X \to X$ is $p - 1$ times differentiable at $y(x_0)$ but otherwise arbitrary and we will make use of the function Y defined in (304a) and (304b).

DEFINITION 307A.

The standard Runge–Kutta method is *of order p* if

$$
Y^{(n)}(x_0) = y^{(n)}(x_0) \tag{307a}
$$

for $n = 1, 2, \ldots, p$.

We emphasize that in this definition the function f is not specified, so that (307a) is supposed to hold for all sufficiently differentiable functions f.

Let $\gamma(t) = \bar{\beta}(t)/\alpha(t) = r(t)\beta(t)/\alpha(t)$ for $t \in T$. Thus $\gamma(t)$ is the density of t introduced in subsection 144. We can now establish a result already informally stated in subsection 223 which becomes the key result of the present section.

THEOREM 307B.

The standard Runge–Kutta method is of order p iff

$$
\Phi(t) = \frac{1}{\gamma(t)} \tag{307b}
$$

for all $t \in T$ such that $r(t) \le p$.

Proof.

For $n \le p$, evaluate $Y^{(n)}(x_0)$ given by (305d) and equate it to $y^{(n)}(x_0)$ given by (302h). We see that definition 307A is equivalent to the statement that, for $n \le p$,

$$
\sum_{r(t)=n} \bar{\beta}(t)\Phi(t)F(t)(y(x_0)) = \sum_{r(t)=n} \alpha(t)F(t)(y(x_0)). \tag{307c}
$$

That (307b) is sufficient for (307c) to hold is clear. The necessity follows by noting that for an appropriate choice of the function f, for example using the construction of theorem 306B, a particular component of $F(t)(y(x_0))$ can be made to take any particular values for any finite set of choices of t in T. Thus, for (307c) to hold, coefficients of $F(t)(y(x_0))$ on the two sides must agree. ∎

We complete this subsection by presenting table 307 which shows the form of (307b) for the various $t \in T$ such that $r(t) \leq 5$. In this table, each subscript $i, j, \ldots$ is summed from 1 to s.

31 LOW-ORDER EXPLICIT METHODS

310 Methods with orders less than four.

It is possible to obtain a method of order 2 with just two stages given by

$$
\begin{array}{c|cc}
0 & & \\
c_2 & a_{21} & \\
\hline
 & b_1 & b_2
\end{array}
$$

with $c_2 = a_{21}$ as long as $b_1 + b_2 = 1$ and $b_2 c_2 = \frac{1}{2}$. The two most well-known examples are those which were quoted in subsection 224,

$$
\begin{array}{c|cc}
0 & & \\
\frac{1}{2} & \frac{1}{2} & \\
\hline
 & 0 & 1
\end{array}
\qquad \text{and} \qquad
\begin{array}{c|cc}
0 & & \\
1 & 1 & \\
\hline
 & \frac{1}{2} & \frac{1}{2}
\end{array}
$$

These are sometimes known as the modified Euler (or improved polygon) method and the improved Euler (or Heun) method respectively.

For order 3, three stages are necessary. From the first four rows of table 307, we see that a method

$$
\begin{array}{c|ccc}
0 & & & \\
c_2 & a_{21} & & \\
c_3 & a_{31} & a_{32} & \\
\hline
 & b_1 & b_2 & b_3
\end{array}
$$

with $c_2 = a_{21}, c_3 = a_{31} + a_{32}$ is of order 3 iff

$$b_1 + b_2 + b_3 = 1, \tag{310a}$$

$$b_2 c_2 + b_3 c_3 = \tfrac{1}{2}, \tag{310b}$$

$$b_2 c_2^2 + b_3 c_3^2 = \tfrac{1}{3}, \tag{310c}$$

$$b_3 a_{32} c_2 = \tfrac{1}{6}. \tag{310d}$$

To solve these equations, we select values of c_2, c_3 such that the linear system (310b) and (310c) can be solved for b_2, b_3 with $b_3 \neq 0, c_2 \neq 0$. We then solve for b_1 from (310a) and a_{32} from (310d) and finally $a_{21} = c_2, a_{31} = c_3 - a_{32}$. We distinguish three cases:

Case 1: $0, c_2, c_3$ all different and $c_2 \neq \frac{2}{3}$. In this case we find:

$$b_2 = \frac{c_3 - \frac{2}{3}}{2c_2(c_3 - c_2)},$$

$$b_3 = \frac{\frac{2}{3} - c_2}{2c_3(c_3 - c_2)},$$

$$b_1 = 1 - b_2 - b_3,$$

$$a_{32} = \frac{1}{6b_3c_2}.$$

Case 2: $c_2 = \frac{2}{3}, c_3 = 0, b_3$ an arbitrary non-zero number. In this case we find:

$$b_2 = \tfrac{3}{4}, \qquad b_1 = \tfrac{1}{4} - b_3, \qquad a_{32} = \frac{1}{4b_3}.$$

Case 3: $c_2 = c_3 = \frac{2}{3}, b_3$ an arbitrary non-zero number. In this case, we find:

$$b_1 = \tfrac{1}{4}, \qquad b_2 = \tfrac{3}{4} - b_3, \qquad a_{32} = \frac{1}{4b_3}.$$

311 The last stage of fourth-order methods.

We continue our study of Runge–Kutta methods by considering the problem of selecting the coefficients $c_2, c_3, c_4, a_{32}, a_{42}, a_{43}, b_1, b_2, b_3, b_4$ for an explicit four-stage fourth-order Runge–Kutta method. It is clear that at least four stages are necessary since if $t = [_3\tau]_3$, then $\Phi(t) = 1/\gamma(t)$ becomes $\Sigma b_i a_{ij} a_{jk} c_k = \frac{1}{24}$ and if there are less than four stages, the left-hand side is zero.

For fourth-order accuracy, the ten coefficients listed above must satisfy the following eight equations:

$$b_1 + b_2 + b_3 + b_4 = 1, \qquad \bullet \qquad \tau \qquad \text{(311a)}$$

$$b_2c_2 + b_3c_3 + b_4c_4 = \tfrac{1}{2}, \qquad [\tau] \qquad \text{(311b)}$$

$$b_2c_2^2 + b_3c_3^2 + b_4c_4^2 = \tfrac{1}{3}, \qquad [\tau^2] \qquad \text{(311c)}$$

$$b_3a_{32}c_2 + b_4a_{42}c_2 + b_4a_{43}c_3 = \tfrac{1}{6}, \qquad [_2\tau]_2 \qquad \text{(311d)}$$

$$b_2 c_2^3 + b_3 c_3^3 + b_4 c_4^3 = \tfrac{1}{4},$$

$[\tau^3]$ (311e)

$$b_3 c_3 a_{32} c_2 + b_4 c_4 a_{42} c_2 + b_4 c_4 a_{43} c_3 = \tfrac{1}{8},$$

$[\tau[\tau]]$ (311f)

$$b_3 a_{32} c_2^2 + b_4 a_{42} c_2^2 + b_4 a_{43} c_3^2 = \tfrac{1}{12},$$

$[_2\tau^2]_2$ (311g)

$$b_4 a_{43} a_{32} c_2 = \tfrac{1}{24}.$$

$[_3\tau]_3$ (311h)

These equations are, of course, equivalent to the first eight of those listed in table 307. To make the correspondence clearer, we have shown the corresponding trees after each of the equations in both diagrammatic and symbolic forms.

The complete solution of these equations will be postponed until the next subsection. For the present we consider only the value of the single parameter c_4 and at the same time we derive formulae that will later be used in the determination of a_{42} and a_{43}. It will be found (lemma 311C) that c_4 necessarily equals 1 and that this fact together with a set of relations connecting the columns of matrix A leads to simplifications to the set of order conditions.

As a first step in this line of attack, we introduce lemma 311A which will also be used in section 32.

LEMMA 311A.

Let U, V be 3×3 matrices such that

$$UV = \begin{bmatrix} w_{11} & w_{12} & 0 \\ w_{21} & w_{22} & 0 \\ 0 & 0 & 0 \end{bmatrix},$$

where $w_{11} w_{22} - w_{21} w_{12} \neq 0$. Then either the third row of U is the zero vector or the third column of V is the zero vector.

Proof.

Since UV is singular, either U or V is singular. In the first case, let x^T be such that $x^T U = 0$ with $x^T \neq 0$. This implies that $x^T UV = 0$, and because of the form assumed for UV, this implies that $x^T = [0\,0\,1]$ or a scalar multiple of this. Hence the last row of U is $[0\,0\,0]$. Similarily, if V is singular, we assume the existence of a non-singular vector y such that $Vy = 0$ and deduce that y must be a scalar multiple of $[0\,0\,1]^T$. Thus the last column of V is $[0\,0\,0]^T$. ∎

Now consider the vectors $u_1^T, u_2^T, u_3^T \in (\mathbb{R}^3)^*$ and $v_1, v_2, v_3 \in \mathbb{R}^3$ defined as follows:

$$u_1^T = [b_2, b_3, b_4],$$

$$u_2^T = [b_2 c_2, b_3 c_3, b_4 c_4],$$

$$u_3^T = \left[\sum_i b_i a_{i2} - b_2(1 - c_2), \sum_i b_i a_{i3} - b_3(1 - c_3), \sum_i b_i a_{i4} - b_4(1 - c_4) \right],$$

$$v_1 = \begin{bmatrix} c_2 \\ c_3 \\ c_4 \end{bmatrix}, \quad v_2 = \begin{bmatrix} c_2^2 \\ c_3^2 \\ c_4^2 \end{bmatrix}, \quad v_3 = \begin{bmatrix} \sum_j a_{2j} c_j - \tfrac{1}{2} c_2^2 \\ \sum_j a_{3j} c_j - \tfrac{1}{2} c_3^2 \\ \sum_j a_{4j} c_j - \tfrac{1}{2} c_4^2 \end{bmatrix}.$$

If the parameters $c_2, c_3, c_4, \ldots, b_4$ are those of a four-stage fourth-order Runge–Kutta method it is possible to compute $u_i^T v_j$ for $i, j = 1, 2, 3$, since this will be a linear combination of the $\Phi(t)$ for various t of order less than five and will thus be equal to a certain number formed from the corresponding $\gamma(t)$. In fact, we find the following result.

LEMMA 311B.

For a four-stage fourth-order method $u_1^T v_1 = \tfrac{1}{2}$, $u_1^T v_2 = u_2^T v_1 = \tfrac{1}{3}$, $u_2^T v_2 = \tfrac{1}{4}$, $u_3^T v_j = u_i^T v_3 = 0$ for $i, j = 1, 2, 3$.

Proof.

Let

$$\alpha^T = \left[\sum_i b_i a_{i2}, \sum_i b_i a_{i3}, \sum_i b_i a_{i4} \right],$$

$$\beta = \left[\sum_j a_{2j} c_j, \sum_j a_{3j} c_j, \sum_j a_{4j} c_j \right]^T,$$

so that $u_3^T = \alpha^T - u_1^T - u_2^T$ and $v_3 = \beta - \frac{1}{2}v_2$. We have

$$u_1^T v_1 = \sum_k b_k c_k = \Phi([\tau]) = \frac{1}{\gamma([\tau])} = \frac{1}{2},$$

$$u_1^T v_2 = u_2^T v_1 = \sum_k b_k c_k^2 = \Phi([\tau^2]) = \frac{1}{\gamma([\tau^2])} = \frac{1}{3},$$

$$u_2^T v_2 = \sum_k b_k c_k^3 = \Phi([\tau^3]) = \frac{1}{\gamma([\tau^3])} = \frac{1}{4},$$

$$\alpha^T v_1 = \sum_{ik} b_i a_{ik} c_k = \Phi([_2\tau]_2) = \frac{1}{\gamma([_2\tau]_2)} = \frac{1}{6},$$

$$\alpha^T v_2 = \sum_{ik} b_i a_{ik} c_k^2 = \Phi([_2\tau^2]_2) = \frac{1}{\gamma([_2\tau^2]_2)} = \frac{1}{12},$$

$$u_1^T \beta = \sum_{kj} b_k a_{kj} c_j = \Phi([_2\tau]_2) = \frac{1}{6},$$

$$u_2^T \beta = \sum_{kj} b_k c_k a_{kj} c_j = \Phi([\tau[\tau]]) = \frac{1}{\gamma([\tau[\tau]])} = \frac{1}{8},$$

$$\alpha^T \beta = \sum_{ijk} b_i a_{ik} a_{kj} c_j = \Phi([_3\tau]_3) = \frac{1}{\gamma([_3\tau]_3)} = \frac{1}{24}$$

and we can immediately compute $u_3^T v_j$, $u_i^T v_3$ ($i, j = 1, 2, 3$) to verify the remaining details of the lemma. ■

We now have the main result of this subsection.

LEMMA 311C.

For a four-stage fourth-order Runge–Kutta method, $c_4 = 1$ and, furthermore, $\Sigma_i b_i a_{ij} = b_j(1 - c_j)$ for $j = 2, 3, 4$.

Proof.

We note first of all that v_3 is not the zero vector, since its first component is $-\frac{1}{2}c_2^2$ (noting that $\Sigma_j a_{2j} c_j = 0$) and if $c_2 = 0$ it would be impossible to satisfy (311h). Let U, V be 3×3 matrices formed respectively from the three rows u_1^T, u_2^T, u_3^T and the three columns v_1, v_2, v_3. From lemma 311B we compute

$$UV = \begin{bmatrix} \frac{1}{2} & \frac{1}{3} & 0 \\ \frac{1}{3} & \frac{1}{4} & 0 \\ 0 & 0 & 0 \end{bmatrix},$$

so that, using lemma 311A and noting that $v_3 \neq 0$, we see that $u_3^T = 0$. Thus

$\sum_i b_i a_{ij} = b_j(1 - c_j)$ for $j = 2, 3, 4$. In the case $j = 4$, since $\sum_i b_i a_{i4} = 0$, this implies that $b_4 = 0$ (contradicting (311h)) or $c_4 = 1$. ∎

312 Fourth-order methods.

Having established that in any four-stage fourth-order method, $c_4 = 1$ and, furthermore, that

$$\sum_i b_i a_{ij} = b_j(1 - c_j), \qquad j = 2, 3, 4, \tag{312a}$$

we proceed to show that such methods do exist. First, we see that we can in effect ignore certain of the equations (311a) to (311h). Specifically, those associated with the trees $[_2\tau]_2$, $[_2\tau^2]_2$, $[_3\tau]_3$ fall into this category since, if $\Phi(t) = 1/\gamma(t)$ holds for all other trees, we have

$$\Phi([_2\tau]_2) = \sum_{ij} b_i a_{ij} c_j = \sum_j b_j(1 - c_j)c_j = \tfrac{1}{2} - \tfrac{1}{3} = \tfrac{1}{6},$$

$$\Phi([_2\tau^2]_2) = \sum_{ij} b_i a_{ij} c_j^2 = \sum_j b_j(1 - c_j)c_j^2 = \tfrac{1}{3} - \tfrac{1}{4} = \tfrac{1}{12},$$

$$\Phi([_3\tau]_3) = \sum_{ijk} b_i a_{ij} a_{jk} c_k = \sum_{jk} b_j(1 - c_j)a_{jk}c_k = \tfrac{1}{6} - \tfrac{1}{8} = \tfrac{1}{24}.$$

It is interesting that as a consequence of (311a) and (311b), (312a) holds when $j = 1$, as well as when $j = 2, 3, 4$. This is because the sum of each side of (312a) over $j = 1, 2, 3, 4$ is $\tfrac{1}{2}$.

The eight trees of order not exceeding four can now be divided into three categories:
(a) $\tau, [\tau], [\tau^2], [\tau^3]$,
(b) $[_2\tau]_2, [_2\tau^2]_2, [_3\tau]_3$,
(c) $[\tau[\tau]]$.
For t in category (a), $\Phi(t) = 1/\gamma(t)$ takes the form

$$\sum_i b_i c_i^{k-1} = \frac{1}{k}, \qquad k = 1, 2, 3, 4, \tag{312b}$$

which enables the determination of b_1, b_2, b_3, b_4 as long as c_2, c_3 are appropriately chosen. The three trees in category (b) can be ignored as we have already indicated and the single tree in category (c) yields the equation

$$\sum_{ij} b_i c_i a_{ij} c_j = \tfrac{1}{8}. \tag{312c}$$

Under the assumption that $\sum_{ij} b_i a_{ij} c_j = \tfrac{1}{6}$ (which would follow, as we have seen, from (312a) and (312b)), (312c) is equivalent to

$$\sum_{ij} b_i(1 - c_i)a_{ij}c_j = \tfrac{1}{24}$$

which simplifies to

$$b_3(1 - c_3)a_{32}c_2 = \tfrac{1}{24}. \tag{312d}$$

Assuming that the choice of c_2, c_3 did not imply that $b_3(1 - c_3)c_2 = 0$, (312d) can be satisfied with an appropriate choice of a_{32}.

Finally, we must ensure that (312a) holds for $j = 2, 3$. Assuming that $b_4 \neq 0$, this can be achieved through the choice of a_{42}, a_{43}. The method determined by the programme that has just been described depends on the choice of c_2, c_3, and if the equations (312b) do not determine them uniquely, depends on the choice of b_1, b_2, b_3, b_4. We now list five cases where the determination can be carried out.

Case 1: $0, c_2, c_3, 1$ all distinct, $c_2 \neq \tfrac{1}{2}$ and $3 - 4(c_2 + c_3) + 6c_2c_3 \neq 0$.

Case 2: $c_2 = c_3 = \tfrac{1}{2}, b_3 \neq 0$.

Case 3: $c_2 = \tfrac{1}{2}, c_3 = 0, b_3 \neq 0$.

Case 4: $c_2 = 1, c_3 = \tfrac{1}{2}, b_4 \neq 0$.

Case 5: $c_2 \neq 0, c_3 = \tfrac{1}{2}, b_2 = 0$.

These cases are not all distinct. For example, case 5 overlaps with each of cases 1, 2, 4. They are, however, exhaustive in that every fourth-order four-stage method is in one of these cases.

The general solution to the problem of determining the coefficients is given in the various cases as follows:

Case 1:

0				
c_2	c_2			
c_3	$\dfrac{c_3(3c_2 - c_3 - 4c_2^2)}{2c_2(1 - 2c_2)}$	$\dfrac{c_3(c_3 - c_2)}{2c_2(1 - 2c_2)}$		
1	a_{41}	a_{42}	a_{43}	
	$\dfrac{1 - 2(c_2 + c_3) + 6c_2c_3}{12c_2c_3}$	$\dfrac{2c_3 - 1}{12c_2(c_3 - c_2)(1 - c_2)}$	$\dfrac{1 - 2c_2}{12c_3(c_3 - c_2)(1 - c_3)}$	$\dfrac{3 - 4(c_2 + c_3) + 6c_2c_3}{12(1 - c_2)(1 - c_3)}$

where

$$a_{41} = \frac{c_3^2(12c_2^2 - 12c_2 + 4) - c_3(12c_2^2 - 15c_2 + 5) + (4c_2^2 - 6c_2 + 2)}{2c_2c_3[3 - 4(c_2 + c_3) + 6c_2c_3]},$$

$$a_{42} = \frac{(-4c_3^2 + 5c_3 + c_2 - 2)(1 - c_2)}{2c_2(c_3 - c_2)[3 - 4(c_2 + c_3) + 6c_2c_3]},$$

$$a_{43} = \frac{(1 - 2c_2)(1 - c_3)(1 - c_2)}{c_3(c_3 - c_2)[3 - 4(c_2 + c_3) + 6c_2c_3]}.$$

Case 2:

$$
\begin{array}{c|cccc}
0 \\
\tfrac{1}{2} & \tfrac{1}{2} \\
\tfrac{1}{2} & \dfrac{3b_3 - 1}{6b_3} & \dfrac{1}{6b_3} \\
1 & 0 & 1 - 3b_3 & 3b_3 \\
\hline
& \tfrac{1}{6} & \tfrac{2}{3} - b_3 & b_3 & \tfrac{1}{6}
\end{array}
$$

Case 3:

$$
\begin{array}{c|cccc}
0 \\
\tfrac{1}{2} & \tfrac{1}{2} \\
0 & -\dfrac{1}{12b_3} & \dfrac{1}{12b_3} \\
1 & -\tfrac{1}{2} - 6b_3 & \tfrac{3}{2} & 6b_3 \\
\hline
& \tfrac{1}{6} - b_3 & \tfrac{2}{3} & b_3 & \tfrac{1}{6}
\end{array}
$$

Case 4:

$$
\begin{array}{c|cccc}
0 \\
1 & 1 \\
\tfrac{1}{2} & \tfrac{3}{8} & \tfrac{1}{8} \\
1 & 1 - \dfrac{1}{4b_4} & -\dfrac{1}{12b_4} & \dfrac{1}{3b_4} \\
\hline
& \tfrac{1}{6} & \tfrac{1}{6} - b_4 & \tfrac{2}{3} & b_4
\end{array}
$$

Case 5:

$$
\begin{array}{c|cccc}
0 \\
c_2 & c_2 \\
\tfrac{1}{2} & \dfrac{4c_2 - 1}{8c_2} & \dfrac{1}{8c_2} \\
1 & \dfrac{1 - 2c_2}{2c_2} & -\dfrac{1}{2c_2} & 2 \\
\hline
& \tfrac{1}{6} & 0 & \tfrac{2}{3} & \tfrac{1}{6}
\end{array}
$$

313   The classical Runge–Kutta method.

If in case 2 of the previous subsection we select the value $b_3 = \frac{1}{3}$, we get the method

$$
\begin{array}{c|cccc}
0 \\
\frac{1}{2} & \frac{1}{2} \\
\frac{1}{2} & 0 & \frac{1}{2} \\
1 & 0 & 0 & 1 \\
\hline
 & \frac{1}{6} & \frac{1}{3} & \frac{1}{3} & \frac{1}{6}
\end{array} \quad .
$$

This method, remarkably simple in that it has zero values for each of a_{31}, a_{41}, a_{42}, was discovered by Kutta (1901). It is widely used and is the most well-known of all Runge–Kutta methods. Although it is sometimes known as *the* Runge–Kutta method, we will refer to it as the *classical Runge–Kutta method*.

Algorithm 313 which we now present carries out a single step of this method. Note that the identifiers y, N, h and *diffeqn* have the same meanings as in algorithm 207. The arrays u, $F1$, $F2$, $F3$, $F4$ all have dimensions $[1 .. N]$ and are used to hold temporary values used in carrying out the step.

Algorithm 313: Classical Runge–Kutta step

```
diffeqn (y, F1);
for i := 1 to N do u[i] := y[i] + 0.5*h*F1[i];
diffeqn (u, F2);
for i := 1 to N do u[i] := y[i] + 0.5*h*F2[i];
diffeqn (u, F3);
for i := 1 to N do u[i] := y[i] + h*F3[i];
diffeqn (u, F4);
for i := 1 to N do y[i] := y[i] + h*(F1[i] + F4[i] + 2*(F2[i] + F3[i]))/6
```

314   Gill's Runge–Kutta method.

Algorithm 313 was written in such a way that five arrays u, $F1$, $F2$, $F3$, $F4$ each requiring N memory positions are necessary to hold intermediate values used in the computation. If N is sufficiently large for this to be prohibitive, it is a simple matter to redesign the algorithm such that only three of these arrays are required. Denoting these arrays u, v and w, we illustrate in algorithm 314 one way in which this may be done for an arbitrary four-stage Runge–Kutta method. Note that $a21$, $a31$, $a32$, ... denote the constants $a_{21}, a_{31}, a_{32}, \ldots$.

Algorithm 314: Arbitrary four-stage Runge–Kutta step

diffeqn (y, u);
for $i := 1$ **to** N **do** $y[i] := y[i] + h*a21*u[i]$;
diffeqn (y, v);
for $i := 1$ **to** N **do** $y[i] := y[i] + h*((a31 - a21)*u[i] + a32*v[i])$;
diffeqn (y, w);
for $i := 1$ **to** N **do**
begin
    $y[i] := y[i] + h*((a41 - a31)*u[i] + (a42 - a32)*v[i] + a43*w[i])$;
    $u[i] := (b1 - a41)*u[i] + (b2 - a42)*v[i]$
end;
diffeqn (y, v);
for $i := 1$ **to** N **do**
    $y[i] := y[i] + h*(u[i] + (b3 - a43)*w[i] + b4*v[i])$

In the search for a fourth-order method that was even more efficient in its use of memory, Gill (1951) proposed the use of methods that could be implemented as in algorithm 314'. Note that only the working arrays u, v are now needed.

Algorithm 314': Gill-type Runge–Kutta step

diffeqn (y, u);
for $i := 1$ **to** N **do** $y[i] := y[i] + h*a21*u[i]$;
diffeqn (y, v);
for $i := 1$ **to** N **do**
begin
    $y[i] := y[i] + h*((a31 - a21)*u[i] + a32*v[i])$;
    $u[i] := P*u[i] + Q*v[i]$
end;
diffeqn (y, v);
for $i := 1$ **to** N **do**
begin
    $y[i] := y[i] + h*(R*u[i] + a43*v[i])$;
    $u[i] := S*u[i] + T*v[i]$
end;
diffeqn (y, v);
for $i := 1$ **to** N **do** $y[i] := y[i] + h*(U*u[i] + b4*v[i])$

In this algorithm, P, Q, R, S, T, U are numerical constants. This method will be identical to the method

$$
\begin{array}{c|cccc}
0 & & & & \\
c_2 & a_{21} & & & \\
c_3 & a_{31} & a_{32} & & \\
1 & a_{41} & a_{42} & a_{43} & \\
\hline
& b_1 & b_2 & b_3 & b_4
\end{array}
$$

if and only if P, Q, R, S, T and U are related to the parameters of the method by

$$PR = a_{41} - a_{31}, \tag{314a}$$

$$QR = a_{42} - a_{32}, \tag{314b}$$

$$PSU = b_1 - a_{41}, \tag{314c}$$

$$QSU = b_2 - a_{42}, \tag{314d}$$

$$TU = b_3 - a_{43}. \tag{314e}$$

Thus a given method can be implemented this special way if and only if P, Q, R, S, T, U can be chosen to satisfy (314a) to (314e). It is easy to see that this is possible if and only if

$$(a_{41} - a_{31})(b_2 - a_{42}) = (a_{42} - a_{32})(b_1 - a_{41}), \tag{314f}$$

and it is specifically for methods satisfying this requirement that Gill sought. He observed that this was not possible for fourth-order methods in what we called cases 3, 4 and 5 in subsection 312. However, in case 2 a method satisfying (314f) occurs when $b_3 = (2 + \sqrt{2})/6$. His method then becomes

$$
\begin{array}{c|cccc}
0 & & & & \\
\dfrac{1}{2} & \dfrac{1}{2} & & & \\
\dfrac{1}{2} & \dfrac{-1 + \sqrt{2}}{2} & \dfrac{2 - \sqrt{2}}{2} & & \\
1 & 0 & \dfrac{-\sqrt{2}}{2} & \dfrac{2 + \sqrt{2}}{2} & \\
\hline
& \dfrac{1}{6} & \dfrac{2 - \sqrt{2}}{6} & \dfrac{2 + \sqrt{2}}{6} & \dfrac{1}{6}
\end{array}
$$

With the choice $R = -a_{43}$, $u = -b_4$ we can solve (314a) to (314e) for P, Q, S, T and substitute numerical values to obtain algorithm 314″ for Gill's version of the Runge–Kutta method.

Note that the following constants appear in this algorithm:

$(2 - \sqrt{2})/2 = 0.2928 \ldots, (3\sqrt{2} - 4)/2 = 0.1213 \ldots, 2 - \sqrt{2} = 0.5857 \ldots,$
$(2 + \sqrt{2})/2 = 1.7071 \ldots, 4 + 2\sqrt{2} = 6.8284 \ldots, 4 + 3\sqrt{2} = 8.2426 \ldots,$
$1/6 = 0.1666 \ldots.$

Algorithm 314": Gill–Runge–Kutta step

```
diffeqn (y, u);
for i := 1 to N do y[i] := y[i] + h*0.5*u[i];
diffeqn (y, v);
for i := 1 to N do
begin
    y(i] := y[i] + h*0.292893218813*(v[i] − u[i]);
    u[i] := 0.121320343560*u[i] + 0.585786437627*v[i]
end;
diffeqn (y, v);
for i := 1 to N do
begin
    y[i] := y[i] + h*1.707106781187*(v[i] − u[i]);
    u[i] := 6.828427124746*v[i] − 8.242640687120*u[i]
end;
diffeqn (y, v);
for i := 1 to N do y[i] := y[i] + h*0.166666666667*(v[i] − u[i])
```

32   THE ATTAINABLE ORDER OF EXPLICIT METHODS

320   Introductory comments.

In an s stage method, the parameters to be chosen are $a_{21}, a_{31}, a_{32}, a_{41}, a_{42}$, $a_{43}, \ldots, a_{s1}, a_{s2}, a_{s3}, \ldots, b_1, b_2, b_3, \ldots, b_s$, a total of $s(s+1)/2$ parameters. On the other hand, for a method to be of order p there are $\#\{t \in T : r(t) \leq p\}$ conditions of the form $\Phi(t) = 1/\gamma(t)$ imposed on these parameters. It is interesting to compare these numbers of parameters and of conditions for various s, p, and such a comparison is given in table 320.

We are interested in the relationship between the value of s that is required to achieve a particular order p and the value of p itself. The following conjecture seems to be a reasonable one and is consistent with the facts about the existence of methods of orders less than five brought out in the previous section.

Table 320

s	$s(s+1)/2$	p	$\#\{t \in T : r(t) \leq p\}$
1	1	1	1
2	3	2	2
3	6	3	4
4	10	4	8
5	15	5	17
6	21	6	37
7	28	7	85
8	36	8	200
9	45	9	486

CONJECTURE 320A.

An order p is attainable with an s-stage Runge–Kutta method if and only if $s(s + 1)/2 \geq \#\{t \in T : r(t) \leq p\}$.

A consequence of this conjecture would be that order 5 methods exist with six stages but not with five stages; this is, in fact, true. The conjecture is, however, quite false: we shall show in the next section, for example, that order 6 methods exist with seven stages even though $37 > 28$ and that order 7 methods exist with nine stages even though $85 > 45$.

As a step towards undermining the credulity of the reader concerning such statements as conjecture 320A, an observation will now be made about the 17 conditions that would have to be satisfied by the 15 constants defining a five-stage method if its order were five.

Suppose, as for fourth-order methods, that extra conditions are imposed as follows:

$$\sum_i b_i a_{ij} = b_j(1 - c_j), \qquad j = 1, 2, 3, 4, 5. \tag{320a}$$

The effect of this is to enable us to remove from consideration the eight trees $[\tau]$, $[_2\tau]_2$, $[_2\tau^2]_2$, $[_3\tau]_3$, $[_2\tau^3]_2$, $[_2[\tau]\tau]_2$, $[_3\tau^2]_3$, $[_4\tau]_4$. Thus, although we are imposing five extra conditions as given by (320a), eight of the conditions $\Phi(t) = 1/\gamma(t)$ are removed so that the effective number of constraints on the parameters is reduced to 14. Since there are 15 parameters, one might now suspect that five-stage fifth-order methods really do exist. However, as we have pointed out, and as we will prove later in this section, this is not the case.

321 An elementary bound.

In this subsection we give a generalization of theorem 300A as follows.

THEOREM 321A.

A Runge–Kutta method with s stages has order not exceeding s.

Proof.

Consider the tree $t = [_s\tau]_s$ of order $s + 1$. If a method were of order $s + 1$, it would hold that $\Phi(t) = 1/\gamma(t)$. This means that

$$\sum_{i_1, i_2, \ldots, i_s} b_{i_1} a_{i_1 i_2} a_{i_2 i_3} \cdots c_{i_s} = \frac{1}{(s + 1)!}. \tag{320a}$$

The subscripts in a possible non-zero term on the left-hand side of (321a) would satisfy

$$s \geq i_1 > i_2 > \cdots > i_s > 1$$

which is impossible for integers $i_1, i_2, \ldots, i_s$. Thus (321a) reduces to

$$0 = \frac{1}{(s+1)!},$$

a contradiction. ■

The fact that we have already (section 31) constructed methods of order s with s stages up to $s = 4$ enables us to state the following theorem without further proof.

THEOREM 321B.

For $s \leq 4$, the greatest attainable order of an s-stage method is s.

322 Impossibility of five-stage fifth-order methods.

In this subsection (and the next) we make frequent use of lemma 311A. Our object is to prove that no method with five stages can have order 5. The argument proceeds in three steps:

Step 1: It is shown that if such a method exists then $c_4 = 1$ (lemma 322A).
Step 2: It is further shown that for such a method $c_5 = 1$ (lemma 322B).
Step 3: It is shown that no five-stage fifth-order method exists with $c_4 = c_5 = 1$ (theorem 322C).

We now complete the subsection by carrying out this programme.

LEMMA 322A.

Let

0					
c_2	a_{21}				
c_3	a_{31}	a_{32}			
c_4	a_{41}	a_{42}	a_{43}		
c_5	a_{51}	a_{52}	a_{53}	a_{54}	
	b_1	b_2	b_3	b_4	b_5

be a Runge–Kutta method of order 5. Then $c_4 = 1$.

Proof.

Let

$$
U = \begin{bmatrix}
\sum_i b_i a_{i2} & \sum_i b_i a_{i3} & \sum_i b_i a_{i4} \\[2mm]
\sum_i b_i a_{i2} c_2 & \sum_i b_i a_{i3} c_3 & \sum_i b_i a_{i4} c_4 \\[2mm]
\sum_{i,j} b_i a_{ij} a_{j2} - \tfrac{1}{2} \sum_i b_i a_{i2}(1 - c_2) & \sum_{i,j} b_i a_{ij} a_{j3} - \tfrac{1}{2} \sum_i b_i a_{i3}(1 - c_3) & \sum_{i,j} b_i a_{ij} a_{j4} - \tfrac{1}{2} \sum_i b_i a_{i4}(1 - c_4)
\end{bmatrix},
$$

$$
V = \begin{bmatrix}
c_2 & c_2^2 & \sum_k a_{2k} c_k - \tfrac{1}{2} c_2^2 \\[2mm]
c_3 & c_3^2 & \sum_k a_{3k} c_k - \tfrac{1}{2} c_3^2 \\[2mm]
c_4 & c_4^2 & \sum_k a_{4k} c_k - \tfrac{1}{2} c_4^2
\end{bmatrix},
$$

so that the product UV can be formed making use of the order conditions for fifth-order methods. It is found that

$$
UV = \begin{bmatrix}
\tfrac{1}{6} & \tfrac{1}{12} & 0 \\[2mm]
\tfrac{1}{12} & \tfrac{1}{20} & 0 \\[2mm]
0 & 0 & 0
\end{bmatrix},
$$

so that, by lemma 311A, $b_5 a_{54}(1 - c_4) = 0$ or $c_2 = 0$. However, we find by considering the tree $t = [_4\tau]_4$ and the corresponding equation $\Phi(t) = 1/\gamma(t)$ that

$$
b_5 a_{54} a_{43} a_{32} c_2 = \tfrac{1}{120},
$$

so that $b_5 a_{54} c_2 \neq 0$. Hence, $c_4 = 1$. ∎

LEMMA 322B.

If the method of lemma 322A is of fifth order, then $c_5 = 1$.

Proof.

Let

$$
= \begin{bmatrix}
b_2(1 - c_2) & b_3(1 - c_3) & b_5(1 - c_5) \\
b_2 c_2(1 - c_2) & b_3 c_3(1 - c_3) & b_5 c_5(1 - c_5) \\[2mm]
\left[\sum_i b_i a_{i2} - b_2(1 - c_2)\right](1 - c_2) & \left[\sum_i b_i a_{i3} - b_3(1 - c_3)\right](1 - c_3) & \left[\sum_i b_i a_{i5} - b_5(1 - c_5)\right](1 - c_5)
\end{bmatrix},
$$

$$V = \begin{bmatrix} c_2 & c_2^2 & \sum_j a_{2j}c_j - \frac{1}{2} c_2^2 \\[2ex] c_3 & c_3^2 & \sum_j a_{3j}c_j - \frac{1}{2} c_3^2 \\[2ex] c_5 & c_5^2 & \sum_j a_{5j}c_j - \frac{1}{2} c_5^2 \end{bmatrix},$$

and compute

$$UV = \begin{bmatrix} \frac{1}{6} & \frac{1}{12} & 0 \\[1ex] \frac{1}{12} & \frac{1}{20} & 0 \\[1ex] 0 & 0 & 0 \end{bmatrix}.$$

Note that in evaluating the elements of UV, the terms with subscript 4 do not occur but their contribution would, in fact, be zero because of the factor $1 - c_4$ which vanishes by lemma 322A.

By lemma 311A, we find $b_5(1 - c_5)^2 = 0$ or $c_2 = 0$. Since, as we have seen in the proof of lemma 322A, $b_5 a_{54} a_{43} a_{32} c_2 \neq 0$, it follows that $c_5 = 1$. ∎

THEOREM 322C.

The method of lemma 322A does not have order 5.

Proof.

If the method were of order 5, it would follow from lemmas 322A and 322B that $c_4 = c_5 = 1$. Now compute $\sum_{i,j,k} b_i(1 - c_i)a_{ij}a_{jk}c_k$. Since $1 - c_i = 0$ for $i = 4$ or 5, any non-vanishing term must have $i \leq 3$, $j \leq 2$, $k = 1$. But $c_1 = 0$ and hence the expression vanishes.

On the other hand,

$$\sum_{i,j,k} b_i(1 - c_i)a_{ij}a_{jk}c_k = \Phi([_3\tau]_3) - \Phi([\tau[_2\tau]_2])$$

$$= \frac{1}{\gamma([_3\tau]_3)} - \frac{1}{\gamma([\tau[_2\tau]_2])}$$

$$= \tfrac{1}{24} - \tfrac{1}{30} \neq 0.$$

This contradication shows that the method cannot have order 5. ∎

323 Impossibility of p-stage pth-order methods with $p \geq 5$.

In this subsection, we generalize the result of the previous subsection. Let

$m = p - 5$ and define $b_i^{(k)}$ $(i = 1, 2, \ldots, s; \ k = 0, 1, \ldots, m)$ by the recursion

$$b_i^{(0)} = b_i, \qquad\qquad i = 1, 2, \ldots, s,$$

$$b_i^{(k)} = \sum_{j=1}^{s} b_j^{(m-1)} a_{ji}, \qquad k = 1, 2, \ldots, m; \ i = 1, 2, \ldots, s,$$

where $a_{21}, a_{31}, \ldots, a_{s,s-1}, b_1, b_2, \ldots, b_s$ are the parameters of an s-stage method

$$
\begin{array}{c|ccccc}
0 & & & & & \\
c_2 & a_{21} & & & & \\
c_3 & a_{31} & a_{32} & & & \\
\vdots & \vdots & \vdots & \ddots & & \\
c_s & a_{s1} & a_{s2} & \cdots & a_{s,s-1} & \\
\hline
& b_1 & b_2 & \cdots & b_{s-1} & b_s
\end{array}
\tag{323a}
$$

It is easy to see that $b_i^{(m)} = 0$ if $i > s - m$. For $t \in T$, define $\Phi^{(m)}(t)$ as $\sum_{i=1}^{s} b_i^{(m)} \chi_i$, where $\chi_1, \chi_2, \ldots, \chi_s$ are such polynomials in $a_{21}, a_{31}, \ldots, a_{s,\,s-1}$ that $\Phi(t) = \sum_{i=1}^{s} b_i \chi_i$. Also define $\gamma^{(m)}(t) = [m + r(t)]! \, \gamma(t)/r(t)!$

LEMMA 323A.

A necessary condition for the Runge–Kutta method (323a) to have order $p = m + 5$ is

$$\Phi^{(m)}(t) = \frac{1}{\gamma^{(m)}(t)} \tag{323b}$$

for all $t \in T$ such that $r(t) \le 5$.

Proof.

In the case $m = 0$, there is nothing to prove. If $m > 0$, let $u = [_m t]_m$ so that $\Phi^{(m)}(t) = \Phi(u), \gamma^{(m)}(t) = \gamma(u)$. Hence, (323b) is implied by $\Phi(u) = 1/\gamma(u)$ where the order of u is $r(u) = m + r(t) \le p$. ∎

THEOREM 323B.

There is no p-stage pth-order method with $p \ge 5$.

Proof.

If there were a method of the form of (323a) with $s = p$ stages, we can apply lemma 311A with

$$
U = \begin{bmatrix}
b_2^{(m+1)} & b_3^{(m+1)} & b_4^{(m+1)} \\
b_2^{(m+1)} c_2 & b_3^{(m+1)} c_3 & b_4^{(m+1)} c_4 \\
b_2^{(m+2)} - \dfrac{1}{m+2} b_2^{(m+1)}(1-c_2) & b_3^{(m+2)} - \dfrac{1}{m+2} b_3^{(m+1)}(1-c_3) & b_4^{(m+2)} - \dfrac{1}{m+2} b_4^{(m+1)}(1-c_4)
\end{bmatrix}
$$

and V as in the proof of lemma 322A. This time it is found that

$$
UV = \begin{bmatrix}
\dfrac{1}{(m+3)!} & \dfrac{2}{(m+4)!} & 0 \\[2mm]
\dfrac{2}{(m+4)!} & \dfrac{6}{(m+5)!} & 0 \\[2mm]
0 & 0 & 0
\end{bmatrix}
$$

and we deduce that $c_4 = 1$ just as in lemma 322A.

Next we apply lemma 311A with

$$
U = \begin{bmatrix}
b_2^{(m)}(1-c_2) & b_3^{(m)}(1-c_3) & b_5^{(m)}(1-c_5) \\
b_2^{(m)} c_2(1-c_2) & b_3^{(m)} c_3(1-c_3) & b_5^{(m)} c_5(1-c_5) \\
\left[b_2^{(m+1)} - \dfrac{1}{m+1} b_2^{(m)}(1-c_2) \right](1-c_2) & \left[b_3^{(m+1)} - \dfrac{1}{m+1} b_3^{(m)}(1-c_3) \right](1-c_3) & \left[b_5^{(m+1)} - \dfrac{1}{m+1} b_5^{(m)}(1-c_5) \right](1-c_5)
\end{bmatrix}
$$

and V as in the proof of lemma 322B.

We find that

$$
UV = \begin{bmatrix}
\dfrac{m+1}{(m+3)!} & \dfrac{2(m+1)}{(m+4)!} & 0 \\[2mm]
\dfrac{2(m+1)}{(m+4)!} & \dfrac{6(m+1)}{(m+5)!} & 0 \\[2mm]
0 & 0 & 0
\end{bmatrix}
$$

and we deduce that $c_5 = 1$ just as in lemma 322B.

Finally, we see that $\Sigma_{i,j,k} \, b_i^{(m)}(1-c_i) a_{ij} a_{jk} c_k$ vanishes because $b_i^{(m)} = 0$ if $i > 5$, $1 - c_i = 0$ if $i = 4$ or 5 and $a_{ij} a_{jk} c_k = 0$ if $i < 4$. On the other hand, since the method is of order $m + 5$,

$$
\Sigma_{i,j,k} \, b_i^{(m)}(1-c_i) a_{ij} a_{jk} c_k = \Phi([_{m+3}\tau]_{m+3}) - \Phi([_{m+1}\tau[_2\tau]_2]_{m+1})
$$

$$
= \frac{1}{\gamma([_{m+3}\tau]_{m+3})} - \frac{1}{\gamma([_{m+1}\tau[_2\tau]_2]_{m+1})}
$$

$$
\neq 0.
$$

This contradiction completes the proof. ∎

324 A sequence of bounds.

In an effort to understand further the relationship between the attainable order and the number of stages, an investigation was undertaken in Butcher (1975) aimed at generalizing theorem 323B, and we report this without proof here as theorem 324A. Let the sequence $u_0, u_1, u_2, \ldots$ be defined by

$$u_0 = 5, \tag{324a}$$

$$u_n = [(4u_{n-1} + 2n + 1)/3], \qquad n = 1, 2, 3, \ldots, \tag{324b}$$

where $[\cdot]$ denotes the integer part, so that

$$u_0 = 5, \qquad u_1 = 7, \qquad u_2 = 11, \qquad u_3 = 17, \qquad u_4 = 25.$$

It is known from theorem 321A that obtaining order $p \geq u_0$ is not possible unless $s > p + 0$ and from theorem 323B that for order $p \geq u_1$ it is necessary that $s > p + 1$. In the result which we now state, these statements are generalized.

THEOREM 324A.

Let the sequence $u_0, u_1, \ldots$ be defined by (324a) and (324b). Then for $p \geq u_n$ $(n = 0, 1, 2, \ldots)$ there does not exist an explicit s-stage method with order p unless $s > p + n$.

325 A lower bound.

In the preceding parts of this section we have been considering upper bounds on the order that could be obtained with various numbers of stages. We now look at the opposite question where a lower bound is sought. It will be convenient to look at this question in another way. That is, given an order p, we ask for a choice of s so that we can guarantee that a method of the required order exists with this many stages. The following theorem is given in Verner (1978) and we refer the reader there for the proof.

THEOREM 325A.

For given p there exists a method of order p with s stages where

$$s = \frac{p^2 - 7p + 20}{2}.$$

Futhermore, if $p \geq 10$ the method exists with the required order with

$$s = \frac{p^2 - 7p + 10}{2}$$

stages.

Using a different approach, it is possible to lower the coefficient of p^2 slightly, and we will show how this can be done below. What seems to be more difficult is to obtain a formula for s which grows less than quadratically.

To obtain our family of methods, we will need to foreshadow results from section 34, where it is shown that implicit Runge–Kutta methods exist of the forms:

Radau I method:

0	0	0	$\cdots$	0
c_2	a_{21}	a_{22}	$\cdots$	a_{2m}
c_3	a_{31}	a_{32}	$\cdots$	a_{3m}
$\vdots$	$\vdots$	$\vdots$		$\vdots$
c_m	a_{m1}	a_{m2}	$\cdots$	a_{mm}
	b_1	b_2	$\ldots$	b_m

Lobatto method:

0	0	0	$\cdots$	0	0
c_2	a_{21}	a_{22}	$\cdots$	a_{2m}	0
c_3	a_{31}	a_{32}	$\cdots$	a_{3m}	0
$\vdots$	$\vdots$	$\vdots$		$\vdots$	$\vdots$
c_m	a_{m1}	a_{m2}	$\cdots$	a_{mm}	0
1	$a_{m+1,1}$	$a_{m+1,2}$	$\cdots$	$a_{m+1,m}$	0
	b_1	b_2	$\cdots$	b_m	b_{m+1}

where $c_2, c_3, \ldots, c_m$ are distinct numbers lying in $(0, 1)$ which are zeros of $(d/dx)^{m-1}[x^m(x-1)^{m-1}]$ for the Radau I method and of $(d/dx)^{m-1}[x^m(x-1)^m]$ for the Lobatto method, and where the orders are $2m-1$ and $2m$ respectively.

Let $Y_1, Y_2, \ldots, Y_m$ (and Y_{m+1} in the Lobatto case) denote the results computed by the stages of these methods. It is known that, if $y_{n-1} = y(x_{n-1})$, then

$$Y_i = y(x_{n-1} + hc_i) + O(h^{m+1}), \qquad i = 1, 2, \ldots, m(, m+1).$$

We now describe the various steps in our explicit method built around obtaining approximations to $Y_1, Y_2, \ldots, Y_m(, Y_{m+1})$:

Step 1: The first stage computes Y_1 just as for the implicit method.

Step 2: For $i = 2, 3, \ldots, m$ and $j = 2, 3, \ldots, i$ construct $Y_j^{(i)}$ as an approximation to Y_j using the formula

$$Y_j^{(i)} = y_{n-1} + h \sum_{k=1}^{i-1} \alpha_k f(Y_k^{(i-1)}),$$

where, for each i, $Y_1^{(i)}$ is identical with Y_1. The numbers $\alpha_1, \alpha_2, \ldots, \alpha_{i-1}$ are chosen to satisfy the equations

$$\sum_{k=1}^{i-1} \alpha_k c_k^{l-1} = \frac{c_j^l}{l}, \qquad l = 1, 2, \ldots, i-1.$$

It is easy to see that

$$Y_j^{(i)} = y(x_{n-1} + hc_j) + O(h^i)$$
$$= Y_j + O(h^i),$$

for each i and j pair.

Step 3: For $i = m+1, m+2, \ldots, 2m-1$ and $j = 2, 3, \ldots, m$ construct $Y_j^{(i)}$ using the formula

$$Y_j^{(i)} = y_{n-1} + h \sum_{k=1}^m a_{jk} f(Y_k^{(i-1)}),$$

where $a_{21}, a_{22}, \ldots, a_{mm}$ are as in the Radau I or Lobatto methods. It is easy to see that

$$Y_j^{(i)} = Y_j + O(h^i)$$

so that

$$Y_j^{(2m-1)} = Y_j + O(h^{2m-1}).$$

Step 4: In the case of the Lobatto method construct a stage $\bar{Y}_{m+1}$ approximating Y_{m+1} using the formula

$$\bar{Y}_{m+1} = y_{n-1} + \sum_{k=1}^m a_{m+1,k} \, f(Y_k^{(2m-1)}).$$

Step 5: Compute the final result by

$$y_n = y_{n-1} + h \sum_{j=1}^m b_j f(Y_j^{(2m-1)}),$$

with the addition of the further term $hb_{m+1} f(\bar{Y}_{m+1})$ in the Lobatto case.

It is easy to see that the order is $2m-1$ (Radau I case) or $2m$ (Lobatto case), and by adding the number of stages in steps 1, 2, 3 (and 4 in the Lobatto case), we find a total of $1 + m(m-1)/2 + (m-1)^2$ ($+1$ in the Lobatto case). This comes to $(3m^2 - 5m + 4)/2$ (or $(3m^2 - 5m + 6)/2$ in the Lobatto case). Thus, if p is odd we can substitute $m = (p+1)/2$ in the total for the Radau I approximation and if p is even we can substitute $m = p/2$ in the Lobatto approximation to obtain the results of the following theorem which we state without a formal proof.

THEOREM 325B.

For given p there exists a method of order p with s stages where

$$s = \begin{cases} \dfrac{3p^2 - 10p + 24}{8}, & p \text{ even,} \\[2em] \dfrac{3p^2 - 4p + 9}{8}, & p \text{ odd.} \end{cases}$$

326 Summary of known results.

In addition to the results we have presented, there are two further sources of information on the attainable order of methods. In the first place, methods to be discussed in section 33 take orders up to 10. This does not include a method of order 9 in its own right and although we can regard the tenth-order method of Hairer (1978) as being such a method we should assume with confidence that ninth order can be achieved with fewer than the 17 stages of the Hairer method.

The second source of information is a recent result (Butcher [1985]) that order p is not possible with $p + 2$ stages for $p \geq 8$. Besides giving the exact result for order 8 this also gives lower bounds for 9 and 10.

Table 326 summarizes the situation up to order 10.

Table 326   Minimum number of stages for various orders.

| Order | Minimum number of stage | |
	Lower bound	Upper bound
1	1	
2	2	
3	3	
4	4	
5	6	
6	7	
7	9	
8	11	
9	12	17
10	13	17

33   HIGH-ORDER EXPLICIT METHODS

330   Row-simplifying assumptions.

In this section we will consider methods of order five or more and we introduce the discussion by writing down the order conditions corresponding to the three trees $[[\tau]^2]$, $[[\tau]\tau^2]$ and $[\tau^4]$. These conditions are, respectively,

$$\sum_i b_i \left(\sum_j a_{ij}c_j \right)^2 = \tfrac{1}{20}, \tag{330a}$$

$$\sum_i b_i \left(\sum_j a_{ij}c_j \right) c_i^2 = \tfrac{1}{10}, \tag{330b}$$

$$\sum_i b_i c_i^4 = \tfrac{1}{5}. \tag{330c}$$

If (330c) is multiplied by $\tfrac{1}{4}$ and the result added to (330a) minus (330b) it is found that

$$\sum_i b_i \left(\sum_j a_{ij}c_j - \tfrac{1}{2} c_i^2 \right)^2 = 0, \tag{330d}$$

which implies that at least one of $b_2, b_3, \ldots$ is negative or else that, for each i, either $\sum_j a_{ij}c_j = \tfrac{1}{2} c_i^2$ or $b_i = 0$.

Negative values of $b_1, b_2, \ldots$ are computationally undesirable, since they lead to coarse bounds on the rounding error produced in a single step, so we consider the alternative that $b_i(\sum_j a_{ij}c_j - \tfrac{1}{2} c_i^2) = 0$ for all i. The second factor cannot vanish for $i = 2$, since this would imply $-\tfrac{1}{2} c_2^2 = 0$ which could only hold if Y_2 and Y_1 were identically equal and we could then find an equivalent method with one less stage. We will consider the possibility that

$$\sum_j a_{ij}c_j = \tfrac{1}{2} c_i^2, \qquad i \neq 2, \tag{330e}$$

$$b_2 = 0. \tag{330f}$$

We shall refer to the pair (330e), (330f) as a row-simplifying assumption. If these conditions are satisfied, then we can immediately see that (330a) and (330b) hold if and only if (330c) holds. Certain other pairs of order conditions also have the same type of relationship and we list these here together with the corresponding trees

$[_2\tau]_2$:	$\sum b_i a_{ij}c_j = \tfrac{1}{6}$,		$[\tau^2]$:	$\sum b_i c_i^2 = \tfrac{1}{3}$,	(330g)
$[[\tau]\tau]$:	$\sum b_i c_i a_{ij}c_j = \tfrac{1}{8}$,		$[\tau^3]$:	$\sum b_i c_i^3 = \tfrac{1}{4}$,	(330h)
$[_3\tau]_3$:	$\sum b_i a_{ij}a_{jk}c_k = \tfrac{1}{24}$,		$[_2\tau^2]_2$:	$\sum b_i a_{ij}c_j^2 = \tfrac{1}{12}$,	(330i)
$[_2[\tau]\tau]_2$:	$\sum b_i a_{ij}c_j a_{jk}c_k = \tfrac{1}{40}$,		$[_2\tau^3]_2$:	$\sum b_i a_{ij}c_j^3 = \tfrac{1}{20}$,	(330j)
$[[_2\tau]_2\tau]$:	$\sum b_i c_i a_{ij}a_{jk}c_k = \tfrac{1}{30}$,		$[[\tau^2]\tau]$:	$\sum b_i c_i a_{ij}c_j^2 = \tfrac{1}{15}$,	(330k)
$[_4\tau]_4$:	$\sum b_i a_{ij}a_{jk}a_{kl}c_l = \tfrac{1}{120}$,		$[_3\tau^2]_3$:	$\sum b_i a_{ij}a_{jk}c_k^2 = \tfrac{1}{60}$.	(330l)

We have restricted ourselves to trees of order not exceeding five but, of course, other pairs of conditions exist for higher orders. From the form of these condition pairs, we immediately see that if (330e) held for all i, including $i = 2$, then the equations in the pairs (330g) to (330l) are equivalent. Since the $i = 2$ case of (330e) is excluded, but (330f) holds, the pairs in (330g) and (330h) are still equivalent whereas those in (330i) and (330j) are equivalent iff it also holds that

$$\Sigma \, b_i a_{i2} = 0, \qquad (330\text{m})$$

and the pairs in (330k) and (330l) are equivalent iff

$$\Sigma \, b_i c_i a_{i2} = 0 \qquad (330\text{n})$$

and

$$\Sigma \, b_i a_{ij} a_{j2} = 0. \qquad (330\text{o})$$

It could well be asked what is gained by imposing these simplifying assumptions. By assuming (330e) ($s - 2$ linear constraints on the rows of the A matrix), (330f), (330m), (330n) and (330o), a total of $s + 2$ algebraic conditions, we have, in effect, removed 8 of the 17 trees of order less than or equal to five from consideration when specifying the restraints on a method so that its order is five. For higher orders than five, the simplification brought about by assuming (330e) together with appropriate subsidiary conditions is even greater and we will invariably make these assumptions.

The equation $\Sigma_j a_{ij} c_j = \frac{1}{2} c_i^2$ is actually the first member of a sequence of such assumptions

$$\sum_j a_{ij} c_j^{k-1} = \frac{1}{k} c_i^k, \qquad k = 2, 3, 4, \ldots, \qquad (330\text{p})$$

which can be made to hold except for some low values of i. These, together with subsidiary conditions, are also used as row-simplifying assumptions. To appreciate the reason that (330p) plays an important role in the simplification of Runge–Kutta order conditions, consider the trees t_1 on the left and t_2 on the right in figure 330. These two trees are supposed to be identical except that, in t_1, $k - 1$ terminal vertices branch from the vertex labelled j and, in t_2, k terminal vertices branch from the vertex labelled i. For these two trees $\Phi(t_1), \Phi(t_2)$ take the forms

$$\Phi(t_1) = \Sigma \cdots a_{ij} c_j^{k-1} \cdots, \qquad \Phi(t_2) = \Sigma \cdots c_i^k \cdots,$$

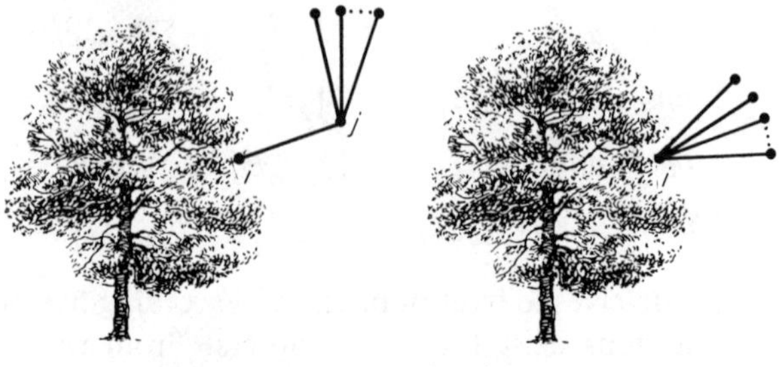

Figure 330 Two related trees.

where the dots represent identical factors for $\Phi(t_1)$ and $\Phi(t_2)$. Furthermore, in computing $\gamma(t_1), \gamma(t_2)$ using the rule given in subsection 144, we find that

$$\gamma(t_1) = 1^{k-1} \cdot k \cdots, \qquad \gamma(t_2) = 1^k \cdots,$$

where the dots indicate identical numerical factors, so that $\gamma(t_1) = k\gamma(t_2)$. Thus, if (330p) holds, then $\Phi(t_1) = 1/\gamma(t_1)$ iff $\Phi(t_2) = 1/\gamma(t_2)$.

As we have already remarked, (330p) cannot actually hold for all values of i, at least in explicit methods, and subsidiary conditions are necessary.

331 Column-simplifying assumptions.

We saw in subsection 312 that the linear condition on the columns of the A matrix

$$\sum_i b_i a_{ij} = b_j(1 - c_j), \qquad j = 1, 2, \ldots, s, \tag{331a}$$

had the effect of removing certain trees from consideration when specifying the order conditions. Specifically, the trees that can be so disregarded are of the form $t = [u]$, where u is a tree. That is to say, the tree t has only a single arc branching from the root. We shall call (331a) a column-simplifying assumption.

In the same way, if it were possible for it to hold that

$$\sum_i b_i c_i^{k-1} a_{ij} = \frac{1}{k} b_j(1 - c_j^k), \tag{331b}$$

for all $j = 1, 2, \ldots, s$ with $k > 1$, then this would mean that trees of the form $t = [\tau^{k-1}u]$ could be removed from consideration also. To see why this is the case, let $u = [v_1 v_2 \ldots]$, where $v_1, v_2, \ldots$ are trees and define $t_1 = u, t_2 = [\tau^k v_1 v_2 \cdots]$. The three equations corresponding to t, t_1, t_2 then become

$$\Phi(t) = \sum b_i c_i^{k-1} a_{ij} \cdots = \frac{1}{r(t)[r(t) - k] \cdots} = \frac{1}{\gamma(t)},$$

$$\Phi(t_1) = \sum b_j \cdots = \frac{1}{[r(t) - k] \cdots} = \frac{1}{\gamma(t_1)},$$

$$\Phi(t_2) = \sum b_j c_j^k \ldots = \frac{1}{r(t) \cdots} = \frac{1}{\gamma(t_2)},$$

so that (331b) implies that $\Phi(t) = (1/k)[\Phi(t_1) - \Phi(t_2)]$ and since $1/\gamma(t) = (1/k)[1/\gamma(t_1) - 1/\gamma(t_2)]$, it follows that $\Phi(t_1) = 1/\gamma(t_1)$ and $\Phi(t_2) = 1/\gamma(t_2)$ together imply $\Phi(t) = 1/\gamma(t)$. Hence, t can be removed completely from further consideration.

For explicit methods, it is not possible to satisfy (331b) with $k = 2$ as well as $k = 1$ and obtain an order greater than unity. To see this, we take the difference of these two instances of (331b) and we find that

$$\sum_{i>j} b_i(1 - c_i)\, a_{ij} = \tfrac{1}{2} b_j(1 - c_j)^2,$$

implying (by induction on $s - j$) that $b_i(1 - c_j) = 0$ for all j and hence that $\sum b_j c_j = \sum b_j$, which would give the contradiction $\tfrac{1}{2} = 1$ if the order of the method were two.

Even though the general case of (331b) will not play any important role in the development of explicit methods, the special case (331a) plays a part of central importance in methods of high order.

There is an interesting relationship between row- and column- simplifying assumptions which we examine in the particular case that (330e), (330f) and (331a) all hold. Sum (331a) for $j = 1, 2, \ldots, s$ and also sum the product of (331a) and c_j over these same j values. In each case, make use of (330e) and (330f) and we find that

$$\sum b_i = 2 \sum b_i c_i = 3 \sum b_i c_i^2, \tag{331c}$$

so that consistency ($\sum b_i = 1$) would imply an order of accuracy of at least three.

We will use (331a) as a means of determining a_{sj} ($j = 1, 2, \ldots, s - 1$) once other elements of the A matrix have been found. Since $b_1, b_2, \ldots, b_s$ will have been selected in accordance with (331c), the row-simplifying assumption on row s will automatically hold.

Another interconnection between row- and column-simplifying assumptions is that if (331a) holds, then the subsidiary conditions (330m) and (330o) can be omitted as they hold automatically. We will also, for convenience, replace (330n) by the difference of (330m) and (330n), that is

$$\sum b_i(1 - c_i)\, a_{i2} = 0. \tag{331d}$$

332   Fifth-order methods.

From the remarks made in the previous two subsections, we see that a six-stage method of order 5 can be found if we can satisfy (330e), (330f), (331a), (331d) together with the remaining order conditions

$$\sum b_i c_i^{k-1} = \frac{1}{k}, \qquad k = 1, 2, 3, 4, 5, \tag{332a}$$

$$\sum b_i c_i a_{ij} c_j^2 = \tfrac{1}{15}, \tag{332b}$$

but, as we have remarked, there is a relationship between (332a) with $k = 1, 2, 3$ and the simplifying assumptions.

For convenience, we will replace (332b) by the following linear combination of this equation and several that are automatically satisfied by (332a) and the simplifying assumptions:

$$\sum b_i(1 - c_i)\, a_{ij}c_j(c_j - c_3) = \frac{1}{60} - \frac{c_3}{24} \qquad (332c)$$

Since $c_6 = 1$, by (331a), the left-hand side reduces to the single term of the following equivalent equation:

$$b_5(1 - c_5)\, a_{54}c_4(c_4 - c_3) = \frac{1}{60} - \frac{c_3}{24}. \qquad (332d)$$

The steps that are now taken to determine the method completely are:

Step 1: Select c_2, c_3, c_4, c_5 arbitrarily and $c_6 = 1$ (certain choices of these parameters have to be excluded if they cause any later steps to be impossible because of zero divisors).

Step 2: Select b_1, b_3, b_4, b_5, b_6 with $b_2 = 0$ to satisfy the linear system (332a).

Step 3: Select a_{54} to satisfy (332d).

Step 4: Select a_{43} arbitrarily.

Step 5: Select a_{32}, a_{42} to satisfy (330e) with $i = 3, 4$.

Step 6: Select a_{52} to satisfy (331d).

Step 7: Select a_{53} to satisfy (330e) with $i = 5$.

Step 8: Select $a_{62}, a_{63}, a_{64}, a_{65}$ to satisfy (331a) with $j = 2, 3, 4, 5$.

Step 9: Select $a_{21}, a_{31}, a_{41}, a_{51}, a_{61}$ to satisfy the condition $\sum_j a_{ij} = c_i$, $i = 2, 3, 4, 5, 6$.

For example, choosing $(c_3, c_4, c_5) = (\frac{1}{4}, \frac{1}{2}, \frac{3}{4})$, $c_2 = u\ (\neq 0)$, $a_{43} = v$, and carrying out this programme we find the method:

0						
u	u					
$\dfrac{1}{4}$	$\dfrac{1}{4} - \dfrac{1}{32u}$	$\dfrac{1}{32u}$				
$\dfrac{1}{2}$	$\left(\dfrac{1}{2} - \dfrac{1}{8u}\right)(1 - 2v)$	$\dfrac{1}{8u}(1 - 2v)$	v			
$\dfrac{3}{4}$	$\dfrac{3}{16}\left(\dfrac{1-v}{u} - 1\right)$	$-\dfrac{3}{8}(1 - 2v) - \dfrac{3}{16u}(1 - v)$	$\dfrac{3}{4}(1 - v)$	$\dfrac{9}{16}$		
1	$\dfrac{11 - 12v}{7} - \dfrac{7 - 6v}{14u}$	$\dfrac{7 - 6v}{14u}$	$\dfrac{12v}{7}$	$\dfrac{-12}{7}$	$\dfrac{8}{7}$	
	$\dfrac{7}{90}$	0	$\dfrac{32}{90}$	$\dfrac{12}{90}$	$\dfrac{32}{90}$	$\dfrac{7}{90}$

Selecting the values $u = \frac{2}{5}$ and $v = \frac{1}{2}$ this becomes:

0					
$\dfrac{2}{5}$	$\dfrac{2}{5}$				
$\dfrac{1}{4}$	$\dfrac{11}{64}$	$\dfrac{5}{64}$			
$\dfrac{1}{2}$	0	0	$\dfrac{1}{2}$		
$\dfrac{3}{4}$	$\dfrac{3}{64}$	$\dfrac{-15}{64}$	$\dfrac{3}{8}$	$\dfrac{9}{16}$	
1	0	$\dfrac{5}{7}$	$\dfrac{6}{7}$	$\dfrac{-12}{7}$	$\dfrac{8}{7}$
	$\dfrac{7}{90}$	0	$\dfrac{32}{90}$	$\dfrac{12}{90}$	$\dfrac{32}{90}$ $\dfrac{7}{90}$

$$(332\mathrm{e})$$

333 A generalization of fifth-order methods.

In this subsection we consider the problem of deriving methods with six stages such that row- but not column-simplifying assumptions hold. Also, for a reason that will become clear in the next subsection, we will generalize the order conditions for a tree t of order $r(t) \leq 5$ to the following

$$\Phi(t) = \frac{r(t)!}{\gamma(t)[r(t) + m]!}, \tag{333a}$$

where m is a non-negative integer. In the case $m = 0$, we arrive at the usual order conditions.

On the assumption that (330e), (330f), (330m), (330n), (330o) all hold, the remaining order conditions are

$$\sum b_i c_i^{k-1} = \frac{(k-1)!}{(k+m)!}, \qquad k = 1, 2, 3, 4, 5, \tag{333b}$$

from which it follows that

$$b_5(c_6 - c_5)\, c_5(c_5 - c_3)(c_5 - c_4) = -\frac{24}{(m+5)!} + \frac{6(c_3 + c_4 + c_6)}{(m+4)!}$$

$$-\frac{2(c_3c_4 + c_4c_6 + c_6c_3)}{(m+3)!} + \frac{c_3c_4c_6}{(m+2)!}, \tag{333c}$$

together with the conditions

$$\sum_{i=5,6} \sum_{j=4,5} b_i a_{ij} c_j (c_j - c_3) = \frac{2}{(m+4)!} - \frac{c_3}{(m+3)!}, \tag{333d}$$

$$b_6 a_{65} c_5 (c_5 - c_3)(c_5 - c_4) = \frac{6}{(m+5)!} - \frac{2(c_3 + c_4)}{(m+4)!} + \frac{c_3 c_4}{(m+3)!}, \tag{333e}$$

$$b_5 (c_6 - c_5) a_{54} c_4 (c_4 - c_3) = \frac{2c_6}{(m+4)!} - \frac{8}{(m+5)!} - \frac{c_3 c_6}{(m+3)!} + \frac{3c_3}{(m+4)!}, \tag{333f}$$

$$b_6 a_{65} a_{54} c_4 (c_4 - c_3) = \frac{2}{(m+5)!} - \frac{c_3}{(m+4)!}. \tag{333g}$$

If the right-hand sides of (333c), (333e), (333f) and (333g) are denoted by C, E, F and G then for consistency of these equations, we must have $CG = EF$ which, after some simplification, leads to the conclusion that $c_6 = 1$ or

$$c_4 = \frac{2c_3}{4 - 4(m+4)c_3 + (m+4)(m+5)c_3^2}. \tag{333h}$$

We will reject the possibility that $c_6 = 1$ since this leads to the column-simplifying assumption when $m = 0$, or to a slightly altered assumption when $m > 0$.

To determine the coefficients of a method of the style considered here, we carry out the following steps:

Step 1: Select c_2, c_3, c_5, c_6 arbitrarily (under the usual proviso that certain special values have to be excluded).

Step 2: Select c_4 by (333h).

Step 3: Select $b_1, b_2, b_3, b_4, b_5, b_6$ with $b_2 = 0$ to satisfy the linear system (333b).

Step 4: Solve for a_{54} and a_{65} from the self-consistent equations (333e), (333f) and (333g). (Note that we cannot necessarily use just (333e) and (333f) since it is possible that $b_5 = 0$.)

Step 5: Solve for a_{64} from (333d).

Step 6: Set a_{43} equal to a parameter λ and solve for a_{32} and a_{42} by (330e).

Step 7: Solve for a_{52} and a_{62} from the linear system (330m) and (330n).

Step 8: Solve for a_{53}, a_{63} from (330e). Note that $a_{42}, a_{43}, a_{52}, a_{53}, a_{62}, a_{63}$ are all linear functions of λ.

Step 9: Use (330o), for which the left-hand side is a linear function of λ, to solve for λ and substitute this value into $a_{42}, a_{43}, a_{52}, a_{53}, a_{62}$ and a_{63}.

We give four examples of methods derived under this programme, two with

$m = 0$ and two with $m = 1$:

$m = 0$:

0					
$\dfrac{2}{5}$	$\dfrac{2}{5}$				
$\dfrac{4}{5}$	0	$\dfrac{4}{5}$			
$\dfrac{2}{5}$	$\dfrac{1}{10}$	$\dfrac{2}{5}$	$\dfrac{-1}{10}$		
$\dfrac{1}{5}$	$\dfrac{3}{10}$	-1	$\dfrac{3}{20}$	$\dfrac{3}{4}$	
$\dfrac{3}{5}$	$\dfrac{-3}{70}$	$\dfrac{9}{5}$	$\dfrac{-47}{140}$	$\dfrac{-15}{28}$	$\dfrac{-2}{7}$
	$\dfrac{19}{144}$	0	$\dfrac{85}{144}$	$\dfrac{5}{6}$	$\dfrac{-5}{72}$ $\quad \dfrac{-35}{72}$

$m = 0$:

0					
$\dfrac{6-\sqrt{6}}{10}$	$\dfrac{6-\sqrt{6}}{10}$				
$\dfrac{6+\sqrt{6}}{10}$	$\dfrac{6-9\sqrt{6}}{100}$	$\dfrac{54+19\sqrt{6}}{100}$			
$\dfrac{6-\sqrt{6}}{10}$	$\dfrac{-282+127\sqrt{6}}{500}$	$\dfrac{-18+23\sqrt{6}}{100}$	$\dfrac{168-73\sqrt{6}}{125}$		
$\dfrac{6+\sqrt{6}}{10}$	$\dfrac{36-4\sqrt{6}}{300}$	0	$\dfrac{24-\sqrt{6}}{120}$	$\dfrac{168+73\sqrt{6}}{600}$	
$\dfrac{6-\sqrt{6}}{10}$	$\dfrac{6+9\sqrt{6}}{100}$	0	$\dfrac{6504-4969\sqrt{6}}{15\,000}$	$\dfrac{24+\sqrt{6}}{120}$	$\dfrac{-96+131\sqrt{6}}{625}$
	$\dfrac{1}{9}$	$0 \qquad 0$	0	$\dfrac{16-\sqrt{6}}{36}$	$\dfrac{16+\sqrt{6}}{36}$

$m = 1$:

0					
$\dfrac{1}{3}$	$\dfrac{1}{3}$				
$\dfrac{2}{3}$	0	$\dfrac{2}{3}$			
$\dfrac{1}{3}$	$\dfrac{1}{12}$	$\dfrac{1}{3}$	$\dfrac{-1}{12}$		
$\dfrac{5}{6}$	$\dfrac{25}{48}$	$\dfrac{-55}{24}$	$\dfrac{35}{48}$	$\dfrac{15}{8}$	
$\dfrac{1}{6}$	$\dfrac{3}{20}$	$\dfrac{-11}{24}$	$\dfrac{-1}{8}$	$\dfrac{1}{2}$	$\dfrac{1}{10}$
	$\dfrac{13}{200}$	0	$\dfrac{11}{120}$	$\dfrac{11}{60}$	$\dfrac{2}{75}$ $\dfrac{2}{15}$

$m = 1$:

0					
$\dfrac{5-\sqrt{5}}{10}$	$\dfrac{5-\sqrt{5}}{10}$				
$\dfrac{5+\sqrt{5}}{10}$	$\dfrac{-\sqrt{5}}{10}$	$\dfrac{5+2\sqrt{5}}{10}$			
$\dfrac{5-\sqrt{5}}{10}$	$\dfrac{-15+7\sqrt{5}}{20}$	$\dfrac{-1+\sqrt{5}}{4}$	$\dfrac{15-7\sqrt{5}}{10}$		
$\dfrac{5+\sqrt{5}}{10}$	$\dfrac{5-\sqrt{5}}{60}$	0	$\dfrac{1}{6}$	$\dfrac{15+7\sqrt{5}}{60}$	
$\dfrac{5-\sqrt{5}}{10}$	$\dfrac{5+\sqrt{5}}{60}$	0	$\dfrac{9-5\sqrt{5}}{12}$	$\dfrac{1}{6}$	$\dfrac{-5+3\sqrt{5}}{10}$
	$\dfrac{1}{12}$	0	0	0	$\dfrac{5-\sqrt{5}}{24}$ $\dfrac{5+\sqrt{5}}{24}$

Note that the first of each pair of methods is chosen for simple rational values of c_2, c_3, c_4, c_5, c_6, while the second of each pair of methods is chosen so that only two distinct values occur among c_2, c_3, c_4, c_5, c_6 and so that $b_3 = b_4 = 0$.

334 Sixth-order methods.

Methods of this order require seven stages. Suppose we had such a method and we formed the vector of numbers

$$b_i^{(1)} = \Sigma_{j=1}^{7} b_j a_{ji}, \quad \text{for } i = 1, 2, \ldots, 6. \tag{334a}$$

By lemma 323A, the method

0				
c_2	a_{21}			
c_3	a_{31}	a_{32}		
$\vdots$	$\vdots$	$\vdots$	$\ddots$	
c_6	a_{61}	a_{62}	$\cdots$	a_{65}
	$b_1^{(1)}$	$b_2^{(1)}$	$\cdots b_5^{(1)}$	$b_6^{(1)}$

satisfies the modified order conditions which correspond to $m = 1$ in the sense of subsection 333. On the other hand, given a method of generalized order 5 (with $m = 1$) it is sometimes possible to reverse the procedure and find a method of order 6. Specifically, we have the following result.

THEOREM 334A.

If a standard six-stage Runge–Kutta method is such that $c_i \neq 1$ ($i = 2, 3, 4, 5, 6$), such that (330e) and (330f) hold and such that its generalized order ($m = 1$) is five, then the method

0					
c_2	a_{21}				
c_3	a_{31}	a_{32}			
$\vdots$	$\vdots$	$\vdots$	$\ddots$		
c_6	a_{61}	a_{62} $\cdots$	a_{65}		
1	a_{71}	a_{72} $\cdots$	a_{75}	a_{76}	
	$\bar{b}_1$	0 $\cdots \bar{b}_5$	$\bar{b}_6$	$\bar{b}_7$	

where $\bar{b}_i = b_i/(1 - c_i)$ $(i = 1, 2, \ldots, 6)$, $\bar{b}_7 = 1 - \bar{b}_1 - \bar{b}_3 - \cdots - \bar{b}_6 \neq 0$ and

$$a_{7i} = \frac{1}{\bar{b}_7}\left[\bar{b}_i(1 - c_i) - \sum_{j<7} \bar{b}_j a_{ji}\right], \tag{334b}$$

is or order 6.

Proof.

By (334b), the column-simplifying assumptions hold. Also, by the generalized ($m = 1$)-order conditions for the original fifth-order method, the order condition for (334a) holds for all trees of the form $t = [u]$ where u is a tree.

By the row-simplifying assumption, we can disregard the tree $[[\tau]^2]$ so that there remain only trees of the form $[\tau^n u]$ to consider. We will show that $\Phi(t) = 1/\gamma(t)$ by induction on n. The case $n = 0$ has already been considered so for $n > 0$ we have, using the quantities $\Phi_i(u)$ introduced in subsection 304,

$$\Phi(t) = \Sigma \, \bar{b}_i \, c_i^n \, \Phi_i(u)$$

$$= \Sigma \bar{b}_i c_i^{n-1} \Phi_i(u) - \Sigma \bar{b}_i (1 - c_i) \, c_i^{n-1} \Phi_i(u)$$

$$= \Sigma \bar{b}_i c_i^{n-1} \Phi_i(u) - \Sigma b_i c_i^{n-1} \Phi_i(u)$$

$$= \frac{r(t)}{[r(t) - 1] \, \gamma(t)} - \frac{1}{[r(t) - 1] \, \gamma(t)}$$

$$= \frac{1}{\gamma(t)}. \quad \blacksquare$$

Using this theorem, we can readily derive sixth-order methods from the $m = 1$ methods of subsection 333. The last row of the A matrix, together with the b^{T} vector for these two methods are respectively

$$
\begin{array}{c|cccccc}
1 & \dfrac{-261}{260} & \dfrac{33}{13} & \dfrac{43}{156} & \dfrac{-118}{39} & \dfrac{32}{195} & \dfrac{80}{39} \\
\hline
 & \dfrac{13}{200} & 0 & \dfrac{11}{40} & \dfrac{11}{40} & \dfrac{4}{25} & \dfrac{4}{25} & \dfrac{13}{200}
\end{array}
$$

and

$$
\begin{array}{c|cccccc}
1 & \dfrac{1}{6} & 0 & \dfrac{-55 + 25\sqrt{5}}{12} & \dfrac{-25 - 7\sqrt{5}}{12} & 5 - 2\sqrt{5} & \dfrac{5 + \sqrt{5}}{2} \\
\hline
 & \dfrac{1}{12} & 0 & 0 & 0 & \dfrac{5}{12} & \dfrac{5}{12} & \dfrac{1}{12}
\end{array}
$$

335 Methods of orders seven or more.

In this subsection we will present the derivation of a method of order 7 with 9 stages. Known methods of order 8 have 11 stages and it is now known that this many stages are necessary. We will present a method of this order due to Cooper and Verner (1972); a further method of order 8 has been derived by Curtis (1970). A method also due to Curtis (1975) with order 10 has 18 stages and one of this order with only 17 stages has been found by Hairer (1978).

In deriving a seventh-order method with nine stages, we assume that

$$3c_2 = \tfrac{3}{2}c_3 = c_4 = c_8 = \frac{7 + \sqrt{21}}{14}, \quad c_5 = c_7 = \tfrac{1}{2}, \quad c_6 = \frac{7 - \sqrt{21}}{14} \quad c_9 = 1, \quad (335a)$$

$$b_2 = b_3 = b_4 = b_5 = 0, \tag{335b}$$

$$\sum b_i c_i^{k-1} = \frac{1}{k}, \qquad k = 1, 2, \ldots, 5, \tag{335c}$$

$$\sum_i b_i a_{ij} = b_j(1 - c_j), \qquad j = 1, 2, \ldots, 9, \tag{335d}$$

$$\sum_j a_{ij} c_j = \tfrac{1}{2} c_i^2, \qquad i \neq 2, \tag{335e}$$

$$\sum_j a_{ij} c_j^2 = \tfrac{1}{3} c_i^3, \qquad i \neq 2, 3, \tag{335f}$$

$$a_{i2} = 0, \qquad i \neq 3, \tag{335g}$$

$$\sum b_i(1 - c_i)\, a_{i3} = \sum b_i(1 - c_i)\, c_i a_{i3} = \sum b_i(1 - c_i)\, a_{ij} a_{j3} = 0, \tag{335h}$$

$$\sum b_i(1 - c_i)\, a_{ij} a_{jk} c_k (c_k - c_4)(c_k - c_5) = \frac{1}{4 \cdot 5 \cdot 6 \cdot 7} - \frac{c_4 + c_5}{3 \cdot 4 \cdot 5 \cdot 6}$$
$$+ \frac{c_4 c_5}{2 \cdot 3 \cdot 4 \cdot 5}, \tag{335i}$$

$$\sum b_i(1 - c_i)\, a_{ij} c_j (c_j - c_4)(c_j - c_5) = \frac{1}{4 \cdot 5 \cdot 6} - \frac{c_4 + c_5}{3 \cdot 4 \cdot 5} + \frac{c_4 c_5}{2 \cdot 3 \cdot 4}, \tag{335j}$$

$$\sum b_i(1 - c_i)\, c_i a_{ij} c_j (c_j - c_4)(c_j - c_5) = \frac{1}{4 \cdot 6 \cdot 7} - \frac{c_4 + c_5}{3 \cdot 5 \cdot 6} + \frac{c_4 c_5}{2 \cdot 4 \cdot 5}. \tag{335k}$$

That the order conditions corresponding to many of the trees of order 7 or less are satisfied is obvious in the light of detailed derivations we have given for various lower-order methods. However, we still have to prove that (335c) holds for $k = 6, 7$ and that the order condition corresponding to the tree $[\tau[\tau^4]]$ is satisfied. These three conditions are equivalent to

$$\sum b_i c_i(1 - c_i)(c_i - c_4)(c_i - c_5)(c_i - c_6) =$$
$$\int_0^1 c(1 - c)(c - c_4)(c - c_5)(c - c_6)\, dc, \tag{335l}$$

$$\sum b_i c_i^2(1 - c_i)(c_i - c_4)(c_i - c_5)(c_i - c_6) =$$
$$\int_0^1 c^2(1 - c)(c - c_4)(c - c_5)(c - c_6)\, dc, \tag{335m}$$

$$\sum b_i(1-c_i)\,a_{ij}c_j(c_j-c_4)(c_j-c_5)(c_j-c_6) =$$

$$\tfrac{1}{2}\int_0^1 (1-c)^2\,c(c-c_4)(c-c_5)(c-c_6)\,dc, \qquad (335\mathrm{n})$$

and are satisfied because the left-hand sides vanish and because

$$c(1-c)(c-c_4)(c-c_5)(c-c_6) = -\tfrac{1}{14}c + \tfrac{5}{7}c^2 - \tfrac{15}{7}c^3 + \tfrac{5}{2}c^4 - c^5$$

is orthogonal over the interval $[0,1]$ to any polynomial of degree less than 3. The actual evaluation of the coefficients of the method now proceeds very much as for the lower-order cases previously considered. Note that (335j) and (335k) give a pair of linear equations for a_{86} and a_{87} and, once these are solved, a_{76} can be found from (335i).

We now display the solution to this system using an abbreviation for numbers of the form $(a + b\sqrt{21})/c$ where a, b, c are integers. The numerator of such a number is written as a, b.

c								
0								
$\dfrac{7,1}{42}$	$\dfrac{7,1}{42}$							
$\dfrac{7,1}{21}$	$\dfrac{7,1}{21}$	0						
$\dfrac{7,1}{14}$	$\dfrac{7,1}{56}$	0	$\dfrac{21,3}{56}$					
$\dfrac{1}{2}$	$\dfrac{8,-1}{16}$	0	$\dfrac{-21,6}{16}$	$\dfrac{21,-5}{16}$				
$\dfrac{7,-1}{14}$	$\dfrac{-1687,374}{196}$	0	$\dfrac{969,-210}{28}$	$\dfrac{-381,83}{14}$	$\dfrac{84,-20}{49}$			
$\dfrac{1}{2}$	$\dfrac{583,-131}{128}$	0	$\dfrac{-2373,501}{128}$	$\dfrac{4221,-914}{288}$	$\dfrac{-9,4}{18}$	$\dfrac{189,35}{576}$		
$\dfrac{7,1}{14}$	$\dfrac{-623,169}{392}$	0	$\dfrac{435,-81}{56}$	$\dfrac{-1437,307}{252}$	$\dfrac{-2028,-1468}{7497}$	$\dfrac{-21,-4}{126}$	$\dfrac{384,80}{833}$	
1	$\dfrac{579,-131}{24}$	0	$\dfrac{-791,167}{8}$	$\dfrac{8099,-1765}{108}$	$\dfrac{-1976,784}{459}$	$\dfrac{70,7}{54}$	$\dfrac{160,-80}{153}$	$\dfrac{49,-7}{18}$
	$\dfrac{1}{20}$	0	0	0	0	$\dfrac{49}{180}$	$\dfrac{16}{45}$	$\dfrac{49}{180}$

$$(335\mathrm{o})$$

Finally, we present the eighth-order method of Cooper and Verner (1972). The various coefficients of this method are rational functions of $\sqrt{21}$ and we use the same abbreviated notation as we used in the method (335o).

0											
$\frac{1}{2}$	$\frac{1}{2}$										
$\frac{1}{2}$	$\frac{1}{4}$	$\frac{1}{4}$									
$\frac{7,1}{14}$	$\frac{1}{7}$	$\frac{-7,-3}{98}$	$\frac{21,5}{49}$								
$\frac{7,1}{14}$	$\frac{11,1}{84}$	0	$\frac{18,4}{63}$	$\frac{21,-1}{252}$							
$\frac{1}{2}$	$\frac{5,1}{48}$	0	$\frac{9,1}{36}$	$\frac{-231,14}{360}$	$\frac{63,-7}{80}$						
$\frac{7,-1}{14}$	$\frac{10,-1}{42}$	0	$\frac{-432,92}{315}$	$\frac{633,-145}{90}$	$\frac{-504,115}{70}$	$\frac{63,-13}{35}$					
$\frac{7,-1}{14}$	$\frac{1}{14}$	0	0	0	$\frac{14,-3}{126}$	$\frac{13,-3}{63}$	$\frac{1}{9}$				
$\frac{1}{2}$	$\frac{1}{32}$	0	0	0	$\frac{91,-21}{576}$	$\frac{11}{72}$	$\frac{-385,-75}{1152}$	$\frac{63,13}{128}$			
$\frac{7,1}{14}$	$\frac{1}{14}$	0	0	0	$\frac{1}{9}$	$\frac{-733,-147}{2205}$	$\frac{515,111}{504}$	$\frac{-51,-11}{56}$	$\frac{132,28}{245}$		
1	0	0	0	0	$\frac{-42,7}{18}$	$\frac{-18,28}{45}$	$\frac{-273,-53}{72}$	$\frac{301,53}{72}$	$\frac{28,-28}{45}$	$\frac{49,-7}{18}$	
	$\frac{1}{20}$	0	0	0	0	0	0	$\frac{49}{180}$	$\frac{16}{45}$	$\frac{49}{180}$	$\frac{1}{20}$

34 IMPLICIT RUNGE—KUTTA METHODS

340 Introduction.

In the search for methods with acceptable computational properties for a variety of problems, we consider the natural generalization of the Runge–Kutta type of method, by defining $Y_1, Y_2, \ldots, Y_s$, the internal approximations used in computing y_n from y_{n-1}, by the formula

$$Y_i = y_{n-1} + h \sum_{j=1}^{s} a_{ij} f(Y_j), \qquad i = 1, 2, \ldots, s, \tag{340a}$$

without requiring that the numbers a_{ij}, for all i, j such that $j \geq i$, are zero. Methods of this general type will be referred to as *implicit* Runge–Kutta methods to distinguish them from the methods of sections 31, 32 and 33 which will now be known as *explicit* Runge–Kutta methods.

A special case is of some interest. This is when $a_{ij} = 0$ for $j > i$ but where some at least of $a_{11}, a_{22}, \ldots, a_{ss}$ are non-zero. Since these methods fall intermediate in ease of use between explicit and implicit methods, they are known as *semi-implicit* methods, and will be discussed in detail in subsection 347.

There are at least six basic reasons for taking a serious interest in implicit Runge–Kutta methods. We list some of these as follows:
(a) Higher orders of accuracy can be obtained than for explicit methods.
(b) For linear systems of differential equations implicit methods can be implemented explicitly.
(c) For stiff problems explicit methods are never satisfactory whereas some implicit methods are.
(d) Implicit methods are closely related to Rosenbrock methods.
(e) The structure of certain high-order explicit methods can be derived directly from some related implicit methods.
(f) Implicit methods have an algebraic nicety not possessed by explicit methods in that the set of implicit methods under a very natural operation is homomorphic to a certain group whereas the subset corresponding to explicit methods is only a semigroup.

As we saw in section 33, when we were studying high-order Runge–Kutta methods, the so-called simplifying assumptions,

$$\sum_{j} a_{ij}c_j = \tfrac{1}{2} c_i^2, \tag{340b}$$

$$\sum_{i} b_i a_{ij} = b_j(1 - c_j), \tag{340c}$$

played an important part even though we saw that (340b) could not hold for all i. At the same time we considered the further assumption

$$\sum_{i} b_i c_i a_{ij} = \tfrac{1}{2} b_j(1 - c_j^2), \tag{340d}$$

which we saw could not hold for explicit methods at the same time as (340c).

By allowing just one diagonal element of the A matrix, namely a_{22}, to be non-zero, we can satisfy (340b) for all $i = 2, 3, \ldots$, and by allowing $a_{s-1, s-1}$ to be non-zero, we can satisfy both (340c) and (340d) for all j. Hence we have, for example, the following two five-stage fifth-order methods:

$$
\begin{array}{c|ccccc}
0 \\
\frac{1}{4} & \frac{1}{8} & \frac{1}{8} \\
\frac{1}{2} & 0 & \frac{1}{2} \\
\frac{3}{4} & \frac{3}{16} & 0 & \frac{9}{16} \\
1 & -\frac{3}{7} & 2 & -\frac{12}{7} & \frac{8}{7} \\
\hline
& \frac{7}{90} & \frac{32}{90} & \frac{12}{90} & \frac{32}{90} & \frac{7}{90}
\end{array}
\quad \text{and} \quad
\begin{array}{c|ccccc}
0 \\
\frac{1}{4} & \frac{1}{4} \\
\frac{1}{4} & \frac{1}{8} & \frac{1}{8} \\
\frac{7}{10} & -\frac{1}{100} & -\frac{7}{25} & \frac{21}{25} & \frac{3}{20} \\
1 & \frac{2}{7} & \frac{4}{5} & -\frac{4}{5} & \frac{5}{7} \\
\hline
& \frac{1}{14} & 0 & \frac{32}{81} & \frac{250}{567} & \frac{5}{54}
\end{array}
$$

These methods do not serve any real purpose except to support the reason (a) above. In the same spirit we present a further example, this time with order 6, in which each of (340b), (340c) and (340d) holds:

$$
\begin{array}{c|ccccc}
0 \\
\frac{1}{5} & \frac{1}{10} & \frac{1}{10} \\
\frac{1}{2} & 0 & \frac{5}{12} & \frac{1}{12} \\
\frac{4}{5} & \frac{7}{50} & \frac{2}{15} & \frac{32}{75} & \frac{1}{10} \\
1 & -\frac{1}{9} & \frac{35}{54} & 0 & \frac{25}{54} \\
\hline
& \frac{1}{16} & \frac{125}{432} & \frac{8}{27} & \frac{125}{432} & \frac{1}{16}
\end{array}
$$

341 Implementation.

Since $Y_1, Y_2, \ldots, Y_s$ are defined only implicitly by (340a), it is appropriate to consider whether this set of equations does actually have a solution, whether or not the solution is unique, and, finally, how it can be computed. Under the assumption that f satisfies a Lipschitz condition with constant L, these three questions can be looked at together and it will be shown that if $|h|L$ is small enough, each of them can be answered satisfactorily. Unfortunately, for stiff problems, where the value of L is large, this result is not a useful one since it requires an unreasonable restriction on the magnitude of h. For stiff problems, we also discuss algorithms for computing $Y_1, Y_2, \ldots, Y_s$, but in this case (that is the case of large $|h|L$) it is not possible to present convenient criteria for their existence and uniqueness.

In the case of small $|h|L$, we consider two types of iteration function. Let $\varphi : X^s \to X^s$ be defined by

$$
\varphi(Y_1, Y_2, \ldots, Y_s) =
\begin{bmatrix}
\varphi_1(Y_1, Y_2, \ldots, Y_s) \\
\varphi_2(Y_1, Y_2, \ldots, Y_s) \\
\vdots \\
\varphi_s(Y_1, Y_2, \ldots, Y_s)
\end{bmatrix},
$$

where

$$
\varphi_i(Y_1, Y_2, \ldots, Y_s) = y_{n-1} + h \sum_{j=1}^{s} a_{ij} f(Y_j), \qquad i = 1, 2, \ldots, s,
$$

and let $\bar{\varphi} : X^s \to X^s$ be defined by

$$
\bar{\varphi}(Y_1, Y_2, \ldots, Y_s) =
\begin{bmatrix}
\bar{\varphi}_1(Y_1, Y_2, \ldots, Y_s) \\
\bar{\varphi}_2(Y_1, Y_2, \ldots, Y_s) \\
\vdots \\
\bar{\varphi}_s(Y_1, Y_2, \ldots, Y_s)
\end{bmatrix},
$$

where

$$
\bar{\varphi}_i(Y_1, Y_2, \ldots, Y_s) = y_{n-1} + h \sum_{j<i} a_{ij} f(\bar{\varphi}_j(Y_1, Y_2, \ldots, Y_s)) + h \sum_{j \ge i} a_{ij} f(Y_j),
$$

$$
i = 1, 2, \ldots, s.
$$

Throughout this subsection A will denote the $s \times s$ matrix with elements $a_{ij}(i, j = 1, 2, \ldots, s)$.

Let $\|\cdot\|$ denote a norm on $\mathbb{R}^s$ with the property that $\|u\| = \|v\|$ for all $u, v \in \mathbb{R}^s$ such that u and v differ only in the signs of their components. If $\|\cdot\|$ denotes also the corresponding operator norm then it follows that

$$
\|A\| = \left\|
\begin{bmatrix}
a_{11} & a_{12} & \cdots & a_{1s} \\
a_{21} & a_{22} & \cdots & a_{2s} \\
\vdots & \vdots & & \vdots \\
a_{s1} & a_{s2} & \cdots & a_{ss}
\end{bmatrix}
\right\| = \left\|
\begin{bmatrix}
|a_{11}| & |a_{12}| & \cdots & |a_{1s}| \\
|a_{21}| & |a_{22}| & \cdots & |a_{2s}| \\
\vdots & \vdots & & \vdots \\
|a_{s1}| & |a_{s1}| & \cdots & |a_{ss}|
\end{bmatrix}
\right\|.
$$

In the context where u denotes a vector in X, then $\|u\|$ will denote the value of an arbitrary norm on this space and if

$$u = \begin{bmatrix} u_1 \\ u_2 \\ \vdots \\ u_s \end{bmatrix} \in X^s,$$

then $\|u\|$ will denote $\|v\|$ where $v \in \mathbb{R}^s$ is defined by

$$v = \begin{bmatrix} \|u_1\| \\ \|u_2\| \\ \vdots \\ \|u_s\| \end{bmatrix}.$$

THEOREM 341A.

If f satisfies a Lipschitz condition with constant L and if $|h| < 1/L\|A\|$, then φ is a contraction mapping and $\bar{\varphi}$ is a generalized contraction mapping. If the norm on $\mathbb{R}^s$ is the l_∞ norm then $\bar{\varphi}$ is a contraction mapping.

Proof.

Let

$$Y = \begin{bmatrix} Y_1 \\ Y_2 \\ \vdots \\ Y_s \end{bmatrix} \quad \text{and} \quad Z = \begin{bmatrix} Z_1 \\ Z_2 \\ \vdots \\ Z_s \end{bmatrix}$$

be any two points in X^s, where $Y_1, Y_2, \ldots, Z_1, Z_2, \ldots, \in X$. Then

$$\|\varphi_i(Y) - \varphi_i(Z)\| \le \left\| \left[y_{n-1} + h \sum_{j=1}^{s} a_{ij} f(Y_j) \right] - \left[y_{n-1} + h \sum_{j=1}^{s} a_{ij}(Z_j) \right] \right\|$$

$$\le |h| \left\| \sum_{j=1}^{s} |a_{ij}[f(Y_j) - f(Z_j))]| \right\|$$

$$\le |h| L \sum_{j=1}^{s} |a_{ij}| \|Y_j - Z_j\|,$$

so that

$$\|\varphi(Y) - \varphi(Z)\| \le |h| L \|A\| \|Y - Z\|$$

and the first result follows.

In the case of $\bar{\varphi}$, let B, C denote the strictly lower triangular part and the weakly upper triangular part of the matrix with elements $|h| L |a_{ij}|$ so that if

v, w denote vectors with components $\| Y_i - Z_i \|$ and $\| \bar{\varphi}_i(Y) - \bar{\varphi}_i(Z) \|$ respectively, then

$$w \leq Bw + Cv,$$

where the vector inequality is to be interpreted in a component by component sense.

The matrix B has spectral radius less than unity, since $\| B \| \leq \| B + C \| < 1$ and the property of $\bar{\varphi}$ being a generalized contraction will then follow if it can be proved that $(I - B)^{-1}C$ has spectral radius less than 1. If, on the contrary, there existed a vector v such that $w = \lambda v$, with $| \lambda | \geq 1$, it would follow that $B + \lambda^{-1} C$ had a unit eigenvalue. However, $\| B + \lambda^{-1} C \| \leq \| B + |\lambda^{-1}| C \| \leq \| B + C \| < 1$.

We omit the straightforward proof that $\bar{\varphi}$ is actually a contraction in the l_∞ case, since this result is established for a more general class of methods in Butcher (1966a). ∎

COROLLARY 341B.

Under the conditions of theorem 341A, $Y_1, Y_2, \ldots, Y_s$ satisfying (340a) exist and are unique. They are the limits of the sequences generated by either of the iteration functions φ or $\bar{\varphi}$.

For situations in which $| h | L$ is very much greater than unity, as for the solution of stiff problems, the sequences of corollary 341B do not converge and the existence and uniqueness of the solution to the algebraic equations come into doubt. It is convenient to consider the existence and uniqueness questions in subsection 359 since they are closely related to stability questions. For the present we will assume that these equations always have solutions and we look for methods of computing $Y_1, Y_2, \ldots, Y_s$ which do not break down for stiff problems. In particular, we consider the Newton–Raphson method for the solution of algebraic equations. In this method, if $g : \mathbb{R}^N \to \mathbb{R}^N$ is given, then a solution of $g(x) = 0$ is computed as the limit of the sequence $x^{(0)}, x^{(1)}, \ldots$, where the initial approximation $x^{(0)}$ is determined in a manner which is not part of the method itself and for $n = 1, 2, 3, \ldots, x^{(n)}$ is determined from the linear system

$$g'(x^{(n-1)})(x^{(n)} - x^{(n-1)}) + g(x^{(n-1)}) = 0. \tag{341a}$$

It is, of course, assumed that $g'(x^{(n-1)})$ is never singular.

Conditions under which $x^{(n)} \to x$, where x is the required solution to the problem, have been studied, for example, by Ostrowski [1966].

In the case where $g : X^s \to X^s$ is given by

$$g(Y_1, Y_2, \ldots, Y_s) = \begin{bmatrix} g_1(Y_1, Y_2, \ldots, Y_s) \\ g_2(Y_1, Y_2, \ldots, Y_s) \\ \vdots \\ g_s(Y_1, Y_2, \ldots, Y_s) \end{bmatrix},$$

where

$$g_i(Y_1, Y_2, \ldots, Y_s) = Y_i - y_{n-1} - h \sum_{j=1}^{s} a_{ij} f(Y_j),$$

we find that the matrix of $g'(Y_1, Y_2, \ldots, Y_s)$ is given by

$$\begin{bmatrix} I - ha_{11} J_1 & -ha_{12} J_2 & \cdots & -ha_{1s} J_s \\ -ha_{21} J_1 & I - ha_{22} J_2 & \cdots & -ha_{2s} J_s \\ \vdots & \vdots & & \vdots \\ -ha_{s1} J_1 & -ha_{s2} J_2 & \cdots & I - ha_{ss} J_s \end{bmatrix}, \qquad (341b)$$

where $J_i = f(Y_i)$, $i = 1, 2, \ldots, s$.

For large systems of differential equations, a considerable proportion of the work in implementing these methods in this way is taken up with the evaluation of $J_1, J_2, \ldots, J_s$ and also with the solution of the linear system for $x^{(n)} - x^{(n-1)}$ given by (341a). In fact, if the differential equation system is in N dimensions, then the matrix (341b) is $sN \times sN$ and its LU factorization requires approximately $s^3 N^3/3$ multiplicative operations (with the same number of additive operations) and the back-substitutions require approximately $s^2 N^2$ multiplicative operations. As it stands, this work has to be performed anew for each iteration unless some modification is made to the Newton iteration scheme.

Such a modification is to replace each of $J_1, J_2, \ldots, J_s$ by a matrix J equal to f' evaluated at some point not too distant from y_{n-1}. If f' is a slowly varying function, this often turns out to be a quite satisfactory alternative to the Newton–Raphson method, although the quadratic convergence property is lost.

If J is kept constant, not only over the sequence of iterations in computing y_n from y_{n-1} but also over several steps, then further savings are made in both the number of Jacobian (f') evaluations and in the linear algebra work, since the LU factorization has to be performed only when J is reevaluated or h is changed.

In Butcher (1976) a technique is described for further reducing the linear algebra work, by taking into account the spectral structure of A.

342 The effect of simplifying assumptions.

As we saw in section 33, simplifying assumptions of the form

$$\sum_{j=1}^{s} a_{ij} c_j^{k-1} = \frac{1}{k} c_i^k, \qquad k = 1, 2, \ldots, m, \; i = 1, 2, \ldots, s, \qquad (342a)$$

have the effect of allowing the equation

$$\Phi(t) = \frac{1}{\gamma(t)} \qquad (342b)$$

to be removed from consideration whenever t has some non-terminal vertex, other than the root, from which less than m other vertices branch and all these are terminals.

Similarly, the simplifying assumptions of the form

$$\sum_{i=1}^{s} b_i c_i^{k-1} a_{ij} = \frac{1}{k} b_j (1 - c_j^k), \qquad k = 1, 2, \ldots, m, \ j = 1, 2, \ldots, s, \qquad (342c)$$

have the effect of removing (342b) from consideration, whenever t has less than m terminals branching from the root and only one non-terminal branching from the root.

If these two types of simplifying assumption are satisfied for m close to s, in each case, and an order close to $2s$ is required, there are very few trees left to consider. In the next subsection we will elaborate this remark further by applying it to the derivation of methods of order exactly $2s$.

First, however, we examine a relationship between the simplifying assumptions, (342a), (342c), and the order conditions (342b) when t is a tree of the form $[\tau^{k-1}]$, $k = 1, 2, \ldots, m$, or of the form $[\tau^{k-1}[\tau^{l-1}]]$, $k = 1, 2, \ldots, m$, $l = 1, 2, \ldots, m$. In these cases the order conditions become

$$\sum_{i=1}^{s} b_i c_i^{k-1} = \frac{1}{k}, \qquad k = 1, 2, \ldots, m, \qquad (342d)$$

$$\sum_{i,j=1}^{s} b_i c_i^{k-1} a_{ij} c_j^{l-1} = \frac{1}{(k+l)l}, \qquad k = 1, 2, \ldots, m, \ l = 1, 2, \ldots, n, \qquad (342e)$$

respectively.

For convenience in stating later results, we write $A(m)$ for (342a), $D(m)$ for (342c), $B(m)$ for (342d) and $E(m, n)$ for (342e). In each case m and n are supposed to be positive integers.

THEOREM 342A.

If $B(m + n)$ and $A(n)$ then $E(m, n)$.

THEOREM 342B.

If $b_1, b_2, \ldots, b_s$ are all non-zero, $c_1, c_2, \ldots, c_s$ are distinct and $B(s + n)$ and $E(s, n)$ then $A(n)$.

Proofs.

To prove theorem 342A, multiply the difference of the two sides of (342a) by $b_i c_i^{l-1}$ where $l \le m$ and $k \le n$ and sum for $i = 1, 2, \ldots, s$. The result is

$$\sum_{i,j=1}^{s} b_i c_i^{l-1} a_{ij} c_j^{k-1} - \frac{1}{k} \sum_{i=1}^{s} b_i c_i^{k+l-1} = \sum_{i,j=1}^{s} b_i c_i^{l-1} a_{ij} c_j^{k-1} - \frac{1}{k(k+l)},$$

because of $B(m + n)$, and $E(m, n)$ follows. To prove theorem 342B, note that when $m = s$, the square matrix with the (i, l) element equal to $b_i c_i^{l-1}$ is non-singular (since we have assumed $b_1, b_2, \ldots$ to be non-zero and $c_1, c_2, \ldots$ to be distinct). Hence, given $B(s + n)$, $E(s, n)$ is both necessary and sufficient for $A(n)$. ∎

THEOREM 342C.

If $B(m + n)$ and $D(m)$ then $E(m, n)$.

THEOREM 342D.

If $c_1, c_2, \ldots, c_s$ are distinct and $B(m + s)$ and $E(m, s)$ then $D(m)$.

Proofs.

To prove theorem 342C, multiply the difference of the two sides of (342c) by c_j^{l-1}, where $k \le m$ and $l \le n$, and sum for $j = 1, 2, \ldots, s$. The result is

$$\sum_{i,\,j=1}^{s} b_i c_i^{k-1} a_{ij} c_j^{l-1} - \frac{1}{k} \sum_{j=1}^{s} b_j(c_j^{l-1} - c_j^{k+l-1})$$

$$= \sum_{i,\,j=1}^{s} b_i c_i^{k-1} a_{ij} c_j^{l-1} - \frac{1}{k}\left(\frac{1}{l} - \frac{1}{k+l}\right)$$

$$= \sum_{i,\,j=1}^{s} b_i c_i^{k-1} a_{ij} c_j^{l-1} - \frac{1}{l(k+l)}$$

where we have again made use of $B(m + n)$. Thus, given $B(m + n)$, $D(m)$ is sufficient for $E(m, n)$. To prove theorem 342D, that if $n = s$ then $D(m)$ is also necessary for $E(m, n)$, we proceed as for the proof of theorem 342B. ∎

These four results can be interpreted to mean that, not only do the simplifying assumptions result in a simplification of the order conditions but, for orders greater than s, some of these assumptions must necessarily hold.

Use will be made of these observations in deriving particular implicit Runge–Kutta methods, especially of the type studied in the next subsection.

343 Methods based on Gauss–Legendre quadrature.

We begin this subsection by identifying a number of key statements that are related to the properties of many implicit Runge–Kutta methods. These statements will be referred to by the names A, B, C, D, E, F, G, and it will be noted that A, B, D, E, F are the same as $A(s)$, $B(s)$, $D(s)$, $E(s, s)$, $B(2s)$ respectively of the last subsection. In the full wording of these statements which we now give, an attempt has been made to choose the initial letter of the text to match the name of the statement and at the same time to provide a mnemonic.

A: $a_{ij}(i, j = 1, 2, \ldots, s)$ satisfy

$$\sum_{j=1}^{s} a_{ij}c_j^{k-1} = \frac{1}{k} c_i^k, \qquad i, k = 1, 2, \ldots, s.$$

B: $b_i(i = 1, 2, \ldots, s)$ satisfy

$$\sum_{i=1}^{s} b_i c_i^{k-1} = \frac{1}{k}, \qquad k = 1, 2, \ldots, s.$$

C: $c_i(i = 1, 2, \ldots, s)$ are the zeros of the polynomial $c \mapsto P_s(2c - 1)$ where P_s is the Legendre polynomial of degree s.

D: Down column j of the a_{ij} matrix $(j = 1, 2, \ldots, s)$ we have

$$\sum_{i=1}^{s} b_i c_i^{k-1} a_{ij} = \frac{1}{k} b_j (1 - c_j^k), \qquad k = 1, 2, \ldots, s.$$

E: Every pair of k and $l(k, l = 1, 2, \ldots, s)$ is such that

$$\sum_{i,j=1}^{s} b_i c_i^{k-1} a_{ij} c_j^{l-1} = \frac{1}{l(k+l)}$$

F: For each $k(k = 1, 2, \ldots, 2s)$ we have

$$\sum_{i=1}^{s} b_i c_i^{k-1} = \frac{1}{k}.$$

G: Good! For each tree t with no more than $2s$ vertices, the order condition $\Phi(t) = 1/\gamma(t)$ is satisfied.

We shall state the next two results in a diagrammatic form using the notation exemplified by

meaning F implies B, and

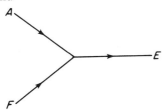

meaning A and F together imply E, and

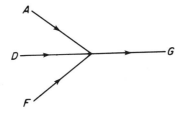

meaning A and D and F together imply G.

218

THEOREM 343A.

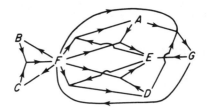

Proof.

We consider first the relationships between B, C and F. That F implies B is trivial. To show that F implies C, we use the theory of Gauss–Legendre quadrature. Let $\varphi(z) = (z - c_1)(z - c_2) \cdots (z - c_s)$ and let ψ be an arbitrary polynomial of degree less than $2s$. Making use of F, we see that

$$\Sigma b_i \psi(c_i) = \int_0^1 \psi(z)\, dz \tag{343a}$$

so that, if $\varphi = \varphi q$, where q is a polynomial of degree less than s, we see that

$$\int_0^1 \varphi(z)q(z)\, dz = 0, \tag{343b}$$

since $\varphi(c_1) = \varphi(c_2) = \cdots = \varphi(c_s) = 0$. Thus, φ is orthogonal on the interval $[0, 1]$ to all lower-degree polynomials, and is therefore a scalar multiple of the polynomial which takes z to $P_s(2z - 1)$. Thus C follows from F. Similarly, if B and C both hold we wish to prove (343a) for all polynomials ψ of degree less than $2s$. Let q and r be the quotient and remainder when ψ is divided by φ so that (343a) will follow from

$$\Sigma b_i \varphi(c_i)q(c_i) = \int_0^1 \varphi(z)q(z)\, dz \tag{343c}$$

and

$$\Sigma b_i r(c_i) = \int_0^1 r(z)\, dz. \tag{343d}$$

However, (343c) is equivalent to (343b) since φ is chosen in accordance with C and (343d) is equivalent to B.

To prove that, with F given, A is equivalent to E and D is equivalent to E, we note that, by F and the consequent properties of Gauss–Legendre quadrature, $b_1, b_2, \ldots, b_s$ are non-zero and $c_1, c_2, \ldots, c_s$ are all distinct. This being the case we deduce these results immediately from theorems 342A, 342B, 342C and 342D.

Finally, we consider the relation between G and the other statements. That G implies E and F follows by noting that E and F are simply order conditions for particular subsets of the trees or orders not exeeding $2s$. That A, D and F imply G follows from the discussion in subsection 342 in which it was noted that the simplifying assumptions A and D remove all trees from consideration except those covered by F. ■

COROLLARY 343B.

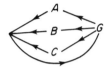

Proof.

By theorem 343A, G implies E and F, which together imply A. Also F implies each of B and C. Conversely, if B and C, then F. Hence, since F is true, A implies in turn E and D. Hence, from theorem 343A, G follows. ■

The importance of corollary 343B lies in the fact that it expresses the existence and uniqueness for each positive integer s of an implicit Runge–Kutta method with order $2s$. To find such a method, we fix $c_1, c_2, \ldots, c_s$ by C, solve for $b_1, b_2, \ldots, b_s$ from B and solve for $a_{11}, a_{12}, \ldots, a_{ss}$ from A.

Below are given the methods in this class for $s = 1, 2$ and 3. Details of the methods with $s = 4$ and 5 are given in a similar form in Butcher (1964) whereas numerical values of the various coefficients up to $s = 20$ are given in Glasmacher and Sommer (1966).

$s = 1$:

$$
\begin{array}{c|c}
\frac{1}{2} & \frac{1}{2} \\
\hline
 & 1
\end{array}
$$

$s = 2$:

$$
\begin{array}{c|cc}
\frac{3 - \sqrt{3}}{6} & \frac{1}{4} & \frac{3 - 2\sqrt{3}}{12} \\
\frac{3 + \sqrt{3}}{6} & \frac{3 + 2\sqrt{3}}{12} & \frac{1}{4} \\
\hline
 & \frac{1}{2} & \frac{1}{2}
\end{array}
$$

$$s = 3: \quad
\begin{array}{c|ccc}
\dfrac{5-\sqrt{15}}{10} & \dfrac{5}{36} & \dfrac{10-3\sqrt{15}}{45} & \dfrac{25-6\sqrt{15}}{180} \\[2ex]
\dfrac{1}{2} & \dfrac{10+3\sqrt{15}}{72} & \dfrac{2}{9} & \dfrac{10-3\sqrt{15}}{72} \\[2ex]
\dfrac{5+\sqrt{15}}{10} & \dfrac{25+6\sqrt{15}}{180} & \dfrac{10+3\sqrt{15}}{45} & \dfrac{5}{36} \\[2ex]
\hline
& \dfrac{5}{18} & \dfrac{4}{9} & \dfrac{5}{18}
\end{array}$$

344 Reflected methods.

Given any implicit Runge–Kutta method

$$
\begin{array}{c|cccc}
c_1 & a_{11} & a_{12} & \cdots & a_{1s} \\
c_2 & a_{21} & a_{22} & \cdots & a_{2s} \\
\vdots & \vdots & \vdots & & \vdots \\
c_s & a_{s1} & a_{s2} & \cdots & a_{ss} \\
\hline
& b_1 & b_2 & \cdots & b_s
\end{array}
$$

we will consider a method which exactly undoes the work of the given method. That is, if y_n is computed by

$$y_n = y_{n-1} + h \sum_{j=1}^{s} b_j f(Y_j), \tag{344a}$$

where

$$Y_i = y_{n-1} + h \sum_{j=1}^{s} a_{ij} f(Y_j), \qquad i = 1, 2, \ldots, s, \tag{344b}$$

we would need to find a method which computes y_{n-1} in terms of y_n. By solving (344a) for y_{n-1} and substituting into (344b) we find this new method to be

$$y_{n-1} = y_n + h \sum_{j=1}^{s} (-b_j) f(Y_j), \tag{344c}$$

where

$$Y_i = y_n + h \sum_{j=1}^{s} (a_{ij} - b_j) f(Y_j), \qquad i = 1, 2, \ldots, s. \tag{344d}$$

If $\sum_j b_j = b$, so that $b = 1$ for a consistent method, the method given by (344c) and (344d) has the coefficient system

$$
\begin{array}{c|cccc}
c_1 - b & a_{11} - b_1 & a_{12} - b_2 & \cdots & a_{1s} - b_s \\
c_2 - b & a_{21} - b_1 & a_{22} - b_2 & \cdots & a_{2s} - b_s \\
\vdots & \vdots & \vdots & & \vdots \\
c_s - b & a_{s1} - b_1 & a_{s2} - b_2 & \cdots & a_{ss} - b_s \\
\hline
 & -b_1 & -b_2 & \cdots & -b_s
\end{array}
$$

We now consider instead, the method with all signs changed, so that we have the method given by (344c) and (344d) but with h replaced by $-h$. Following the terminology of Scherer (1977, 1978) this will be called the *reflection* of the original method that was given by (344a) and (344b).

It is often convenient to arrange a method so that $c_1, c_2, \ldots, c_s$ are in non-decreasing order. If this is the case, then, to ensure that the reflected method is in a similar style, we would number subscripts in the opposite order. Thus, we can write the reflection of the standard method as

$$
\begin{array}{c|cccc}
b - c_s & b_s - a_{ss} & b_{s-1} - a_{s,s-1} & \cdots & b_1 - a_{s1} \\
b - c_{s-1} & b_s - a_{s-1,s} & b_{s-1} - a_{s-1,s-1} & \cdots & b_1 - a_{s-1,1} \\
\vdots & \vdots & \vdots & & \vdots \\
b - c_1 & b_s - a_{1s} & b_{s-1} - a_{1,s-1} & \cdots & b_1 - a_{11} \\
\hline
 & b_s & b_{s-1} & \cdots & b_1
\end{array}
$$

Our first result dealing with this concept is stated without its trivial proof.

THEOREM 344A.

The reflection of the reflection of an implicit Runge–Kutta method is the original method.

Recalling the notation of subsection 342, we let $A(m)$, $B(m)$, $D(m)$, $E(m, n)$ denote the properties introduced there as applying to a particular method, and $A'(m)$, $B'(m)$, $D'(m)$, $E'(m, n)$ the same properties but as applying to the reflection of this method.

THEOREM 344B.

If m, n are positive integers then

$$B(m) \Rightarrow B'(m), \tag{344e}$$

$$B(m) \wedge A(m) \Rightarrow A'(m), \tag{344f}$$

$$B(m) \wedge D(m) \Rightarrow D'(m), \tag{344g}$$

$$B(m + n) \wedge E(m, n) \Rightarrow E'(m, n). \tag{344h}$$

222

Proof.

$B(m)$, $A(m)$, $D(m)$, $E(m, n)$ are respectively equivalent to the statements that, whenever P is a polynomial of degree less that m and Q is a polynomial of degree less than n,

$$\sum_j b_j P(c_j) = \int_0^1 P(x)\, dx,$$

$$\sum_j a_{ij} P(c_j) = \int_0^{c_i} P(x)\, dx, \qquad i = 1, 2, \ldots, s,$$

$$\sum_i b_i P(c_i) a_{ij} = b_j \int_{c_j}^1 P(x)\, dx,$$

$$\sum_{i,j} b_i P(c_i) a_{ij} Q(c_j) = \int_0^1 P(x) \left[\int_0^x Q(x)\, dx \right] dx.$$

Since $B(1)$ holds on the left-hand sides of each of (344e) to (344h), we will write $b = 1$. Throughout this proof, k and l are positive integers satisfying $k \leq m$, $l \leq n$. To prove (344e), we compute

$$\sum_i b_i (1 - c_i)^{k-1} = \int_0^1 (1 - x)^{k-1}\, dx = \frac{1}{k}.$$

To prove (344f), we compute

$$\sum_j (b_j - a_{ij})(1 - c_j)^{k-1} = \int_0^1 (1 - x)^{k-1}\, dx - \int_0^{c_i} (1 - x)^{k-1}\, dx = \frac{(1 - c_i)^k}{k}.$$

To prove (344g), we compute

$$\sum_i b_i (1 - c_i)^{k-1}(b_j - a_{ij}) = b_j \left[\int_0^1 (1 - x)^{k-1}\, dx - \int_{c_j}^1 (1 - x)^{k-1}\, dx \right]$$

$$= \frac{1}{k} b_j [1 - (1 - c_j)^k].$$

Finally, to prove (344h), we compute

$$\sum_i b_i (1 - c_i)^{k-1}(b_j - a_{ij})(1 - c_j)^{l-1}$$

$$= \int_0^1 (1 - x)^{k-1} \left[\int_x^1 (1 - y)^{l-1}\, dy \right] dx$$

$$= \frac{1}{l(k + l)}. \quad \blacksquare$$

Before dealing with a result on the orders of reflected methods, we prove a preliminary lemma.

LEMMA 344C.

If, using a particular implicit Runge–Kutta method, y_n is computed from y_{n-1} and z_n is computed from z_{n-1} each using the same stepsize h, then there are positive constants h_0 and C such that, if $|h| \leq h_0$, then

$$\| y_n - z_n \| \leq (1 + Ch) \| y_{n-1} - z_{n-1} \| .$$

Proof.

Without loss of generality, we can assume that $b_1 = a_{s1}$, $b_2 = a_{s2}$, ..., $b_s = a_{ss}$ since we could, if necessary, add an extra stage. If $(Y_1, Y_2, \ldots, Y_s) \in X^s$ then we define $\| (Y_1, Y_2, \ldots, Y_s) \| = \max(\| Y_1 \|, \| Y_2 \|, \ldots, \| Y_s \|)$. Let $Y_1, Y_2, \ldots, Y_s = y_n$ and $Z_1, Z_2, \ldots, Z_s = z_n$ denote the internal approximations computed within the steps of each of the two applications of the method and let $Y = (Y_1, Y_2, \ldots, Y_s)$, $Z = (Z_1, Z_2, \ldots, Z_s)$. We have for $i = 1, 2, \ldots, s$,

$$Y_i - Z_i = y_{n-1} - z_{n-1} + h \sum_{j=1}^{s} a_{ij} [f(Y_j) - f(Z_j)],$$

so that

$$\| Y - Z \| \leq \| y_{n-1} - z_{n-1} \| + |h| \| A \| L \| Y - Z \| ,$$

where $\| A \|$ is the row-sum norm of A and L the Lipschitz constant for the function f. Hence,

$$\| y_n - z_n \| \leq \| Y - Z \| \leq (1 - |h| L \| A \|)^{-1} \| y_{n-1} - z_{n-1} \|$$

and the result follows. ∎

THEOREM 344D.

If a method is of order p then the reflection of this method is also of order p.

Proof.

We assume that the function f satisfies sufficient smoothness conditions. Let $y_{n-1}, y_n, z_{n-1}, z_n$ be as in lemma 344C where the method by which y_n and z_n are computed is the reflected method under study and where the exact solution y satisfies $y(x_{n-1}) = y_{n-1}$ and $y(x_n) = z_n$. Since the original method, with stepsize $-h$ computes z_{n-1} from an initial value z_n, the local truncation error for that method is $\| z_{n-1} - y_{n-1} \| = O(h^{p+1})$. Hence, the local truncation error of the reflected method is

$$\| y_n - z_n \| \leq (1 + Ch) \| y_{n-1} - z_{n-1} \| = O(h^{p+1}). \ ∎$$

Finally, in this subsection, we show that a certain class of methods are their own reflections.

THEOREM 344E.

If an s-stage method has order $2s$ then its reflection is this same method.

Proof.

This follows from corollary 343B (that such a method is unique) and from theorem 344D that its reflection has this same order. ∎

345   Methods of order $2s - 1, 2s - 2$.

As we shall see in section 35, there are some advantages possessed by methods in which the order is slightly lower than $2s$. Specifically, we will consider the cases where the order is $2s - d$ with $d = 1$ or 2. Making use of the notation introduced in subsection 342, we see that the conditions for this order imply that $B(2s - d)$, $E(s - d, s)$, $E(s, s - d)$ (and in the case where $d = 2$, $E(s - 1, s - 1)$) must all hold. If $b_1, b_2, \ldots, b_s$ are non-zero and $c_1, c_2, \ldots, c_s$ are distinct, this immediately implies from theorems 342B and 342D that $A(s - d)$ and $D(s - d)$ hold. In the case of $d = 1$, these two statements, together with $B(2s - d)$, are easily seen to be sufficient for order $2s - d$. In the case $d = 2$, we must also require that $E(s - 1, s - 1)$ holds. We will discuss in this subsection how $c_1, c_2, \ldots, c_s$ might be chosen and how the remaining parameters are evaluated to obtain the required order.

THEOREM 345A.

If an implicit Runge–Kutta method has order $2s - 1$ then $c_1, c_2, \ldots, c_s$ are zeros of a polynomial

$$x \mapsto P_s(2x - 1) + \lambda P_{s-1}(2x - 1) \tag{345a}$$

and if the method has order $2s - 2$ then $c_1, c_1, \ldots, c_s$ are zeros of a polynomial

$$x \mapsto P_s(2x - 1) + \lambda P_{s-1}(2x - 1) + \mu P_{s-2}(2x - 1) \tag{345b}$$

where λ (respectively λ, μ) are arbitrary constants.

THEOREM 345B.

If $c_1, c_2, \ldots, c_s$ are chosen as zeros of the polynomial (345a) (respectively (345b)), then $b_1, b_2, \ldots, b_s$ can be chosen so as to satisfy $B(2s - 1)$ (respectively $B(2s - 2)$).

Proofs.

Since the polynomial (345a) is orthogonal to all polynomials of degree less than $s - 1$ and (345b) is orthogonal to all polynomials of degree less than $s - 2$, in each case on the interval $[0, 1]$, the result of theorems 345A and 345B follows in a similar way to the first part of the proof of theorem 343A. ∎

Important special cases of theorems 345A and 345B are, for the $B(2s - 1)$ case, when $\lambda = 1$ and $\lambda = -1$. These will be denoted by *Radau I* and *Radau II* respectively since they correspond to Radau quadrature methods with one of $c_1, c_2, \ldots, c_s$ fixed at 0 (in the Radau I case) or 1 (in the Radau II case). The important special case when only $B(2s - 2)$ is achieved is when $\lambda = 0$ and $\mu = -1$, and this will be denoted by the name *Lobatto III* since it corresponds to Lobatto quadrature where two of $c_1, c_2, \ldots, c_s$ are fixed, one at 0 and the other at 1.

In discussions of these special cases we will assume that $c_1, c_2, \ldots, c_s$ are numbered in non-decreasing order.

THEOREM 345C.

For the Radau I case, $c_1, c_2, \ldots, c_s$ are the distinct roots of $(\mathrm{d}/\mathrm{d}x)^{s-1} x^s (x - 1)^{s-1} = 0$, with $c_1 = 0$ and $c_s < 1$. For the Radau II case, $c_1, c_2, \ldots, c_s$ are the distinct roots of $(\mathrm{d}/\mathrm{d}x)^{s-1} x^{s-1} (x - 1)^s = 0$, with $c_1 > 0$ and $c_s = 1$. For the Lobatto III case, $c_1, c_2, \ldots, c_s$ are the distinct roots of $(\mathrm{d}/\mathrm{d}x)^{s-2} x^{s-1} (x - 1)^{s-1} = 0$, with $c_1 = 0$ and $c_s = 1$.

Proof.

We consider only the Radau I case in detail since the others follow in a similar way. Since $P_s(x) = (1/2^s s!)(\mathrm{d}/\mathrm{d}x)^s (x^2 - 1)^s$, we find that

$$P_s(2x - 1) = \frac{1}{s!} \left(\frac{\mathrm{d}}{\mathrm{d}x}\right)^s x^s (x - 1)^s,$$

which takes on the value $(-1)^s$ when $x = 0$. Since $P_s(2x - 1)$ and $P_{s-1}(2x - 1)$ are each $(s - 1)$-order derivatives of polynomials of degrees less than $2s$ which have at least $(s - 1)$-fold zeros at both 0 and 1, there are numbers a and b such that

$$P_s(2x - 1) + P_{s-1}(2x - 1) = \left(\frac{\mathrm{d}}{\mathrm{d}x}\right)^{s-1} [x^{s-1}(x - 1)^{s-1}(ax + b)]. \quad (345c)$$

However, when $x = 0$, the value of this polynomial is $(-1)^s + (-1)^{s-1} = 0$, so that $b = 0$ and the zeros of (345c) are the zeros of

$$\left(\frac{\mathrm{d}}{\mathrm{d}x}\right)^{s-1} x^s (x - 1)^{s-1}.$$

That the roots are distinct and lie in $[0, 1)$ follows by applying $s - 1$ times the result that differentiation of a real polynomial reduces by one the multiplicity of repeated zeros and gives further zeros between adjacent zeros in the original polynomial. ∎

Although we will not deal in detail with implicit Runge–Kutta methods with orders $2s - 1$ or $2s - 2$ where the underlying quadrature formulae are not of the Radau type or of the Lobatto type, it is perhaps interesting to consider for which other cases covered by theorems 345A and 345B the numbers $b_1, b_2, \ldots, b_s, c_1, c_2, \ldots, c_s$ satisfy certain desirable conditions.

THEOREM 345D.

Consider an implicit Runge–Kutta method in which $c_1, c_2, \ldots, c_s$, $b_1, b_2, \ldots, b_s$ are selected in accordance with theorems 345A and 345B. If $B(2s - 1)$ then $c_1, c_2, \ldots, c_s$ are distinct and $b_1, b_2, \ldots, b_s$ are all positive. If $B(2s - 2)$ and if μ in (345b) satisfies $\mu \neq (s - 1)/s$ then $c_1, c_2, \ldots, c_s$ are distinct and if $\mu < (s - 1)/s$ then $b_1, b_2, \ldots, b_s$ are all positive.

Proof.

In the case $B(2s - 1)$, if two of $c_1, c_2, \ldots, c_s$ were equal or one of $b_1, b_2, \ldots, b_s$ was zero, then there would exist an $(s - 1)$-point Gaussian quadrature formula over the interval $[0, 1]$ with order greater than $2(s - 1)$. That $b_1, b_2, \ldots, b_s$ are always positive can be seen by observing that $\min(b_1, b_2, \ldots, b_s)$ is a continuous function of λ, is positive in the $\lambda = 0$ (order $2s$) case and never vanishes. In the case $B(2s - 2)$, the only way that two of $c_1, c_2, \ldots, c_s$ could be equal or one of $b_1, b_2, \ldots, b_s$ could be zero would be for the polynomial (345b) to be a linear factor multiplied by $P_{s-1}(2x - 1)$. Since

$$P_s(2x - 1) - \frac{2s - 1}{s}(2x - 1)P_{s-1}(2x - 1) + \frac{s - 1}{s} P_{s-2}(2x - 1) = 0$$

identically for Legendre polynomials, it would then follow that $\mu = (s - 1)/s$. To show that $b_1, b_2, \ldots, b_s$ are always positive, we keep λ fixed and note that $\min(b_1, b_2, \ldots, b_s)$ is continuous in μ and (as we have seen in the $B(2s - 1)$ case) is positive when $\mu = 0$. ∎

Returning now to the two Radau cases and the Lobatto case, we consider how the A matrix can be chosen. Among a variety of possibilities, we distinguish the following.

Radau I methods: Choose $c_1, c_2, \ldots, c_s, b_1, b_2, \ldots, b_s$ in accordance with the Radau I quadrature case and choose $a_{11}, a_{12}, \ldots, a_{ss}$ in accordance with $A(s)$.

Radau IA methods: The reflections of the Radau II methods.

Radau II methods: Choose $c_1, c_2, \ldots, c_s, b_1, b_2, \ldots, b_s$ in accordance with the Radau II quadrature case and choose $a_{11}, a_{12}, \ldots, a_{ss}$ in accordance with $D(s)$.

Radau IIA methods: The reflections of the Radau I methods.

Lobatto III methods: Choose $c_1, c_2, \ldots, c_s, b_1, b_2, \ldots, b_s$ in accordance with the Lobatto quadrature case and choose $a_{11}, a_{12}, \ldots, a_{ss}$ in accordance with $a_{1s} = a_{2s} = \cdots = a_{ss} = 0$ and $A(s - 1)$.

Lobatto IIIA methods: Choose $c_1, c_2, \ldots, c_s, b_1, b_2, \ldots, b_s$ as for Lobatto methods and $a_{11}, a_{12}, \ldots, a_{ss}$ in accordance with $A(s)$.

Lobatto IIIB methods: Choose $c_1, c_2, \ldots, c_s, b_1, b_2, \ldots, b_s$ as for Lobatto methods and $a_{11}, a_{12}, \ldots, a_{ss}$ in accordance with $D(s)$.

Lobatto IIIC methods: The reflections of the Lobatto III methods.

Note that in Butcher (1964a) Radau I, Radau II and Lobatto III methods were designated simply as I, II and III methods. The terminologies IA, IIA, IIIA, IIIB and IIIC were introduced by Ehle [1969] and Chipman (1971). We now establish the order of these special methods.

THEOREM 345E.

The methods Radau I, Radau IA, Radau II and Radau IIA are of order $2s - 1$ whereas the methods Lobatto III, Lobatto IIIA, Lobatto IIIB and Lobatto IIIC are of order $2s - 2$.

Proof.

By theorems 342A, 342B, 342C, 342D, Radau I satisfies $D(s - 1)$ and Radau II satisfies $A(s - 1)$. From the discussion in subsection 342, $A(s - 1)$ and $D(s - 1)$ allow all trees to be removed from consideration except those covered by $B(2s - 1)$. By theorem 344D, Radau IA and Radau IIA are also of order $2s - 1$.

In the case of Lobatto III methods we note that

$$\sum_{i=1}^{s} b_i c_i^{k-1} a_{ij} - \frac{1}{k} b_j (1 - c_j^k)$$

vanishes when $j = s$ so that the proof of theorems 342C and 342D can be modified by restricting $l \leq n - 1$ and summing $j = 1, 2, \ldots, s - 1$. The modified form of theorem 342D states that $B(m + s - 1)$ and $E(m, s - 1)$ together imply $D(m)$ and from this result together with theorem 342A we deduce $D(s - 1)$. Referring again to the discussion in subsection 342, $B(2s - 1)$, $A(s - 2)$ and $D(s - 2)$ are sufficient for order $2s - 2$ if the order condition holds for the single tree $[\tau^{s-2}[\tau^{s-2}]]$. However, this follows immediately from either of $A(s - 1)$, $D(s - 1)$.

For Lobatto IIIA methods (respectively Lobatto IIIB methods) we can use theorems 342A, 342B, 342C and 342D to prove $D(s - 2)$ (respectively $A(s - 2)$) and again the order conditions hold. Finally, Lobatto IIIC methods have order $2s - 2$ since they are the reflections of Lobatto III methods. ∎

Although Radau I, Radau II and Lobatto III methods were originally proposed by Butcher (1964a) as having as few as possible *implicit* steps to achieve a particular order, stability considerations favour the other five types of method based on the same quadrature formulae. These stability questions will be discussed in subsection 352.

Some examples of the methods considered here are given below:

Radau IA, $s = 2$, order $= 3$:

$$
\begin{array}{c|cc}
0 & \dfrac{1}{4} & -\dfrac{1}{4} \\[2ex]
\dfrac{2}{3} & \dfrac{1}{4} & \dfrac{5}{12} \\[2ex]
\hline
 & \dfrac{1}{4} & \dfrac{3}{4}
\end{array}
$$

Radau IA, $s = 3$, order $= 5$:

$$
\begin{array}{c|ccc}
0 & \dfrac{1}{9} & \dfrac{-1-\sqrt{6}}{18} & \dfrac{-1+\sqrt{6}}{18} \\[2ex]
\dfrac{6-\sqrt{6}}{10} & \dfrac{1}{9} & \dfrac{88+7\sqrt{6}}{360} & \dfrac{88-43\sqrt{6}}{360} \\[2ex]
\dfrac{6+\sqrt{6}}{10} & \dfrac{1}{9} & \dfrac{88+43\sqrt{6}}{360} & \dfrac{88-7\sqrt{6}}{360} \\[2ex]
\hline
 & \dfrac{1}{9} & \dfrac{16+\sqrt{6}}{36} & \dfrac{16-\sqrt{6}}{36}
\end{array}
$$

Radau IIA $s = 2$, order $= 3$:

$$
\begin{array}{c|cc}
\dfrac{1}{3} & \dfrac{5}{12} & -\dfrac{1}{12} \\[2ex]
1 & \dfrac{3}{4} & \dfrac{1}{4} \\[2ex]
\hline
 & \dfrac{3}{4} & \dfrac{1}{4}
\end{array}
$$

Radau IIA, $s = 3$, order = 5:

$$
\begin{array}{c|ccc}
\dfrac{4 - \sqrt{6}}{10} & \dfrac{88 - 7\sqrt{6}}{360} & \dfrac{296 - 169\sqrt{6}}{1800} & \dfrac{-2 + 3\sqrt{6}}{225} \\[2ex]
\dfrac{4 + \sqrt{6}}{10} & \dfrac{296 + 169\sqrt{6}}{1800} & \dfrac{88 + 7\sqrt{6}}{360} & \dfrac{-2 - 3\sqrt{6}}{225} \\[2ex]
1 & \dfrac{16 - \sqrt{6}}{36} & \dfrac{16 + \sqrt{6}}{36} & \dfrac{1}{9} \\[2ex]
\hline
& \dfrac{16 - \sqrt{6}}{36} & \dfrac{16 + \sqrt{6}}{36} & \dfrac{1}{9}
\end{array}
$$

Lobatto IIIC, $s = 3$, order = 4:

$$
\begin{array}{c|ccc}
0 & \dfrac{1}{6} & -\dfrac{1}{3} & \dfrac{1}{6} \\[2ex]
\dfrac{1}{2} & \dfrac{1}{6} & \dfrac{5}{12} & -\dfrac{1}{12} \\[2ex]
1 & \dfrac{1}{6} & \dfrac{2}{3} & \dfrac{1}{6} \\[2ex]
\hline
& \dfrac{1}{6} & \dfrac{2}{3} & \dfrac{1}{6}
\end{array}
$$

Lobatto IIIC, $s = 4$, order = 6:

$$
\begin{array}{c|cccc}
0 & \dfrac{1}{12} & -\dfrac{\sqrt{5}}{12} & \dfrac{\sqrt{5}}{12} & -\dfrac{1}{12} \\[2ex]
\dfrac{5 - \sqrt{5}}{10} & \dfrac{1}{12} & \dfrac{1}{4} & \dfrac{10 - 7\sqrt{5}}{60} & \dfrac{\sqrt{5}}{60} \\[2ex]
\dfrac{5 + \sqrt{5}}{10} & \dfrac{1}{12} & \dfrac{10 + 7\sqrt{5}}{60} & \dfrac{1}{4} & -\dfrac{\sqrt{5}}{60} \\[2ex]
1 & \dfrac{1}{12} & \dfrac{5}{12} & \dfrac{5}{12} & \dfrac{1}{12} \\[2ex]
\hline
& \dfrac{1}{12} & \dfrac{5}{12} & \dfrac{5}{12} & \dfrac{1}{12}
\end{array}
$$

Lobatto IIIC, $s = 5$, order $= 8$:

0	$\dfrac{1}{20}$	$-\dfrac{7}{60}$	$\dfrac{2}{15}$	$-\dfrac{7}{60}$	$\dfrac{1}{20}$
$\dfrac{7 - \sqrt{21}}{14}$	$\dfrac{1}{20}$	$\dfrac{29}{180}$	$\dfrac{47 - 15\sqrt{21}}{315}$	$\dfrac{203 - 30\sqrt{21}}{1260}$	$-\dfrac{3}{140}$
$\dfrac{1}{2}$	$\dfrac{1}{20}$	$\dfrac{329 + 105\sqrt{21}}{2880}$	$\dfrac{73}{360}$	$\dfrac{329 - 105\sqrt{21}}{2880}$	$\dfrac{3}{160}$
$\dfrac{7 + \sqrt{21}}{14}$	$\dfrac{1}{20}$	$\dfrac{203 + 30\sqrt{21}}{1260}$	$\dfrac{47 + 15\sqrt{21}}{315}$	$\dfrac{29}{180}$	$-\dfrac{3}{140}$
1	$\dfrac{1}{20}$	$\dfrac{49}{180}$	$\dfrac{16}{45}$	$\dfrac{49}{180}$	$\dfrac{1}{20}$
	$\dfrac{1}{20}$	$\dfrac{49}{180}$	$\dfrac{16}{45}$	$\dfrac{49}{180}$	$\dfrac{1}{20}$

346 Collocation methods.

A wide range of numerical problems require the evaluation of a function y on some set I such that $\phi(y)(x) = 0$ for all $x \in I$ where ϕ is an operator on an appropriate function space. By a collocation method for the solution of this type of problem, we mean the approximation of y by such a member $\bar{y}$ of some finite-dimensional linear space of functions that $\phi(\bar{y})(x) = 0$ for all x in some finite subset of I.

For example, in the solution of $y'(x) = f(y(x))$ in the interval $[x_{n-1}, x_n]$ with $y(x_{n-1})$ given as y_{n-1}, an approximation of the form

$$y(x_{n-1} + h\xi) \approx y_{n-1} + h(\xi v_1 + \xi^2 v_2 + \cdots + \xi^s v_s) \tag{346a}$$

can be used, where the vectors $v_1, v_2, \cdots, v_s$ are to be determined. In this case, the function ϕ can be chosen as $\phi(y)(x) = y'(x) - f(y(x))$ so that the values of $v_1, v_2, \cdots, v_s$ would be determined by requiring that

$$v_1 + 2\xi v_2 + \cdots + s\xi^{s-1} v_s = f(y_{n-1} + h(\xi v_1 + \xi^2 v_2 + \ldots + \xi^s v_s)) \tag{346b}$$

at a finite set of ξ values. The corresponding value of $x = x_{n-1} + h\xi$ for a given choice of ξ is usually known as a collocation point, although we will find it more convenient to refer to ξ itself as a collocation point.

THEOREM 346A.

Let $\{c_1, c_2, \ldots, c_s\}$ be a set of s distinct collocation points and let C be the number

$$\| M^{-1} \|_\infty \max_{i=1}^{s} (\, |c_i| + |c_i|^2 + \cdots + |c_i|^s\,),$$

where M is the matrix

$$\begin{bmatrix} 1 & 2c_1 & \cdots & sc_1^{s-1} \\ 1 & 2c_2 & \cdots & sc_2^{s-1} \\ \vdots & \vdots & & \vdots \\ 1 & 2c_s & \cdots & sc_s^{s-1} \end{bmatrix}.$$

Then if f satisfies a Lipschitz condition with constant L and h is a constant such that $|h|LC < 1$, then there is a unique choice of $v_1, v_2, \ldots, v_s$ satisfying (346b) for each $\xi \in \{c_1, c_2, \ldots, c_s\}$.

Proof.

Rewrite (346b) in the form

$$\begin{bmatrix} v_1 \\ v_2 \\ \vdots \\ v_s \end{bmatrix} = (\tilde{M}^{-1} \circ \tilde{f} \circ \tilde{N}) \left(\begin{bmatrix} v_1 \\ v_2 \\ \vdots \\ v_s \end{bmatrix} \right), \tag{346c}$$

where $\tilde{M} = M \otimes I$ (I being the unit operator in the space where values of $y(x)$ lie),

$$\tilde{f} \left(\begin{bmatrix} v_1 \\ v_2 \\ \vdots \\ v_s \end{bmatrix} \right) = \begin{bmatrix} f(y_{n-1} + v_1) \\ f(y_{n-1} + v_2) \\ \vdots \\ f(y_{n-1} + v_s) \end{bmatrix}$$

and $\tilde{N} = N \otimes I$ with

$$N = h \begin{bmatrix} c_1 & c_1^2 & \cdots & c_1^s \\ c_2 & c_2^2 & \cdots & c_2^s \\ \vdots & \vdots & & \vdots \\ c_s & c_s^2 & \cdots & c_s^s \end{bmatrix}.$$

The result now follows by noting that the function on the right-hand side of (346c) is a contraction. ∎

It was pointed out by Wright (1970) that this type of collocation method, in which $v_1, v_2, \ldots, v_s$ are determined by (346b) and y_n is then given by setting $\xi = 1$ in the right-hand side of (346a), is precisely equivalent to an implicit Runge–Kutta method and we now give a proof of this result.

THEOREM 346B.

If f satisfies a Lipschitz condition and $|h|$ is so small that the conditions of theorems 346A and 341A are satisfied, in the latter case as applied to the Runge–Kutta method

$$
\begin{array}{c|cccc}
c_1 & a_{11} & a_{12} & \cdots & a_{1s} \\
c_2 & a_{21} & a_{22} & \cdots & a_{2s} \\
\vdots & \vdots & \vdots & & \vdots \\
c_s & a_{s1} & a_{s2} & \cdots & a_{ss} \\
\hline
 & b_1 & b_2 & \cdots & b_s
\end{array}
\tag{346d}
$$

with $c_1, c_2, \ldots, c_s$ the same as the collocation points and $a_{11}, a_{12}, \ldots, a_{ss}$ are defined by $A(s)$ of subsection 342 and $b_1, b_2, \ldots, b_s$ are defined by $B(s)$ of subsection 342, then the result computed by the collocation and Runge–Kutta methods are the same.

Proof.

Let $Y_1, Y_2, \ldots, Y_s, y_n$ be the internal approximations and the value at the end of the step as computed by (346d) and $Z_1, Z_2, \ldots, Z_s$ be the values of the right-hand side of (346a) with $\xi = c_1, c_2, \ldots, c_s$ respectively. Also let z_n be the result computed at the end of the step by the collocation method. Substitute $\xi = c_1, c_2, \ldots, c_s$ in (346b) and we find

$$
v_1 + 2c_j v_2 + \cdots + s c_j^{s-1} v_s = f(Z_j).
\tag{346e}
$$

If (346e) is multiplied by a_{ij} and summed, $j = 1, \ldots, s$, then making use of (342a) for $k = 1, 2, \ldots, s$ (that is to say, making use of $A(s)$), we find

$$
c_i v_1 + c_i^2 v_2 + \cdots + c_i^s v_s = \sum_{j=1}^{s} a_{ij} f(Z_j)
$$

so that

$$
Z_i = y_{n-1} + h \sum_{j=1}^{s} a_{ij} f(Z_j),
$$

implying that the unique values of $Y_1, Y_2, \ldots, Y_s$ are identified with $Z_1, Z_2, \ldots, Z_s$. Now multiply (346e) by b_j, sum, make use of (342d) for $k = 1, 2, \ldots, s$ (the condition $B(s)$), and rearrange to find

$$
z_n = y_{n-1} + h \sum_{j=1}^{s} b_j f(Z_j)
$$

$$
= y_{n-1} + h \sum_{j=1}^{s} b_j f(Y_j)
$$

$$
= y_n. \quad \blacksquare
$$

Evidently, the methods of subsection 343 (based on $c_1, c_2, \ldots, c_s$ being the Gauss–Legendre points on $[0, 1]$) are collocation methods as are the Radau I, Radau IIA and Lobatto IIIA methods of subsection 345. Collocation methods can be studied in their own right, however, and we will see in subsection 348 that choosing the collocation points in a way that depends on Laguerre zeros, rather than on Legendre zeros, gives another interesting type of method.

Even though we have emphasized the equivalence of collocation methods with certain implicit Runge–Kutta methods, so that the quantities $v_1, v_2, \ldots, v_s$ need not enter explicitly into computations with these methods, there is an advantage in actually evaluating them. This is because the approximation given by (346a) can be used at other values than the collocation points or at $\xi = 1$. It may, for example, be used to interpolate results within an integration step and it may also be used to extrapolate beyond a step to yield initial approximations for the iterations in the next integration step.

347 Semi-implicit methods.

The special class of methods in which $a_{ij} = 0$ whenever $j > i$ is known variously by the names 'semi-implicit' and 'semi-explicit'. Although examples of such methods appear in papers of Butcher (1964, 1964a) the first systematic studies of semi-implicit methods did not appear until the works of Alt (1972), Nørsett (1974), Crouziex (1976), and Alexander (1977). There is a particular interest in methods of this type in which a_{ii} is constant for all $i = 1, 2, \ldots, s$. These were considered in the works of Nørsett and of Alexander and we will use Alexander's name, 'diagonally implicit', for them.

The importance of semi-implicit methods is that the computation of $Y_1, Y_2, \ldots, Y_s$ can be carried out in sequence as s systems of N algebraic equations rather than as one system of sN equations. As we will see when we consider the efficient solution of these algebraic equations, this special structure leads to cost reductions and there is a further improvement in the diagonally implicit case.

In the next subsection we will consider a closely related type of method in which the matrix of coefficients is *similar* to one of diagonally implicit form. This more general class of methods has been called 'singly implicit' by Burrage (1978) emphasizing the fact that the coefficient matrix has only a single distinct eigenvalue. As we will see, they are almost as economical to implement as are diagonally implicit methods.

In the case of $s = 1$, there is no distinction between semi-implicit and general implicit methods, and all consistent methods take the form

$$
\begin{array}{c|c}
c & c \\
\hline
 & 1
\end{array}
$$

with order 2 rather than 1 only in the case of $c = \frac{1}{2}$.

For $s = 2$, the method

$$
\begin{array}{c|cc}
c_1 & a_{11} & 0 \\
c_2 & a_{21} & a_{22} \\
\hline
 & b_1 & b_2
\end{array}
$$

with $a_{11} = c_1$, $a_{21} = c_2 - a_{22}$ has order 2 if one or both of c_1 and c_2 is $\frac{1}{2}$, $b_1 + b_2 = 1$ and b_i vanishes unless $c_i = \frac{1}{2}$. Apart from this trivial case, we can obtain order 2 if $c_2 \neq c_1$ and $b_1 = (c_2 - \frac{1}{2})/(c_2 - c_1)$, $b_2 = (\frac{1}{2} - c_1)/(c_2 - c_1)$. To obtain order 3, $c_1 c_2 - \frac{1}{2}(c_1 + c_2) + \frac{1}{3}$ must be zero and we must choose a_{22} to satisfy $\Sigma_{i,j} b_i a_{ij} c_j = \frac{1}{6}$. Thus, order 3 methods of this type take the form

$$
\begin{array}{c|cc}
c_1 & c_1 & 0 \\[4pt]
\dfrac{1}{2} + \dfrac{\frac{1}{12}}{\frac{1}{2} - c_1} & \dfrac{\frac{1}{6}}{\frac{1}{2} - c_1} & \dfrac{\frac{1}{6} - \frac{1}{2}c_1}{\frac{1}{2} - c_1} \\[10pt]
\hline
 & \dfrac{\frac{1}{12}}{c_1^2 - c_1 + \frac{1}{3}} & \dfrac{(\frac{1}{2} - c_1)^2}{c_1^2 - c_1 + \frac{1}{3}}
\end{array}.
$$

Clearly we cannot obtain a higher order than this, since the unique two-stage, order 4 method is not semi-implicit. On the other hand, we can maintain order 3 under the added constraint that the method is diagonally implicit. It is found that this is possible if c equals $(3 + \sqrt{3})/6$. Hence, we have the following method of order 3, together with the method in which $\sqrt{3}$ is replaced throughout by $-\sqrt{3}$:

$$
\begin{array}{c|cc}
(3 + \sqrt{3})/6 & (3 + \sqrt{3})/6 & 0 \\
(3 - \sqrt{3})/6 & -\sqrt{3}/3 & (3 + \sqrt{3})/6 \\
\hline
 & \frac{1}{2} & \frac{1}{2}
\end{array}.
$$

With $s = 3$, order 4 can be obtained very much as for four-stage fourth-order explicit methods. In particular (312a) must hold for $j = 1, 2, 3$, implying that $a_{33} = 1 - c_3$. With this simplifying assumption imposed on us, it is sufficient to choose c_1, c_2, c_3, b_1, b_2, b_3 to satisfy fourth-order quadrature conditions

$$
b_1 c_1^{k-1} + b_2 c_2^{k-1} + b_3 c_3^{k-1} = \frac{1}{k}, \qquad k = 1, 2, 3, 4,
$$

with $b_3 \neq 0$, a_{21} to satisfy

$$
\sum_{i,j} b_i (c_3 - c_i) a_{ij} c_j = \frac{c_3}{6} - \frac{1}{8}
$$

and a_{31}, a_{32}, a_{33} to satisfy (312a).

From the different cases of order 4 methods with $s = 3$, which were given by Crouziex (1976) the diagonally implicit methods have been extracted by Alexander (1977) as being

$$
\begin{array}{c|ccc}
(1 + \xi)/2 & (1 + \xi)/2 & & \\
1/2 & -\xi/2 & (1 + \xi)/2 & \\
(1 - \xi)/2 & 1 + \xi & -1 - 2\xi & (1 + \xi)/2 \\
\hline
& 1/6\xi^2 & 1 - 1/3\xi^2 & 1/6\xi^2
\end{array}
$$

where $\xi^3 - \xi = \frac{1}{3}$. The solutions to this cubic can be written as $\xi = (2/\sqrt{3})\cos 10°$, $-(2/\sqrt{3})\cos 70°$ and $-(2/\sqrt{3})\cos 50°$.

As s increases, the existence of methods with order greater than s comes increasingly into doubt.

Although methods with order 5 and $s = 4$ exist, Alexander has shown that necessarily $a_{11} = 0$ or $a_{44} = 0$ so that such a method cannot be diagonally implicit. Nørsett (1974) conjectured and presented some evidence for the belief that for s any even number greater than two, no diagonally implicit method exists with order $s + 1$.

348 Singly implicit methods.

Following Burrage (1978) we consider s-stage methods

$$
\begin{array}{c|c}
c & A \\
\hline
& b^{\mathrm{T}}
\end{array}
$$

in which the characteristic polynomial of A has a single s-fold zero λ. We will restrict our attention to collocation methods so that the order will be at least s.

As it happens, there is a unique method for each s once λ is specified, because a collocation method is characterized completely by the c vector and because of the following result.

THEOREM 348A.

In an s-stage implicit method with $\sigma(A) = \{\lambda\}$, $c_1/\lambda, c_2/\lambda, \ldots, c_s/\lambda$ are the zeros of the s-degree Laguerre polynomial

$$
L_s(x) = \sum_{i=1}^{s} \binom{s}{i} \frac{(-x)^i}{i!}
$$

Proof.

Because A is non-singular, $c_1, c_2, \ldots, c_s$ are distinct. It remains to show that they are equal to λ times the Laguerre zeros. Let $c^k, k = 0, 1, \ldots$, denote the component-by-component kth power of c so that (342a) can be written for $k = 1, 2, \ldots, s$ in the form

$$Ac^{k-1} = \frac{1}{k} c^k,$$

implying that

$$A^k c^0 = \frac{1}{k!} c^k. \qquad (348a)$$

We now premultiply $c°$ by the result of applying the Cayley–Hamilton theorem to the matrix A to obtain

$$\sum_{i=0}^{s} \binom{s}{i} \left(\frac{-1}{\lambda} \right)^i A^i c^0 = 0. \qquad (348b)$$

Now substitute (348a) into (348b) and interpret the result component by component to give

$$L_s \left(\frac{c_j}{\lambda} \right) = 0, \qquad j = 1, 2, \ldots, s. \quad \blacksquare$$

It is interesting that a similarity transformation that relates A to a matrix of diagonally implicit form can be written explicitly and so can its inverse.

THEOREM 348B.

Let A denote the matrix of an s-stage singly implicit collocation method with $\sigma(A) = \{\lambda\}$. Then

$$T^{-1}AT = \tilde{A},$$

where

$$\tilde{A} = \lambda \begin{bmatrix} 1 & 0 & 0 & \cdots & 0 & 0 \\ -1 & 1 & 0 & \cdots & 0 & 0 \\ 0 & -1 & 1 & \cdots & 0 & 0 \\ \vdots & \vdots & \vdots & & \vdots & \vdots \\ 0 & 0 & 0 & \cdots & 1 & 0 \\ 0 & 0 & 0 & \cdots & -1 & 1 \end{bmatrix},$$

T has the (i, j) element equal to $L_{j-1}(c_i/\lambda)$ and T^{-1} has the (i, j) element equal to $c_j L_{i-1}(c_j/\lambda)/[\lambda s^2 L_{s-1}(c_j/\lambda)^2]$. The proof of this result, which makes use of identities relating Laguerre polynomials is omitted here but is given in Butcher (1979).

35 STABILITY PROPERTIES OF RUNGE–KUTTA METHODS

350 Stability analysis.

Since the purpose of numerical analysis is to represent the solution to actual problems, it is important that what could be called qualitative properties of the numerical solution should resemble those of the true solution. By stability analysis we will mean a study of such qualitative properties as boundedness and convergence to zero of numerical solutions when these properties are possessed by the exact solution. Given a slightly different emphasis, this type of analysis is appropriate for studying the growth of numerical errors in a computed solution to a differential equation.

To illustrate the idea of qualitative behaviour, consider the differential equation $y' = qy$, where q is a (possibly complex) constant which we used as a test equation for the Euler method in subsection 206. If $y(0) = 1$ then the exact solution is $y(x) = \exp(qx)$ which is bounded (respectively convergent to zero) iff $\mathrm{Re}(q) \leq 0$ (respectively $\mathrm{Re}(q) < 0$). This differential equation will be known as the *standard test problem*.

If we attempt to solve the standard test problem by the Euler method, using positive step size h, then, as we saw in subsection 206, the result computed after n steps is $y_n = (1 + hq)^n$. This quantity is bounded (respectively convergent to zero) iff hq lies in the closed (respectively open) disc with centre -1 and radius 1. Thus, only if h is so chosen that hq lies in this closed disc can bounded solution sequences be obtained to model the behaviour of the exact solution, and hq must lie in the interior of the disc if the solution sequence is to have values converging to zero.

If we write $hq = z$, then this type of analyis generalizes in the case of explicit Runge–Kutta methods to give a result y_n, computed after n steps from $y(0) = 1$, given by $y_n = r(z)^n$, where r is a particular polynomial determined by the coefficients in the method. In the case of implicit Runge–Kutta methods, r is not in general a polynomial but rather a rational function.

We are interested in studying the regions in which $|r(z)| \leq 1$ (the stability region) or $|r(z)| < 1$ (the strict stability region). In the case of explicit Runge–Kutta methods, where the rational function r is specialized to a polynomial, these regions are necessarily bounded (unless, of course, r is a constant which breaks the requirement that the method be consistent).

DEFINITION 350A.

A Runge–Kutta method is said to be *A-stable* if its stability region contains $\mathbb{C}^-$, the non-positive half-plane.

There are a number of strengthenings and weakenings of this definition that have been proposed, on the one hand, to put more stringent requirements on the qualitative behaviour of numerical solutions and, on the other hand, to make the requirements broad enough so as not to exclude too many otherwise acceptable methods.

Among the stronger requirements is one that requires a method to be such that $|r(z)| \leq 1$ for all $z \in \mathbb{C}^-$ and in addition that $\lim_{|z| \to \infty} |r(z)| = 0$. This property was first studied by Ehle (1973) and is known as L-stability. The point of this concept is that for problems in which q has a very negative real part, it is not enough that the numerical solution should fail to increase in value: it should decrease and tend to zero as rapidly as possible. It is clear that an A-stable method corresponding to r has this property iff the numerator of r has a lower degree than the denominator. Intermediate in strength between A and L stabilities is the requirement that a method be A-stable and that $\lim_{|z| \to \infty} |r(z)| < 1$. This is a reasonable alternative to L-stability and is more frequently realized since it does not require that the coefficient in the numerator corresponding to the highest-degree coefficient in the denominator should vanish but only that its magnitude be smaller than for the denominator.

Among the standard weakenings of A-stability is the requirement that the stability region include the set $\mathbb{C}(\alpha) = \{z \in \mathbb{C} : |\arg(-z)| \leq \alpha\}$ and the requirement that the stability region contains *some* left half-plane together with the intersection of the negative half-plane with some open set containing the real axis. The first of these properties was named $A(\alpha)$ by Widlund (1967) who first studied it and proposed its use. The second was proposed by Gear (1969, 1971) in essentially the form in which we have expressed it and is known as stiff stability.

Although these weak variations of A-stability are of limited interest in the case of Runge–Kutta methods, the stronger property of L-stability is realized in the case of certain large classes of methods. We will also consider a different type of strengthening of this requirement which refers to the qualitative behaviour of numerical solutions to certain non-linear problems.

Specifically, we will consider one-sided Lipschitz continuous problems with the one-sided Lipschitz constant 0. For two particular solutions to such a problem the difference between them is non-increasing and we wish the corresponding property to apply also to numerical solutions. Methods which preserve this property are said to be B-stable (Butcher, 1975a) although it is convenient to also consider a property known as BN-stability, in which the differential equation is allowed to be non-autonomous (Burrage and Butcher, 1979). These properties are implied by algebraic stability, a property which also has a meaning as applied to general linear methods. These properties, and the relationship between them, are explored in subsections 355 to 358. Finally, in subsection 359, we will explore the closely related questions of the existence and uniqueness of solutions to the algebraic equations relating the stages of an implicit Runge–Kutta method.

351 Stability regions for explicit methods.

Consider a Runge–Kutta method given by

$Y_1 = y_{n-1},$

$Y_2 = y_{n-1} + ha_{21}f(Y_1),$

$\vdots$

$Y_s = y_{n-1} + h[a_{s1}f(Y_1) + a_{s2}f(Y_2) + \cdots + a_{s,s-1}f(Y_{s-1})],$

$y_n = y_{n-1} + h[b_1f(Y_1) + b_2f(Y_2) + \cdots + b_sf(Y_s)].$

Using the standard test problem and writing $z = hq$ as usual, we see that these can be rewritten as

$$Y = y_{n-1}e + zAY, \qquad (351a)$$

$$y_n = y_{n-1} + zb^{\mathrm{T}}Y, \qquad (351b)$$

where $e = [1, 1, \ldots, 1]^{\mathrm{T}}$, $Y = [Y_1, Y_2, \ldots, Y_s]^{\mathrm{T}}$ and $b^{\mathrm{T}} = [b_1, b_2, \ldots, b_s]$.

The polynomial r which determines the stability of this method is given by

$$r(z) = \frac{y_n}{y_{n-1}} = 1 + zb^{\mathrm{T}}(y_{n-1}^{-1}Y)$$

and, from (351a),

$$y_{n-1}^{-1}Y = (I + zA + z^2A^2 + \cdots + z^{s-1}A^{s-1})e.$$

Hence, we have

$$r(z) = 1 + zb^{\mathrm{T}}(I + zA + z^2A^2 + \cdots + z^{s-1}A^{s-1})e.$$

If we know that the method in question has order p then, because $b^{\mathrm{T}}A^{k-1}e = \Phi([_{k-1}\tau]_{k-1})$, we see that $b^{\mathrm{T}}A^{k-1}e = 1/k!$ for $k \leq p$. Using this fact, we find that

$$r(z) = 1 + z + \frac{z^2}{2!} + \cdots + \frac{z^p}{p!} + c_{p+1}z^{p+1} + \cdots + c_sz^s,$$

where the coefficients $c_{p+1}, \ldots, c_s$ are equal to $\Phi([_p\tau]_p), \ldots, \Phi([_{s-1}\tau]_{s-1})$.

As we saw in section 31, an order equal to s is possible if $s \leq 4$ so that, if this optimal situation is realized, the polynomial r has a unique form for each of $s = 1, 2, 3, 4$.

Given r, we can determine its stability region by noting, by the maximum modulus principle, that it is the region enclosed by the set of points for which $|r(z)| = 1$. For a particular point z on the boundary of the stability region there must exist an angle θ for which $r(z) = \exp(i\theta)$ and we can trace out this boundary by solving this polynomial equation for values of θ in $[0, 2\pi)$.

Algorithm 351 implements this procedure. Various points on the boundary are located by taking θ in steps of $2\pi/n$ from 0 to $2\pi s$ and then invoking a procedure *print* which is supposed to print a point $x + iy$ on this boundary.

The method used is to solve for $z = x + iy$ by the Newton–Raphson method, taking the value at the previous angle as the initial approximation and taking zero as the initial approximation for $\theta = 0$. The algorithm is designed so that it will deal with a polynomial $r(z) = a[0] + a[1]z + a[2]z^2 + \cdots + a[s]z^s$, rather than just the special case where $a[m] = 1/m!$, $m = 0, 1, \ldots, s$. The variable *eps* is the required accuracy.

Algorithm 351: Plotting of stability region

```
pi := 3.14159265359;
d := 2*pi/n;  x := 0;  y := 0;
for k := 1 to n*s do
begin
     t := k*d;  j := 1;
     repeat
          xn := a[s];  yn := 0;  xd := 0;  yd := 0;
          for l := s - 1 downto 0 do
          begin
               temp := x*xd - y*yd + xn;
               yd := y*  xd + x*yd + yn;  xd := temp;
               temp := x*xn - y*yn + a[l];
               yn := y*xn + x*yn;  xn := temp
          end;
          xn := xn - cos(t);  yn := yn - sin(t);
          temp := xd*xd + yd*yd;
          dx := (xn*xd + yn*yd)/temp;
          dy := (yn*xd - xn*yd)/temp;
          x := x - dx;  y := y - dy;
          j := j + 1
     until dx*dx + dy*dy < eps*eps;
     print(x, y)
end
```

For the methods of optimal order with $s = 1, 2, 3, 4$, the boundaries of the stability regions computed using this algorithm are shown in figure 351 with the values of s attached.

352   *A*-stable methods.

Let r be the rational function determined by a particular implicit Runge–Kutta method as in subsection 350. Since the truth of (351a) is not dependent on the explicitness of the method we can solve for $(Y_1, Y_2, \ldots, Y_s)$ from this equation and substitute into (351b) to give

$$r(z) = 1 + zb^{\mathrm{T}}(I - zA)^{-1}e. \tag{352a}$$

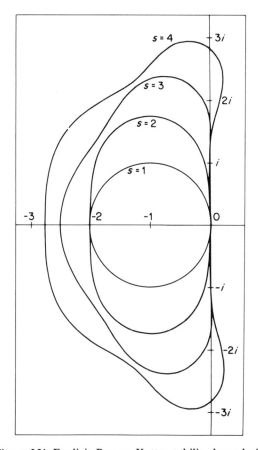

Figure 351 Explicit Runge–Kutta stability boundaries.

If M is a non-singular $s \times s$ matrix, then it is well-known that the elements of M^{-1} are equal to polynomials of degree $s-1$ in the elements of M, divided by $\det(M)$. Hence, taking $M = I - zA$, which is non-singular for $|z|$ sufficiently small, we see that $r(z)$ in (352a) is a rational function

$$r(z) = \frac{n_0 + n_1 z + n_2 z^2 + \cdots + n_s z^s}{1 - d_1 z + d_2 z^2 + \cdots + (-1)^s d_s z^s} \tag{352b}$$

such that $z \to z^s - d_1 z^{s-1} + d_2 z^{s-2} - \cdots + (-1)^s d_s$ is the characteristic polynomial of A. Setting $z = 0$ in (352b) and comparing with (352a), we see that $n_0 = 1$. Furthermore, if the Runge–Kutta method is consistent, so that $b^{\mathrm{T}} e = 1$, we see that $r'(0) = 1$ which implies that $n_1 + d_1 = 1$.

To obtain further information on the coefficients in (352b), when the method is of order p, we consider the differential equation $y' = y$ and the numerical solution after a single step of length $h = z$ from an initial value $y_0 = 1$. Since the exact solution is $\exp(z)$, the numerical solution $r(z)$ and the error $O(z^{p+1})$ as $z \to 0$, we see that

$$|r(z) - \exp(z)| = O(z^{p+1}).\qquad(352c)$$

Since the denominator in (352b) has limit 1 as $z \to 0$, it follows from (352c) that

$$n_0 + n_1 z + n_2 z^2 + \cdots + n_s z^s$$
$$- \exp(z)(1 - d_1 z + d_2 z^2 - \cdots + (-1)^s d_s z^s) = O(z^{p+1})\qquad(352d)$$

so that we obtain $p + 1$ linear relations between $n_0, n_1, n_2, \ldots, d_1, d_2, \ldots$ by equating to zero the coefficients of $1, z, z^2, \ldots, z^p$ in the Taylor series for the left-hand side of (352d).

As we saw in subsection 137 it is possible to obtain a rational function r such that (352c) (or equivalently (352d)) holds with $p = 2s$. Furthermore, if r is restricted to having a numerator of degree only $s - \delta$ (δ a positive integer) so that $n_s = n_{s-1} = \cdots = n_{s-\delta+1} = 0$, then r can be obtained such that (352c) holds with $p = 2s - \delta$. These approximations, the diagonal (s, s) and the subdiagonal $(s - \delta, s)$ Padé approximations to the exponential function, are given by theorem 137E and tabulated in several cases in table 137.

We wish to show that for certain of the implicit Runge–Kutta families considered in subsections 343 and 345 the function r is just one of these Padé approximations. As a preliminary step we find a relationship between the rational function associated with a method and with the corresponding function for its reflection (as in subsection 344).

LEMMA 352A.

If r is the rational function associated with an implicit Runge–Kutta method and $\bar{r}$ the corresponding function for the reflected method then

$$\bar{r}(z) = r(-z)^{-1}.$$

Proof.

The reflection of a method exactly undoes the work or the original method with a change in the sign of h. However, for the equation $y' = qy$, the work of a method consists of multiplying the solution value by $r(z)$. ∎

We are now in a position to find the relation we seek between certain methods and Padé approximations to the exponential function.

THEOREM 352B.

The rational function corresponding to the s-stage Runge–Kutta method of order $2s$ (subsection 343) is the (s, s) Padé approximation to the exponential function; the rational function corresponding to each of the s-stage Radau IA and Radau IIA methods (subsection 345) is the $(s - 1, s)$ Padé approximation to the exponential function; and the rational function corresponding to the

Lobatto IIIC method (subsection 345) is the $(s-2, s)$ Padé approximation to the exponential function.

Proof.

In the case of the methods of order $2s$, the condition $|r(z) - \exp(z)| = O(z^{2s+1})$ identifies r as the (s, s) Padé approximation. The other cases follow similarly with $O(z^{2s+1})$ replaced by $O(z^{2s})$ in the case of Radau IA and Radau IIA and by $O(z^{2s-1})$ in the case of Lobatto IIIC, once it is established that the numerators have degrees $s - 1$ for the Radau IA and Radau IIA cases and $s - 2$ for the Lobatto IIIC case. However, by lemma 352A, this is equivalent to requiring that the reflections of these methods have these lower degrees in their denominators and this in turn follows by noting that the reflections of these methods (respectively Radau II, Radau I and Lobatto III) were specially constructed to have, in their A matrices, zero elements in the first column, first row or both, implying the existence of one or two zero roots in the characteristic polynomial. Since the denominator of the rational function is just the characteristic polynomial with the coefficients in reverse order, the result follows. ∎

Our aim, of course, is to establish that the methods referred to in theorem 352B are A-stable, and according to that theorem we can now shift our attention from the methods themselves to the (s, s), $(s-1, s)$ and $(s-2, s)$ Padé approximations to the exponential function. We formally establish the properties of these approximations that we require in theorem 352C and then state without further proof that the various methods are A-stable, in corollary 352D.

THEOREM 352C.

If l and m are integers such that $0 \le l \le m \le l + 2$, then the (l, m) Padé approximations to the exponential function is bounded in magnitude by 1 in the negative half-plane.

Proof.

For a complex number z with negative real part, consider the vectors $V_0, V_1, \ldots, V_m$ in $\mathbb{C}^2$ given by

$$V_0 = X_{00},$$

$$V_1 = z^{-1}X_{11},$$

$$\vdots$$

$$V_{m-1} = z^{-(m-1)}X_{m-1, m-1},$$

$$V_m = z^{-m}X_{l, m},$$

where, for non-negative integers i, j, X_{ij} denotes the vector introduced in subsection 137 with components formed from the values at z of the numerator and denominator of the (i, j) Padé approximation to the exponential function, which was shown to be normal in theorem 137D.

From theorem 137F and the discussion following it, we see that for $n = 2, 3, \ldots, m$ there are real numbers $\alpha_n, \beta_n, \gamma_n$ satisfying $\alpha_n \leq 0, \beta_n, \gamma_n > 0$ in the exponential case such that

$$V_n = \left(\alpha_n + \frac{\beta_n}{z}\right) V_{n-1} + \gamma_n V_{n-2},$$

and we can also verify that

$$V_0 = \begin{bmatrix} 1 \\ 1 \end{bmatrix}, \quad V_1 = \begin{bmatrix} 2/z + 1 \\ 2/z - 1 \end{bmatrix}.$$

Hence, the number formed by dividing the first component of V_m by the second is given by the continued fraction

$$1 + \cfrac{2}{-1 + 2/z +} \cfrac{\gamma_2}{\alpha_2 + \beta_2/z +} \cdots \cfrac{\gamma_m}{\alpha_m + \beta_m/z}.$$

By induction on $k = m, m - 1, \ldots, 2$ we note that the number

$$\cfrac{\gamma_k}{\alpha_k + \beta_k/z +} \cfrac{\gamma_{k+1}}{\alpha_{k+1} + \beta_{k+1}/z +} \cdots \cfrac{\gamma_m}{\alpha_m + \beta_m/z}$$

has negative real part and accordingly the required Padé approximation at z is given by

$$1 + \frac{2}{-1 + w} = \frac{w + 1}{w - 1}, \tag{352e}$$

where $\operatorname{Re}(w) < 0$ and the magnitude of (352e) is less than 1. ∎

COROLLARY 352D.

The s-stage Runge–Kutta method of order $2s$ and the Radau IA, Radau IIA and Lobatta IIIC methods are A-stable.

It is now straightforward to consider the possible L-stability of the methods which were the subject of corollary 352D. For the methods of order $2s$, $r(z)r(-z) = 1$, since they are their own reflections and, therefore, it is not possible that $\lim_{|z| \to \infty} |r(z)| = 0$. On the other hand, since the other methods have lower degrees in the numerator than in the denominator, we can state the following without further proof.

THEOREM 352E.

Radau IA, Radau IIA and Lobatto IIIC methods are L-stable.

It is interesting to ask whether any of the other Padé approximations to the exponential function besides those covered by theorem 352C also correspond to A-stable methods. Clearly those above the diagonal cannot have this property and the cases dealt with lie on the diagonal or on the first two subdiagonals. It is impossible to obtain A-stable methods for any of the lower diagonals, and this can be proved for the third, fourth, etc., subdiagonals by studying the behaviour of the approximation on the imaginary axis. For the third subdiagonal the result was proved by Ehle [1969] and for the fourth and fifth subdiagonal by Nørsett (1975) but the general result was not settled until Wanner, Hairer and Nørsett (1978) devised the method of 'order stars' which we review briefly in subsection 354.

The proof that has been given here for the A-stability of methods corresponding to the diagonal and first two subdiagonals of the Padé table for the exponential function appeared in Butcher (1977) and is quite atypical of such proofs. Other approaches to the study of A-stable Runge–Kutta methods have been considered by Axelsson (1969), Ehle (1973), Ehle and Picel (1975), Watts and Shampine (1972) and by Wright (1970), and in the work of these authors there are several independent proofs of these and related results.

The usual approach is to note that the rational function r given by (352a) is A-stable iff (a) $|r(z)| \leq 1$ whenever $\mathrm{Re}(z) = 0$ and (b) the only poles of r are in the right half-plane. This fact, which is a consequence of the maximum modulus principle for regular functions, neatly breaks the work of determining whether or not a particular method is A-stable into two parts, each with a quite different flavour.

In the case of methods which are their own reflections, for example collocation methods based on a symmetric arrangement of $c_1, c_2, \ldots, c_s$ about $\frac{1}{2}$, and in particular the order $2s$ method, we necessarily have $r(z) = 1/r(-z)$ so that $|r(z)| = 1$ if $\mathrm{Re}(z) = 0$, so that all that remains is to locate the poles of r. For methods of order $2s - 1$ or $2s - 2$, the analysis is very similar. That is to say, the behaviour of $|r(z)|$ when $\mathrm{Re}(z) = 0$ is easy to analyze but in contrast the proof that r is regular in the negative half-plane is far from trivial.

This approach has been used by Wright (1970) and by Watts and Shampine (1972) to determine the A-stability of a number of collocation methods. In particular, for the choice $c_i = (i - 1)/(s - 1)$ (corresponding to Newton–Cotes quadrature), Wright has verified numerically and Watts and Shampine have confirmed using exact arithmetic that A-stability is achieved for $s \leq 9$ but evidently not for higher s.

For singly implicit methods, the analyticity of r in the left half-plane is a consequence of the structure of A, and in this case all the effort of determining whether or not a method is A-stable reduces to a study of the behaviour of r on the imaginary axis.

In the next subsection we will consider the possible A-stability of singly implicit methods and, in particular, of diagonally implicit methods.

353 Further A-stable methods.

Because of their relatively low implementation costs, diagonally implicit methods are promising as practical methods. These, together with singly implicit methods, are characterized by the fact that the function r given by (352a) has the form

$$r(z) = \frac{P(z/\lambda)}{(1 - z/\lambda)^s} \tag{353a}$$

and it is appropriate that we consider the stability of such methods as this. Note that r has an s-fold pole at λ. The factor $1/\lambda$ is inserted into the argument of the polynomial P for convenience of notation.

By a *restricted Padé approximation* to the exponential function we will mean a function r given by (353a) where P has degree m and where

$$|r(z) - \exp(z)| = O(z^{m+1}). \tag{353b}$$

Although the cases $m = s$ and $m = s - 1$ are the most important, we will also consider the possibility that $m = s + 1$ since this gives us the opportunity of obtaining order $s + 1$ with $\deg(P) = s$ by determining conditions under which the coefficient of degree $s + 1$ in P vanishes.

Substituting λz for z in (353b) and multiplying by $(1 - z)^s$ gives the relation

$$P(z) = (1 - z)^s \exp(z\lambda) + O(z^{m+1}),$$

so that P is simply a truncated power series for $(1 - z)^s \exp(z\lambda)$. We recall that the term $O(z^{m+1})$ denotes a function which is bounded by $C|z^{m+1}|$ as $z \to 0$ but note that the value of C depends on λ. Let L_s denote the Laguerre polynomial of degree s and for $n = 1, 2, \ldots,$ let $L_s^{(n)}$ denote the nth derivative. For $n = -1, -2, \ldots,$ define $L_s^{(n)}$ by $L_s^{(-1)}(0) = L_s^{(-2)}(0) = \cdots = 0$ with the derivative of $L_s^{(n-1)}$ always equal to $L_s^{(n)}$ and $L_s^{(0)}$ identified with L_s.

With this notation we have the following result.

LEMMA 353A. $(1 - z)^s \exp(z\lambda) = (-1)^s \sum_{i=0}^{\infty} L_s^{(s-i)}(\lambda) z^i$.

Proof.

From the formula $L_s(\lambda) = \sum_{j=0}^{s}(-\lambda)^j \binom{s}{j}/j!$, leading to $L_s^{(s-i)}(\lambda) = (-1)^{s-i} \sum_{j=0}^{s}(-\lambda)^{j+i-s}\binom{s}{j}/(j+i-s)!$, with the convention that $1/n! = 0$ for n a negative integer, we obtain

$$(-1)^s \sum_{i=0}^{\infty} L_s^{(s-i)}(\lambda) z^i = \sum_{i=0}^{\infty} (-1)^i z^i \sum_{j=0}^{s} (-\lambda)^{j+i-s} \binom{s}{j} \frac{1}{(j+i-s)!}$$

$$= \sum_{j=0}^{s} \sum_{k=j-s}^{j} \binom{s}{j}(-z)^{s-j} \frac{(z\lambda)^k}{k!},$$

where we have written $k = i + j - s$, leading to the formula for $(1 - z)^s \exp(z\lambda)$.

∎

Using this result, we find an expression for the restricted Padé approximation

$$r(\lambda z) = \frac{(-1)^s \sum_{i=0}^{m} L_s^{(s-i)}(\lambda)z^i}{(1-z)^s}, \tag{353c}$$

with the error term

$$\exp(\lambda z) - r(\lambda z) = (-1)^s L_s^{(s-1-m)}(\lambda)z^{m+1} + O(z^{m+2}).$$

If $m = s$, the factor $(-1)^s L_s^{(s-1-m)}(\lambda)$ can be written as

$$(-1)^s L_s^{(-1)}(\lambda) = (-1)^s \int_0^\lambda [L_s(\mu)] \, d\mu$$

$$= (-1)^s \int_0^\lambda [L_s'(\mu) - L_{s+1}'(\mu)] \, d\mu$$

$$= (-1)^s [L_s(\lambda) - L_{s+1}(\lambda)]$$

$$= (-1)^{s+1} \frac{\lambda}{s+1} L_{s+1}'(\lambda),$$

so that, if λ is chosen as a zero of L_{s+1}', order $s + 1$ can be obtained. Note that in these manipulations we have used standard identities involving the Laguerre polynomials (see, for example, Abramowitz and Stegun [1965]). On the other hand, to obtain order s with $m = s - 1$, it is only necessary to make sure that $L_s(\lambda) = 0$. If such a method is A-stable, it is also L-stable.

Because of the maximum modulus principle, A-stability is determined entirely by the behaviour of the rational function $P(z)/(1 - z)^s$ for z on the imaginary axis. Write $z = iy$ and $u = y^2$, so that the method is A-stable iff $|(1 - iy)^s|^2 - |P(iy)|^2 \geq 0$ for all real y.

Denote this even function of y (the so-called E-polynomial of Nørsett, 1975) by $E(u)$ and we see that, because the order is at least s, $E(u)$ takes the form $u^{[s/2]+1}F(u)$, where F is a polynomial of degree $[(s-1)/2]$ where, for a real number x, $[x]$ denotes the integer part of x. Using the expression for $P(z)/(1-z)^s$ given by (353c), we obtain for $E(u)$ the formula

$$E(u) = (1+u)^s - \left(\sum_{i=0}^{[s/2]} (-u)^i L_s^{(s-2i)}(\lambda) \right)^2 - u \left(\sum_{i=0}^{[(s-1)/2]} (-u)^i L_s^{(s-2i-1)}(\lambda) \right)^2$$

which leads to the following expressions for $F(u)$ with $s = 1, 2, 3, 4$:

$s = 1$:

$$F(u) = \lambda(2 - \lambda),$$

$s = 2$:

$$F(u) = \frac{\lambda(4 - \lambda)(4 - 4\lambda + \lambda^2)}{4},$$

$s = 3$:

$$F(u) = \frac{\lambda(\lambda^3 - 12\lambda^2 + 36\lambda - 24)}{12} - \frac{u\lambda(\lambda - 3)(\lambda - 6)(\lambda^3 - 9\lambda^2 + 18\lambda - 12)}{36},$$

$s = 4$:

$$F(u) = \frac{\lambda(\lambda^5 - 24\lambda^4 + 204\lambda^3 - 768\lambda^2 + 1224\lambda - 576)}{72}$$

$$- \frac{u\lambda(\lambda - 4)(\lambda - 6 + 2\sqrt{3})(\lambda - 6 - 2\sqrt{3})(\lambda^4 - 16\lambda^3 + 72\lambda^2 - 96\lambda + 48)}{576}.$$

Values of positive λ which guarantee that these expressions for $F(u)$ take non-negative values for all $u \geq 0$ are, for $s = 1$, $(0, 2]$, for $s = 2$, $(0, 4]$, for $s = 3$, $[0.935822227, 3]$, where $0.9358\ldots$ is a zero of $\lambda^3 - 12\lambda^2 + 36\lambda - 24$, and for $s = 4$, $[0.780896302, 2.535898385]$, where $0.7808\ldots$ is a zero of $\lambda^5 - 24\lambda^4 + 204\lambda^3 - 768\lambda^2 + 1224\lambda - 576$ and $2.5358\ldots = 6 - 2\sqrt{3}$.

To obtain order $s + 1$ we must choose a zero of L'_{s+1} and to retain A-stability such a zero must lie in the appropriate interval of allowed λ values. It turns out that for each of $s = 1, 2, 3$ the lowest zero of L'_{s+1} achieves this result. For $s = 4$ there is no possible choice, but for $s = 5$ the second zero of L'_6 gives A-stability. It has been proved by Wanner, Hairer and Nørsett (1978) that these five cases are the only ones that combine order $s + 1$ with A-stability.

The alternative goal, of choosing λ as a zero of L_s so as to obtain L-stability, is achievable for the only zero, 1, in the case $s = 1$, for both zeros, $2 \pm \sqrt{2}$ in the case $s = 2$, for the second zero in the cases $s = 3$ and $s = 4$ and for the third zero in the cases $s = 5$ and $s = 6$. According to Wolfbrandt [1977] whose investigations went as far as $s = 15$, there is only one known case of L-stability beyond these and this is for the fourth zero of L_8 in the case $s = 8$.

Although L-stability seems to be possible in only this handful of cases, a good approximation to L-stability exists for a wide range of values of s and usually for a choice of possible zeros of L_s.

354 Order stars and stability.

As pointed out by Wanner, Hairer and Nørsett (1978), the set of points where $r(z)$ defined by (352a) has magnitude less than 1 gives little insight into its behaviour throughout the complex plane. On the other hand, the relative stability function $r(z)\exp(-z)$ turns out to be an excellent link between stability questions and order of accuracy.

To illustrate the idea of using stability functions with the negative exponential factor inserted, consider the contour line maps for both $|r(z)|$ and $|r(z)\exp(-z)|$ for four choices of r shown in figure 354. In each case the contour line corresponding to height 1 is shown together with the location of poles and zeros. The shaded part of the diagram represents the regions where

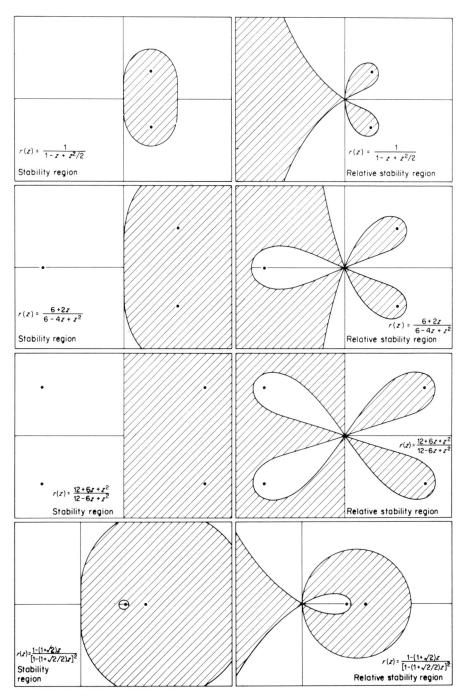

Figure 354 Examples of stability and relative stability regions.

$|r(z)| > 1$ (respectively $|r(z)\exp(-z)| > 1$). The first three choices of r are for Padé approximations whereas the fourth is one of the two L-stable restricted Padé approximations with $s = 2$.

For the relative stability diagrams, the structure near the origin is directly related to the order of accuracy as in lemma 354C below. On the other hand, the behaviour on the imaginary axis and the locations of the poles are the same whether the exponential factor is present or not and, by lemma 354A, this is enough to determine A-stability.

Throughout this subsection 'the method' will refer to a Runge–Kutta method for which r is the corresponding stability rational function. Also S will denote the set in the complex plane where $|r(z)\exp(-z)| > 1$ and I will denote the set of purely imaginary numbers.

LEMMA 354A.

The method is A stable iff r has no poles in the negative half-plane and $S \cap I = \emptyset$.

Proof.

(Only if). Clearly poles in the negative half-plane are impossible for an A-stable method. On I, $|\exp(-z)| = 1$ so that $|r(z)\exp(-z)| > 1$ implies $|r(z)| > 1$.

(If). $|r(z)|$ is bounded by unity on the imaginary axis and therefore, by the maximum modulus principle, on the negative half-plane. ∎

Information on the behaviour of the set S, sufficient to apply lemma 354A in particular cases, is afforded by lemmas 354B, 354C and 354D below. In the outline proofs we will give, we will refer to 'contour lines' (where $|r(z)\exp(-z)|$ is constant) and to 'flux lines' (where $\arg(r(z)\exp(-z))$ is constant so that along these lines $|r(z)\exp(-z)|$ changes most rapidly) which are orthogonal to contour lines. The special contour of height 1 ($|r(z)\exp(-z)| = 1$) is the boundary of S. If the method is consistent, this boundary passes through zero. The components of S whose boundaries intersect at zero are, following the terminology of Wanner, Hairer and Nørsett (1978), called 'fingers'. If the same connected component of S has an m component intersection with a sufficiently small disc centred at zero, it is called a 'finger with multiplicity m'.

In figure 354' the three diagrams contain fingers of multiplicities 1, 2 and 3.

LEMMA 354B.

There is a $\rho_0 > 0$ such that for $\rho \geq \rho_0$ there exist two angles $\theta_1(\rho)$, $\theta_2(\rho)$ such that $\rho \exp(i\theta) \in S$ with $\theta \in [0, 2\pi]$ iff $\theta \in (\theta_1(\rho), \theta_2(\rho))$. Furthermore, $\theta_1(\rho) \to \pi/2$, $\theta_2(\rho) \to 3\pi/2$ as $\rho \to \infty$.

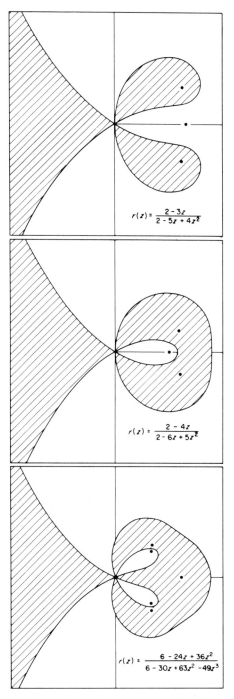

Figure 354′ Order stars containing fingers of various multiplicities.

Proof outline.

For large $|z|$, the contour and flux line system for $r(z)\exp(-z)$ tends to that for $\exp(-z)$, and for this function the contour lines are given by $\mathrm{Re}(z) = \text{constant}$ and the flux lines by $\mathrm{Im}(z) = \text{constant}$. ∎

LEMMA 354C.

If the order of the method is p, then the lines making an angle $(n + \frac{1}{2})\pi/(p + 1)$ $(n = 0, 1, 2, \ldots, 2p + 1)$ with the real axis are tangential to the boundary of S at 0.

Proof.

By the order conditions, $r(z)\exp(-z) = 1 + Cz^{p+1} + O(z^{p+2})$ where $C(\neq 0)$ is a real constant. Hence, for $z = \rho \exp(i\theta)$, with ρ sufficiently small, the condition that $|r(z)\exp(-z)| > 1$ is that

$$\mathrm{Re}(C\rho^{p+1} \exp(i(p + 1)\theta)) < O(\rho^{p+2})$$

or that $\cos((p + 1)\theta)$ is greater (respectively less) than $O(\rho)$ for C positive (respectively negative). ∎

LEMMA 354D.

Each bounded finger of S, with multiplicity m, contains at least m poles (counted with their multiplicities).

Proof outline.

Traverse the boundary of each bounded finger in a counterclockwise direction. By orthogonality, the flux increases and eventually returns to the origin m times. Hence, the total change of flux is at least $2\pi m$. It is easy to see that the change of flux around a pole of multiplicity μ is $2\pi\mu$. ∎

Although the ideas related to order stars have a variety of applications, we will content ourselves here with the proof of the Ehle conjecture due to Wanner, Hairer and Nørsett (1978).

THEOREM 354E.

The only members of the Padé table for the exponential function which correspond to A-stable methods are on the diagonal and on the first two subdiagonals.

Proof.

This result is the converse of theorem 352C. Because the case of superdiagonal approximations is trivial, we consider only (l, m) Padé approximations with $m > l + 2$ and order $p = l + m$. If there are no poles in the negative half-plane, there must be a total of m (including multiplicities) in the positive half-plane. By lemma 354D, a total of m bounded fingers (again including multiplicities) of S must contain these poles. However, by lemma 354C, the angle subtended by the union of all these fingers must be at least $(2m - 1)\pi/(p + 1) > \pi$. Hence, at least one of the fingers intersects with the imaginary axis. ∎

355 *AN*-stability.

The linear differential equation

$$y'(x) = q(x)y(x), \tag{355a}$$

where $q(x)$ takes on the constant value of a complex number with non-positive real part, is used as a test problem in the definition of A-stability. We here consider a generalization in which q in (355a) is allowed to vary with x. If $\mathrm{Re}(q(x)) \le 0$ for all x, the solution to the differential equation is still seen to be bounded as $x \to \infty$ and, for a particular numerical method, we wish to consider whether or not this stability property carries over to computed results.

For a single step of a Runge–Kutta method from x_{n-1} to $x_{n-1} + h$, let $z_i = hq(x_{n-1} + hc_i)$ $(i = 1, 2, \ldots, s)$ so that the stages $Y_1, Y_2, \ldots, Y_s$ and the final result y_n are computed from the given value of y_{n-1} by the equations

$$Y_i = y_{n-1} + \sum_{j=1}^{s} a_{ij}z_j Y_j, \qquad i = 1, 2, \ldots, s \tag{355b}$$

$$y_n = y_{n-1} + \sum_{i=1}^{s} b_i z_i Y_i. \tag{355c}$$

Let $Z = \mathrm{diag}(z_1, z_2, \ldots, z_s)$ so that (355b) and (355c) can be rewritten as

$$(I - AZ)Y = y_{n-1}e \tag{355d}$$

$$y_n = y_{n-1} + b^{\mathrm{T}}ZY, \tag{355e}$$

where $Y = [Y_1, Y_2, \ldots, Y_s]^{\mathrm{T}}$. Assuming that the linear equations occurring in (355d) have a solution, which will be the case if $\det(I - AZ) \ne 0$, the value of y_n/y_{n-1} is found to be

$$\frac{y_n}{y_{n-1}} = R(Z),$$

where $R(Z)$ is given by (355h) below and generalizes the function r given by (352a). Note that $r(z) = R(zI)$.

DEFINITION 355A.

A Runge–Kutta method is said to be *AN-stable* if for all $z_1, z_2, \ldots, z_s \in \mathbb{C}^-$ such that $z_i = z_j$ if $c_i = c_j$,

$$\det(I - AZ) \neq 0, \qquad (355f)$$

$$|R(Z)| \leq 1, \qquad (355g)$$

where

$$R(Z) = 1 + b^{\mathrm{T}} Z (I - AZ)^{-1} e. \qquad (355h)$$

The name *AN*-stability in this definition comes from the observation that it is *A*-stability modified for non-autonomous problems.

We remark that, in the original version of this definition given by Burrage and Butcher (1979) the requirement that $I - AZ$ should be non-singular was omitted. In the third of three illustrative examples of what can happen for particular Runge–Kutta methods, the significance of this requirement will be highlighted.

The two examples which precede this show that *AN*-stability is possible but that it is not necessarily a consequence of *A*-stability. After the three examples, we come to a theorem which characterizes those *AN*-stable methods for which $c_1, c_2, \ldots, c_s$ are all distinct.

Consider the Gauss method with $s = 2$:

$$
\begin{array}{c|cc}
\dfrac{1}{2} - \dfrac{\sqrt{3}}{6} & \dfrac{1}{4} & \dfrac{1}{4} - \dfrac{\sqrt{3}}{6} \\[2ex]
\dfrac{1}{2} + \dfrac{\sqrt{3}}{6} & \dfrac{1}{4} + \dfrac{\sqrt{3}}{6} & \dfrac{1}{4} \\[2ex]
\hline
& \dfrac{1}{2} & \dfrac{1}{2}
\end{array}
\qquad (355i)
$$

We find that $\det(I - AZ) = 1 - (z_1 + z_2)/4 + z_1 z_2/12$ and, generalizing the usual stability function $r(z) = (1 + z/2 + z^2/12)/(1 - z/2 + z^2/12)$,

$$R(Z) = \frac{1 + (z_1 + z_2)/4 + z_1 z_2/12}{1 - (z_1 + z_2)/4 + z_1 z_2/12}. \qquad (355j)$$

If $\mathrm{Re}(z_1) \leq 0$ and $\mathrm{Re}(z_2) \leq 0$, it is not possible that $\det(I - AZ) = 0$, since this would imply that $(1 - z_1/3) + (3 - z_2)^{-1} = 0$, contrary to the observation that each of the two terms has a positive real part. Since the denominator of (355j) never vanishes we can use the maximum modulus principle for a regular function of two complex variables to see that it is sufficient to prove that $|R(Z)| \leq 1$ for z_1 and z_2 purely imaginary. This of course holds because the numerator and denominator are conjugates.

An alternative verification that this method is *AN*-stable comes from dealing separately with the trivial case that $z_1z_2 = 0$ and otherwise noting that

$$R(Z) = 1 + \frac{2}{-1 + 4/(z_1 + z_2) + \frac{1}{3}(1/z_1 + 1/z_2)^{-1}}$$

and that both $4/(z_1 + z_2)$ and $\frac{1}{3}(1/z_1 + 1/z_2)^{-1}$ have non-positive real parts.

Even though (355i) is both A-stable and AN-stable, as is the case also for other methods of the Gauss type, not all A-stable methods are AN-stable. For a counterexample given by Burrage and Butcher (1979)

$$
\begin{array}{c|cc}
\frac{1}{4} & \frac{1}{8} & \frac{1}{8} \\[2mm]
\frac{3}{4} & \frac{3}{8} & \frac{3}{8} \\[2mm]
\hline
& \frac{1}{2} & \frac{1}{2}
\end{array}
$$

it is found that $r(z) = (1 + z/2)/(1 - z/2)$, just as for the A-stable mid-point rule method, whereas

$$R(Z) = \frac{8 + 3z_1 + z_2}{8 - 3z_2 - z_1}$$

which equals -1.4 when $z_1 = -12$, $z_2 = 0$.

The final example from Crouzeix, Hundsdorfer and Spijker [1983] illustrates the crucial role played by the non-singularity of $I - AZ$. For this method

$$
\begin{array}{c|ccc}
\frac{1}{2} - \frac{\sqrt{6}}{6} & \frac{1}{8} & \frac{1}{4} - \frac{\sqrt{6}}{3} & \frac{1}{8} + \frac{\sqrt{6}}{6} \\[2mm]
\frac{1}{2} & \frac{1}{8} + \frac{\sqrt{6}}{6} & \frac{1}{4} & \frac{1}{8} - \frac{\sqrt{6}}{6} \\[2mm]
\frac{1}{2} + \frac{\sqrt{6}}{6} & \frac{1}{8} - \frac{\sqrt{6}}{6} & \frac{1}{4} + \frac{\sqrt{6}}{3} & \frac{1}{8} \\[2mm]
\hline
& \frac{1}{4} & \frac{1}{2} & \frac{1}{4}
\end{array}
$$

it is found that

$$\det(I - AZ) = 1 - \left(\frac{z_1}{8} + \frac{z_2}{4} + \frac{z_3}{8}\right) + \frac{z_2(z_1 + z_3)}{3} + \frac{z_1 z_3}{6} - \frac{3z_1 z_2 z_3}{8},$$

which vanishes when $z_1 = 2z_2 = z_3 = i\sqrt{2}$. The value of $R(Z)$ is found to be $\det(I + AZ)/\det(I - AZ)$ and is bounded in magnitude by 1 for all $[z_1, z_2, z_3]^T \in (\mathbb{C}^-)^3$, except at the two points $\pm[i\sqrt{2}, i\sqrt{2}/2, i\sqrt{2}]^T$, where both numerator and denominator vanish.

If an attempt were made to solve the differential equation system

$$y_1'(x) = -g(x)y_2(x) + 1,$$

$$y_2'(x) = g(x)y_1(x),$$

with $y_1(0) = y_2(0) = 0$ using a step size $h = 1$, where the function g takes the values

$$g\left(\frac{1}{2} - \frac{\sqrt{6}}{6}\right) = g\left(\frac{1}{2} + \frac{\sqrt{6}}{6}\right) = \sqrt{2}, \qquad g\left(\frac{1}{2}\right) = \frac{\sqrt{2}}{2},$$

then in the first integration step non-solvable algebraic equations arise. To see this, write the differential system as a single complex-valued differential equation in the variable $y(x) = y_1(x) + iy_2(x)$. The differential equation becomes $y'(x) = ig(x)y(x) + 1$ with $y(0) = 0$ and the algebraic equations become

$$(I - AZ)Y = c, \tag{355k}$$

where $Z = \text{diag}(i\sqrt{2}, i\sqrt{2}/2, i\sqrt{2})$. Let $v^{\mathrm{T}} = [-\frac{1}{2} + i\sqrt{3}/2, 1, -\frac{1}{2} - i\sqrt{3}/2]$ and we find that $v^{\mathrm{T}}(I - AZ) = 0$ but $v^{\mathrm{T}}c \neq 0$, showing that (355k) has no solution.

Finally, in this subsection, we will introduce a criterion which characterizes AN-stability in a fairly satisfactory way. We will need to make use of the fact that a real symmetric $s \times s$ matrix M can be written in the form

$$M = N^{\mathrm{T}}GN, \tag{355l}$$

where $G = \text{diag}(g_1, g_2, \ldots, g_s)$ and N is a non-singular $s \times s$ matrix with elements n_{ij}. It follows that the quadratic form $Q(x) = x^{\mathrm{T}}Mx$ can be written as a weighted sum of squares

$$Q(x) = \sum_{k=1}^{s} g_k \left(\sum_{i=1}^{s} n_{ki}x_i\right)^2$$

for $x = [x_1, x_2, \ldots, x_k]^{\mathrm{T}}$. It is well known that the numbers of positive, zero and negative values among $g_1, g_2, \ldots, g_s$ are invariants of M. If none of them is negative then M(or Q) is said to be *positive semidefinite* and if all are positive then M (or Q) is said to be *positive definite*. We will also generalize positive definiteness to non-symmetric matrices as follows.

DEFINITION 355B.

An $s \times s$ matrix A is *positive definite* if there exists a diagonal matrix $D = \text{diag}(d_1, d_2, \ldots, d_s)$ with $d_1, d_2, \ldots, d_s > 0$ such that $DA + A^{\mathrm{T}}D$ is positive definite.

It is easy to see that a symmetric matrix is positive definite in the sense of this definition iff it is positive definite in the usual sense. It is also easy to see that if A is positive definite then it cannot be singular because if $Av = 0$

then $v^T(DA + A^TD)v = 0$ and, furthermore, because of the congruence $(A^{-1})^T(DA + A^TD)A^{-1} = DA^{-1} + (A^{-1})^TD$, A^{-1} is also positive definite.

For the characterization of AN-stability, define $B = \text{diag}(b_1, b_2, \ldots, b_s)$ so that $b^T = e^TB$ and define M by (355m) below so that M is a symmetric matrix with the (i, j) element equal to $b_i a_{ij} + b_j a_{ji} - b_i b_j$. Because of its importance in a wider context, M is sometimes known as the *algebraic stability matrix* for the method. In the theorem which follows, from Burrage and Butcher (1979) we restrict ourselves to non-confluent methods in the sense that $c_i \neq c_j$ for $i \neq j$.

THEOREM 355C.

If a non-confluent Runge–Kutta method is AN-stable then

$$M = BA + A^TB - bb^T \tag{355m}$$

and $B = \text{diag}(b_1, b_2, \ldots, b_s)$ are each positive semidefinite matrices.

Proof.

If $b_i < 0$ for some $i \in \{1, 2, \ldots, s\}$ then choose Z so that $z_j = 0$ if $j \neq i$ and $z_i = -\varepsilon$ for ε a positive real number. We see that

$$|R(Z)| = 1 - \varepsilon b_i + O(\varepsilon^2)$$

which exceeds 1 for ε sufficiently small.

Having established that $b_1, b_2, \ldots, b_s \geq 0$, choose $Z = \text{diag}(i\varepsilon y_1, i\varepsilon y_2, \ldots, i\varepsilon y_s)$ for $i = \sqrt{(-1)}$, $y = [y_1, y_2, \ldots, y_s]^T$ an arbitrary real vector and ε a non-zero real number. Expand $(I - AZ)^{-1}$ by the geometric series in (355h) and we find

$$R(Z) = 1 + \varepsilon i b^T y - \varepsilon^2 y^T BAy + O(\varepsilon^3)$$

so that

$$|R(Z)|^2 = R(\bar{Z})R(Z)$$
$$= [1 - \varepsilon i y^T b - \varepsilon^2 y^T A^T By + O(\varepsilon^3)][1 + \varepsilon i b^T y - \varepsilon^2 y^T BAy + O(\varepsilon^3)]$$
$$= 1 + \varepsilon i(b^T y - y^T b) - \varepsilon^2 y^T(BA + A^TB - bb^T)y + O(\varepsilon^3)$$
$$= 1 - \varepsilon^2 y^T My + O(\varepsilon^3)$$

which is greater than 1 for ε sufficiently small unless $y^T My \geq 0$. ∎

356   Non-linear stability.

We consider the stability of methods with a non-linear test problem $y'(x) = f(y(x))$ in which f is one-sided Lipschitz continuous with one-sided Lipschitz constant 0. As we saw from (112g) in the proof of theorem 112G, if y and z are two solutions to such a problem with initial values given at x_0 then for any $x > x_0$ we have the bound $\| y(x) - z(x) \| \le \| y(x_0) - z(x_0) \|$. We consider a stability condition that guarantees that computed solutions have a similar behaviour.

DEFINITION 356A.

A Runge–Kutta method is said to be *B-stable* if for any problem $y'(x) = f(y(x))$ where f is such that $(f(u) - f(v))^{T}(u - v) \le 0$ for all u and v, then $\| y_n - z_n \| \le \| y_{n-1} - z_{n-1} \|$ for two solution sequences $\ldots,$ $y_{n-1}, y_n, \ldots$ and $\ldots, z_{n-1}, z_n, \ldots$.

If the coefficients of the method are

$$
\begin{array}{c|cccc}
c_1 & a_{11} & a_{12} & \cdots & a_{1s} \\
c_2 & a_{21} & a_{22} & \cdots & a_{2s} \\
\vdots & \vdots & \vdots & & \vdots \\
c_s & a_{s1} & a_{s2} & \cdots & a_{ss} \\
\hline
 & b_1 & b_2 & \cdots & b_s
\end{array}
, \qquad (356a)
$$

then it turns out to be appropriate to make use of the matrix

$$
M = \begin{bmatrix}
m_{11} & m_{12} & \cdots & m_{1s} \\
m_{21} & m_{22} & \cdots & m_{2s} \\
\vdots & \vdots & & \vdots \\
m_{s1} & m_{s2} & \cdots & m_{ss}
\end{bmatrix}, \qquad (356b)
$$

defined by (355m). We recall that the individual elements of M are given by

$$
m_{ij} = b_i a_{ij} + b_j a_{ji} - b_i b_j, \qquad i, j = 1, 2, \ldots, s. \qquad (356c)
$$

DEFINITION 356B.

If for a method (356a), M given by (356b) and (356c) is positive semidefinite and none of $b_1, b_2, \ldots, b_s$ is negative, then the method is said to be *algebraically stable*.

Since M is symmetric, it can be written in the form (355l) and if the method is algebraically stable then

$$
g_1, g_2, \ldots, g_s \ge 0. \qquad (356d)
$$

THEOREM 356C.

An algebraically stable Runge–Kutta method is *B-stable*.

Proof.

If the stages used to compute y_n from y_{n-1} (respectively z_n from z_{n-1}) are $Y_1, Y_2, \ldots, Y_s$ (respectively $Z_1, Z_2, \ldots, Z_s$) then

$$\left. \begin{aligned} Y_i &= y_{n-1} + h \sum_j a_{ij} f(Y_j), \\ Z_i &= z_{n-1} + h \sum_j a_{ij} f(Z_j), \end{aligned} \right\} \quad i = 1, 2, \ldots, s,$$

so that subtracting, taking the scalar product with $h b_i [f(Y_i) - f(Z_i)]$ and summing over $i = 1, 2, \ldots, s$, we find

$$h \sum_i b_i [f(Y_i) - f(Z_i)]^T (Y_i - Z_i)$$

$$= h \sum_i b_i [f(Y_i) - f(Z_i)]^T (y_{n-1} - z_{n-1})$$

$$+ h^2 \sum_{i,j} b_i a_{ij} [f(Y_i) - f(Z_i)]^T [f(Y_j) - f(Z_j)]. \tag{356e}$$

However,

$$h \sum_i b_i [f(Y_i) - f(Z_i)] = (y_n - z_n) - (y_{n-1} - z_{n-1}), \tag{356f}$$

which we substitute into the first term on the right-hand side of (356e). We also take the scalar product of (356f) with itself to give, after rearranging, the two equations.

$$(y_n - z_n)^T (y_{n-1} - z_{n-1}) - \| y_{n-1} - z_{n-1} \|^2$$

$$= - h^2 \sum_{i,j} b_i a_{ij} [f(Y_i) - f(Z_i)]^T [f(Y_j) - f(Z_j)]$$

$$+ h \sum_i b_i (f(Y_i) - f(Z_i))^T (Y_i - Z_i) \tag{356g}$$

and

$$\| y_n - z_n \|^2 - 2(y_n - z_n)^T (y_{n-1} - z_{n-1}) + \| y_{n-1} - z_{n-1} \|^2$$

$$= h^2 \sum_{i,j} b_i b_j [f(Y_i) - f(Z_i)]^T [f(Y_j) - f(Z_j)]. \tag{356h}$$

We now add to (356h) two copies of (356g), one of which has i and j interchanged, to give the result

$$\| y_n - z_n \|^2 - \| y_{n-1} - z_{n-1} \|^2$$

$$= - h \sum_{i,j} m_{ij} [f(Y_i) - f(Z_i)]^T [f(Y_j) - f(Z_j)]$$

$$+ h \sum_i b_i [f(Y_i) - f(Z_i)]^T (Y_i - Z_i)$$

which, because M is positive semidefinite, can be written in the form

$$\| y_n - z_n \|^2 - \| y_{n-1} - z_{n-1} \|^2$$

$$= - h^2 \sum_k g_k \| \sum_i n_{ki}[f(Y_i) - f(Z_i)] \|^2$$

$$+ h \sum_i b_i [f(Y_i) - f(Z_i)]^T (Y_i - Z_i)$$

which is non-positive because $g_1, g_2, \ldots, b_1, b_2, \ldots$ are non-negative and because for $i = 1, 2, \ldots, s$, $[f(Y_i) - f(Z_i)]^T(Y_i - Z_i) \leq 0$. ∎

It is interesting to note that if the definition of B-stability were modified to use the test equation

$$y'(x) = f(x, y(x))$$

where, for all x, u, v, $[f(x, u) - f(x, v)]^T(u - v) \leq 0$, then, as we state in theorem 356E below, an algebraically stable method still enjoys the modified B-stability property. Since this modified property has a similar relationship to B-stability as AN-stability has to A-stability it will be known as 'BN-stability'.

DEFINITION 356D.

A Runge–Kutta method is said to be BN-stable if for any problem $y'(x) = f(x, y(x))$, where f is such that $[f(x, u) - f(x, v)]^T(u - v) \leq 0$ for all x and for all u and v, then $\| y_n - z_n \| \leq \| y_{n-1} - z_{n-1} \|$ for two solution sequences $\ldots, y_{n-1}, y_n, \ldots$ and $\ldots, z_{n-1}, z_n, \ldots$.

Just as for theorem 356C we have the following result.

THEOREM 356E.

An algebraically stable Runge–Kutta method is BN-stable.

Proof.

As in the proof of theorem 356C. ∎

Details of the interrelations between algebraic stability and other stability properties are given in Burrage and Butcher (1979) and in Crouzeix (1979). The original definition of B-stability was given by Butcher (1975a).

For examples of algebraically stable and therefore B- and BN-stable methods we have the methods already shown to be A-stable.

THEOREM 356F.

The s-stage methods of order $2s$ and the Radau IA, Radau IIA and Lobatto IIIC methods are algebraically stable.

Proof.

For each of these methods, $b_1, b_2, \ldots, b_s$ are positive so it is only necessary to verify the positive semidefiniteness of M. Since for these methods $c_1, c_2, \ldots, c_s$ are distinct, the (Vandermonde) matrix

$$C = \begin{bmatrix} 1 & c_1 & c_1^2 & \cdots & c_1^{s-1} \\ 1 & c_2 & c_2^2 & \cdots & c_2^{s-1} \\ \vdots & \vdots & \vdots & & \vdots \\ 1 & c_s & c_s^2 & \cdots & c_s^{s-1} \end{bmatrix}$$

is non-singular and $C^\mathsf{T} MC$ and M are either both positive semidefinite or neither is. We can compute the k, l element of $C^\mathsf{T} MC$ $(k, l = 1, 2, \ldots, s)$ making use of the elementary weights for the method and the result is

$$\sum_i \sum_j (b_i a_{ij} + b_j a_{ji} - b_i b_j) c_i^{k-1} c_j^{l-1}$$

$$= \Phi([\tau^{k-1}[\tau^{l-1}]]) + \Phi([\tau^{l-1}[\tau^{k-1}]])$$

$$- \Phi([\tau^{k-1}])\Phi([\tau^{l-1}]).$$

By the order conditions, this can be evaluated as long as $k + l$ does not exceed the order and the result is

$$\frac{1}{l(k+l)} + \frac{1}{k(k+l)} - \frac{1}{kl} = 0.$$

This establishes the result in full for the order $2s$ case but it remains to show that

$$2\Phi([\tau^{s-1}[\tau^{s-1}]]) - \frac{1}{s^2} \geq 0 \tag{356i}$$

for the other cases and that, in addition,

$$\Phi([\tau^{s-2}[\tau^{s-1}]]) + \Phi([\tau^{s-1}[\tau^{s-2}]]) - \frac{1}{s(s-1)} = 0 \tag{356j}$$

in the Lobatto IIIC case.

We first consider (356i) for the Radau IA and Radau IIA cases for which, in the notation of subsection 342, $D(s)$ (in the Radau IA case) or $A(s)$ (in the Radau IIA case) holds. Hence, for Radau IA,

$$\Phi([\tau^{s-1}[\tau^{s-1}]]) = \sum_{i,j} b_i c_i^{s-1} a_{ij} c_j^{s-1}$$

$$= \frac{1}{s} \sum_j b_j (1 - c_j^s) c_j^{s-1}$$

$$= \frac{1}{s^2} - \frac{1}{s} \sum_j b_j c_j^{2s-1},$$

whereas for Radau IIA,

$$\Phi([\tau^{s-1}[\tau^{s-1}]]) = \sum_{i,j} b_i c_i^{s-1} a_{ij} c_j^{s-1}$$

$$= \frac{1}{s} \sum_i b_i c_i^{2s-1},$$

and it remains to prove that for Radau I quadrature

$$\sum_j b_j c_j^{2s-1} \le \frac{1}{2s}, \tag{356k}$$

since the corresponding result for Radau II quadrature with the inequality reversed is equivalent. We note that (356k) is equivalent to the statement that $\sum_j b_j p(c_j) \le \int_0^1 p(c)\,dc$ where $p(c) = c(c-c_2)^2 \cdots (c-c_s)^2$ assuming that $c_1 = 0$, and this is true because the summation is zero and the integrand is positive.

To prove (356i) in the Lobatto IIIC case, we note that for the reflection of this method an inequality of this sort would be reversed. Hence, we consider instead the reverse inequality for the Lobatto case. Thus we must prove that

$$\sum_{i,j} b_i c_i^{s-1} a_{ij} c_j^{s-1} - \frac{1}{2s^2} \le 0,$$

or, what is equivalent, that $L(p) \le 0$, where

$$L(p) = \sum_{i,j} b_i p(c_i) a_{ij} c_j^{s-1} - \frac{1}{s} \int_0^1 p(c) c^s \, dc \tag{356l}$$

and where $p(c) = c^{s-1}$. Let $p = p_1 + p_2 + p_3$, where $p_1(c) = (c-c_2)(c-c_3) \cdots (c-1)$, $p_2(c) = (s/2) c^{s-2}$ and where $\deg(p_3) < s-2$, so that $L(p) = L(p_1) + L(p_2) + L(p_3)$. The first term in $L(p_1)$ vanishes, the first term of $L(p_2)$ is equal to

$$\frac{s}{2(s-1)} \sum_j b_j (1 - c_j^{s-1}) c_j^{s-1} = \frac{1}{2(s-1)} - \frac{s}{2(s-1)} \sum_j b_j c_j^{2s-2}$$

and $L(p_3) = 0$. Hence, we evaluate $L(p)$ given by (356l) as

$$L(p) = \frac{1}{s} \int_0^1 (c-c_2) \cdots (c-c_{s-1})(1-c) c^s \, dc$$

$$+ \frac{1}{2(s-1)} - \frac{s}{2(s-1)} \sum_j b_j c_j^{2s-2} - \frac{1}{2} \int_0^1 c^{2s-2} \, dc$$

$$= \frac{1}{s} \int_0^1 (c-c_2) \cdots (c-c_{s-1})(1-c) c^s \, dc$$

$$- \frac{s}{2(s-1)} \left(\sum_j b_j c_j^{2s-2} - \int_0^1 c^{2s-2} \, dc \right)$$

$$= \sum_j b_j q(c_j) - \int_0^1 q(c) \, dc, \tag{356m}$$

where q is given by

$$q(c) = \frac{1}{s}(c - c_2) \cdots (c - c_{s-1})(c - 1)c^s - \frac{s}{2(s-1)}c^{2s-2}.$$

Denote the expression given in (356m) by $M(q)$ and write $q = q_1 + q_2 + q_3$, where

$$q_1(c) = \frac{1}{s}c(c - c_2)^2 \cdots (c - c_{s-1})^2(c - 1)(c - \tfrac{1}{2}),$$

$$q_2(c) = \left(\frac{s-1}{2s} - \frac{s}{2s-1}\right)c(c - c_2)^2 \cdots (c - c_{s-1})^2(c - 1)$$

and $\deg(q_3) < 2s - 2$ so that

$$L(p) = M(q_1) + M(q_2) + M(q_3).$$

By the order conditions $M(q_3) = 0$, by the oddness of the integrand as a function of $c - \tfrac{1}{2}$, $M(q_1) = 0$, and because the summation vanishes in $M(q_2)$, we have

$$L(p) = M(q) = -\left[\frac{s-1}{2s} - \frac{s}{2(s-1)}\right]\int_0^1 q_2(c)\,dc$$

$$= -\frac{2s-1}{2s(s-1)}\int_0^1 c(c - c_2)^2 \cdots (c - c_{s-1})^2(1 - c)\,dc,$$

which is negative because the integrand is non-negative.

It remains to prove (356j) in the Lobatto IIIC case. We have, using the conditions $A(s-1)$ and $D(s-1)$ of subsection 342,

$$\sum b_i c_i^{s-2} a_{ij} c_j^{s-1} + \sum b_i c_i^{s-1} a_{ij} c_j^{s-2}$$

$$= \frac{1}{s-1}\sum b_j(1 - c_j^{s-1})c_j^{s-1} + \frac{1}{s-1}\sum b_i c_i^{s-1} c_i^{s-1}$$

$$= \frac{1}{s-1}\sum b_i c_i^{s-1}$$

$$= \frac{1}{s(s-1)}$$

by the order conditions. ∎

That the methods referred to in theorem 356F are B-stable is also proved by Wanner (1976) but using an approach which bypasses any consideration of algebraic stability.

357 Generalizations of algebraic stability.

Let k and l be real numbers with $k > 0$. We wish to determine conditions under which

$$\| y_n - z_n \|^2 \le k \| y_{n-1} - z_{n-1} \|^2$$

in the case of differential equations for which

$$[f(x, u) - f(x, v)]^{\mathsf{T}}(u - v) \le \lambda \| u - v \|^2 \tag{357a}$$

for all u, v. For equations which satisfy this one-sided Lipschitz condition, we see from an argument used in the proof of theorem 112G that the difference between two solutions satisfies the condition

$$\| y(x) - z(x) \|^2 \le \exp(2l) \| y(x - h) - z(x - h) \|^2,$$

where $l = h\lambda$.

Hence, to the extent that $k^{1/2}$ can be chosen as close to $\exp(l)$ for l near zero, the stability properties of the numerical solution are more or less related to those of the exact solution. We note that $k = 1$, $l = 0$ corresponds to BN-stability. In the argument which follows we revert to an autonomous problem for convenience only.

Let D denote a diagonal matrix of non-negative real numbers and d the corresponding vector $d = De = [d_1, d_2, \ldots, d_s]^{\mathsf{T}}$. From the sum of two copies of (356e), with b replaced by d in each case and with i and j interchanged in one copy of the last term, we find

$$0 = 2h \sum_i d_i [f(Y_i) - f(Z_i)]^{\mathsf{T}}(Y_i - Z_i)$$

$$- 2h \sum_i d_i [f(Y_i) - f(Z_i)]^{\mathsf{T}}(y_{n-1} - z_{n-1})$$

$$- h^2 \sum_{i,j} (d_i a_{ij} + d_j a_{ji}) [f(Y_i) - f(Z_i)]^{\mathsf{T}}[f(Y_j) - f(Z_j)],$$

so that

$$0 \le 2l \sum_i d_i \| Y_i - Z_i \|^2 - 2h \sum_i d_i [f(Y_i) - f(Z_i)]^{\mathsf{T}}(y_{n-1} - z_{n-1})$$

$$- h^2 \sum_{i,j} (d_i a_{ij} + d_j a_{ji}) [f(Y_i) - f(Z_i)]^{\mathsf{T}}[f(Y_j) - f(Z_j)], \tag{357b}$$

where we have used (357a) to bound the first term with $l = h\lambda$ substituted.

Add (357b) to $\| y_n - z_n \|^2 - k \| y_{n-1} - z_{n-1} \|^2$, expand $y_n - z_n$ and $Y_i - Z_i$ as linear combinations of $y_{n-1} - z_{n-1}$ and of $hf(Y_j) - hf(Z_j)$ for $j = 1, 2, \ldots, s$ and we find

$$\| y_n - z_n \|^2 - k \| y_{n-1} - z_{n-1} \|^2 \le - m_{00} \| y_{n-1} - z_{n-1} \|^2$$

$$- 2h \sum_{i=1}^{s} m_{0i}(y_{n-1} - z_{n-1})^{\mathsf{T}}[f(Y_i) - f(Z_i)]$$

$$- h^2 \sum_{i,j=1}^{s} m_{ij} [f(Y_i) - f(Z_i)]^{\mathsf{T}}[f(Y_j) - f(Z_j)], \tag{357c}$$

where the symmetric $(1 + s) \times (1 + s)$ matrix

$$M = \begin{bmatrix} m_{00} & m_{01} & m_{02} & \cdots & m_{0s} \\ m_{10} & m_{11} & m_{12} & \cdots & m_{1s} \\ m_{20} & m_{21} & m_{22} & \cdots & m_{2s} \\ \vdots & \vdots & \vdots & & \vdots \\ m_{s0} & m_{s1} & m_{s2} & \cdots & m_{ss} \end{bmatrix}$$

is given by

$$M = \begin{bmatrix} k - 1 - 2le^{\mathrm{T}}d & d^{\mathrm{T}} - b^{\mathrm{T}} - 2ld^{\mathrm{T}}A \\ d - b - 2lA^{\mathrm{T}}d & DA + A^{\mathrm{T}}D - bb^{\mathrm{T}} - 2lA^{\mathrm{T}}DA \end{bmatrix}. \tag{357d}$$

Motivated by the inequality (357c) and the observation that the right-hand side is the negative of a quadratic form with matrix of coefficients given by (357d), we give a definition and theorem.

DEFINITION 357A.

An implicit Runge–Kutta method is *(k, l) algebraically stable* if for some non-negative diagonal matrix D, M given by (357d) is positive semidefinite.

THEOREM 357B.

If $\ldots, y_{n-1}, y_n, \ldots$ and $\ldots, z_{n-1}, z_n, \ldots$ are two numerical sequences computed from a (k, l) algebraically stable Runge–Kutta method applied to a one-sided Lipschitz continuous differential equation with one-sided Lipschitz constant l/h, then

$$\| y_n - z_n \| \le k^{1/2} \| y_{n-1} - z_{n-1} \|.$$

It is sometimes convenient to transform M into a form in which l appears only in one of the four blocks.

THEOREM 357C.

If an implicit Runge–Kutta method is such that $e \in \mathrm{Im}(A)$, so that a vector v exists such that $e = Av$, then the method is (k, l) algebraically stable iff $\widetilde{M}$ is non-negative definite, where

$$\widetilde{M} = \begin{bmatrix} k - (1 - b^{\mathrm{T}}v)^2 & -v^{\mathrm{T}}DA - (1 - b^{\mathrm{T}}v)b^{\mathrm{T}} \\ -A^{\mathrm{T}}Dv - (1 - b^{\mathrm{T}}v)b & DA + A^{\mathrm{T}}D - bb^{\mathrm{T}} - 2lA^{\mathrm{T}}DA \end{bmatrix}.$$

Proof.

$$\widetilde{M} = \begin{bmatrix} 1 & -v^{\mathrm{T}} \\ 0 & I \end{bmatrix} M \begin{bmatrix} 1 & 0 \\ -v & I \end{bmatrix}. \quad \blacksquare$$

In particular, if A is non-singular, then

$$\tilde{M} = \begin{bmatrix} k - (1 - b^T A^{-1} e)^2 & -e^T (A^T)^{-1} DA - (1 - b^T A^{-1} e) b^T \\ -A^T DA^{-1} e - (1 - b^T A^{-1} e) b & DA + A^T D - bb^T - 2lA^T DA \end{bmatrix}$$

For a method with s stages of Gauss type, it is known that $1 - b^T A^{-1} e = (-1)^s$ and hence

$$\tilde{M} = \begin{bmatrix} k - 1 & -e^T (A^T)^{-1} DA - (-1)^s b^T \\ -A^T DA^{-1} e - (-1)^s b & DA + A^T D - bb^T - 2lA^T DA \end{bmatrix},$$

so that $k \geq 1$ even for negative l. For a method of Radau IA, Radau IIA or Lobatto IIIC types, $1 - b^T A^{-1} e = 0$ and $\tilde{M}$ takes the form

$$\tilde{M} = \begin{bmatrix} k & -e^T (A^T)^{-1} DA \\ -A^T DA^{-1} e & DA + A^T D - bb^T - 2lA^T DA \end{bmatrix},$$

and it is always possible, by choosing l sufficiently negative, to decrease k to an arbitrarily small positive value.

For $l > 0$ it is clear from the form of (357d) that a necessary condition for (k, l) algebraic stability is that A is positive definite. It is also easy to see that for l sufficiently small the converse is also true and we now present a strengthened form of the converse result on the assumption that a method is $(1, 0)$ algebraically stable.

THEOREM 357D.

If a Runge–Kutta method is positive definite and algebraically stable then there exists θ and $l_0 > 0$ such that for all $l \in [0, l_0]$, the method is $(1 + 2\theta l, l)$ algebraically stable.

Proof.

Since A is positive definite, a positive diagonal matrix F exists such that $FA + A^T F$ is positive definite. Choose $\varphi > 0$ so that $X_0 = \varphi(FA + A^T F) - A^T BA$ is positive definite and l_0 such that $X_1 = X_0 - 2l_0\varphi A^T FA$ is positive definite. Let $v_0 = (\varphi F - A^T B)e$, $v_1 = v_0 - 2l_0\varphi A^T Fe$, $\theta_0 = e^T b + v_0^T X_0^{-1} v_0$ and $\theta_1 = e^T(b + 2l\varphi Fe) + v_1^T X_1^{-1} v_1$. Choose $\theta = \max(\theta_0, \theta_1)$ so that

$$M_1 = \begin{bmatrix} \theta - e^T(b + 2l\varphi Fe) & e^T[\varphi F - (B + 2l\varphi F)A] \\ [\varphi F - A^T(B + 2l\varphi F)]e & \varphi(FA + A^T F) - A^T(B + 2l\varphi F)A \end{bmatrix}$$

is positive semidefinite for $l = 0$ and $l = l_0$, and hence for all $l \in [0, l_0]$.
For $k = 1 + 2\theta l$, $D = B + 2l\varphi F$ we find that M in (357d) is given by

$$M = \begin{bmatrix} 0 & 0 \\ 0 & BA + A^T B - bb^T \end{bmatrix} + 2lM_1,$$

both terms of which are positive semidefinite. ∎

To enable us to apply this result in practice, we will show that the standard types of A-stable methods, the Gauss, Radau IA and Radau IIA methods, are all positive definite. Conspicuously absent is the s-stage Lobatto IIIC method which is not positive definite if $s > 2$. We need two preparatory lemmas.

LEMMA 357E.

For any method with the following properties:

$$c_1, c_2, \ldots, c_s > 0, \tag{357e}$$

$$b_1, b_2, \ldots, b_s > 0, \tag{357f}$$

$$\sum_{j=1}^{s} a_{ij} c_j^{k-1} = \frac{1}{k} c_i^k, \qquad i, k = 1, 2, \ldots, s, \tag{357g}$$

$$\sum_{i=1}^{s} b_i c_i^{k-1} = \frac{1}{k}, \qquad k = 1, 2, \ldots, 2s - 1, \tag{357h}$$

A is positive definite with $D = \operatorname{diag}(b_1 c_1^{-1}, b_2 c_2^{-1}, \ldots, b_s c_s^{-1})$.

Proof.

The diagonal elements of D are positive because of (357e) and (357f). Let U denote the $s \times s$ matrix with the (i, j) element $j c_i^{j-1}$. Since $c_1, c_2, \ldots, c_s$ are distinct because of (357h), U is non-singular. We will show that $U^T(DA + A^T D)U$ is positive definite. The (k, l) element of $U^T DAU$ is given by $kl\sum_{i,j=1}^{s} b_i c_i^{k-2} a_{ij} c_j^{l-1}$, which equals $k\sum_{i=1}^{s} b_i c_i^{k+l-2}$ by (357g), which equals $k/(k + l - 1)$ by (357h). Hence, the (k, l) element of $U^T(DA + A^T D)U$ is $(k + l)/(k + l - 1) = 1 + 1/k + l - 1)$. That is,

$$U^T(DA + A^T D)U = ee^T + H,$$

where H is the $s \times s$ Hilbert matrix. Since ee^T is positive semidefinite and H is positive definite, the result follows. ∎

LEMMA 357F.

For any method with the following properties:

$$c_1, c_2, \ldots, c_s < 1, \tag{357i}$$

$$b_1, b_2, \ldots, b_s > 0, \tag{357j}$$

$$\sum_{i=1}^{s} b_i c_i^{k-1} a_{ij} = \frac{1}{k} b_j(1 - c_j^k), \qquad j, k = 1, 2, \ldots, s, \tag{357k}$$

$$\Sigma b_i c_i^{k-1} = \frac{1}{k}, \qquad k = 1, 2, \ldots, 2s - 1, \tag{357l}$$

A is positive definite with $D = \operatorname{diag}(b_1(1 - c_1), b_2(1 - c_2), \ldots, b_s(1 - c_s))$.

Proof.

From (357k) and the fact that $c_1, c_2, \ldots, c_s$ are distinct because of (357l) it follows that A is non-singular. We will prove that A^{-1} is positive definite. Let $D = \text{diag}(d_1, d_2, \ldots, d_s)$ with $d_1, d_2, \ldots, d_s > 0$ by (357i) and (357j). Let V be the $s \times s$ matrix with the (i, j) element c_i^{j-1}. Since V is non-singular it will be sufficient to prove that $V^T[DA^{-1} + (A^{-1})^T D] V$ is positive definite. Denoting the (i, j) element of A^{-1} by $a_{ij}^{(-1)}$, we have, for the (k, l) element of $V^T DA^{-1} V$,

$$m_{kl} = \sum_{i,j=1}^{s} b_i(1 - c_i)c_i^{k-1}a_{ij}^{(-1)}c_j^{l-1}.$$

If $k = 1$ then using (357k) we see that $m_{kl} = \sum_{j=1}^{s} b_j c_j^{l-1}$, which equals $1/l$ using (357l). If $k > 1$, then using (357k) and (357l) we find

$$m_{kl} = \sum_{i,j=1}^{s} b_i(1 - c_i^k)a_{ij}^{(-1)}c_j^{l-1} - \sum_{i,j=1}^{s} b_i(1 - c_i^{k-1})a_{ij}^{(-1)}c_j^{l-1}$$

$$= k\sum_{j=1}^{s} b_j c_j^{k+l-2} - (k - 1)\sum_{j=1}^{s} b_j c_j^{k+l-3}$$

$$= \frac{k}{k+l-1} - \frac{k-1}{k+l-2}.$$

Adding m_{kl} to m_{lk} to obtain the (k, l) element of $V^T[DA^{-1} + (A^{-1})^T D] V$ we find

$$V^T[DA^{-1} + (A^{-1})^T D] V = e_1 e_1^T + H,$$

where H is the Hilbert matrix and is thus positive definite. As we remarked following definition 355B, A positive definite and A^{-1} positive definite are equivalent. ∎

THEOREM 357G.

The A matrix is positive definite for Gauss, Radau IA and Radau IIA methods.

Proof.

The Radua IA case is covered by lemma 357F and the Radau IIA case by lemma 357E. The result for Gauss methods follows from either of lemmas 357E or 357F. ∎

Combining the results of theorems 357D and 357G we have the following result.

COROLLARY 357H.

For each Gauss, Radau IA and Radau IIA method, there exist θ and $l_0 > 0$ such that for $l \in [0, l_0]$, the method is $(1 + 2\theta l, l)$ algebraically stable.

Because of the relationship between linear and non-linear stability, the value of θ occurring in theorem 357F cannot be less than unity. That this minimum value of θ is achieved, at least for the Gauss and Radau IA methods, we now show.

THEOREM 357I.

For any Gauss or Radau IA method, there exists K and $l_0 > 0$ such that for $l \in [0, l_0]$ the method is $(1 + 2l + 4l^2 K, l)$ algebraically stable.

Proof.

Let $F = \text{diag}(f_1, f_2, \ldots, f_s)$ be chosen with positive diagonal elements so that $FA + A^T F$ is positive definite and define $d_i = b_i + 2lb_i(1 - c_i) + 4l^2 \varphi f_i$ $(i = 1, 2, \ldots, s)$, $D = \text{diag}(d_1, d_2, \ldots, d_s)$, $C = \text{diag}(c_1, c_2, \ldots, c_s)$. If $k = 1 + 2l + 4l^2 K$ then M given by (357d) is equal to

$$M = M_0 + 2lM_1 + 4l^2 M_2 + 8l^3 M_3,$$

where

$$M_0 = \begin{bmatrix} 0 & 0 \\ 0 & BA + A^T B - bb^T \end{bmatrix},$$

$$M_1 = \begin{bmatrix} 1 - e^T b & b^T - b^T C - b^T A \\ b - Cb - A^T b & (B - BC)A + A^T(B - BC) - A^T BA \end{bmatrix},$$

$$M_2 = \begin{bmatrix} K - e^T(b - Cb) & \varphi e^T F - (b^T - b^T C)A \\ \varphi Fe - A^T(b - Cb) & \varphi(FA + A^T F) - A^T(B - BC)A \end{bmatrix},$$

$$M_3 = \begin{bmatrix} -\varphi e^T Fe & -\varphi e^T FA \\ -\varphi A^T Fe & -\varphi A^T FA \end{bmatrix}.$$

By algebraic stability, M_0 is positive semidefinite. By choosing φ and K sufficiently large positive numbers, M_2 can be made positive definite and by choosing l_0 a sufficiently small positive number, $4l^2 M_2 + 8l^3 M_3$ can be made positive semidefinite for all $l \in [0, l_0]$. It remains to show that M_1 is positive semidefinite. From known properties of the Gauss and Radau methods, we see that

$$M_1 = \begin{bmatrix} 0 & 0 \\ 0 & A^T NA \end{bmatrix},$$

where $N = (A^T)^{-1}(B - BC) + (B - BC)A^{-1} - B$. From the matrix $V^T NV$, where the Vandermonde matrix V is as in the proof of theorem 357F. We find that

$$V^{\mathrm{T}}NV = e_1 e_1^{\mathrm{T}} + H - V^{\mathrm{T}}BV = e_1 e_1^{\mathrm{T}},$$

which is positive semidefinite. ∎

358 The W transformation.

It was found to be convenient in the proof of theorem 356F to carry out a congruence transformation on the matrix M, occurring in the definition of algebraic stability, using the Vandermonde matrix C as the transforming matrix. Following the work of Hairer and Wanner (1981, 1982), and in further papers by Hairer (1982a) and by Wanner (1980a), we consider the effect of using instead a particular generalized Vandermonde matrix. To simplify the discussion, we will assume that $c_1, c_2, \ldots, c_s$ are all distinct and that $b_1, b_2, \ldots, b_s$ are all positive.

Under these assumptions, it is possible to construct a sequence of polynomials $P_0, P_1, \ldots, P_{s-1}$ of degrees $0, 1, \ldots, s-1$ respectively such that

$$\sum_{i=1}^{s} b_i P_{k-1}(c_i) P_{l-1}(c_i) = \delta_{kl}, \qquad k, l = 1, 2, \ldots, s, \tag{358a}$$

and to define W as the $s \times s$ matrix

$$W = \begin{bmatrix} P_0(c_1) & P_1(c_1) & \cdots & P_{s-1}(c_1) \\ P_0(c_2) & P_1(c_2) & \cdots & P_{s-1}(c_2) \\ \vdots & \vdots & & \vdots \\ P_0(c_s) & P_1(c_s) & \cdots & P_{s-1}(c_s) \end{bmatrix}.$$

We note that W can be found from a QR factorization

$$B^{\frac{1}{2}}C = (B^{\frac{1}{2}}W)R,$$

where $B^{\frac{1}{2}}$ is the square root of $B = \operatorname{diag}(b_1, b_2, \ldots, b_s)$, with R upper triangular.

If the b_i, c_i are weights and abscissae of a quadrature formula of order p, then for $k + l \le p + 1$, the order conditions imply that

$$\sum_{i=1}^{s} b_i P_{k-1}(c_i) P_{l-1}(c_i) = \int_0^1 P_{k-1}(c) P_{l-1}(c) \, \mathrm{d}c, \tag{358b}$$

so that comparing (358a) with (358b), we see that $P_0, P_1, \ldots, P_{[(p-1)/2]}$ are orthonormal with respect to integration on $[0, 1]$. Conventionally choosing the highest-degree coefficient of each P_i to be positive, we see that this determines $P_0, P_1, \ldots, P_{[(p-1)/2]}$ uniquely as the normalized Legendre polynomials on the interval $[0, 1]$ given by

$$P_k(c) = (2k + 1)^{\frac{1}{2}} \sum_{i=0}^{k} (-1)^{k-i} \binom{k}{i} \binom{k+i}{i} c^i. \tag{358c}$$

In particular, $P_0(c) = 1$ for a consistent method.

For methods of the Gauss type, for which $p = 2s$, the W matrix is known in full and it is interesting to identify the matrix X_G introduced in the following lemma.

LEMMA 358A.

Let X_G denote the $s \times s$ matrix

$$X_G = W^T B A W.$$

Then

$$X_G = \begin{bmatrix}
\frac{1}{2} & -\xi_1 & 0 & 0 & \cdots & 0 & 0 \\
\xi_1 & 0 & -\xi_2 & 0 & \cdots & 0 & 0 \\
0 & \xi_2 & 0 & -\xi_3 & \cdots & 0 & 0 \\
0 & 0 & \xi_3 & 0 & \cdots & 0 & 0 \\
\vdots & \vdots & \vdots & \vdots & & \vdots & \vdots \\
0 & 0 & 0 & 0 & \cdots & 0 & -\xi_{s-1} \\
0 & 0 & 0 & 0 & \cdots & \xi_{s-1} & 0
\end{bmatrix},$$

where $\xi_k = \frac{1}{2}(4k^2 - 1)^{-\frac{1}{2}} (k = 1, 2, \ldots, s - 1)$.

Proof.

From linear combinations of order conditions of the type $\Sigma_{i,j} b_i c_i^{k-1} a_{ij} c_j^{l-1} = 1/l(k + l)$, we see that

$$\sum_{i,j=1}^{s} b_i P_{k-1}(c_i) a_{ij} P_{l-1}(c_j) = \int_0^1 P_{k-1}(u) \int_0^u P_{l-1}(v) \, dv \, du$$

and this is identical with the (k, l) element of X_G.

Using integration by parts, we see that

$$\int_0^1 P_{k-1}(u) \int_0^u P_{l-1}(v) \, dv \, du + \int_0^1 P_{l-1}(u) \int_0^u P_{k-1}(v) \, dv \, du$$

$$= \int_0^1 P_{k-1}(u) \, du \int_0^1 P_{l-1}(u) \, du$$

$$= \begin{cases} 1, & k = l = 1, \\ 0, & \text{otherwise} \end{cases}$$

and the values of the diagonal elements of X_G follow immediately, as does the skew-symmetric form of $X_G - \frac{1}{2} e_1 e_1^T$. It remains to evaluate

$$\int_0^1 P_k(u) \int_0^u P_l(v) \, dv \, du, \qquad k > l. \tag{358d}$$

If $k > l + 1$ the integral of P_l is a polynomial of lower degree than k and the integral is zero by orthogonality. If $k = l + 1$, the integral of P_l is a known constant multiplied by P_k plus a lower-degree polynomial. Evaluating the constant from the highest-degree coefficients in P_k and P_{k-1} as given by (358c), we have

$$\int_0^1 P_k(u) \int_0^u P_{k-1}(v) \, dv \, du$$

$$= k^{-1}(2k-1)^{\frac{1}{2}} \binom{2k-2}{k-1}(2k+1)^{-\frac{1}{2}} \binom{2k}{k}^{-1} \int_0^1 P_k(u)^2 \, du = \xi_k. \quad \blacksquare$$

Since only order conditions up to order p are involved in the computation of the principal $[(p+1)/2] \times [(p+1)/2]$ submatrix of X in the following corollary, we can state this result without detailed proof.

COROLLARY 358B.

For a Runge–Kutta method with order p, let

$$X = W^T B A W,$$

then the elements in the principal $[(p+1)/2] \times [(p+1)/2]$ submatrix of X are identical to those in X_G.

It is interesting to note that the simplifying assumptions discussed in subsection 342 also have a role in determining the form of X. The assumptions that we denoted there by $A(m)$ and $D(m)$ can be written equivalently as

$$A(m): \qquad \sum_{j=1}^s a_{ij} P(c_j) = \int_0^{c_i} P(c) \, dc, \qquad \deg(P) < m,$$

$$D(m): \qquad \sum_{i=1}^s b_i P(c_i) a_{ij} = b_j \int_{c_j}^1 P(c) \, dc, \qquad \deg(P) < m.$$

If $m \le [(p-1)/2]$, where p is the 'quadrature order' of the method (that is $\sum_{i=1}^s b_i P(c_i) = \int_0^1 P(c) \, dc$ for $\deg(P) < p$), then these observations enable us to evaluate the first m columns of X if $A(m)$ is true and the first m rows of X if $D(m)$ is true. In each case the specified elements of X coincide in value with the corresponding elements of X_G. If the method is of full order p, then under our assumption that $b_i > 0$, for $i = 1, 2, \ldots, s$, $A([(p-1)/2])$ is a necessary consequence because of the identity for methods of this order

$$\sum_{i=1}^s b_i \left(\sum_{j=1}^s a_{ij} P(c_j) - \int_0^{c_i} P(c) \, dc \right)^2 = 0, \qquad \deg(P) < \left[\frac{p-1}{2}\right].$$

The main result of Hairer and Wanner (1981) gives a characterization of high-order algebraically stable methods.

THEOREM 358C.

An implicit Runge–Kutta method of order p, in which $b_1, b_2, \ldots, b_s > 0$, is algebraically stable if and only if $X = M^{\mathrm{T}}BAM$ is given by

$$
X = \left[
\begin{array}{ccccc:cc}
\frac{1}{2} & -\xi_1 & 0 & \cdots & 0 & 0 & 0 \cdots 0 \\
\xi_1 & 0 & -\xi_2 & \cdots & 0 & 0 & 0 \cdots 0 \\
0 & \xi_2 & 0 & \cdots & 0 & 0 & 0 \cdots 0 \\
\vdots & \vdots & \vdots & & \vdots & \vdots & \vdots \\
0 & 0 & 0 & \cdots 0 & -\xi_{k-1} & 0 & 0 \cdots 0 \\
0 & 0 & 0 & \cdots \xi_{k-1} & 0 & -\xi_k & 0 \cdots 0 \\
\hdashline
0 & 0 & 0 & \cdots 0 & \xi_k & & \\
0 & 0 & 0 & \cdots 0 & 0 & & \widetilde{X} \\
\vdots & \vdots & \vdots & & \vdots & & \\
0 & 0 & 0 & \cdots 0 & 0 & &
\end{array}
\right], \quad (358e)
$$

with $\xi_i = \frac{1}{2}(4i^2 - 1)^{-1/2}$, where the symmetric part of the $(s - k) \times (s - k)$ matrix $\widetilde{X}$ is positive semidefinite with $k = [(p - 1)/2]$, and is such that $e_1^{\mathrm{T}}\widetilde{X}e_1 = 0$ if p is even.

Proof.

Algebraic stability is equivalent to $M = BA + A^{\mathrm{T}}B - Be \cdot e^{\mathrm{T}}B$ is positive semidefinite, which is equivalent to the same statement applied to the transformed matrix

$$
W^{\mathrm{T}}MW = W^{\mathrm{T}}BAW + W^{\mathrm{T}}A^{\mathrm{T}}BW - W^{\mathrm{T}}Be \cdot e^{\mathrm{T}}BW = X + X^{\mathrm{T}} - e_1 e_1^{\mathrm{T}}.
$$

From the known form of the first k columns of X it follows that

$$
W^{\mathrm{T}}MW = \begin{bmatrix} 0 & Y \\ Y^{\mathrm{T}} & \widetilde{X} + \widetilde{X}^{\mathrm{T}} \end{bmatrix}
$$

on the assumption that X differs from the form given in (358e) by a matrix Y in the upper right submatrix position. It is now clear that $W^{\mathrm{T}}MW$ positive semidefinite is equivalent to $Y = 0$ and $\widetilde{X} + \widetilde{X}^{\mathrm{T}}$ positive semidefinite. ∎

The very specific form of X required to satisfy the requirements of this theorem is exploited by Hairer and Wanner to characterize various high-order classes of algebraically stable methods, including those that are required to be singly implicit.

359 Solvability of the algebraic equations.

We explore the questions of the existence and uniqueness only in the case that A is positive definite. We recall that this requires the existence of a positive-valued diagonal matrix D such that $DA + A^{\mathrm{T}}D$ is positive definite in the usual sense and that if A is positive definite then A^{-1} is also positive definite. We will consider the solution of the algebraic equations arising when a method with this property is applied to a problem satisfying a one-sided Lipschitz condition. We will then consider the application of our conclusions to the special classes of methods covered by theorem 357G. The results presented here arise directly out of the study of B-convergence due to Frank, Schneid and Ueberhuber (1981). Further contributions are due to Crouzeix, Hundsdorfer and Spijker [1983] and to Spijker [1985]. Finally, the recent book by Dekker and Verwer [1984] surveys and further develops the topic in this subsection as well as other questions related to stability of Runge–Kutta methods.

THEOREM 359A.

If f is continuously differentiable and satisfies (357a) and A is positive definite, then the algebraic equations

$$Y_i = y_{n-1} + h \sum_{j=1}^{s} a_{ij} f(Y_j), \qquad i = 1, 2, \ldots, s, \tag{359a}$$

have a unique solution for $h \le h_0$, where h_0 depends only on the method and on the value of λ.

Proof.

Choose D so that $M = DA^{-1} + (A^{-1})^{\mathrm{T}}D$ is positive definite and choose h_0 so that $2\lambda h_0 < \mu$, where μ is the least eigenvalue of $D^{-\frac{1}{2}}MD^{-\frac{1}{2}}$.

Let $Y, Z \in X^s$ be made up from subvectors $Y_1, Y_2, \ldots, Y_s \in X$ and $Z_1, Z_2, \ldots, Z_s \in X$. Define $F: X^s \to X^s$ so that $F(Y)$ is made up from subvectors $d_i h f(Y_i) - d_i \sum_{j=1}^{s} a_{ij}^{(-1)}(Y_j - Y_{n-1})$, $i = 1, 2, \ldots, s$, where $a_{ij}^{(-1)}$ is a typical element of A^{-1}. Thus, $F(Y) = 0$ is equivalent to the statement that (359a) is satisfied.

We now make use of a classical result on uniformly monotonic functions (see, for example, Ortega and Rheinboldt [1970]) and note that the conclusion of our theorem would follow if F is such a function. It is found that

$$[F(Y) - F(Z)]^{\mathrm{T}}(Y - Z) = \sum_{i=1}^{s} d_i h [f(Y_i) - f(Z_i)]^{\mathrm{T}}(Y_i - Z_i)$$

$$- \tfrac{1}{2} \sum_{i,j=1}^{s} d_i a_{ij}^{(-1)}(Y_j - Z_j)^{\mathrm{T}}(Y_i - Z_i)$$

$$- \tfrac{1}{2} \sum_{i,j=1}^{s} d_j a_{ij}^{(-1)}(Y_j - Z_j)^{\mathrm{T}}(Y_i - Z_i)$$

$$\leq \lambda h \sum_{i=1}^{s} d_i (Y_i - Z_i)^{\mathrm{T}} (Y_i - Z_i)$$

$$-\tfrac{1}{2} \sum_{i,j=1}^{s} m_{ij} (Y_i - Z_i)^{\mathrm{T}} (Y_j - Z_j)$$

$$\leq -\tfrac{1}{2} \sum_{i,j=1}^{s} \widetilde{m}_{ij} (Y_i - Z_i)^{\mathrm{T}} (Y_j - Z_j)$$

$$\leq -\gamma \| Y - Z \|^2,$$

where m_{ij} is a typical element of M and $\widetilde{m}_{ij}$ is a typical element of $\widetilde{M} = M - 2\lambda h D$ which is positive definite if $2\lambda h < \mu$. The constant $\gamma > 0$ can be chosen as the least eigenvalue of $\tfrac{1}{2} \widetilde{M}$. ∎

We note in passing that if $\lambda \leq 0$ then h_0 can be chosen arbitrarily large.

Finally, we use theorem 357G to enable us to apply theorem 359A to important classes of methods.

THEOREM 359B.

If f is continuously differentiable and satisfies (357a) then the algebraic equations (359a) have a unique solution for $h \leq h_0$ for any Gauss, Radau IA and Radau IIA method, where for each particular method h_0 depends only on the value of λ.

36   ERROR PROPOGATION IN RUNGE–KUTTA METHODS

360   Introduction to error propagation.

In this section we will examine the way in which errors, generated in a single step of integration, propagate over a series of steps. If we refer to the error in a single step as the 'local truncation error', then this means that we wish to know how the global truncation error is related to the individual local truncation errors committed from step to step. Having obtained a rather crude bound by simply using the perturbation result for differential equations, theorem 112J, repeatedly for each step as well as approaching the same question with a difference inequality, we examine the magnitude of the local truncation errors themselves through the use of the so-called 'principal error function'.

A sharper type of bound can be obtained based on a variational differential equation as for the Euler method in theorem 205B. For details in the Runge–Kutta case, see Stetter (1973).

For the special case of algebraically stable Runge–Kutta methods, a problem satisfying the dissipativity condition

$$[f(x, u) - f(x, v)]^{\mathrm{T}} (u - v) \leq 0$$

gives a bound that does not depend explicitly on the Lipschitz constant and this furnishes an appropriate approach in the case of stiff problems.

361 Local truncation error.

We will assume that we are dealing with a particular fixed Runge–Kutta method and that the step size h is constant from step to step. For the differential equation

$$y'(x) = f(y(x)), \qquad y(x_0) = y_0, \qquad (361a)$$

let E denote the function that maps y_0 into the solution at $x_0 + h$. For the particular numerical method with the given initial value y_0, let φ denote the mapping taking y_0 into the result computed at the end of the step. In the sense that φ is supposed to approximate E, it is natural to think of the difference function $E - \varphi$ as representing the error generated in a single step.

If y_{n-1} is the approximation computed to $y(x_{n-1})$, then $y_n = \varphi(y_{n-1})$ and $E(y_{n-1})$ is the *exact* solution at x_n to the differential equation given by (361a) with its initial value replaced by $y(x_{n-1}) = y_{n-1}$. Thus $(E - \varphi)(y_{n-1})$ is the additional amount by which the computed result falls short of the exact result as a consequence of the nth integration step. To understand how such local errors contribute to the error after many steps, one may study figure 361.

In this figure, the full lines denote trajectories which exactly satisfy the differential equation and the broken line connecting $y_0, y_1, y_2, \ldots, y_{n-2}, y_{n-1}, y_n$ denotes the solution curve computed by the Runge–Kutta method. At the

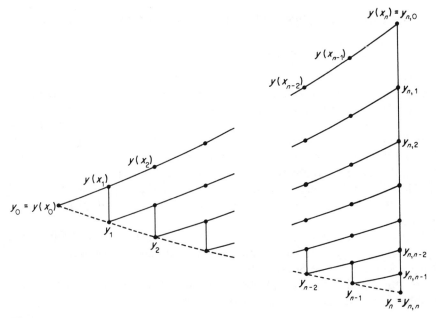

Figure 361 Growth of exact solution from points on the approximate trajectory.

point x_n, we distinguish n approximations to $y(x_n)$. For $i = 1, 2, \ldots, n - 1$, we denote by $y_{n,i}$ the exact solution to the differential equation defined by the initial condition $y(x_i) = y_i$. For convenience, we will also write $y_{n,n}$ as an alternative name for y_n and $y_{n,0}$ as an alternative name for $y(x_n)$ corresponding to the given initial condition at x_0.

To estimate the overall error at x_n, we decompose $y(x_n) - y_n$ into n terms thus:

$$y(x_n) - y_n = y_{n,0} - y_{n,n}$$

$$= (y_{n,0} - y_{n,1}) + (y_{n,1} - y_{n,2}) + \cdots + (y_{n,n-1} - y_{n,n}). \tag{361b}$$

Looking at the terms on the right-hand side (361b), we see that the first is the distance apart, at x_n, of two trajectories in which their values at x_1 differ by the local truncation error $E(y_0) - \varphi(y_0)$ in the first integration step. Similarly, $(y_{n,1} - y_{n,2})$ is the difference between two trajectories at x_n, in which the solution values at x_2 differ by the local truncation error $E(y_1) - \varphi(y_1)$ in the second integration step. The same pattern continues until the final term $(y_{n,n-1} - y_{n,n})$ which is exactly the local truncation error in the last integration step given by $E(y_{n-1}) - \varphi(y_{n-1})$.

Taking norms in (361b) and using the triangle inequality repeatedly, together with the result of theorem 112J to bound the growth of the separation between two trajectories, we find that

$$\| y(x_n) - y_n \| \leq \exp(hL(n-1)) \| E(y_0) - \varphi(y_0) \|$$
$$+ \exp(hL(n-2)) \| E(y_1) - \varphi(y_1) \|$$
$$+ \cdots + \| E(y_{n-1}) - \varphi(y_{n-1}) \|.$$

If each of the n local truncation errors can be bounded by a constant K, then we have the bound

$$\| y(x_n) - y_n \| \leq K[1 + \exp(hL) + \exp(2hL) + \cdots + \exp((n-1)hL)]$$

$$\leq \frac{K}{e^{hL} - 1} [\exp(nhL) - 1]$$

$$\leq \frac{K}{hL} [\exp(L(x_n - x_0)) - 1],$$

assuming, of course, that $hL > 0$.

In spite of the simplicity of this type of analysis, it is customary to interrelate local and global errors in a somewhat different way. In the definition of the local truncation error associated with the nth integration step, the function $E - \varphi$ is applied not to the numerical solution but rather to the exact solution at x_{n-1}. Thus, the local truncation error is defined as $(E - \varphi)(y(x_{n-1}))$ $= y(x_n) - \varphi(y(x_{n-1}))$, and we will estimate global errors as suggested not by figure 361 but by 361'.

In this figure, we have used the same convention as in figure 361 of denoting the exact solution by a full line and various approximate solution curves by

278

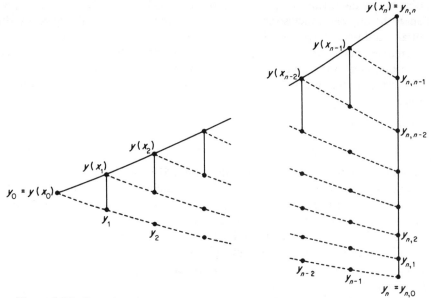

Figure 361' Growth of computed solution from points on the exact trajectory.

broken lines. The role of the notation for $y_{n,i}$ is now reversed in that it now denotes the approximation formed by carrying out a further $n - i$ steps with the Runge–Kutta method under consideration, using the exact value $y(x_i)$ as the approximation at x_i. Estimation of the total error using the inequality

$$\| y(x_n) - y_n \| \leq \| y_{n,n} - y_{n,n-1} \| + \| y_{n,n-1} - y_{n,n-2} \| + \cdots + \| y_{n,1} - y_{n,0} \|$$

(361c)

is still possible, but now we have to bound the individual terms in accordance with the growth of perturbations in the computed rather than the exact solution.

Although this approach is slightly less natural, it is the preferred one because it generalizes more easily to the case of multistep methods. Accordingly we conclude this subsection with a formal definition of the local truncation error $(E - \varphi)y(x_{n-1})$.

DEFINITION 361A.

Consider a Runge–Kutta method

$$
\begin{array}{c|cccc}
c_1 & a_{11} & a_{12} & \cdots & a_{1s} \\
c_2 & a_{21} & a_{22} & \cdots & a_{2s} \\
\vdots & \vdots & \vdots & & \vdots \\
c_s & a_{s1} & a_{s2} & \cdots & a_{ss} \\
\hline
 & b_1 & b_2 & \cdots & b_s
\end{array}
$$

and let $h = x_n - x_{n-1}$ be sufficiently small so as to guarantee that the system (361e) below has a unique solution. The *local truncation error* associated with the nth integration step is defined by

$$l_n = y(x_n) - y(x_{n-1}) - h \sum_{i=1}^{s} b_i f(Y_i), \tag{361d}$$

where $Y_1, Y_2, \ldots, Y_s$ satisfy

$$Y_i = y(x_{n-1}) + h \sum_{j=1}^{s} a_{ij} f(Y_j), \qquad i = 1, 2, \ldots, s. \tag{361e}$$

Throughout section 36, we will continue to use the functions φ and E so that the expression in (361d) could be written as $y(x_n) - \varphi(y(x_{n-1}))$ or as $E(y(x_{n-1})) - \varphi(y(x_{n-1}))$.

362 The principal error function.

Assuming that f has a sufficient number of bounded derivatives, the local truncation error can be written in the form

$$l_n = h^{p+1} \sum_{r(t) \leq p+1} \sigma(t)^{-1} \left[\gamma(t)^{-1} - \Phi(t) \right] F(t)(y(x_{n-1})) + O(h^{p+2}).$$

For small h, l_n can be accurately approximated by the h^{p+1} terms, which motivates the following definitions.

DEFINITION 362A.

The *principal error coefficient* for a given Runge–Kutta method of order p at the point x is defined as

$$C(x) = \sum_{r(t) = p+1} \sigma(t)^{-1} [\gamma(t)^{-1} - \Phi(t)] F(t)(y(x)). \tag{362a}$$

DEFINITION 362B.

The *principal error function* for step number n in the Runge–Kutta method of definition 362A is defined as

$$P_n = h^{p+1} C(x_{n-1}).$$

For large p, $C(x)$ is usually a complicated combination of many elementary differentials and it is difficult to unravel them, for example, to choose between two alternative Runge–Kutta methods of the same order from the point of view of accuracy.

Various approaches to the question of minimizing $\| C(x) \|$ in some sense have been put forward by Bieberbach (1930), Henrici (1962), Lotkin (1951), Ralston (1962), and others. We present our own approach in the next subsection. This is, however, put forward with little enthusiasm, since empirical evidence does not support the view that a method selected as optimal in this sense is actually more accurate than other methods for realistic problems.

363 Optimal methods.

For a method of order p, let M denote a bound along the solution trajectory of $\| f(y(x)) \|$ and N the maximum for $m = 1, 2, \ldots, p$ of a bound on $\| f^{(m)}(y(x)) \|^{1/m} M^{1-1/m}$. Thus, for $m = 1, 2, \ldots, p$, $\| f^{(m)}(y(x)) \|$ is bounded by $N^m M^{1-m}$. Under these assumptions, for any tree t with $r(t) = p + 1$, the corresponding elementary differential is bounded by

$$\| F(t)(y(x)) \| \leq MN^p. \tag{363a}$$

Substituting (363a) into (362a) gives the bound

$$\| C(x) \| \leq \sum_{r(t)=p+1} \sigma(t)^{-1} |\gamma(t)^{-1} - \Phi(t)| MN^p, \tag{363b}$$

so that, by use of an artificial assumption, all terms in the principal error function have been forced into a comparable form. Evidently an optimal choice of the right-hand side of (363b) is thus dependent only on the method rather than the problem.

In the case of a three-stage third-order method, for example, (363b) becomes

$$\| C(x) \| \leq MN^3 (\tfrac{1}{6} |\tfrac{1}{4} - \Sigma b_i c_i^3| + |\tfrac{1}{8} - \Sigma b_i c_i a_{ij} c_j| + \tfrac{1}{2} |\tfrac{1}{12} - \Sigma b_i a_{ij} c_j^2| + \tfrac{1}{24})$$

$$= MN^3 [\tfrac{1}{6} |\tfrac{1}{4} - \tfrac{1}{3}(c_2 + c_3) + \tfrac{1}{6} c_2 c_3| + \tfrac{1}{6} |c_3 - \tfrac{3}{4}| + \tfrac{1}{12} |c_2 - \tfrac{1}{2}| + \tfrac{1}{24}].$$

It is easily seen that the bound is minimized by $c_2 = \tfrac{1}{2}$, $c_3 = \tfrac{3}{4}$, giving the method

$$
\begin{array}{c|ccc}
0 & & & \\
\tfrac{1}{2} & \tfrac{1}{2} & & \\
\tfrac{3}{4} & -\tfrac{3}{4} & \tfrac{3}{2} & \\
\hline
 & \tfrac{1}{9} & \tfrac{2}{3} & \tfrac{2}{9}
\end{array}
$$

A similar analysis for four-stage fourth-order methods leads to an 'optimal' method given by case 1 in subsection 312 with $c_2 \approx 0.371615$ and $c_3 = \tfrac{3}{5}$.

364 Global truncation error.

In theorem 112J, we considered the growth of a perturbation in the solution to a differential equation. It can be seen from the result that when x increases by h, $\| y(x) - z(x) \|$ increases by no more than a factor $\exp(hL)$. If f satisfies a one-sided Lipschitz condition with constant λ, it follows from (112g) in the proof of theorem 112G, that the factor can be replaced by $\exp(h\lambda)$.

We now consider an analogous question concerning the growth of a perturbation in the solution computed in a single step of a Runge–Kutta method. For a method characterized by A, b^T, let $|A|$, $|b|^T$ denote the matrix and vector formed by taking the absolute value of each element of A and of b^T respectively.

Let y_n, z_n denote results computed for the same problem and this same method but using different initial values y_{n-1}, z_{n-1} at the start of the step. That is, $y_n = \varphi(y_{n-1})$, $z_n = \varphi(z_{n-1})$. We will obtain a bound for $\| \varphi(y_{n-1}) - \varphi(z_{n-1}) \|$.

THEOREM 364A.

Let $Y_1, Y_2, \ldots, Y_s, Z_1, Z_2, \ldots, Z_s, y_n, z_n$ be defined by

$$Y_i = y_{n-1} + h \sum_{j=1}^{s} a_{ij} f(Y_j) \qquad i = 1, 2, \ldots, s, \qquad (364a)$$

$$Z_i = z_{n-1} + h \sum_{j=1}^{s} a_{ij} f(Z_j), \qquad i = 1, 2, \ldots, s, \qquad (364b)$$

$$y_n = y_{n-1} + h \sum_{i=1}^{s} b_i f(Y_i), \qquad (364c)$$

$$z_n = z_{n-1} + \sum_{i=1}^{s} b_i f(Z_i), \qquad (364d)$$

where f satisfies a Lipschitz condition with constant L and $hL \| \, |A| \, \| < 1$; then

$$\| y_n - z_n \| \le [1 + hL |b|^T (I - hL |A|)^{-1} e] \, \| y_{n-1} - z_{n-1} \|.$$

Proof.

Let $v = [v_1, v_2, \ldots, v_s]^T$, where $v_i = \| Y_i - Z_i \|$; then by subtracting (364b) from (364a), taking norms and using the Lipschitz condition, we see that

$$v_i \le \| y_{n-1} - z_{n-1} \| + hL \sum_{j=1}^{s} |a_{ij}| \, v_j$$

or

$$(I - hL |A|) v \le \| y_{n-1} - z_{n-1} \| e.$$

From (364d) and (364c) it follows that

$$\| y_n - z_n \| \le \| y_{n-1} - z_{n-1} \| + hL \sum_{i=1}^{s} |b_i| \, v_i$$

$$\le \| y_{n-1} - z_{n-1} \| + hL |b|^T v$$

and the result follows. ∎

After a number of integration steps, with constant step size h, the growth factor $1 + hL \, |b|^{\mathrm{T}} (I - hL \, |A|)^{-1} e$ applies at every step. Let h_0 be chosen such that $h_0 L \, \| \, |A| \, \| < 1$, then it can be seen that

$$1 + hL \, |b|^{\mathrm{T}} (I - hL \, |A|)^{-1} e \le \exp(hL \sum_{i=1}^{s} |b_i| + h^2 C), \qquad (364e)$$

where C depends on L, $|b|^{\mathrm{T}}$, $|A|$ and h_0. Thus for integration over m integration steps, the error grows by a factor not exceeding

$$\exp\left(mh\left(L \sum_{i=1}^{s} |b_i| + hC \right)\right).$$

If $b_1, b_2, \ldots, b_s \ge 0$, as is the case for most Runge–Kutta methods in practical use, $\sum_i |b_i| = \sum_i b_i = 1$, and we see that error growth for Runge–Kutta methods is bounded by essentially the same factor as for the exact solution.

We now consider the global truncation error resulting from the repeated use of a Runge–Kutta method. We will assume that f has sufficient smoothness to obtain appropriate local truncation error behaviour and that f satisfies a Lipschitz condition with $L > 0$.

THEOREM 364B.

Consider a Runge–Kutta method used with constant step size h such that the local truncation error is bounded in norm by $Ah^{p+1} + Bh^{p+2}$ for $h < h_0$. Let $L^* = L \sum_{i=1}^{s} |b_i|$, then the global truncation error is bounded by

$$\| y(x_n) - y_n \| \le \frac{Ah^p [\exp(L^*(x_n - x_0)) - 1]}{L^*} + Dh^{p+1}$$

for $h < h_0$ for some constant D.

Proof.

Using the inequality (361c) and bounding each term by a combination of (364e) and the bound assumed for the local truncation error, we find that

$$\| y(x_n) - y_n \| \le (Ah^{p+1} + Bh^{p+2}) \sum_{j=0}^{s} \exp(jhL \sum_{i=1}^{s} |b_i| + jh^2 C)$$

$$\le \frac{(Ah^{p+1} + Bh^{p+2})\exp((x_n - x_0)Ch)[\exp(nhL^*) - 1]}{\exp(hL^*) - 1}$$

The factor $[\exp(hL^*) - 1]^{-1}$ can be bounded by $(hL^*)^{-1}$ and $\exp((x_n - x_0)Ch)$ by $1 + (x_n - x_0)Ch \exp((x_n - x_0)Ch_0)$, and the result follows using standard inequalities. ∎

We emphasize that for a consistent method with $b_1, b_2, \ldots, b_s > 0$, $L^* = L$. This result, as it stands, is entirely unsatisfactory for stiff problems.

However, for methods which are (k, l) algebraically stable, where $k^{1/2} = 1 + \gamma l + O(l^2)$ for $l > 0$, L^* may be replaced by $\lambda^* = \gamma \lambda$ where the constant γ is characteristic of the method. We will look again at error analysis for stiff problems in subsections 366 and 367.

In a variable stepsize setting, theorem 364B is not directly applicable. We consider briefly how it should be generalized to allow for this and for a principal error coefficient which varies widely along the trajectory.

Suppose that H is a bound on the greatest step size permitted in integrating from x_0 to $\bar{x}$ and that a function $h(x, H)$ is defined to determine the step size that would be used from a point x on the solution trajectory. We will assume that h is continuous in its first variable and continuous and monotonically decreasing as H decreases. We note that $0 < h(x, H) \leq H$.

We will seek a bound on the error in terms of H. Since the rate at which errors are generated per step is $\| C(x) \| h(x, H)^{p+1} + O(h(x, H)^{p+2})$, where $C(x)$ is the principal error coefficient, and the rate at which x increases per step is $h(x, H)$, we see that $\| C(x) \| h(x, H)^p + O(H^{p+1})$ is the rate of error generation per increase in x.

Allowing for a growth of the effect of this error by the appropriate exponential factor, we state the following theorem without detailed proof.

THEOREM 364C.

Under the assumptions made in this discussion, the global truncation error at $\bar{x}$ is bounded by

$$\int_{x_0}^{\bar{x}} \exp(L^*(\bar{x} - x)) \, \| C(x) \| \, h(x, H)^p \, dx + O(H^{p+1}).$$

365 Optimal order and stepsize sequences.

A major goal in the design of differential equations software is the automatic selection among methods of varying orders and the automatic choice of step size in accordance with the observed behaviour of the solution and with the user's requirements. What is usually available is a collection of methods for which the local error estimates can be conveniently made. At the same time it is desirable to be able to estimate not only the error in a method actually in use, but also the error in alternative methods that might be considered for use in succeeding steps. A variety of techniques for obtaining local error estimates, for a method currently in use, are discussed in section 37. Although less progress has been made on the estimation of errors for methods of higher or lower order, there is at least one important paper, by Verner (1979) on this topic.

Leaving questions of technique aside, we will deal here with an idealized situation in which, for a particular problem, the principal error is assumed to be known for a set of methods of differing orders and it is required to resolve the optimization question of balancing the two conflicting aims of low total error and low total computing cost.

Assuming that the global truncation error can be accurately approximated by the integral occurring in theorem 364C, it is easy to obtain a more general result in which the numerical method varies within the interval of integration. Suppose that at a step beginning at x, a step size $h(x)$ is to be used and that $p(x)$ is the order of this method. If $C(x, p(x))$ denotes the magnitude of the error constant with the factor $\exp(L(\bar{x} - x))$ of theorem 364C absorbed into it, then the approximation to the global error becomes

$$E = \int_{x_0}^{\bar{x}} C(x, p(x)) h(x)^{p(x)} \, dx. \tag{365a}$$

We further assume that a method of order $p(x)$ incurs a computational cost of $w(p(x))$ in carrying out a single step. Thus, the rate at which computing work is done as the integration proceeds is $w(p(x))h(x)^{-1}$. Hence, we will approximate the total work done by

$$W = \int_{x_0}^{\bar{x}} w(p(x)) h(x)^{-1} \, dx. \tag{365b}$$

Our aim will now be to understand how $p(x)$ and $h(x)$ should be selected so that E and W should together be made as low as possible. Specifically, we will seek to achieve optimality in the following sense.

DEFINITION 365A.

The order selection function p and the stepsize selection function h are said to be *optimal* if it is not possible to replace p and h by different functions so that both E and W given by (365a) and (365b) are simultaneously reduced.

There are two obvious ways in which an optimal sequence may be interpreted. We may regard it as providing a minimal value of E for a given W or as providing minimal W for given E.

In practice, it is not usually possible to make use of optimal sequences in either of these ways. However, with an optimal sequence, we are assured that resources are not wasted in the sense that effort put into particular steps is matched by appropriate effort put into other steps.

Our analysis of optimal sequences of order and step size will proceed in two phases. For $p(x)$ given at each point, we will determine conditions on $h(x)$ for it to correspond to such a sequence. The second phase will be to determine how $p(x)$ should be selected at each point. To avoid unnecessary complications we will assume that for $p(x)$ equal to the constant value p_1, $C(x, p_1)$ is a piecewise continuous function of x. We will also assume that the function p in an optimal sequence switches between different orders a finite number of times and that the function h is piecewise continuous.

THEOREM 365B.

In an optimal sequence, the value of

$$T(x) = \frac{p(x)C(x, p(x))h(x)^{p(x)+1}}{w(p(x))} \tag{365c}$$

is constant.

Proof.

This result follows from a calculus of variations argument as in Gear (1971, p. 77). We will here give a proof using elementary methods. If T were not constant, let x_1, x_2 be such that $x_1 \in O_1$, $x_2 \in O_2$, where O_1, O_2 are open sets in $[x_0, \bar{x}]$ for which $p(x) = p_1$ for $x \in O_1$, $p(x) = p_2$ for $x \in O_2$ and for which $h(x)$ and $C(x, p(x))$ are each continuous in O_1 and O_2 and $T(x_1) > T(x_2)$. Let $T_0 = \frac{1}{2}[T(x_1) + T(x_2)]$ so that $T(x_1) > T_0 > T(x_2)$ and let $w_1 = w(p_1)$, $w_2 = w(p_2)$. By continuity, a positive number δ may be chosen so that $I_1 = [x_1 - \delta, x_1 + \delta] \subset O_1$, $I_2 = [x_2 - \delta, x_2 + \delta] \subset O_2$ and such that $T(x) > T_0$ for $x \in I_1$, $T(x) < T_0$ for $x \in I_2$.

For $i = 1, 2$ define

$$\alpha_i = \int_{I_i} w_i(T(x) + T_0)h(x)^{-2}\left(1 - \frac{|x - x_i|}{\delta}\right) dx,$$

and, for $|\varepsilon|$ sufficiently small to ensure that $\bar{h}(x, \varepsilon)$ is never negative,

$$\bar{h}(x, \varepsilon) = \begin{cases} h(x) - \alpha_2\varepsilon\left(1 - \dfrac{|x - x_1|}{\delta}\right), & x \in I_1, \\[2mm] h(x) + \alpha_1\varepsilon\left(1 - \dfrac{|x - x_2|}{\delta}\right), & x \in I_2, \\[2mm] h(x), & x \notin I_1 \cup I_2. \end{cases}$$

If

$$\bar{E}(\varepsilon) = \int_{x_0}^{\bar{x}} C(x, p(x))\bar{h}(x, \varepsilon)^{p(x)} dx$$

$$\bar{W}(\varepsilon) = \int_{x_0}^{\bar{x}} w(p(x))\bar{h}(x, \varepsilon)^{-1} dx,$$

so that $\bar{E}(0) = E$, $\bar{W}(0) = W$, it is found that $\bar{E}'(0) < 0$, $\bar{W}'(0) < 0$. Thus, by selecting a small positive value for ε, it is possible to reduce both E and W by replacing $h(x)$ by $\bar{h}(x, \varepsilon)$. ∎

Having established the constancy of T given by (365c) in an optimal sequence, we note that $h(x) = H(x, p(x))$, where

$$H(x, q) = \left[\frac{Tw(q)}{qC(x, q)}\right]^{1/(q+1)} \tag{365d}$$

and that

$$E = \int_{x_0}^{\bar{x}} \alpha(x, p(x))\, dx,$$

$$W = \int_{x_0}^{\bar{x}} \beta(x, p(x))\, dx,$$

where the rates at which global truncation error (respectively total computing costs) are accumulated using order q are $\alpha(x, q)$ (respectively $\beta(x, q)$), where these functions are given by

$$\alpha(x, q) = [C(x, q)w(q)^q]^{1/(q+1)}(T/q)^{q/(q+1)}, \tag{365e}$$

$$\beta(x, q) = [C(x, q)w(q)^q]^{1/(q+1)}(T/q)^{-1/(q+1)}. \tag{365f}$$

We now characterize the optimal choice of $p(x)$ at a given value of x.

THEOREM 365C.

In an optimal sequence $p(x)$ is the value of q which satisfies each of the equivalent criteria

$\quad (q + 1)\alpha(x, q) \qquad$ is minimized,

$\quad \dfrac{q + 1}{q} \beta(x, q) \qquad$ is minimized,

$\quad \dfrac{q}{q + 1} \dfrac{H(x, q)}{w(q)} \qquad$ is maximized,

where $\alpha(x, q)$, $\beta(x, q)$, $H(x, q)$ are given by (365e), (365f), (365d).

Proof.

It is easy to verify the equivalence of the three criteria.

We suppose that an optimal sequence exists for which some $x \in [x_0, \bar{x}]$, there is a possible order p_1 such that

$$(p_1 + 1)\alpha(x, p_1) < [p(x) + 1]\alpha(x, p(x)). \tag{365g}$$

By continuity, we may assume that (365g) holds throughout an open interval O in which $p(x)$ has constant value p_0 and such that each of $C(x, p_0)$ and $C(x, p_1)$ are continuous in O. Let x be a point in O and let ε_0 denote a positive number such that $x_1 + \varepsilon_0 \in O$ and such that $T + \theta\varepsilon_0 > 0$, where θ is defined by

$$2\theta \int_{x_0}^{\bar{x}} \frac{\beta(x, p(x))}{p(x) + 1}\, dx = \alpha(x_1, p_1)(p_1 - 1) - \alpha(x_1, p_0)(p_0 - 1). \tag{365h}$$

For $\varepsilon \in [0, \varepsilon_0]$, let an alternative sequence be defined by the functions

$\bar{p}(x, \varepsilon)$ and $\bar{h}(x, \varepsilon)$, where

$$\bar{p}(x, \varepsilon) = \begin{cases} p_1, & x \in [x_1, x_1 + \varepsilon], \\ p(x), & x \notin [x_1, x_1 + \varepsilon], \end{cases}$$

$$\bar{h}(x, \varepsilon) = \left[\frac{(T + \theta\varepsilon)w(\bar{p}(x, \varepsilon))}{\bar{p}(x, \varepsilon)C(x, \bar{p}(x, \varepsilon))} \right]^{1/(p(x, \varepsilon) + 1)},$$

so that $\bar{p}(x, 0) = p(x)$ and $\bar{h}(x, 0) = h(x)$. Define the corresponding error and cost functions

$$\bar{E}(\varepsilon) = \int_{x_0}^{\bar{x}} C(x, \bar{p}(x, \varepsilon))\bar{h}(x, \varepsilon)^{p(x, \varepsilon)} \, dx,$$

$$\bar{W}(\varepsilon) = \int_{x_0}^{\bar{x}} w(\bar{p}(x, \varepsilon))\bar{h}(x, \varepsilon)^{-1} \, dx$$

and compute the right-handed derivatives $\bar{E}'_+(0)$, $\bar{W}'_+(0)$. It is found that $\bar{E}'_+(0) < 0$ and that $\bar{W}'_+(0) < 0$, implying the existence of $\varepsilon > 0$ such that $\bar{E}(\varepsilon) < E$ and $\bar{W}(\varepsilon) < W$. ∎

Informally we may regard the results of this subsection as requiring that the order multiplied by the 'error per work done' should be maintained constant from step to step and that we should choose the order p which minimizes $(p + 1)$ multiplied by the 'error per unit step'.

366 S-stability and stiff order.

We will motivate the work of Prothero and Robinson (1974) by considering the numerical solution of the equation

$$y'(x) = \lambda(y(x) - \exp(\mu x)) + \mu \exp(\mu x), \qquad y(0) = 1, \tag{366a}$$

with exact solution $y(x) = \exp(\mu x)$, using the s-stage implicit Runge–Kutta method of order $2s$. It is easily verified that the global error after n steps is

$$y_n - \exp(nh\mu) = [\exp(nh\mu) - r(h\lambda)^n] \left[\frac{h(\mu - \lambda)b^{\mathrm{T}}(I - h\lambda A)^{-1}v}{\exp(h\mu) - r(h\lambda)} - 1 \right], \tag{366b}$$

where $v = [\exp(h\mu c_1), \exp(h\mu c_2), \ldots, \exp(h\mu c_s)]^{\mathrm{T}}$ and r is the usual rational function associated with this method. That is r is the (s, s) diagonal entry of the Padé table for the exponential function. For n constant, the first factor on the right-hand side of (366b) converges to zero as $h \to 0$, but is bounded and bounded away from zero if $\mu \neq \lambda$ and $h \to 0$ with nh constant. Thus, from the known behaviour of both local and global truncation error, the last factor tends to zero as h^{2s}. We write this factor as

$$\varphi(h) = \frac{h(\mu - \lambda)b^{\mathrm{T}}(I - h\lambda A)^{-1}v - \exp(h\mu) + r(h\lambda)}{\exp(h\mu) - r(h\lambda)}$$

If $|\lambda|$ is large compared with $|\mu|$, it will be possible to choose values of h for which $\varphi(h)$ is quite accurately approximated as the limit when $\lambda \to \infty$. This gives the approximation

$$\varphi(h) \approx \frac{b^{\mathrm{T}} A^{-1} v - \exp(h\mu) + (-1)^s}{\exp(h\mu) - (-1)^s} \tag{366c}$$

and we will consider the behaviour of this for small $h\mu$. Write the components of v in power series and we find

$$b^{\mathrm{T}} A^{-1} v = \sum_{i=0}^{\infty} \frac{b^{\mathrm{T}} A^{-1} c^i}{i!} h^i,$$

where c^i denotes the componentwise power. It is easy to verify that $b^{\mathrm{T}} A^{-1} c^0 = b^{\mathrm{T}} A^{-1} e = 1 - (-1)^s$, $b^{\mathrm{T}} A^{-1} c^i = 1$ (for $i = 1, 2, \ldots, s$) and that $b^{\mathrm{T}} A^{-1} c^{s+1} \neq 1$. Thus the numerator on the right-hand side of (366c) equals a constant multiple of h^{s+1} plus $O(h^{s+2})$. If s is odd the denominator is $2 + O(h)$ and if s is even the denominator is $h\mu + O(h^2)$. Thus, for $h\mu$ small, but $h\lambda$ large, $\varphi(h)$ approximates a constant multiplied by h^s (s even) or h^{s+1} (s odd). For $s > 1$, this represents an effective reduction of order in the sense that, for a very stiff problem, the size to which h must be reduced before the benefits of the h^{2s} behaviour of error is realized, is much smaller than if stiffness were not present.

To look at the question in more detail, we consider the value of $h\varphi'(h)/\varphi(h)$ as representing the exponent E in the approximate behaviour $\varphi(\bar{h}) \approx C\bar{h}^E$ for $\bar{h} \approx h$. We confine ourselves to the special case $s = 2$ and present in table 366 values of this quantity for $\mu = 1$ and a variety of values of h and λ.

We see from this table that the greater the ratio $|\lambda|/|\mu|$, the further the order 2 behaviour persists as h becomes smaller.

In the Prothero and Robinson paper, (366a) is generalized to the problem

$$y'(x) = \lambda(y(x) - g(x)) + g'(x), \tag{366d}$$

with the exact solution $y(x) = g(x)$, and the relative error growth in a single step

$$\frac{\| y_n - g(x_n) \|}{\| y_{n-1} - g(x_{n-1}) \|} \tag{366e}$$

is studied on the assumption that $y_{n-1} \neq g(x_{n-1})$. If for any negative number λ_0, a positive h_0 exists such that (366e) is bounded by 1, wherever $0 < h < h_0$ and $\mathrm{Re}(\lambda) \leq \lambda_0$, then the method is said to be S-stable. It is found that Gauss methods do not have this property although Radau IA, Radau IIA and Lobatto IIIC methods do.

Assuming sufficient smoothness for g, it is possible to obtain an asymptotic result for the local truncation error if h is small but λh is large. This analysis, which greatly extends the considerations which we confined to the problem (366a), highlights serious difficulties in obtaining accurate results with several large classes of implicit Runge–Kutta methods. Besides the importance of this

Table 366   Values of $h\phi'(h)/\varphi(h)$ for $\mu = 1$

h λ	-10^4	-10^5	-10^6	-10^7	-10^8
10	0.55	0.55	0.55	0.55	0.55
5	1.23	1.23	1.23	1.23	1.23
2	1.82	1.82	1.82	1.82	1.82
1	1.95	1.95	1.95	1.95	1.95
0.5	2.00	1.99	1.99	1.99	1.99
0.2	2.06	2.00	2.00	2.00	2.00
0.1	2.21	2.02	2.00	2.00	2.00
0.05	2.65	2.09	2.01	2.00	2.00
0.02	3.50	2.46	2.06	2.01	2.00
0.01	3.85	3.09	2.21	2.02	2.00
0.005	3.96	3.66	2.65	2.09	2.01
0.002	3.99	3.94	3.50	2.46	2.06
0.001	4.00	3.98	3.85	3.09	2.21
0.0005	4.00	4.00	3.96	3.66	2.65
0.0002	4.00	4.00	3.99	3.93	3.50
0.0001	4.00	4.00	4.00	3.98	3.84

work in its own right, it is also significant as a forerunner to the work of Frank, Schneid and Ueberhuber (1981) which will be discussed in subsection 367.

367   B-consistency and B-convergence.

In an algebraically stable implicit Runge–Kutta method, assume that the internal order conditions

$$\sum_{j=1}^{s} a_{ij}c_j^{k-1} = \frac{1}{k}c_i^k, \qquad i = 1, 2, \ldots, s, \tag{367a}$$

hold for $k = 1, 2, \ldots, r$. These conditions have previously appeared in this volume as row-simplifying assumptions but we will highlight the importance they will now assume by referring to the highest r such that (367a) holds whenever $k \leq r$ as the *stage order*.

For a method with stage order r, let the *quadrature order* q be defined as the highest integer such that the quadrature conditions on the final result,

$$\sum_{i=1}^{s} b_i c_i^{k-1} = \frac{1}{k}, \tag{367b}$$

hold for all $k \leq q$. It is easy to see that the order p satisfies the inequality

$$q \geq p \geq \min(q, r+1).$$

We are particularly interested in methods such as the Gauss methods, for which p is much greater than r.

We will show how to obtain bounds on the local truncation error in terms of $h^{\min(r,q)+1}$ on condition that $y^{(\min(r,q))}$ is a continuous bounded function, that the problem satisfies (357a) and that A is positive definite (in the sense of definition 355A). In contrast, if p is much greater than the stage order, while bounds in terms of h^{p+1} can still be obtained, the bound will invariably depend on quantities which depend directly on the magnitude of the Lipschitz constant. The order of B-consistency $\bar{p}$ can be defined informally as the maximum integer such that the local truncation error can be bounded by $h^{\bar{p}+1}$ multiplied by a quantity which can be written in terms of derivatives to the solution to the problem but *not* on f' or on anything else that would become large as a problem became more stiff in such a way that the smooth solution itself did not change in any way.

We will not attempt to discuss B-consistency and related concepts in the full detail to be found in the paper by Frank, Schneid and Ueberhuber (1981) and in the other works by these and other authors cited in subsection 359, but will confine ourselves to the result given in theorem 367A and its immediate consequences. Among the concepts we have avoided using are BS- and BSI-stability. These are now recognized as being of fundamental significance in assessing the applicability of implicit Runge–Kutta methods to stiff problems, but it is convenient in this brief introduction to the theory of B-consistency to rely instead on the closely related criterion that A is positive definite.

To simplify the presentation we will write r as the lower of the stage order and the quadrature order for a given method.

THEOREM 367A.

Consider a Runge–Kutta method characterized by A, b^{T}, c such that A is positive definite and such that the stage order and the quadrature order are each at least r. Consider also a problem satisfying the one-sided Lipschitz condition (357a) for which $y^{(r+1)}$ is continuous on a closed interval I which contains x_{n-1}, x_n and each of $x_{n-1} + hc_i$ for $i = 1, 2, \ldots, s$. If the method is used with the problem from x_{n-1} to $x_n = x_{n-1} + h$, for $h \leq h_0$, then the local truncation error is bounded by $m \| y^{(r+1)} \| h^{r+1}$, where $\| y^{(r+1)} \| = \sup_{x \in I} \| y^{(r+1)}(x) \|$ and m and h_0 depend only on the method.

Proof.

Without loss of generality assume that the one-sided Lipschitz constant satisfies $\lambda \geq 0$. Since A is positive definite, choose $D = \mathrm{diag}(d_1, d_2, \ldots, d_s)$ so that $DA^{-1} + (A^{\mathrm{T}})^{-1}D$ is positive definite and choose $\mu > 0$, $h_0 > 0$ in such a way that $DA^{-1} + (A^{\mathrm{T}})^{-1}D - (2\lambda h_0 + \mu)D$ is positive semidefinite. Let $Y_1, Y_2, \ldots, Y_s, y_n$ denote stage values and the result at the end of the step starting from an initial value $y_{n-1} = y(x_{n-1})$. Also let $Z_1, Z_2, \ldots, Z_s, z_n$ denote the corresponding exact quantities $Z_i = y(X_i)$ $(i = 1, 2, \ldots, s)$, $z_n = y(x_n)$, where $X_i = x_{n-1} + hc_i (i = 1, 2, \ldots, s)$. We thus have

$$Y_i = y_{n-1} + h \sum_{j=1}^{s} a_{ij} f(X_j, Y_j), \qquad\qquad i = 1, 2, \ldots, s, \qquad (367c)$$

$$Z_i = y_{n-1} + h \sum_{j=1}^{s} a_{ij} f(X_j, Z_j) + \delta_i, \qquad i = 1, 2, \ldots, s, \qquad (367d)$$

$$y_n = y_{n-1} + h \sum_{i=1}^{s} b_i f(X_i, Y_i),$$

$$z_n = y_{n-1} + h \sum_{i=1}^{s} b_i f(X_i, Z_i) + \delta,$$

where the quadrature errors $\delta_1, \delta_2, \ldots, \delta_s, \delta$ satisfy $\| \delta_i \| \le \alpha_i \| y^{(r+1)} \| h^{r+1}$ $(i = 1, 2, \ldots, s)$, $\| \delta \| \le \alpha \| y^{(r+1)} \| h^{r+1}$. The local truncation error is $\| y_n - z_n \| \le \sum_{i=1}^{s} | b_i | \| hf(X_i, Y_i) - hf(X_i, Z_i) \| + \| \delta \|$ and the result of the theorem will follow if we can show that, for some $\bar{m}$,

$$\| hf(X_i, Y_i) - hf(X_i, Z_i) \| \le \bar{m} \| y^{(r+1)} \| h^{r+1}, \qquad i = 1, 2, \ldots, s.$$

Subtract (367c) from (367d) and form linear combinations of the s differences using A^{-1} with elements $a_{ij}^{(-1)}$ and we find

$$hf(X_i, Z_i) - hf(X_i, Y_i) = \sum_{j=1}^{s} a_{ij}^{(-1)} (Z_j - Y_j) + \sum_{j=1}^{s} a_{ij}^{(-1)} \delta_j. \qquad (367e)$$

Form the inner product with $Z_i - Y_i$, multiply by d_i and sum from $i = 1$ to s. Add the resulting sum to itself with i and j interchanged in the first term on the right-hand side and we find

$$2 \sum_{i=1}^{s} d_i [hf(X_i, Z_i) - hf(X_i, Y_i)]^{\mathrm{T}} (Z_i - Y_i)$$

$$= \sum_{i,j=1}^{s} [d_i a_{ji}^{(-1)} + d_j a_{ji}^{(-1)}] (Z_i - Y_i)^{\mathrm{T}} (Z_j - Y_j) + 2 \sum_{i,j=1}^{s} d_i a_{ij}^{(-1)} (Z_i - Y_i)^{\mathrm{T}} \delta_j.$$

Bound the left-hand side using (357a), write g_{ij} as the i, j element of $G = DA^{-1} + (A^{\mathrm{T}})^{-1} D - (2\lambda h_0 + \mu) D$ and we find

$$\sum_{i,j=1}^{s} g_{ij} (Z_i - Y_i)^{\mathrm{T}} (Z_j - Y_j) + \mu \sum_{i=1}^{s} d_i \| Z_i - Y_i \|^2$$

$$+ 2 \sum_{i,j=1}^{s} d_i a_{ij}^{(-1)} (Z_i - Y_i)^{\mathrm{T}} \delta_j \le 0,$$

so that, using the non-negative property of G, it follows that

$$\mu \sum_{i=1}^{s} d_i \| Z_i - Y_i \|^2 \le -2 \sum_{i,j=1}^{s} d_i a_{ij}^{(-1)} (Z_i - Y_i)^{\mathrm{T}} \delta_j$$

$$\le \theta \mu \left(\sum_{i=1}^{s} d_i \| Z_i - Y_i \|^2 \right)^{1/2} \left(\sum_{i=1}^{s} d_i^{-1} \| \delta_i \|^2 \right)^{1/2},$$

where $\theta = (2/\mu) \| D^{1/2} A^{-1} D^{1/2} \|_2$. Hence,

$$\left(\sum_{i=1}^{s} d_i \| Z_i - Y_i \|^2 \right)^{1/2} \le \theta \left(\sum_{i=1}^{s} d_i^{-1} \| \delta_i \|^2 \right)^{1/2}$$

Because $\| \delta \|_i$ is bounded in terms of $\| y^{(r+1)} \| h^{r+1}$, a similar bound exists for $\| Z_i - Y_i \|$ and, using (367e), for $\| hf(X_i, Z_i) - hf(X_i, Y_i) \|$ $(i = 1, 2, \ldots, s)$. ∎

368 Runge–Kutta methods with even error expansions.

We consider Runge–Kutta methods for which the reflection, in the sense of subsection 344, is identical with the method itself. For such a method, let $\phi(h)$ denote the function which maps y_{n-1} to y_n using a step size h and a particular differential equation system. Thus, $\phi(h)^n(y_0) = y_n \approx y(x_0 + nh)$, where the exponent denotes iterated composition of $\phi(h)$ with itself. Because $\phi(h)^{-1} = \phi(-h)$, the approximations at $\bar{x}$ given by $\phi((\bar{x} - x_0)/n)^n(y_0)$ and $\phi((\bar{x} - x_0)/(-n))^{-n}(y_0)$ are identical. This means that if the function f defining the differential equation is sufficiently smooth as to guarantee that the computed result has an asymptotic error expansion in n^{-1} then only even terms will occur.

Among methods that have this even error expansion property are the Gauss methods. This was shown in theorem 344E.

It seems appropriate to explore the use of extrapolation for these methods. The approaches based on the Neville interpolation formula or its rational variants does not generalize in a simple way to the situation in which the computed result for a step size h has an asymptotic expansion $C_0 + C_m h^{2m} + C_{m+1} h^{2m+2} + \cdots$ with $m > 1$. However, if the sequence of step sizes is a geometric progression $(h_0, \rho h_0, \rho^2 h_0, \ldots)$ then by forming linear combinations from a pair of adjacent members of the sequence, a new sequence can be constructed of exactly the same form but with m replaced by $m + 1$.

Let $\eta_i^{(0)}$ denote the approximate solution at $\bar{x}$ computed using a step size $h_i = h_0 \rho^i$ $(i = 0, 1, 2, \ldots)$. Thus $\eta_i^{(0)} \sim C_0 + \sum_{j=m}^{\infty} C_j h_0^{2j} \rho^{2ij}$. To annihilate the h_0^{2m} term in this asymptotic expansion, $\eta_i^{(1)}$ is formed from $\eta_i^{(0)}$ and $\eta_{i+1}^{(0)}$ using the formula

$$\eta_i^{(1)} = \eta_{i+1}^{(0)} - \rho^{2m} (1 - \rho^{2m})^{-1} (\eta_i^{(0)} - \eta_{i+1}^{(0)})$$

to yield the asymptotic formula

$$\eta_i^{(1)} = C_0 + \sum_{j = m+1}^{\infty} C_j h_0^{2j} \rho^{2ij} (1 - \rho^{2m})^{-1} (\rho^{2j} - \rho^{2m}).$$

In a similar way, we may form further sequences $\eta_i^{(2)}$, $\eta_i^{(3)}$, $\ldots$, where

$$\eta_i^{(k)} = \eta_{i+1}^{(k-1)} - \rho^{2(m+k-1)} (1 - \rho^{2(m+k-1)})^{-1} (\eta_i^{(k-1)} - \eta_{i+1}^{(k-1)}),$$

$$k = 1, 2, 3, \ldots, i = 0, 1, 2, \ldots,$$

Table 368 Extrapolated sequences from fourth-order Gauss method

i	k 0	1	2	3	4
0	0.945846660357688	0.945988971016537	0.945983879897505	0.945983777857574	0.945983777842598
1	0.945980076600359	0.945983389133740	0.945983777865542	0.945983777842613	0.945983777842554
2	0.945978696600403	0.945983780041608	0.945983777842702	0.945983777842554	0.945983777842554
3	0.94598346264033	0.945983777845810	0.945983777842555	0.945983777842554	0.945983777842554
4	0.945983758719949	0.945983777842606	0.945983777842554	0.945983777842554	
5	0.945983777719440	0.945983777842555	0.945983777842554		
6	0.945983777834861	0.945983777842554			
7	0.945983777842073				

and for $k = 1, 2, \ldots$ each sequence may be expected to converge more rapidly than its predecessor.

To illustrate the behaviour of extrapolated sequences of approximations formed in this way, results are presented for the Gauss method with $s = 2$ applied to the problem used for illustrative purposes in subsection 225 using the values $h_0 = 1$, $\rho = \frac{1}{2}$. In table 368 which gives $\eta_0^{(0)}$, $\eta_1^{(0)}$, $\ldots$, $\eta_7^{(0)}$ in the first column and extrapolated sequences in later columns, digits are italicized where they differ from the exact answer $y(1) = 0.945988377842554$.

37 RUNGE–KUTTA METHODS WITH ERROR ESTIMATES

370 Introduction.

The practical use of Runge–Kutta methods requires convenient techniques for estimating local errors. These techniques are used in adaptive implementations of the methods for assessing the appropriateness of the step size being used in the light of the accuracy requirements being imposed.

Thus, if the local truncation error for the method which computes y_n by the formula $y_n = \phi(y_{n-1})$ is given by

$$y(x_n) - \phi(y(x_{n-1})) = c(x_n)h^{p+1} + O(h^{p+2}),$$

where $h = x_n - x_{n-1}$, then we require an approximation to $c(x_n)h^{p+1}$.

A possible procedure for obtaining such an approximation was proposed by Richardson (1927) and called 'the deferred approach to the limit'. This consists in repeating the integration from x_{n-1} to x_n but with two half-size steps instead of a single full-size step. A comparison of the results obtained furnishes the error estimate.

A more modern approach is to devise special methods which are actually two methods built into one. One of the constituent methods has an order p, say, and the second has an order $p + 1$. The difference of the results computed by these methods provides an error estimate for the order p method.

Methods of this type have been devised by E. Fehlberg and by a number of other authors.

Finally, we consider a less costly approach to error estimation but an approach which requires results from several successive steps to be used together in furnishing the error estimate. This approach, which is based on the use of a high-order quadrature formula, was first suggested by Stoller and Morrison (1958) and proposed, in a more general form, by Ceschino and Kuntzmann (1963) and by Lawson and Ehle (1970).

371 Richardson error estimates.

The so-called 'deferred approach to the limit' consists in repeating a calculation with some parameter altered. If the error depends on this parameter and vanishes for some limiting value, the two results are used to estimate this limit.

In the present situation, the first calculation of $y(x_{n-1} + h)$ is the evaluation of y_n from y_{n-1} using a Runge–Kutta method:

$$
\begin{array}{c|ccccc}
0 & & & & & \\
c_2 & a_{21} & & & & \\
c_3 & a_{31} & a_{32} & & & \\
\vdots & \vdots & \vdots & \ddots & & \\
c_s & a_{s1} & a_{s2} & \cdots & a_{s,s-1} & \\
\hline
& b_1 & b_2 & \cdots & b_{s-1} & b_s
\end{array}
$$

and the second is the evaluation of a second approximation to $y(x_{n-1} + h)$ by carrying out two steps by the same method with step size $h/2$ in each case. Thus the second computation is equivalent to the use of a Runge–Kutta method with $2s$ stages given by

$$
\begin{array}{c|cccccccc}
0 & & & & & & & & \\
c_2/2 & a_{21}/2 & & & & & & & \\
c_3/2 & a_{31}/2 & a_{32}/2 & & & & & & \\
\vdots & \vdots & \vdots & \ddots & & & & & \\
c_s/2 & a_{s1}/2 & a_{s2}/2 & \cdots & a_{s,s-1}/2 & & & & \\
1/2 & b_1/2 & b_2/2 & \cdots & b_{s-1}/2 & b_s/2 & & & \\
(1+c_2)/2 & b_1/2 & b_2/2 & \cdots & b_{s-1}/2 & b_s/2 & a_{21}/2 & & \\
(1+c_3)/2 & b_1/2 & b_2/2 & \cdots & b_{s-1}/2 & b_s/2 & a_{31}/2 & a_{32}/2 & \\
\vdots & \vdots & \vdots & \vdots & \vdots & \vdots & \vdots & \vdots & \ddots \\
(1+c_s)/2 & b_1/2 & b_2/2 & \cdots & b_{s-1}/2 & b_s/2 & a_{s1}/2 & a_{s2}/2 & \cdots & a_{s,s-1}/2 \\
\hline
& b_1/2 & b_2/2 & \cdots & b_{s-1}/2 & b_s/2 & b_1/2 & b_2/2 & \cdots & b_{s-1}/2 & b_s/2
\end{array}
$$

This second method has the same order, p, as the original, but the truncation error is

$$
2^{-p}c(x_n)h^{p+1} + O(h^{p+2}).
$$

Hence, at the expense of carrying out both the s stages of the original method and the $2s$ stages of the second method, it is possible to obtain asymptotically correct error estimates.

If $\bar{y}_n$ is the result computed by the original method and $\tilde{y}_n$ the result computed by the second method, then the difference is

$$
\bar{y}_n - \tilde{y}_n = (1 - 2^{-p})c(x_n)h^{p+1} + O(h^{p+2})
$$

so that

$$
(1 - 2^{-p})^{-1}(\bar{y}_n - \tilde{y}_n) = c(x_n)h^{p+1} + O(h^{p+2})
$$

approximates the error in $\bar{y}_n$.

At first sight, this appears like a tripling of the work just to obtain an error estimate but, looked at another way, the work is multiplied by only $(3s - 1)/2s$. This is because the first derivative evaluation is common to each of the two methods and needs to be carried out only only once, and because we can use the value of $\tilde{y}_n$ as the value of y_n. In this case, the error estimate becomes

$$(2^p - 1)^{-1}(\bar{y}_n - \tilde{y}_n).$$

Since this is just the original method repeated over a pair of steps of length $h/2$, the total computational cost is of $3s - 1$ evaluations of f compared with $2s$ if no error estimate had been provided.

We illustrate this idea by presenting algorithm 371 for the execution of a single h step followed by two $h/2$ steps of the classical (order 4) Runge–Kutta method. The two short steps are used to advance the solution a distance h and the difference between this result and the result computed by the larger step furnishes the error estimate, represented here by the vector *error*. Other variables and the procedure *diffeqn* have the same meanings as in the algorithms of subsections 313 and 314.

Algorithm 371: Runge–Kutta step with error estimate

```
H := h;
for j := 1 to 3 do
begin
      if j = 2 then H := h/2 else diffeqn(y,F1);
      for i := 1 to n do Y0[i] := y[i] + 0.5*H*F1[i];
      diffeqn(Y0, F2);
      for i := 1 to n do Y0[i] := y[i] + 0.5*H*F2[i];
      diffeqn(Y0,F3);
      for i := 1 to n do Y0[i] := y[i] + H*F3[i];
      diffeqn(Y0,F4);
      for i := 1 to n do
      begin
            temp := y[i] + (H/6)*(F1[i] + F4[i] + 2*(F2[i] + F3[i]));
            if j = 1 then Y1[i] := temp else y[i] := temp
      end
end;
for i := 1 to n do error[i] := (Y1[i] − y[i])/15;
```

372 Methods with built-in estimates.

Rather than use the Richardson technique of the last subsection it is possible to devise Runge–Kutta schemes in which there are two alternative choices of the vector $(b_1, b_2, \ldots, b_s)$, such that each of the two resulting methods have different orders. The lower-order method provides the result for continuing the integration while the difference of the two results provides an error estimate.

The earliest proposed method of this type is due to Merson (1957). We will write this method in the form

$$
\begin{array}{c|ccccc}
0 & & & & & \\
\frac{1}{3} & \frac{1}{3} & & & & \\
\frac{1}{3} & \frac{1}{6} & \frac{1}{6} & & & \\
\frac{1}{2} & \frac{1}{8} & 0 & \frac{3}{8} & & \\
1 & \frac{1}{2} & 0 & -\frac{3}{2} & 2 & \\
\hline
 & \frac{1}{6} & 0 & 0 & \frac{2}{3} & \frac{1}{6} \\
\hline
 & \frac{1}{10} & 0 & \frac{3}{10} & \frac{2}{5} & \frac{1}{5}
\end{array}
$$

indicating that it combines the two methods

$$
\begin{array}{c|ccccc}
0 & & & & & \\
\frac{1}{3} & \frac{1}{3} & & & & \\
\frac{1}{3} & \frac{1}{6} & \frac{1}{6} & & & \\
\frac{1}{2} & \frac{1}{8} & 0 & \frac{3}{8} & & \\
1 & \frac{1}{2} & 0 & -\frac{3}{2} & 2 & \\
\hline
 & \frac{1}{6} & 0 & 0 & \frac{2}{3} & \frac{1}{6}
\end{array}
\qquad (372a)
$$

and

$$
\begin{array}{c|ccccc}
0 & & & & & \\
\frac{1}{3} & \frac{1}{3} & & & & \\
\frac{1}{3} & \frac{1}{6} & \frac{1}{6} & & & \\
\frac{1}{2} & \frac{1}{8} & 0 & \frac{3}{8} & & \\
1 & \frac{1}{2} & 0 & -\frac{3}{2} & 2 & \\
\hline
 & \frac{1}{10} & 0 & \frac{3}{10} & \frac{2}{5} & \frac{1}{5}
\end{array}
\qquad (372b)
$$

Merson's proposal was that (327a) be used to compute the approximation at the end of the step and the difference between (327a) and (327b) be used to estimate the local truncation error in this approximation. Although (327b) is only a third-order method, for linear differential equations with constant co-efficients it becomes effectively fifth order.

While computational success has been claimed for the Merson method, it is easy to devise problems which defeat its error-estimating property.

In considering possible alternatives to Merson's approach, we might ask whether it is possible to obtain both a fourth-order method and an error estimator in just five stages. That it cannot, follows from theorem 322C, since combining the error estimator with the result would yield a fifth-order method. Thus fourth-order error-estimating methods require at least six stages.

Important contributions to the search for error-estimating methods have been made by a number of authors, particularly E. Fehlberg and J. H. Verner. In subsection 374 and 375 some of the methods due to Fehlberg and to Verner will be touched on. In the meantime, we look at a possible extension to the order conditions for a standard Runge–Kutta method to provide convenient error estimates.

373 A class of error-estimating methods.

Continuing our study of methods with a built-in error estimate, we consider methods of the form

$$
\begin{array}{c|cccccc}
0 \\
c_2 & a_{21} \\
c_3 & a_{31} & a_{32} \\
\vdots & \vdots & \vdots & \ddots \\
\vdots & \vdots & \vdots & \ddots \\
c_s & a_{s1} & a_{s2} & \cdots & a_{s,s-1} \\
c_{s+1} & a_{s+1,1} & a_{s+1,2} & \cdots & a_{s+1,s-1} & a_{s+1,s} \\
\hline
& b_1 & b_2 & \cdots & b_{s-1} & b_s & b_{s+1}
\end{array}
\tag{373a}
$$

which have order $p + 1$ and which are such that the embedded method

$$
\begin{array}{c|ccccc}
0 \\
c_2 & a_{21} \\
c_3 & a_{31} & a_{32} \\
\vdots & \vdots & \vdots & \ddots \\
\vdots & \vdots & \vdots & \ddots \\
c_s & a_{s1} & a_{s2} & \cdots & a_{s,s-1} \\
\hline
& a_{s+1,1} & a_{s+1,2} & \cdots & a_{s+1,s-1} & a_{s+1,s}
\end{array}
\tag{373b}
$$

has order p. Since stage number $s + 1$ in (373a) is required for subsequent steps, we need not count the f evaluation associated with it both in the current step and as the first stage of the next step. Thus, if advantage of this fact is taken in a program, then from the point of view of computational cost there are effectively only s stages in this method. This assumes, of course, that the method (373b) is used to compute y_n from a given value of y_{n-1}, whereas the difference of the results computed by the two methods is of use only for error-estimation purposes.

To analyse the order of those combined methods, it is convenient to write $B = b_{s+1}$ and to write $\Phi(t)$ for an elementary weight corresponding to the tree t for the fictitious Runge–Kutta method:

$$
\begin{array}{c|ccccc}
0 & & & & & \\
c_2 & a_{21} & & & & \\
c_3 & a_{31} & a_{32} & & & \\
\vdots & \vdots & \vdots & \ddots & & \\
c_s & a_{s1} & a_{s2} & \cdots & a_{s,s-1} & \\
\hline
 & b_1 & b_2 & \cdots & b_{s-1} & b_s
\end{array}
\tag{373c}
$$

Also write $\overline{\Phi}(t)$ for the corresponding elementary weight of (373b) and $\widetilde{\Phi}(t)$ in the case of (373a). With these conventions we have the following result.

THEOREM 373A.

If (373b) is of order p and (373a) of order $p + 1$ and $B = b_{s+1}$, then

$$
c_{s+1} = 1, \tag{373d}
$$

$$
\Phi(t) = \frac{1}{\gamma(t)} [1 - Br(t)], \qquad r(t) \le p + 1. \tag{373e}
$$

Conversely, if (373d) and (373e) hold with $c_s \ne 1$ and $B \ne 0$ and, in addition,

$$
b_{s+1} = B, \tag{373f}
$$

$$
a_{s+1,s} = \frac{b_s(1 - c_s)}{B}, \tag{373g}
$$

$$
a_{s+1,i} = \frac{b_i(1 - c_i) - \sum_{j \le s} b_j a_{ji}}{B}, \qquad 1, 2, \ldots, s - 1, \tag{373h}
$$

then (373b) is of order p and (373a) is of order $p + 1$.

Proof.

The first result, (373d), is just the consistency condition, $\Phi(\tau) = 1$, for the method (373b). To prove (373e), let $t = [t_1 t_2 \ldots t_m]$, where $t_1, t_2, \ldots, t_m$ have lower orders than t, so that

$$
\widetilde{\Phi}(t) = \Phi(t) + B\overline{\Phi}(t_1)\overline{\Phi}(t_2) \ldots \overline{\Phi}(t_m). \tag{373i}
$$

Using the order conditions for (373a) and (373b) and the relation $\gamma(t) = r(t)\gamma(t_1)\gamma(t_2) \cdots \gamma(t_m)$, the result follows immediately.

The converse result will be proved by induction. For an integer $k = 1, 2, \ldots, p + 1$ we will show that (373a) has order k and that (373b) has order $k - 1$ (in the case $k = 1$, this is to be interpreted as a vacuous statement concerning (373b)).

For $k = 1$, the only tree is $t = \tau$ and the result is obtained by rearranging (373e). For $k > 1$, let $t = [t_1 t_2 \ldots t_m]$ be a tree of this order. If $m > 1$, then $t_1, t_2, \ldots, t_m$ have lower orders than $k - 1$, and $\widetilde{\Phi}(t) = 1/\gamma(t)$ follows from (373i). If $m = 1$, the column simplifying assumptions given by (373g) and (373h) guarantee that $\widetilde{\Phi}(t) = 1/\gamma(t)$ whereas if $t = [t_1]$ and t_1 is an arbitrary tree of order $k - 1$, the relation (373i) together with the known values of $\widetilde{\Phi}(t)$ and $\Phi(t)$ yield the results $\overline{\Phi}(t_1) = 1/\gamma(t_1)$. ∎

We now see that (373e) may be interpreted as an extension of the conventional order conditions. These modified conditions, together with (373f), (373g) and (373h), are used to derive several particular methods which we present below. In each case, the coefficients will be displayed thus:

$$
\begin{array}{c|ccccc}
0 & & & & & \\
c_2 & a_{21} & & & & \\
c_3 & a_{31} & a_{32} & & & \\
\vdots & \vdots & \vdots & \ddots & & \\
c_s & a_{s1} & a_{s2} & \cdots & a_{s,s-1} & \\
\hline
1 & a_{s+1,1} & a_{s+1,2} & \cdots & a_{s+1,s-1} & a_{s+1,s} \\
\hline
& b_1 & b_2 & \cdots & b_{s-1} & b_s & b_{s+1} \\
& [d_1 & d_2 & \cdots & d_{s-1} & d_s & d_{s+1}]
\end{array}
$$

so that $y_{n-1} + h \sum_{i=1}^{s} a_{s+1,i} f(Y_i)$ is the result of order p, $y_{n-1} + h \sum_{i=1}^{s+1} b_i f(Y_i)$ is the order $p + 1$ result to be used for comparison and, finally, $h \sum_{i=1}^{s+1} d_i f(Y_i)$ is the error estimate itself where $d_i = b_i - a_{s+1,i}$ $(i = 1, 2, \ldots, s)$, $d_{s+1} = b_{s+1}$.

Methods for orders 1 and 2 are easy to derive and we present an example of each:

$$
\begin{array}{c|cc}
0 & & \\
\hline
1 & 1 & \\
\hline
& \frac{1}{2} & \frac{1}{2} \\
& [-\frac{1}{2} & \frac{1}{2}]
\end{array}
$$

and

$$
\begin{array}{c|cccc}
0 & & & & \\
\frac{1}{2} & \frac{1}{2} & & & \\
\frac{1}{2} & 0 & \frac{1}{2} & & \\
\hline
1 & 0 & 0 & 1 & \\
\hline
& \frac{1}{6} & \frac{1}{3} & \frac{1}{3} & \frac{1}{6} \\
& [\frac{1}{6} & \frac{1}{3} & -\frac{2}{3} & \frac{1}{6}]
\end{array}
$$

Note that, for the second-order method, the error-estimating method in which it is embedded is actually the classical fourth-order method.

For order 3 (with order 4 error estimate) 4 stages are necessary (together with the essentially free additional error-estimating stage). From the eight equations of the form given by (373d), with $s = 4$, we write down four expressions involving b_3, b_4, a_{32}, a_{43}, c_2, c_3, c_4, B as follows:

$$b_4 a_{43} c_3 (c_3 - c_2) = \Sigma\, b_i a_{ij} c_j (c_j - c_2) = \frac{1}{12} - \frac{B}{3} - c_2 \left(\frac{1}{6} - \frac{B}{2} \right), \qquad (373j)$$

$$b_3 (c_3 - c_4) a_{32} c_2 = \Sigma\, b_i (c_i - c_4) a_{ij} c_j = \frac{1}{8} - \frac{B}{2} - c_4 \left(\frac{1}{6} - \frac{B}{2} \right), \qquad (373k)$$

$$b_4 a_{43} a_{32} c_2 = \Sigma\, b_i a_{ij} a_{jk} c_k = \frac{1}{24} - \frac{B}{6}, \qquad (373l)$$

$$b_3 (c_3 - c_4)(c_3 - c_2) c_3 = \Sigma\, b_i (c_i - c_4)(c_i - c_2) c_i$$

$$= \left(\frac{1}{4} - B \right) - (c_2 + c_4) \left(\frac{1}{3} - B \right) + c_2 c_4 \left(\frac{1}{2} - B \right). \qquad (373m)$$

Since the products of the left-hand sides of (373j) and (373k) and of (373l) and (373m) are equal, we obtain consistency conditions on c_2, c_4 and B by equating the corresponding products of the right-hand sides. After simplification and removal of a factor c_2, which cannot be zero, we find the following relation between B and c_4:

$$c_4 = 1 - \frac{B}{1 - 6B + 12B^2}.$$

The simplest choices seem to be $B = \frac{1}{6}$ giving $c_4 = \frac{1}{2}$ and $B = \frac{1}{12}$ giving $c_4 = \frac{6}{7}$. In each case, the choices of c_2 and c_3 are somewhat arbitrary and they are given convenient values in the methods which follow:

0				
$\frac{1}{4}$	$\frac{1}{4}$			
$\frac{3}{4}$	$-\frac{9}{4}$	3		
$\frac{1}{2}$	$\frac{1}{18}$	$\frac{5}{12}$	$\frac{1}{36}$	
1	$\frac{7}{9}$	$-\frac{5}{3}$	$-\frac{1}{9}$	2
	$\frac{1}{6}$	0	0	$\frac{2}{3}$	$\frac{1}{6}$
	$[-\frac{11}{18}$	$\frac{5}{3}$	$\frac{1}{9}$	$-\frac{4}{3}$	$\frac{1}{6}]$

$$
\begin{array}{c|cccccc}
0 \\
\tfrac{2}{7} & \tfrac{2}{7} \\
\tfrac{4}{7} & -\tfrac{8}{35} & \tfrac{4}{5} \\
\tfrac{6}{7} & \tfrac{29}{42} & -\tfrac{2}{3} & \tfrac{5}{6} \\
1 & \tfrac{1}{6} & \tfrac{1}{6} & \tfrac{5}{12} & \tfrac{1}{4} \\
\hline
 & \tfrac{11}{96} & \tfrac{7}{24} & \tfrac{35}{96} & \tfrac{7}{48} & \tfrac{1}{12} \\
 & [-\tfrac{5}{96} & \tfrac{1}{8} & -\tfrac{5}{96} & -\tfrac{5}{48} & \tfrac{1}{12}]
\end{array}
$$

For fourth order (embedded within fifth order), we present a single method with $s = 6$. It is based on the simplifying assumption

$$\Sigma a_{ij}c_j = \tfrac{1}{2}c_i^2, \qquad i \neq 2,$$

together with the subsidiary conditions

$$b_2 = \Sigma b_i a_{i2} = \Sigma b_i c_i a_{i2} = \Sigma b_i a_{ij} a_{j2} = 0,$$

and uses expressions for $\Sigma b_i(c_i - c_6)a_{ij}c_j(c_j - c_3)$, $\Sigma b_i a_{ij}c_j(c_j - c_3)(c_j - c_4)$, $\Sigma b_i a_{ij}a_{jk}c_k(c_k - c_3)$, $\Sigma b_i c_i(c_i - c_3)(c_i - c_4)(c_i - c_6)$ involving B, c_3, c_4 and c_6 to obtain, after some manipulation, the following consistency condition:

$$c_6 = 1 - B[c_3 - (2 - 8c_3 + 10c_3^2)c_4] / \{c_3(1 - 8B + 20B^2)$$
$$- [(2 - 16B + 40B^2) - (8 - 56B + 120B^2)c_3 + (10 - 60B + 120B^2)c_3^2]c_4\},$$

from which we have made the particular choice, $c_3 = \tfrac{1}{4}$, $c_4 = \tfrac{1}{2}$, $c_6 = \tfrac{4}{5}$, $B = \tfrac{1}{15}$. Selecting also the values $c_2 = \tfrac{1}{4}$, $c_5 = \tfrac{12}{30}$ we obtain the method

$$
\begin{array}{c|ccccccc}
0 \\
\tfrac{1}{4} & \tfrac{1}{4} \\
\tfrac{1}{4} & \tfrac{1}{8} & \tfrac{1}{8} \\
\tfrac{1}{2} & 0 & -\tfrac{1}{2} & 1 \\
\tfrac{13}{20} & \tfrac{13}{200} & -\tfrac{299}{1000} & \tfrac{78}{125} & \tfrac{13}{50} \\
\tfrac{4}{5} & \tfrac{548}{7475} & \tfrac{688}{2875} & \tfrac{572}{2875} & -\tfrac{88}{575} & \tfrac{132}{299} \\
1 & \tfrac{37}{312} & 0 & \tfrac{4}{33} & \tfrac{8}{9} & \tfrac{100}{117} & \tfrac{575}{792} \\
\hline
 & \tfrac{41}{520} & 0 & \tfrac{58}{165} & \tfrac{16}{135} & \tfrac{50}{351} & \tfrac{575}{2376} & \tfrac{1}{15} \\
 & [-\tfrac{31}{780} & 0 & \tfrac{38}{165} & -\tfrac{104}{135} & \tfrac{350}{351} & -\tfrac{575}{1188} & \tfrac{1}{15}]
\end{array}
$$

Since the derivation of methods of higher order, of the type considered here, becomes increasingly more involved we will stop at fifth order (embedded within sixth order). In the fifth-order case, $s = 8$ appears to be necessary and

303

Table 373 Coefficients of a fifth-order Runge–Kutta method with error estimate

	Numerator	Denominator	
c_2	1	9	0.111111111111111
a_{21}	1	9	0.111111111111111
c_3	1	9	0.111111111111111
a_{31}	1	18	0.055555555555556
a_{32}	1	18	0.055555555555556
c_4	1	6	0.166666666666667
a_{41}	1	24	0.041666666666667
a_{43}	1	8	0.125
c_5	1	3	0.333333333333333
a_{51}	1	6	0.166666666666667
a_{53}	−1	2	−0.5
a_{54}	2	3	0.666666666666667
c_6	1	2	0.5
a_{61}	15	8	1.875
a_{63}	−63	8	−7.875
a_{64}	7	1	7.0
a_{65}	−1	2	−0.5
c_7	3	4	0.75
a_{71}	−93	22	−4.227272727272727
a_{73}	24921	1408	17.699573863636364
a_{74}	−10059	704	−14.288352272727273
a_{75}	735	1408	0.522017045454545
a_{76}	735	704	1.044034090909091
c_8	17	19	0.894736842105263
a_{81}	86547313055	10295619642	8.406226731797519
a_{83}	−96707067	2867062	−33.730371718504867
a_{84}	15526951598	571978869	27.146023112962238
a_{85}	27949088	81711267	0.342046929709216
a_{86}	−452648800	245133801	−1.846537679232575
a_{87}	270189568	467982711	0.577349465373733
c_9	1	1	1.0
a_{91}	98	765	0.128104575163399
a_{94}	−9	83	−0.108433734939759
a_{95}	1071	1600	0.669375
a_{96}	−11	75	−0.146666666666667
a_{97}	64	225	0.284444444444444
a_{98}	390963	2257600	0.173176381998583
b_1	188	3315	0.056711915535445
b_4	1593	7553	0.210909572355356
b_5	2943	20800	0.141490384615385
b_6	197	975	0.202051282051282
b_7	576	2275	0.253186813186813
b_8	2476099	29348800	0.084367980973668
b_9	2	39	0.051282051282051
d_1	−142	1989	−0.071392659627954
d_4	2412	7553	0.319343307295115
d_5	−549	1040	−0.527884615384615
d_6	68	195	0.348717948717949
d_7	−128	4095	−0.031257631257631
d_8	−130321	1467440	−0.088808401024914
d_9	2	39	0.051282051282051

we can make simplifying and subsidiary assumptions as follows:

$$b_2 = 0,$$
$$a_{i2} = 0, \qquad i \neq 3,$$
$$\Sigma a_{ij}c_j = \tfrac{1}{2}c_i^2, \qquad i \neq 2,$$
$$\Sigma a_{ij}c_j^2 = \tfrac{1}{3}c_i^3, \qquad i \neq 2, 3,$$
$$b_3 = \Sigma b_i a_{i3} = \Sigma b_i c_i a_{i3} = \Sigma b_i a_{ij} a_{j3} = 0.$$

Consistency conditions based on expressions for $\Sigma b_i a_{ij} c_j (c_j - c_4)(c_j - c_5)$ $(c_j - c_6)$, $\Sigma b_i (c_i - c_8) a_{ij} c_j (c_j - c_4)(c_j - c_5)$, $\Sigma b_i a_{ij} a_{jk} c_k (c_k - c_4)(c_k - c_5)$ and $\Sigma b_i c_i (c_i - c_4)(c_i - c_5)(c_i - c_6)(c_i - c_8)$ lead to a complicated relationship between c_4, c_5, c_6, c_8 and B. One of the solutions is $c_4 = \tfrac{1}{6}$, $c_5 = \tfrac{1}{3}$, $c_6 = \tfrac{1}{2}$, $c_8 = \tfrac{17}{19}$, $B = \tfrac{2}{39}$ and together with the choice $c_2 = \tfrac{1}{9}$, $c_3 = \tfrac{1}{9}$, $c_7 = \tfrac{3}{4}$, the method is uniquely determined. Writing d_1, d_2, ..., d_9 as $b_1 - a_{91}$, $b_2 - a_{92}$, ..., b_9, the error-estimating coefficients, so that the method is given by

$$
\begin{array}{c|ccccccc}
0 \\
c_2 & a_{21} \\
c_3 & a_{31} & a_{32} \\
\vdots & \vdots & \vdots & \ddots \\
c_8 & a_{81} & a_{82} & \cdots & a_{87} \\
c_9 & a_{91} & a_{92} & \cdots & a_{97} & a_{98} \\
\hline
& b_1 & b_2 & \cdots & b_7 & b_8 & b_9 \\
& [d_1 & d_2 & \cdots & d_7 & d_8 & d_9]
\end{array}
$$

we have displayed the non-zero coefficients in table 373 both as numerator and denominator of rational numbers and as decimal approximations.

374 The methods of E. Fehlberg.

In this subsection we review the methods of Fehlberg (1968, 1969, 1970) of orders up to eight (with an error estimate of one order higher in each case). Methods of orders 1, 2 and 3 are proposed in Fehlberg (1970) as follows:
Order 1:

$$
\begin{array}{c|ccc}
0 \\
\tfrac{1}{2} & \tfrac{1}{2} \\
1 & \tfrac{1}{256} & \tfrac{255}{256} \\
\hline
& \tfrac{1}{512} & \tfrac{255}{256} & \tfrac{1}{512} \\
& [-\tfrac{1}{512} & 0 & \tfrac{1}{512}]
\end{array}
$$

Order 2:

$$
\begin{array}{c|cccc}
0 \\
\frac{1}{4} & \frac{1}{4} \\
\frac{27}{40} & -\frac{189}{800} & \frac{729}{800} \\
\hline
1 & \frac{214}{891} & \frac{1}{33} & \frac{650}{891} \\
\hline
& \frac{41}{162} & 0 & \frac{800}{1053} & -\frac{1}{78} \\
& [\frac{23}{1782} & -\frac{1}{33} & \frac{350}{11583} & -\frac{1}{78}]
\end{array}
$$

$$
\begin{array}{c|ccc}
0 \\
1 & 1 \\
\frac{1}{2} & \frac{1}{4} & \frac{1}{4} \\
\hline
1 & \frac{1}{2} & \frac{1}{2} & 0 \\
\hline
& \frac{1}{6} & \frac{1}{6} & \frac{2}{3} \\
& [-\frac{1}{3} & -\frac{1}{3} & \frac{2}{3}]
\end{array}
$$

Order 3:

$$
\begin{array}{c|ccccc}
0 \\
\frac{1}{4} & \frac{1}{4} \\
\frac{4}{9} & \frac{4}{81} & \frac{32}{81} \\
\frac{6}{7} & \frac{57}{98} & -\frac{432}{343} & \frac{1053}{686} \\
\hline
1 & \frac{1}{6} & 0 & \frac{27}{52} & \frac{49}{156} \\
\hline
& \frac{43}{288} & 0 & \frac{243}{416} & \frac{343}{1872} & \frac{1}{12} \\
& [-\frac{5}{288} & 0 & \frac{27}{416} & -\frac{245}{1872} & \frac{1}{12}]
\end{array}
$$

$$
\begin{array}{c|ccccc}
0 \\
\frac{2}{7} & \frac{2}{7} \\
\frac{7}{15} & \frac{77}{900} & \frac{343}{900} \\
\frac{35}{38} & \frac{805}{1444} & -\frac{77175}{54872} & \frac{97125}{54872} \\
\hline
1 & \frac{79}{490} & 0 & \frac{2175}{3626} & \frac{2166}{9065} \\
\hline
& \frac{229}{1470} & 0 & \frac{1125}{1813} & \frac{13718}{81585} & \frac{1}{18} \\
& [-\frac{4}{735} & 0 & \frac{75}{3626} & -\frac{5776}{81585} & \frac{1}{18}]
\end{array}
$$

We see that each of these methods is of the same style as those of subsection 373 in that the error-estimating stage uses the derivative computed as the first stage in the following step. This is not true of Fehlberg's fourth-order methods, one of which is as follows:

$$
\begin{array}{c|cccccc}
0 \\
\frac{2}{9} & \frac{2}{9} \\
\frac{1}{3} & \frac{1}{12} & \frac{1}{4} \\
\frac{3}{4} & \frac{69}{128} & -\frac{243}{128} & \frac{135}{64} \\
1 & -\frac{17}{12} & \frac{27}{4} & -\frac{27}{5} & \frac{16}{15} \\
\frac{5}{6} & \frac{65}{432} & -\frac{5}{16} & \frac{13}{16} & \frac{4}{27} & \frac{5}{144} \\
\hline
 & \frac{1}{9} & 0 & \frac{9}{20} & \frac{16}{45} & \frac{1}{12} \\
\hline
 & \frac{47}{450} & 0 & \frac{12}{25} & \frac{32}{225} & \frac{1}{30} & \frac{6}{25} \\
\hline
 & [-\frac{1}{150} & 0 & \frac{3}{100} & -\frac{16}{75} & -\frac{1}{20} & \frac{6}{25}]
\end{array}
$$

The coefficients of this method have been displayed in the same way as for Merson's method in subsection 372 so that the coefficients b_1, b_2, b_3, b_4, b_5 for the embedded fourth-order method are $1/9$, 0, $9/20$, $16/45$, $1/12$, the coefficients for the fifth-order approximation are given by the line beginning $47/450$, and the difference of these, used for error-estimating purposes, by the line beginning $-1/150$.

For methods of orders greater than four, the approach used in Fehlberg (1968), (1969) is a little different. In order to make use of the column simplifying assumption

$$\sum_i b_i a_{ij} = b_j(1 - c_j)$$

for both the main method and the error-estimating method, the overall method takes the form

$$
\begin{array}{c|ccccccc}
0 \\
c_2 & a_{21} \\
c_3 & a_{31} & a_{32} \\
\vdots & \vdots & \vdots & \ddots \\
c_{s-3} & a_{s-3,1} & a_{s-3,2} & \cdots & a_{s-3,s-4} \\
1 & a_{s-2,1} & a_{s-2,2} & \cdots & a_{s-2,s-4} & a_{s-2,s-3} \\
0 & a_{s-1,1} & a_{s-1,2} & \cdots & a_{s-1,s-4} & a_{s-1,s-3} & 0 \\
1 & a_{s1} & a_{s2} & \cdots & a_{s,s-4} & a_{s,s-3} & 0 & 1 \\
\hline
 & b_1 & b_2 & \cdots & b_{s-4} & b_{s-3} & b_{s-2} & 0 & 0 \\
\hline
 & b_1 - b_{s-2} & b_2 & \cdots & b_{s-4} & b_{s-3} & 0 & b_{s-2} & b_{s-2} \\
 & [\quad -b_{s-2} & 0 & \cdots & 0 & 0 & -b_{s-2} & b_{s-2} & b_{s-2}]
\end{array}
$$

In the case of fifth order, $s = 8$ is chosen, the row-simplifying assumption

$$\sum a_{ij} c_j = \tfrac{1}{2} c_i^2, \qquad i \neq 2,$$

is imposed as well as the subsidiary conditions

$$b_2 = \Sigma b_i(1 - c_i)a_{i2} = \Sigma b_i(1 - c_i)c_i a_{i2} = \Sigma b_i(1 - c_i)a_{ij}a_{j2} = 0,$$

which is to hold for both of the embedded methods.

The derivation of the actual parameters now proceeds very much as for sixth-order methods in subsection 334. In the choice proposed by Fehlberg, the relationship between c_3 and c_4 as required by (333h) with $m = 1$ is settled as $c_3 = \frac{4}{15}$ and $c_4 = \frac{2}{3}$. This leads to the method

0								
$\frac{1}{6}$	$\frac{1}{6}$							
$\frac{4}{15}$	$\frac{4}{75}$	$\frac{16}{75}$						
$\frac{2}{3}$	$\frac{5}{6}$	$-\frac{8}{3}$	$\frac{5}{2}$					
$\frac{4}{5}$	$-\frac{8}{5}$	$\frac{144}{25}$	-4	$\frac{16}{25}$				
1	$\frac{361}{320}$	$-\frac{18}{5}$	$\frac{407}{128}$	$-\frac{11}{80}$	$\frac{55}{128}$			
0	$-\frac{11}{640}$	0	$\frac{11}{256}$	$-\frac{11}{160}$	$\frac{11}{256}$	0		
1	$\frac{93}{640}$	$-\frac{18}{5}$	$\frac{803}{256}$	$-\frac{11}{160}$	$\frac{99}{256}$	0	1	
	$\frac{31}{384}$	0	$\frac{1125}{2816}$	$\frac{9}{32}$	$\frac{125}{768}$	$\frac{5}{66}$	0	0
	$\frac{7}{1408}$	0	$\frac{1125}{2816}$	$\frac{9}{32}$	$\frac{125}{768}$	0	$\frac{5}{66}$	$\frac{5}{66}$
	$[-\frac{5}{66}$	0	0	0	0	$-\frac{5}{66}$	$\frac{5}{66}$	$\frac{5}{66}]$

Methods similar to this have been derived by Fehlberg up to order 8. Unfortunately, the methods above order 4 have a feature which makes them inappropriate with certain problems. Consider a problem written in non-autonomous form $y'(x) = f(x, y(x))$, where f depends much more strongly on its first than on its second argument. That is, the problem behaves very much like the integration of a known function of x in which $f(x_{n-1} + hc_i, Y_i)$ is effectively equal to $y'(x_{n-1} + hc_i)$ and hb_i is a weight in a quadrature formula. In such an extreme case, the result computed by each of the two embedded methods would be identical and we would obtain a zero estimate. Hence, we should expect misleading stepsize controls.

In the next subsection we examine some methods without this characteristic.

375 The methods of J. H. Verner

The modification to the Fehlberg type of method which Verner uses to overcome the difficulty metnioned in the last subsection is to choose $c_{s-1} \neq 0$ but to retain the requirement that $a_{s-1,s-2} = a_{s,s-2} = b_{s-1} = \bar{b}_{s-2} = 0$ where $b_1, b_2, \ldots, b_s$ correspond to the lower-order embedded method and $\bar{b}_1, \bar{b}_2, \ldots, \bar{b}_s$ to the higher-order estimating method. It is still possible in this case to satisfy the simplifying assumptions

$$\sum_i b_i a_{ij} = b_j(1 - c_j),$$

$$\sum_i \bar{b}_i a_{ij} = \bar{b}_j(1 - c_j)$$

and to obtain the required order with the same numbers of stages as in the work of Fehlberg. In Verner (1978) methods are given for orders 5 to 8 and we present here just one case, the eight-stage method with order 5 with a sixth-order error-estimating stage:

$$
\begin{array}{c|cccccccc}
0 \\
\frac{1}{18} & \frac{1}{18} \\
\frac{1}{6} & -\frac{1}{12} & \frac{1}{4} \\
\frac{2}{9} & -\frac{2}{81} & \frac{4}{27} & \frac{8}{81} \\
\frac{2}{3} & \frac{40}{33} & -\frac{4}{11} & -\frac{56}{11} & \frac{54}{11} \\
1 & -\frac{369}{73} & \frac{72}{73} & \frac{5380}{219} & -\frac{12285}{584} & \frac{2695}{1752} \\
\frac{8}{9} & -\frac{8716}{891} & \frac{656}{297} & \frac{39520}{891} & -\frac{416}{11} & \frac{52}{27} & 0 \\
1 & \frac{3015}{256} & -\frac{9}{4} & -\frac{4219}{78} & \frac{5985}{128} & -\frac{539}{384} & 0 & \frac{693}{3328} \\
\hline
 & \frac{3}{80} & 0 & \frac{4}{25} & \frac{243}{1120} & \frac{77}{160} & \frac{73}{700} & 0 & 0 \\
\hline
 & \frac{57}{640} & 0 & -\frac{16}{65} & \frac{1377}{2240} & \frac{121}{320} & 0 & \frac{891}{8320} & \frac{2}{35} \\
\big[& \frac{33}{640} & 0 & -\frac{132}{325} & \frac{891}{2240} & -\frac{33}{320} & -\frac{73}{700} & \frac{891}{8320} & \frac{2}{35} \big]
\end{array}
$$

376 Further error-estimating methods.

By combining the results computed over several successive steps, it is possible to obtain error estimates requiring no additional derivative calculations. If, for example, a fourth-order method is being used, the step sizes over three successive steps can be forced to be in the ratio $(6 - \sqrt{6})/10 : \sqrt{6}/5 : (4 - \sqrt{6})/10$. Let the total of these three step sizes be H, so that the results computed at the start and at the end of these steps, say $y_{n-3}, y_{n-2}, y_{n-1}, y_n$ are fourth-order approximations to $y(x_n - H)$, $y(x_n - H(4 + \sqrt{6})/10)$, $y(x_n - H(4 - \sqrt{6})/10)$ and $y(x_n)$. Furthermore, $Hf(y_{n-3}), Hf(y_{n-2}), Hf(y_{n-1})$ and $Hf(y_n)$ are fifth-order approximations to $Hy'(x_n - H)$, $Hy'(x_n - H(4 + \sqrt{6})/10)$, $Hy'(x_n - H(4 - \sqrt{6})/10)$ and $Hy'(x_n)$. Thus the fifth-order Radau quadrature formula

$$y(x_n) - y(x_n - H) \approx \frac{H}{9} y'(x_n - H) + \frac{16 + \sqrt{6}}{36} Hy'\left(x_n - \frac{H(4 + \sqrt{6})}{10}\right)$$

$$+ \frac{16 - \sqrt{6}}{36} Hy'\left(x_n - \frac{H(4 - \sqrt{6})}{10}\right)$$

leads to a proposal of using as an error estimate the expression

$$(H/36)[4f(y_{n-3}) + (16 + \sqrt{6})f(y_{n-2}) + (16 - \sqrt{6})f(y_{n-1})] + y_{n-3} - y_n.$$

This suggestion of Stoller and Morrison (1958) has been generalized and improved in the book by Ceschino and Kuntzmann (1963). The necessity of forcing the sequence of consecutive steps over which the error estimate is made to an unequal spacing is removed and orders up to five can be dealt with. Although equal spacing is not an absolute requirement of this approach, it will be described in this way for simplicity.

The idea is to choose numbers $\alpha_0, \alpha_1, \ldots, \alpha_k, \beta_0, \beta_1, \ldots, \beta_k$ such that, for any polynomial p of degree not exceeding $m + 1$ say,

$$\sum_{i=0}^{k} \alpha_i p(-i) + \sum_{i=0}^{k} \beta_i p'(-i) = 0,$$

and such that $\sum_{i=1}^{k} i\alpha_i = 1$. Such choices of $\alpha_0, \alpha_1, \ldots$ can always be made with $k = [m/2] + 1$ by forming the $(m + 1)$-order divided differences of the function p on the set of arguments $-k, -k, -k+1, -k+1, \ldots, 0$ where 0 occurs twice when m is odd and once when m is even. Normalizing such that $\sum_{i=1}^{k} i\alpha_i = 1$ implies that

$$\sum_{i=0}^{k-1} (k - i)\alpha_i = -1. \tag{376a}$$

By expanding each term by Taylor series about $x = x_n$, it follows that

$$\sum_{i=0}^{k} [\alpha_i y(x_{n-i}) + h\beta_i y'(x_{n-i})] = O(h^{m+2}) \tag{376b}$$

if y is the solution to a sufficiently smooth differential equation.

If the local truncation error is $Ch^{m+1} + O(h^{m+2})$ in each of the steps from x_{n-k} to x_n, so that, assuming y_{n-k} exactly equals $y(x_{n-k})$,

$$y_{n-i} = y(x_{n-i}) - (k - i)Ch^{m+1} + O(h^{m+2}), \quad i = 0, 1, \ldots, k \tag{376c}$$

$$hf(y_{n-i}) = hy'(x_{n-i}) + O(h^{m+2}), \quad i = 0, 1, \ldots, k. \tag{376d}$$

Using the coefficients $\alpha_0, \alpha_1, \ldots, \beta_0, \beta_1, \ldots$ we form the expression

$$\varepsilon_m = \alpha_0 y_n + \alpha_1 y_{n-1} + \cdots + \alpha_k y_{n-k}$$
$$+ h[\beta_0 f(y_n) + \beta_1 f(y_{n-1}) + \cdots + \beta_k f(y_{n-k})] \tag{376e}$$

and use this as an estimator for Ch^{m+1}. This is easily verified by substituting (376c) and (376d) into (376e) and using the equations (376a) and (376b) to find that

$$\varepsilon_m = Ch^{m+1} + O(h^{m+2}).$$

These estimators for orders $m = 1, 2, \ldots, 5$, as given by Ceschino and

Kuntzmann (1963), are

$$\varepsilon_1 = \frac{h}{2} \left[f(y_{n-1}) + f(y_n) \right] + y_{n-1} - y_n,$$

$$\varepsilon_2 = \frac{h}{3} \left[f(y_{n-2}) + 2f(y_{n-1}) \right] + \frac{1}{6} (5 y_{n-2} - 4 y_{n-1} - y_n),$$

$$\varepsilon_3 = \frac{h}{6} \left[f(y_{n-2}) + 4f(y_{n-1}) + f(y_n) \right] + \frac{1}{2}(y_{n-2} - y_n),$$

$$\varepsilon_4 = \frac{h}{10} \left[f(y_{n-3}) + 6f(y_{n-2}) + 3f(y_{n-1}) \right]$$

$$+ \frac{1}{30} (10 y_{n-3} + 9 y_{n-2} - 18 y_{n-1} - y_n),$$

$$\varepsilon_5 = \frac{h}{20} \left[f(y_{n-3}) + 9f(y_{n-2}) + 9f(y_{n-1}) + f(y_n) \right]$$

$$+ \frac{1}{60} (11 y_{n-3} + 27 y_{n-2} - 27 y_{n-1} - 11 y_n).$$

377 Embedded implicit methods.

With the fully implicit methods of orders at least $2s - 2$ introduced in subsections 343 and 345 it is unrealistic to expect to find a method of $s + 1$ stages in which the given method is embedded such that the new method has higher order than the original. If such a feat were possible for a method of order $2s - d (d = 0, 1$ or $2)$ then there would exist $\bar{b}_1, \bar{b}_2, \ldots, \bar{b}_{s+1}$ corresponding to the new method such that

$$\sum_{i=1}^{s+1} \bar{b}_i p(c_i) = \int_0^1 p(x) \, dx$$

for any polynomial p of degree not exceeding $2s - d + 1$.

From the order conditions for the original method we find that

$$\sum_{i=1}^{s} b_i p(c_i) = \int_0^1 p(x) \, dx,$$

for any polynomial of degree up to $2s - d$. Conventionally writing $b_{s+1} = 0$, it then follows that

$$\sum_{i=1}^{s+1} (\bar{b}_i - b_i) p(c_i) = 0$$

if the degree of p does not exceed $2s - d$. Choose $p(x) = \Pi_{j \neq k}(x - c_j)$, where $k \in \{1, 2, \ldots, s + 1\}$, so that in each case p has degree s. If $s \geq d$, as is always the case for these methods, $\bar{b}_k = b_k$ for all $k = 1, 2, \ldots, s + 1$. Hence, the new method is the same as the old and does not provide an error estimate. Only by adding approximately s additional stages can we have any hope of overcoming this difficulty.

On the other hand, for methods of order s, embedding in a method with an extra stage is completely successful. We assume, of course, that the order of each stage is also s, so that the original method is equivalent to a collocation method. It is only necessary to choose c_{s+1} different from each of $c_1, c_2, \ldots, c_s$, choose $a_{s+1, s+1}$ arbitrarily and then choose $a_{s+1,1}, a_{s+1,2}, \ldots, a_{s+1,s}$ so that

$$\sum_{i=1}^{s+1} a_{s+1,i} c_i^{k-1} = \frac{1}{k} c_{s+1}^k, \qquad k = 1, 2, \ldots, s;$$

$\bar{b}_1, \bar{b}_2, \ldots, \bar{b}_{s+1}$ for the error-estimating method are then chosen to satisfy

$$\sum_{i=1}^{s+1} \bar{b}_i c_i^{k-1} = \frac{1}{k}, \qquad k = 1, 2, \ldots, s+1.$$

In the particular case of the singly implicit methods based on Laguerre zeros, as described in subsection 348, the value of $a_{s+1,s+1} = \lambda$, where $\{\lambda\}$ is the spectrum of A, is the appropriate choice to maintain the cheap implementation property. Details of this embedding are given in Burrage (1978).

38 ALGEBRAIC PROPERTIES OF RUNGE–KUTTA METHODS

380 The Runge–Kutta space.

In this subsection, we will be concerned with a generalization of Runge–Kutta methods. Specifically, we consider methods respresentable by a tableau of numbers

$$
\begin{array}{c|cccc}
c_1 & a_{11} & a_{12} \cdots & & a_{1s} \\
c_2 & a_{21} & a_{22} \cdots & & a_{2s} \\
\vdots & \vdots & \vdots & & \vdots \\
c_s & a_{s1} & a_{s2} \cdots & & a_{ss} \\
\hline
 & b_0 & b_1 & b_2 \cdots & b_s
\end{array}
\qquad (380a)
$$

in which $c_i = \sum_{j=1}^{s} a_{ij} (i = 1, 2, \ldots, s)$ and the 'result' y_1 is computed from the 'initial value' y_0 by

$$y_1 = b_0 y_0 + h \sum_{j=1}^{s} b_j f(Y_j), \qquad (380b)$$

where

$$Y_i = y_0 + h \sum_{j=1}^{s} a_{ij} f(Y_j), \qquad i = 1, 2, \ldots, s. \qquad (380c)$$

Just as for standard Runge–Kutta methods, in which $b_0 = 1$ and y_1 represents a numerical result after a single Runge–Kutta step starting from the initial value y_0, the values of $Y_1, Y_2, \ldots, Y_s$, and therefore of y_1 given by (380b) and (380c), are uniquely defined if f satisfies a Lipschitz condition and $|h|$ is sufficiently small.

There are two reasons for allowing the slight generalization of inserting a factor b_0 in (380b). The first is that the method can be used not only to represent quantities appropriate for the approximation of $y(x)$ for x values near x_0, where it is assumed that $y(x_0) = y_0$, but also for representing quantities like $hy'(x), h^2 y''(x), \ldots$ or the difference of two approximations of different orders or a variety of other quantities. The second is that, with this generalization, we can view our methods as a linear space, as we shall now see.

If, in addition to the method m, say, given by (380a), we have a second method $\bar{m}$ given by

$$
\begin{array}{c|cccc}
\bar{c}_1 & \bar{a}_{11} & \bar{a}_{12} & \cdots & \bar{a}_{1\bar{s}} \\
\bar{c}_2 & \bar{a}_{21} & \bar{a}_{22} & \cdots & \bar{a}_{2\bar{s}} \\
\vdots & \vdots & \vdots & & \vdots \\
\bar{c}_{\bar{s}} & \bar{a}_{\bar{s}1} & \bar{a}_{\bar{s}2} & \cdots & \bar{a}_{\bar{s}\bar{s}} \\
\hline
 & \bar{b}_0 & \bar{b}_1 & \bar{b}_2 & \cdots & \bar{b}_{\bar{s}}
\end{array}
\tag{380d}
$$

then it is natural to think of the method given by

$$
\begin{array}{c|ccccc|cccc}
c_1 & a_{11} & a_{12} & \cdots & a_{1s} & 0 & 0 & \cdots & 0 \\
c_2 & a_{21} & a_{22} & \cdots & a_{2s} & 0 & 0 & \cdots & 0 \\
\vdots & \vdots & \vdots & & \vdots & \vdots & \vdots & & \vdots \\
c_s & a_{s1} & a_{s2} & \cdots & a_{ss} & 0 & 0 & \cdots & 0 \\
\hline
\bar{c}_1 & 0 & 0 & \cdots & 0 & \bar{a}_{11} & \bar{a}_{12} & \cdots & \bar{a}_{1\bar{s}} \\
\bar{c}_2 & 0 & 0 & \cdots & 0 & \bar{a}_{21} & \bar{a}_{22} & \cdots & \bar{a}_{2\bar{s}} \\
\vdots & \vdots & \vdots & & \vdots & \vdots & \vdots & & \vdots \\
\bar{c}_{\bar{s}} & 0 & 0 & \cdots & 0 & \bar{a}_{\bar{s}1} & \bar{a}_{\bar{s}2} & \cdots & \bar{a}_{\bar{s}\bar{s}} \\
\hline
 & \alpha b_0 + \beta \bar{b}_0 & \alpha b_1 & \alpha b_2 & \cdots & \alpha b_s & \beta \bar{b}_1 & \beta \bar{b}_2 & \cdots & \beta \bar{b}_{\bar{s}}
\end{array}
\tag{380e}
$$

as being $\alpha m + \beta \bar{m}$, since it computes a result $\alpha y_1 + \beta y_2$ from an initial value y_0, where y_2 is the result computed by (380d), just as y_1 is the result computed by (380a).

Although we will wish to follow the spirit of this idea in our construction of a linear space of Runge–Kutta methods, we have to allow for the possibility that m and $\bar{m}$ have common stages and therefore that a simpler method than that represented by (380e) would equally well serve as $\alpha m + \beta \bar{m}$. This leads us to consider the relationship between two methods that are essentially the same but where one might differ by having redundant stages, or collections of stages that can be combined into fewer stages. More simply, the two methods might differ only by having their stages numbered in a different order.

To deal with these situations, we introduce further concepts.

DEFINITION 380A.

If f satisfies a Lipschitz condition and the difference of the results computed by each of two methods is zero, for $|h|$ sufficiently small, then the two methods are said to be *equivalent*. The equivalence classes under this equivalence relation are said to be *Runge–Kutta points*.

Note that the tacit assumption, that the relation defined here is an equivalence, is easily verified and the details will not be given. Note also that Runge–Kutta points are closely related to what Stetter (1973) calls Runge–Kutta schemes.

DEFINITION 380B.

The set of Runge–Kutta points, together with the linear structure in which $\alpha \xi + \beta \bar{\xi}$ is the point containing (380e) if ξ (respectively $\bar{\xi}$) denotes the point containing (380a) (respectively (380d)), will be known as the *Runge–Kutta space*.

It follows from definition 380A that the choices of representatives in definition 380B do not affect the meaning of $\alpha \xi + \beta \bar{\xi}$.

In the special case of the method given by (380a) when $s = 0$, we are restricted to results of the form given by $y_1 = b_0 y_0$, and it is natural to denote the Runge–Kutta point containing this method by the name b_0. In particular, we consider the cases $b_0 = 0$ and $b_0 = 1$ for which the corresponding points, denoted by 0 and 1 respectively, have a special significance.

The point 0 is just the identity element of the linear space of Runge–Kutta points, while the point 1, which contains methods that leave the initial value unaltered, is the identity element with respect to a multiplication operation to be introduced in the next subsection.

The way we have defined Runge–Kutta points as classes of methods raises the question as to what properties of a method are determined by virtue of its being in the same class as another method. As a matter of fact, there is an interesting set of invariants over each class, but for the moment we consider just two.

THEOREM 380C.

If two methods given by (380a) and (380d) are equivalent then

$$\bar{b}_0 = b_0, \tag{380f}$$

$$\sum_{i=1}^{s} \bar{b}_i = \sum_{i=1}^{s} b_i. \tag{380g}$$

Proof.

To prove (380f), we consider $f(u) = 0$, $y_0 = 1$. For any value of h, we obtain the results b_0 and $\bar{b}_0$ respectively. To prove (380g), we consider $f(u) = 1$, $y_0 = 0$ and any positive value of h. The results are h times the two sides of (380g). ∎

The value of b_0 in a method, (380a), representing a particular point ξ will be denoted by $\varphi(\xi)$. We omit the simple verification of the following theorem.

THEOREM 380D.

The function φ is a linear functional on the Runge–Kutta space.

Among all members of the Runge–Kutta space, those members ξ for which $\varphi(\xi) = 1$ have a special role since they are equivalence classes of the usual type of Runge–Kutta methods with the coefficient of y_0 always equal to unity.

We will denote the full Runge–Kutta space by the name K and $\varphi^{-1}(\{1\})$ by K_1.

381 Multiplication in the Runge–Kutta space.

Given two points in K, of which the first is in K_1 and is represented by (380a) (with $b_0 = 1$) and the second is represented by (380d), we consider the point containing

$$
\begin{array}{c|cccc|cccc}
c_1 & a_{11} & a_{12} & \cdots & a_{1s} & 0 & 0 & \cdots & 0 \\
c_2 & a_{21} & a_{22} & \cdots & a_{2s} & 0 & 0 & \cdots & 0 \\
\vdots & \vdots & \vdots & & \vdots & \vdots & \vdots & & \vdots \\
c_s & a_{s1} & a_{s2} & \cdots & a_{ss} & 0 & 0 & \cdots & 0 \\
\hline
\sum_{i=1}^{s} b_i + \bar{c}_1 & b_1 & b_2 & \cdots & b_s & \bar{a}_{11} & \bar{a}_{12} & \cdots & \bar{a}_{1\bar{s}} \\
\sum_{i=1}^{s} b_i + \bar{c}_2 & b_1 & b_2 & \cdots & b_s & \bar{a}_{21} & \bar{a}_{22} & \cdots & \bar{a}_{2\bar{s}} \\
\vdots & \vdots & \vdots & & \vdots & \vdots & \vdots & & \vdots \\
\sum_{i=1}^{s} b_i + \bar{c}_{\bar{s}} & b_1 & b_2 & \cdots & b_s & \bar{a}_{\bar{s}1} & \bar{a}_{\bar{s}2} & \cdots & \bar{a}_{\bar{s}\bar{s}} \\
\hline
\bar{b}_0 & \bar{b}_0 b_1 & \bar{b}_0 b_2 & \cdots & \bar{b}_0 b_s & \bar{b}_1 & \bar{b}_2 & \cdots & \bar{b}_{\bar{s}}
\end{array}
\tag{381a}
$$

To see the numerical significance of this generalized method and thus of the point containing it, we suppose that f and h are given and that g denotes the mapping taking y_0 into y_1 given by (380b). Similarly, let $\bar{g}$ denote the corresponding mapping associated with the method (380d). The result computed by the method (381a), subject as usual to the proviso that $|h|$ is sufficiently small, is $(\bar{g} \circ g)(y_0)$, as we will now verify.

Let the $s + \bar{s}$ stages computed by (381a) be $Y_1, Y_2, \ldots, Y_s, \bar{Y}_1, \bar{Y}_2, \ldots, \bar{Y}_{\bar{s}}$ and let the final result for the first submethod be $y_1 = y_0 + h \sum b_i f(Y_i)$ and, for the method as a whole, y_2. We then have

$$Y_i = y_0 + h \sum_{j=1}^{s} a_{ij} f(Y_j), \qquad i = 1, 2, \ldots, s, \qquad (381b)$$

$$y_1 = y_0 + h \sum_{j=1}^{s} b_j f(Y_j), \qquad (381c)$$

$$\bar{Y}_i = y_0 + h \sum_{j=1}^{s} b_j f(Y_j) + h \sum_{j=1}^{\bar{s}} \bar{a}_{ij} f(\bar{Y}_j), \qquad i = 1, 2, \ldots, \bar{s}, \qquad (381d)$$

$$y_2 = \bar{b}_0 y_0 + h \sum_{j=1}^{s} \bar{b}_0 b_j f(Y_j) + h \sum_{j=1}^{\bar{s}} \bar{b}_j f(\bar{Y}_j). \qquad (381e)$$

To see that y_2 is equal to $(\bar{g} \circ g)(y_0)$, subtract (381c) from (381d) and b_0 times (381c) from (381e). We find that $y_1 = g(y_0)$, $y_2 = \bar{g}(y_1)$.

DEFINITION 381A.

If $\xi \in K_1$ is the Runge–Kutta point containing (380a) and $\bar{\xi} \in K$ is the Runge–Kutta point containing (380d), then the *product,* $\xi \bar{\xi}$, of ξ and $\bar{\xi}$ is the point containing (381a).

By the composition property of this product, it is clear that the choices of representative methods for ξ and $\bar{\xi}$ do not alter the set of methods constituting $\xi \bar{\xi}$.

It is interesting that we can choose (380d) so that the point containing it is the inverse, in the sense of the multiplicative operation of definition 381A, to the point containing (380a). That is to say, we can choose the method (380d) so that it exactly undoes the work of (380a). One way of doing this is described in subsection 344. With such a choice, $\bar{g} \circ g$ becomes the identity mapping, subject as always to a proviso on the size of $|h|$ for the given Lipschitzian f.

382 Algebraic properties of the Runge–Kutta space.

In this subsection, we consider purely algebraic properties of the Runge–Kutta space K, together with the subset K_1 defined in subsection 380. The linear functional φ, also introduced there, will be made use of.

THEOREM 382A.

$$K \text{ is a real linear space.} \tag{382a}$$

$$\text{If } \xi \in K_1, \eta, \zeta \in K, \text{ then } \xi(\eta + \zeta) = \xi\eta + \xi\zeta. \tag{382b}$$

$$\text{If } \xi \in K_1, \eta \in K, c \in \mathbb{R}, \text{ then } \xi(c\eta) = c(\xi\eta). \tag{382c}$$

$$\text{If } \xi, \eta \in K_1, \zeta \in K, \text{ then } (\xi\eta)\zeta = \xi(\eta\zeta). \tag{382d}$$

$$K_1 \text{ is a multiplicative group with identity } 1. \tag{382e}$$

$$\text{If } \xi \in K_1, \eta \in K, \text{ then } \varphi(\xi\eta) = \varphi(\eta). \tag{382f}$$

Proof.

The first result, (382a) is a consequence of definitions in subsection 380. To prove (382b), suppose that a Lipschitzian function f is given and that h is given as a sufficiently small non-zero number. Let g_1, g_2, g_3 denote the functions relating the result to the initial value for members of the points ξ, η, ζ respectively. The result computed from an initial value y_0, by a method in $\xi(\eta + \zeta)$, is $g_2(g_1(y_0)) + g_3(g_1(y_0))$, the same as for a method in $\xi\eta + \xi\zeta$. The truth of (382c) follows in a similar way and (382d) is a consequence of the associativity of compositions of functions. The group property (382e) requires verification of associativity (already proved as a special case of (382d)), the characteristic property of 1 (clear from the fact that, independent of f and h, the result computed by a method in 1 is identical to the initial value) and the existence of inverses (following, as we remarked in the last subsection, from a construction in subsection 344). Finally, (382f) is a consequence of the form of (381a). ∎

Having constructed an algebraic system which is somewhat reminiscent of a ring, it is natural to ask whether entitites corresponding to ideals exist, and, not forgetting our context, whether they have any computational significance. Although the second as well as the first of these questions has an affirmative answer, we begin our discussion on a purely algebraic level.

In the definition which follows, K_0 will denote the subspace of K given by $\varphi^{-1}(\{0\})$. Thus, if $\xi \in K_0$ then $\varphi(\xi) = 0$.

DEFINITION 382B.

A linear subspace I of K_0 is an *ideal* of K if, whenever $\xi \in K_1, \eta \in I$ and $\zeta \in K$, $\xi\eta$ and $(\xi + \eta)\zeta - \xi\zeta$ are each members of I.

DEFINITION 382C.

For an ideal I and $\xi \in K$, we write $\xi + I$ as the set of points $\bar{\xi}$ in K such that $\bar{\xi} - \xi \in I$, and we call this a *coset* of K.

If $\xi + I$, $\eta + I$ are cosets of K, then $(\xi + I) + (\eta + I)$ is defined as the coset containing $\xi + \eta$; if also $c \in \mathbb{R}$ then $c(\xi + I)$ is defined as the coset containing $c\xi$ and if ξ is actually a member of K_1, then $(\xi + I)(\eta + I)$ is defined as the coset containing $\xi\eta$.

That the particular representatives of the ideals in this definition do not affect the sum, the product by a scalar and the product by an ideal in K_1 follows from definition 382B.

DEFINITION 382D.

If I is an ideal of K, then the set of cosets of K will be written K/I. This set, together with its multiplier subset K_1/I, will be called the *quotient* of the Runge–Kutta space *relative* to I. The mapping $K \to K/I$ (which also takes K_1 to K_1/I) in which $\xi \in K$ maps to $\xi + I$ will be called the *natural homomorphism*.

THEOREM 382E.

If I is an ideal of K and θ is the natural homomorphism from K to K/I, then θ is a homomorphism in the sense that for all $\xi, \eta \in K, \theta(\xi) + \theta(\eta) = \theta(\xi + \eta)$; for all $\xi \in K$, $c \in \mathbb{R}$, $c\theta(\xi) = \theta(c\xi)$; and for all $\xi \in K_1$, $\eta \in K$, $\theta(\xi)\theta(\eta) = \theta(\xi\eta)$. Furthermore, if $\bar{\varphi}$ is defined by $\bar{\varphi}(\theta(\xi)) = \varphi(\xi)$ for all $\xi \in K$, then $\bar{y}$ is a linear functional on K/I and K_1/I is equal to $\bar{\varphi}^{-1}(\{1\})$.

Proof.

This result follows from definitions 382B, 382C and 382D. ∎

This last theorem enables us to extend the applicability of theorem 382A to any quotient space of K. That is to say, statements corresponding to (382a), (382b), ..., (382f) in the statement of that theorem would still hold if K, K_1 and φ were replaced at each occurrence by K/I, K_1/I and $\bar{\varphi}$ respectively.

As examples of ideals, and therefore of quotient spaces, consider I_p, for p a non-negative integer, defined as the set of Runge–Kutta points whose members, when used to solve a differential equation for which f is continuously differentiable arbitrarily often, give a result y_1 which is $O(|h|^{p+1})$ as $|h| \to 0$. Members of points in the same coset in K/I_p give numerical results which are identical except for $O(|h|^{p+1})$. This is our promised computational significance of the otherwise abstract ideas arising from the study of ideals of K.

383 Graph–theoretic representation.

We recall the properties of the space G and its subset G_1 studied in subsection 146 and, in particular, in theorem 146E. The similarity between that algebraic system and the space K together with its subset K_1 can hardly go unnoticed, especially when we realize that the functional φ also has a counterpart in the previously studied theory. If we define χ, say, to be that linear functional on G such that $\chi(\alpha) = \alpha(\emptyset)$ for all $\alpha \in G$, then χ becomes this counterpart. A member α of G_1 is characterized by $\chi(\alpha) = 1$, and for any $\beta \in G$, we then have $\chi(\alpha\beta) = \chi(\beta)$ exactly corresponding to (382f) of theorem 382A.

What we will now do is construct a homomorphism $\Theta : K \to G$ for which $\Theta(K_1) = G_1$ and such that $\chi \circ \Theta = \varphi$. When we have carried out this construction, we will have a mechanism for representing what are naturally thought of as analytic properties of numerical methods in terms of the purely algebraic entities studied in subsection 146. It will turn out that the images of the ideals I_p of the last subsection have a very natural role in G, as also do the corresponding quotient sets.

As a first step in the construction of Θ, we define, for any generalized Runge–Kutta method, a corresponding member of G. Let the generalized Runge–Kutta method be m given by (380a), let $\Phi(t)$ for $t \in T$ denote the elementary weight for the closely related Runge–Kutta method

$$
\begin{array}{c|cccc}
\sum_{j=1}^{s} a_{1j} & a_{11} & a_{12} \cdots & & a_{1s} \\
\sum_{j=1}^{s} a_{2j} & a_{21} & a_{22} \cdots & & a_{2s} \\
\vdots & \vdots & \vdots & & \vdots \\
\sum_{j=1}^{s} a_{sj} & a_{s1} & a_{s2} \cdots & & a_{ss} \\
\hline
& b_1 & b_2 & \cdots & b_s
\end{array}
$$

and let $g : T^{\#} \to \mathbb{R}$ denote the corresponding member of G we are defining. The values of $g(t)$ for $t \in T^{\#}$ are given by

$$
\begin{aligned}
g(\emptyset) &= b_0, \\
g(t) &= \Phi(t), \qquad t \in T,
\end{aligned}
$$

and we write $\bar{\Theta}$ for the mapping taking the method m to g.

If we are given two generalized methods, they may or may not be related by yielding the same $g \in G$ under the mapping $\bar{\Theta}$, just as they may or may not be equivalent in the sense of definition 380A. What we will show, in theorem 383C and in the two results that precede it, is that the equivalence relations defined in these two different ways exactly coincide.

LEMMA 383A.

Let m and $\bar{m}$ be generalized Runge–Kutta methods. If m and $\bar{m}$ are equivalent, then $\bar{\Theta}(m) = \bar{\Theta}(\bar{m})$.

Proof.

Suppose, on the contrary, that $\bar{\Theta}(m) \neq \bar{\Theta}(\bar{m})$ so that there is a tree t such that $\Phi(t)$ takes on different values for m and $\bar{m}$ but such that for any tree u with order less than that of t, $\Phi(u)$ takes on the same value for m and $\bar{m}$. If $t = \emptyset$, select $f = 0$ on the vector space $\mathbb{R}$ and the initial value $y_0 = 1$. The results computed by m and $\bar{m}$ are constants for all h, and since they are different constants we obtain a contradiction to an assertion that m and $\bar{m}$ are equivalent. Now suppose that $t \in T$. Apply the results of theorem 306B where T_0 is the set of all trees with order not exceeding r, where r is the order of t; and θ takes on the value 1 at t and 0 for all other members of T_0. The numerical results computed from the differential equation construction in this way satisfy

$$p^* y_1 = \frac{\beta(t)\bar{\Theta}(m)(t)h^r}{r!} + O(h^{r+1})$$

and

$$p^* y_2 = \frac{\beta(t)\bar{\Theta}(\bar{m})(t)h^r}{r!} + O(h^{r+1})$$

and these are different for h sufficiently small. ∎

Before moving any closer to the stronger statement of this lemma, with 'if' replaced by 'if and only if', it is convenient to make a study of equivalence relations beween the *stages* of one particular method. Thus, we consider the s stages of a method

$$
\begin{array}{c|cccc}
c_1 & a_{11} & a_{12} & \cdots & a_{1s} \\
c_2 & a_{21} & a_{22} & \cdots & a_{2s} \\
\vdots & \vdots & \vdots & & \vdots \\
c_s & a_{s1} & a_{s2} & \cdots & a_{ss} \\
\end{array}
$$

where we take no interest in the values of the coefficients giving the final result, since these are not relevant to our discussion.

Let $\Phi_i(t)$, $i = 1, 2, \ldots, s$, denote the elementary weight for the tree t associated with the ith stage of this method. Also, for some particular differential equation, let $Y_1, Y_2, \ldots, Y_s$ denote the results computed in these stages using step size h. We define three relations on the set $\{1, 2, \ldots, s\}$ of stages.

Relation 1: i is equivalent to j. For any problem satisfying a Lipschitz condition, there exists $h_0 > 0$, such that, if $|h| < h_0$, then $Y_i = Y_j$.

This is the same as saying that the following two generalized methods are equivalent:

$$
\begin{array}{c|cccc}
c_1 & a_{11} & a_{12} & \cdots & a_{1s} \\
c_2 & a_{21} & a_{22} & \cdots & a_{2s} \\
\vdots & \vdots & \vdots & & \vdots \\
c_s & a_{s1} & a_{s2} & \cdots & a_{ss} \\
\hline
1 & a_{i1} & a_{i2} & \cdots & a_{is} \\
\end{array}
\quad \text{and} \quad
\begin{array}{c|cccc}
c_1 & a_{11} & a_{12} & \cdots & a_{1s} \\
c_2 & a_{21} & a_{22} & \cdots & a_{2s} \\
\vdots & \vdots & \vdots & & \vdots \\
c_s & a_{s1} & a_{s2} & \cdots & a_{ss} \\
\hline
1 & a_{j1} & a_{j2} & \cdots & a_{js} \\
\end{array}
$$

Relation 2: i is Φ-*equivalent* to j. For all $t \in T$, $\Phi_i(t) = \Phi_j(t)$.

Relation 3: i is P-*equivalent* to j. There is a partition P of $\{1, 2, \ldots, s\}$ into disjoint subsets $P_1, P_2, \ldots, P_q$ such that, for each $m = 1, 2, \ldots, q$, $\sum_{l \in P_m} a_{kl}$ takes the same value for all k in the same component of P; furthermore, i and j are in the same component of P.

That these three relations are justifiably called equivalences in that they are transitive, symmetric and reflexive is easy to verify. Our key result concerning these relations between the stages of a method is that they are all the same relation. We state this formally.

THEOREM 383B.

Two stages of a Runge–Kutta method are equivalent iff they are Φ-equivalent and iff they are P-equivalent.

Proof.

We will prove the result in three cyclic steps:

Step 1 (equivalence implies Φ-equivalence): This is an immediate application of lemma 383A.

Step 2 (Φ-equivalence implies P-equivalence): Given an equivalence relation based on Φ, let P be the corresponding partition into equivalence classes. We will prove that, if i and j are in the same class, then

$$\sum_{l \in P_m} a_{il} = \sum_{l \in P_m} a_{jl}, \tag{383a}$$

for $m = 1, 2, \ldots, q$.

For some choice of m, let $t_1, t_2, \ldots, t_{m-1}, t_{m+1}, \ldots, t_q \in T$ be chosen so that $\Phi_l(t_k)$ takes on a different value for $l \in P_m$ from the value it takes when $l \in P_k$. Now consider the vector in $\mathbb{R}^s$ whose lth component is

$$\prod_{\substack{k=1 \\ k \neq m}}^{q} [\Phi_l(t_k) - \Phi_{p_k}(t_k)], \tag{383b}$$

where, for $k \neq m$, p_k is an arbitrary member of P_k.

If this quantity is multiplied by each of a_{il} and a_{jl} and summed in each case for $l = 1, 2, \ldots, s$, the vanishing of (383b) for $l \notin P_m$ and its constant non-zero value for $l \in P_m$ implies (383a), since we have the same linear combinations of various elementary weights at the ith and jth stages.

Step 3 (P-equivalence implies equivalence): Consider the iterative computation of $Y_1, Y_2, \ldots Y_s$ where the iteration number zero has constant values over P-equivalent stages. This constancy over P-equivalent stages is preserved in each iteration and, since the space of vectors on $\{1, 2, \ldots, s\}$ which have this property is a closed subspace of $V^{\{1,2,\ldots,s\}}$, where V is the vector space in which solution values lie, the result follows. ∎

We are finally in a position to restate lemma 383A with its converse built into it.

THEOREM 383C.

Let m and $\bar{m}$ be generalized Runge–Kutta methods. Then m and $\bar{m}$ are equivalent if and only if $\bar{\Theta}(m) = \bar{\Theta}(\bar{m})$.

Proof.

The 'only if' part has been proved as lemma 383A, so it remains to prove the 'if' part.

By choosing $f = 0$, we see that the equivalence of m and $\bar{m}$ implies that $b_0 = \bar{b}_0$. Hence, by adding a multiple of y_0 to the results produced by each generalized method, we may assume that $b_0 = \bar{b}_0 = 1$. We can now combine the computed results of the two methods into a single Runge–Kutta method with $(s + 1) + (\bar{s} + 1)$ stages. The result now follows from theorem 383B. ∎

We are now in a position to define the mapping $\Theta : K \to G$.

DEFINITION 383D.

Let ξ be given in K and let m be any member of ξ. Then $\Theta(\xi)$ is defined as $\bar{\Theta}(m)$.

The fact that the choice of $m \in \xi$ does not affect the value of the result obtained for $\Theta(\xi)$ follows immediately from theorem 383C.

The usefulness of the representation of Runge–Kutta points by members of G hinges on the fact that manipulations of members of G can be carried out by straightforward algebraic rules. To justify the carrying out of such manipulations as a means of obtaining information about K, we will show in the next subsection that Θ is a homomorphism.

384 The Runge–Kutta homomorphism theorem.

We introduce a result concerning the mapping Θ from K, the space of Runge–Kutta points, to G, the algebraic structure studied in subsection 146. The result simply states that there is an exact correspondence between algebraic properties of the elements of K and the corresponding elements in G. This detailed result, which will be stated as theorem 384A and referred to as the Runge–Kutta homomorphism theorem, will be used to provide a convenient means of studying computational properties of methods, as represented by elements of K, by their counterparts in what is conceptually a simpler mathematical system. Because of this relationship, it will be convenient to refer to either K or G as the Runge–Kutta space, relying on the context to make clear which representation is intended.

THEOREM 384A.

If $c \in \mathbb{R}, \xi \in K_1, \eta, \zeta \in K$, then

$$\Theta(c\eta) = c\Theta(\eta), \tag{384a}$$

$$\Theta(\eta + \zeta) = \Theta(\eta) + \Theta(\zeta), \tag{384b}$$

$$\Theta(0) = 0, \tag{384c}$$

$$\Theta(1) = e, \tag{384d}$$

$$\varphi(\eta) = \chi(\Theta(\eta)), \tag{384e}$$

$$\Theta(\xi) \in G_1, \tag{384f}$$

$$\Theta(\xi\eta) = \Theta(\xi)\Theta(\eta). \tag{384g}$$

Proof.

To prove (384a) and (384b), let m, $\bar{m}$ be the generalized methods given by (380a) and (380d) with $m \in \zeta$ where, without loss of generality (since we may combine stages of two different methods if necessary), we may assume that $\bar{s} = s$ and that $\bar{a}_{ij} = a_{ij}$ for $i, j = 1, 2, \ldots, s$. For methods in $c\eta$ and $\eta + \zeta$, we may use the method with the same A matrix but with cb and $b + \bar{b}$ respectively in place of the b vector. It is then clear that for any tree t, $\Theta(c\eta)(t) = c\Theta(\eta)(t)$, $\Theta(\eta + \zeta)(t) = \Theta(\eta)(t) + \Theta(\zeta)(t)$.

The next four results, (384c), (384d), (384e) and (384f), are all consequences of the fact that if m given by (380a) is a member of the class η, then $\chi(\Theta(\eta)) = \bar{\Theta}(m)(\emptyset) = b_0$.

To prove (384g), let m (respectively $\bar{m}$) be in ξ (respectively η) and consider the $s + \bar{s}$ stages of the product method (381a). We need to prove that $\bar{\Theta}(m\bar{m})(t) = [\bar{\Theta}(m)\bar{\Theta}(\bar{m})](t)$ and, in the case $t = \emptyset$, the result follows by both sides having the value $\bar{b}_0$. Hence, without loss of generality, we assume that $t \in T$.

Let $\Phi(t)$ be the elementary weight associated with m, so that $\Phi(t) = \Sigma_i b_i \alpha_i(t)$, where $\alpha_i(\tau) = 1$, and for $t = uv$ $(u, v \in T)$, $\alpha_i(t) = \alpha_i(u)\Phi_i(v)$ and $\Phi_i(v) = \Sigma_j a_{ij}\alpha_j(v)$. Similarly, define $\bar{\Phi}(t), \beta_i(t)$ and $\bar{\Phi}_i(t)$ for the method $\bar{m}$ and $\gamma_i(t)$ and $\tilde{\Phi}_i(t)$ for the last $\bar{s}$ stages of the product method $m\bar{m}$. Also, $\tilde{\Phi}(t)$ denotes the elementary weight for $m\bar{m}$.

We now prove that

$$\gamma_i(t) = \lambda(\Phi, t)(\beta_i), \tag{384h}$$

for $i = 1, 2, \ldots, \bar{s}$ and $t \in T$, where the function λ was introduced in subsection 146. If $t = \tau$, both sides equal 1, so we use an inductive argument on the order of $t = uv$, where u and v have lower orders than t. Let

$$\lambda(\Phi, u) = \sum_{U \in T} p(U)\hat{U}, \tag{384i}$$

$$\lambda(\Phi, v) = \sum_{V \in T} q(V)\hat{V}, \tag{384j}$$

where p, q vanish except for a finite number of trees. We then have

$$\lambda(\Phi, t)(\beta_i) = \Phi(v) \sum_{U \in T} p(U)\beta_i(U) + \sum_{U, V \in T} p(U)q(V)\beta_i(UV), \tag{384k}$$

where we have used a property of the function λ from theorem 146F. We now evaluate the left-hand side of (384h):

$$\gamma_i(t) = \gamma_i(uv)$$

$$= \gamma_i(u)\tilde{\Phi}_i(v)$$

$$= \gamma_i(u)\left[\sum_{j=1}^{s} b_j \alpha_j(v) + \sum_{j=1}^{s} \bar{a}_{ij}\gamma_j(v)\right]$$

$$= \Phi(v)\lambda(\Phi, u)(\beta_i) + \sum_{j=1}^{s} \bar{a}_{ij}\lambda(\Phi, u)(\beta_i)\lambda(\Phi, v)(\beta_j)$$

using the induction hypothesis. Using the linearity of $\lambda(\Phi, v)$ and the fact that $\sum_j \bar{a}_{ij}\beta_j = \bar{\Phi}_i$ we find that

$$\gamma_i(t) = \Phi(v)\lambda(\Phi, u)(\beta_i) + \lambda(\Phi, u)(\beta_i)\lambda(\Phi, v)(\bar{\Phi}_i),$$

which, together with the formulas for λ given by (384i) and (384j), gives

$$\gamma_i(t) = \Phi(v) \sum_{U \in T} p(U)\beta_i(U) + \sum_{U, V \in T} p(U)\beta_i(U)q(V)\bar{\Phi}_i(V)$$

$$= \Phi(v) \sum_{U \in T} p(U)\beta_i(U) + \sum_{U, V \in T} p(U)q(V)\beta_i(UV),$$

which, by (384k), gives (384h).

Finally, to prove that $\bar{\Theta}(m\bar{m})(t) = [\bar{\Theta}(m)\bar{\Theta}(\bar{m})](t)$, or in other words that

$$\tilde{\Phi}(t) = \bar{b}_0\Phi(t) + \lambda(\Phi, t)(\bar{\Phi}),$$

we evaluate

$$\tilde{\Phi}(t) = \bar{b}_0 \sum_{i=1}^{s} b_i \alpha_i(t) + \sum_{i=1}^{\bar{s}} \bar{b}_i \gamma_i(t)$$

$$= \bar{b}_0\Phi(t) + \lambda(\Phi, t)\left(\sum_{i=1}^{\bar{s}} b_i\beta_i(t)\right)$$

$$= \bar{b}_0\Phi(t) + \lambda(\Phi, t)(\bar{\Phi}). \quad \blacksquare$$

385 The image of the ideals I_p.

In subsection 382, the ideals I_p (p a non-negative integer) were introduced as the sets of Runge–Kutta points for which, for a sufficiently smooth f defining a differential equation, the computed result is $O(|h|^{p+1})$ as $h \to 0$. For a Runge–Kutta method in such a point, say the method given by (380a), it is clear that $b_0 = 0$ and that the elementary weight $\Phi(t)$ is zero for each tree t with no more than p vertices. This follows easily from theorems 305C and 306B and it furthermore follows that any method with $b_0 = 0$ and $\Phi(t) = 0$ for $r(t) \le p$ is a member of a point in I_p. The image of I_p is thus the set of members of G such that the empty tree and all trees with no more than p vertices all map to zero.

To avoid a profusion of notation, I_p will also be used to denote its homomorphic counterpart in G. The quotient space G/I_p consists of cosets of mappings from $T^{\#}$ to $\mathbb{R}$ in which two elements α, β are in the same coset if $\alpha(t) = \beta(t)$ unless $r(t) > p$. We will find it convenient to represent the quotient space as the set of mappings from a finite set of trees, that is only those which are relevant, to $\mathbb{R}$.

We illustrate these ideas in the case $p = 3$. When referring to a point in K, the Runge–Kutta space, we will give a single representative generalized Runge–Kutta method.

The Runge–Kutta point represented by the generalized method

$$
\begin{array}{c|ccccc}
0 \\
\frac{1}{3} & \frac{1}{3} \\
\frac{2}{3} & 0 & \frac{2}{3} \\
1 & 0 & 0 & 1 \\
\hline
0 & -1 & 3 & -3 & 1
\end{array}
$$

is a member of I_3 because

$$b_0 = 0,$$

$$\sum_i b_i = 0,$$

$$\sum_i b_i c_i = 0,$$

$$\sum_i b_i c_i^2 = 0,$$

$$\sum_{i,j} b_i a_{ij} c_j = 0,$$

and the result computed after a single step is therefore $O(|h|^4)$.

The two Runge–Kutta methods

$$
\begin{array}{c|cc}
0 & 0 & 0 \\
\frac{2}{3} & \frac{1}{3} & \frac{1}{3} \\
\hline
& \frac{1}{4} & \frac{3}{4}
\end{array}
\qquad \text{and} \qquad
\begin{array}{c|ccc}
0 & 0 & 0 & 0 \\
\frac{1}{2} & \frac{1}{2} & 0 & 0 \\
1 & -1 & 2 & 0 \\
\hline
& \frac{1}{6} & \frac{2}{3} & \frac{1}{6}
\end{array}
$$

represent points in the same member of K/I_3 because they are both of third order of accuracy and therefore give results after a single step which differ by $O(|h|^4)$.

The ideal I_3 in G consists of the set of all mappings from $T^\#$ to $\mathbb{R}$ which map $\emptyset$, τ, $[\tau]$, $[\tau^2]$ and $[_2\tau]_2$ each to zero. The quotient space G/I_3 can be represented by the set of functions on this set of five trees to the real numbers.

386   The element E.

We consider generalizing the type of method given by (380a) in such a way that the number of stages becomes infinite with the sums in (380b) and (380c) replaced by integrals. We will not attempt to encompass a wide variety of situations but instead look in detail at a method in which the set $\{1, 2, \ldots, s\}$ is replaced by the interval $[0, 1]$ and in which the matrix A is replaced by the indefinite integration operator. The vector $[b_1, b_2, \ldots, b_s]$ is replaced by integration from 0 to 1 and the value of b_0 is set at 1.

We now examine what it means to solve a differential equation system

$$y'(x) = f(y(x)), \qquad y(x_0) = y_0, \tag{386a}$$

by this method. Instead of Y_i indexed on the set $\{1, 2, \ldots, s\}$, we consider $Y(u)$ indexed on $[0, 1]$. That is, $u \in [0, 1]$. Because of the form selected for the operator taking the role of A, $Y(u)$ is defined by

$$Y(u) = y_0 + h \int_0^u f(Y(v))\, dv,$$

and y_1, the result after a single step, is simply

$$y_1 = y_0 + h \int_0^1 f(Y(v))\, dv$$

$$= Y(1).$$

Comparing this with the existence theorem of Picard, theorem 112D, we see that y_1 is the solution to the differential equation (386a) at $x = x_0 + h$. It is natural, then, to regard this limiting type of Runge–Kutta method as the method to which all finite-stage methods are approximations. Since we are interested in representing Runge–Kutta methods by their counterparts in G under the mapping $\bar{\Theta}$, we naturally ask what the elementary weights become for this method. For a tree t, let $\Phi(t)$ denote the elementary weight for this method, and, corresponding to the 'stage' $u \in [0, 1]$, let $\Phi(u)(t)$ denote the stage elementary weight analogous to $\Phi_i(t)$.

Using the analogy of simply replacing sums by integrals, we can define $\Phi(t)$ and $\Phi(u)(t)$ recursively as follows:

$$\Phi(\tau) = \int_0^1 1 \, dv, \tag{386b}$$

$$\Phi(u)(\tau) = \int_0^u 1 \, dv, \tag{386c}$$

$$\Phi([t_1 t_2 \cdots t_k]) = \int_0^1 \prod_{i=1}^k \Phi(v)(t_i) \, dv, \tag{386d}$$

$$\Phi(u)([t_1 t_2 \cdots t_k]) = \int_0^u \prod_{i=1}^k \Phi(v)(t_i) \, dv. \tag{386e}$$

THEOREM 386A.

For $t \in T, u \in [0, 1], \Phi(t)$ and $\Phi(u)(t)$ defined by (386b), (386c), (386d) and (386e) are given by

$$\Phi(t) = \frac{1}{\gamma(t)}, \tag{386f}$$

$$\Phi(u)(t) = \frac{u^{r(t)}}{\gamma(t)}, \tag{386g}$$

where $\gamma(t)$ is the density of t defined in subsection 144.

Proof.

Because (386f) is an instance of (386g), it will be sufficient to verify (386g). For $t = \tau$ the result $\Phi(u)(\tau) = u$ is immediate. We now complete the proof by induction by writing $t = [t_1 t_2 \cdots t_k]$ and assuming that

$$\Phi(u)(t_i) = \frac{u^{r(t_i)}}{\gamma(t_i),} \qquad i = 1, 2, \ldots, k.$$

From (386e) we have

$$\Phi(u)(t) = \int_0^u \prod_{i=1}^k \left[\frac{v^{r(t_i)}}{\gamma(t_i)} \right] dv$$

$$= \frac{\int_0^u v^{r(t)-1} \, dv}{\prod_{i=1}^k \gamma(t_i)}$$

Table 386

t	$E^a(t)$	$E^b(t)$	$(E^a \cdot E^b)(t)$	$E^{a+b}(t)$
$\emptyset$	1	1	1	1
τ	a	b	$a \cdot 1 + b$	$a + b$
$[\tau]$	$a^2/2$	$b^2/2$	$(a^2/2) \cdot 1 + a \cdot b + b^2/2$	$(a+b)^2/2$
$[\tau^2]$	$a^3/3$	$b^3/3$	$(a^3/3) \cdot 1 + a^2 \cdot b + 2a \cdot b^2/2 + b^3/3$	$(a+b)^3/3$
$[_2\tau]_2$	$a^3/6$	$b^3/6$	$(a^3/6) \cdot 1 + (a^2/2) \cdot b + a \cdot b^2/2 + b^3/6$	$(a+b)^3/6$

$$= \frac{u^{r(t)}}{r(t) \prod\limits_{i=1}^{k} \gamma(t_i)}$$

$$= \frac{u^{r(t)}}{\gamma(t)}. \quad \blacksquare$$

In addition to E, it is convenient to consider a family of members of G_1 denoted by E^α for α a real number which are defined such that a tree t maps to $\alpha^{r(t)}/\gamma(t)$. This corresponds to a Runge–Kutta method on the same index set $[0, 1]$ as E but with h replaced by αh. Thus, this limiting Runge–Kutta method gives a result after a single step at $x_0 + \alpha h$. Since E^α is homomorphic to the shift operator, the following result follows immediately.

THEOREM 386B.

For $a, b \in \mathbb{R}$,

$$E^a \cdot E^b = E^{a+b}.$$

We illustrate the result of theorem 386B and the use of the ideals I_m by verifying in detail that $(E^a + I_3)(E^b + I_3) = E^{a+b} + I_3$. This is done in tabular fashion in table 386 where the multiplication is performed using theorem 146F where $\lambda(\alpha, t_i)$ is found from table 146.

387 The element D and the order of Runge–Kutta methods.

Consider the Runge–Kutta point containing

$$\begin{array}{c|c} 0 & 0 \\ \hline & \\ 0 & 1 \end{array} \tag{387a}$$

giving the numerical result after a single step equal to $hf(y_0)$. Its image in G is easily found to be the function which maps the single tree τ to 1 and every other tree, including the empty tree, to 0. Because of its connection with differentiation, this member of G will be denoted by D.

Let α denote any member of G_1 corresponding to a Runge–Kutta method m. If y_1 is the numerical result found by applying a single step of m to an initial value y_0, and the method (387a) is then applied to y_1, the result produced is now $hf(y_1)$ with the corresponding member in G equal to αD. If α corresponds not to a Runge–Kutta method but to E^a, then αD corresponds to the formation of $hf(y(x_0 + ah))$, where the function y satisfies the initial value problem $y'(x) = f(y(x))$, $y(x_0) = y_0$.

These two examples illustrate the significance of the product αD and accordingly we seek to find a simple expression for it.

THEOREM 387A.

If $\alpha \in G_1$, then

$$(\alpha D)(\emptyset) = 0, \tag{387b}$$

$$(\alpha D)(\tau) = 1, \tag{387c}$$

$$(\alpha D)([t_1 t_2 \cdots t_k]) = \alpha(t_1)\,\alpha(t_2) \cdots \alpha(t_k), \qquad t_1, t_2, \ldots, t_k \in T, \tag{387d}$$

$$(\alpha D)(t_0 t_1) = (\alpha D)(t_0)\,\alpha(t_1), \qquad t_0, t_1 \in T. \tag{387e}$$

Proof.

The first two results, (387b) and (387c), follow from the formulae $(\alpha D)(\emptyset) = D(\emptyset)$ and $(\alpha D)(\tau) = D(\tau) + \alpha(\tau)\,D(\emptyset)$, whereas (387d) is a consequence of (387e) together with the observations that $[t_1 t_2 \cdots t_k] = [t_1 t_2 \cdots t_{k-1}] t_k$ and $[t_1] = \tau t_1$. To prove the remaining result, (387e), we make use of the function λ and write

$$\lambda(\alpha, t) = \sum_{u \in T} L(\alpha, t, u)\,\hat{u},$$

where the sum vanishes except for a finite set of members $u \in T$. Since $(\alpha D)(t) = L(\alpha, t, \tau)$, this is the only coefficient which interests us.

Using the recursive definition of λ,

$$\lambda(\alpha, \tau) = \hat{\tau},$$

$$\lambda(\alpha, t_0 t_1) = \lambda(\alpha, t_0)\,\lambda(\alpha, t_1) + \alpha(t_1)\,\lambda(\alpha, t_0),$$

and picking out the coefficient of $\hat{\tau}$, we find that

$$L(\alpha, \tau, \tau) = 1,$$

$$L(\alpha, t_0 t_1, \tau) = \alpha(t_1)\,L(\alpha, t_0, \tau),$$

which is equivalent to the required result. ∎

We are now in a position to restate the order conditions for Runge–Kutta methods using the formalism we have constructed.

THEOREM 387B.

Consider the s-stage Runge–Kutta method with the coefficient matrix A and vector b^T. Then $\alpha_1, \alpha_2, \ldots, \alpha_s, \alpha \in G$, satisfying

$$\alpha_i = 1 + \sum_{j=1}^{s} a_{ij}(\alpha_j D), \qquad i = 1, 2, \ldots, s, \tag{387f}$$

$$\alpha = 1 + \sum_{j=1}^{s} b_j(\alpha_j D), \tag{387g}$$

exist and are unique and the Runge–Kutta method is of order p iff

$$\alpha \in E + I_p.$$

Proof.

To prove the existence and uniqueness of $\alpha_1, \alpha_2, \ldots, \alpha_s$ and α satisfying (387f) and (387g), we operate each side on a tree t. If $t = \emptyset$, then $\alpha_i(t) = \alpha(t) = 1$. If $t = \tau$ then $\alpha_i(t) = \sum_{j=1}^{s} a_{ij}$ and $\alpha(t) = \sum_{j=1}^{s} b_j$. If $t = [t_1 t_2 \cdots t_k]$ then

$$\alpha_i(t) = \sum_{j=1}^{s} a_{ij} \prod_{l=1}^{k} \alpha_j(t_l), \qquad i = 1, 2, \ldots, s, \tag{387h}$$

$$\alpha(t) = \sum_{j=1}^{s} b_j \prod_{l=1}^{k} \alpha_j(t_l). \tag{387i}$$

In each case, for each t there is a uniquely defined expression for $\alpha_i(t)$ and $\alpha(t)$. To verify that the order conditions agree with those given by theorem 307B we note that with $\alpha_i(\tau) = c_i = \Phi_i(\tau)$ and $\alpha(\tau) = \sum_{j=1}^{s} b_j = \Phi(\tau)$, the recursion formulae for $\Phi_i(t)$ and $\Phi(t)$ exactly agree with (387h) and (387i) if $\alpha_i(t) = \Phi_i(t)$, $\alpha(t) = \Phi(t)$ where we note that $\alpha \in E + I_p$ means that $\alpha(t) = E(t) = 1/\gamma(t)$ for $r(t) \le p$. ∎

388 Miscellaneous constructs

Since D has the effect of generating the scaled first derivative, it is natural to consider some type of generalization of a Runge–Kutta method which generates higher derivatives. For convenience, we will scale the nth derivative by $h^n/n!$ so that it corresponds to a term in E written as a Taylor series.

We denote the Runge–Kutta point, and its image in G, which represents this scaled nth derivative by T_n. Among generalized methods which represent T_n is the method

$$\frac{c \mid A}{0 \mid b^*} \tag{388a}$$

where the index set for the stages is the interval $I = [0, \varepsilon]$ with $\varepsilon > 0$, A is the linear operator on $C^\infty(I)$ defined by

$$(AY)(x) = \int_0^x Y(u)\, du, \qquad (388b)$$

c is the identity function on I, and b^* is the linear functional defined by

$$b^*(Y) = \frac{1}{n!}\, Y^{(n-1)}(0). \qquad (388c)$$

It is easy to verify the formulae for the stage weights $\Phi(u)(t)$ and the elementary weights for the method as a whole, $\Phi(t)$. In fact, $\Phi(u)(t)$ is essentially the same as that given by (386c) and (386e) and we obtain for $\Phi(t)$ the formula given in the following theorem which we state without proof.

THEOREM 388A.

For the method defined by (388a), (388b) and (388c), $\Phi(t)$ is given for $t \in T^\#$ by

$$\Phi(t) = \frac{1}{\gamma(t)}, \qquad r(t) = n,$$

$$\Phi(t) = 0, \qquad r(t) \neq n.$$

The application of T_n is to the order analysis of methods using the Nordsieck vector approach. If it is required to compute $(h^n/n!)y^{(n)}(x_0 + mh)$ by a generalized Runge–Kutta method whose counterpart in G_1 is α, then, to within the order being sought, we must have $\alpha = E^m T_n$.

The next quantity we will introduce is a linear operator on the subspace of G corresponding to members of K_0. If $G_0 = \{\alpha \in G: \alpha(\emptyset) = 0\}$ is this subspace then we define $J: G_0 \to G_0$ by the formulae

$$(J\alpha)(\tau) = 0,$$

$$(J\alpha)([t]) = \alpha(t), \qquad t \in T,$$

$$(J\alpha)([t_1 t_2 \cdots t_k]) = 0, \qquad t_1, t_2, \ldots, t_k \in T, \qquad k > 1.$$

The significance of J is that it corresponds to multiplication of a quantity computed by a generalized Runge–Kutta method, for which $b_0 = 0$, by $hf'(y_0)$. The infinite series $(1 - qJ)^{-1} = 1 + qJ + q^2 J^2 + \cdots$, where q is a number, operating on α is defined so that

$$[(1 - qJ)^{-1}\alpha]\,(t)$$

is computed using only those terms that can possibly affect the answer. The role of this type of linear operator is in the analysis of Rosenbrock methods and their generalizations.

Table 388 The effect of various operations on $\alpha \in G/I_5$.

i	t_i	$\alpha(t_i)$	$(E^{\pm 1}\alpha)(t_i)$	$(D\alpha)(t_i)$	$(J\alpha)(t_i)(\alpha_0=0)$
0	$\emptyset$	α_0	α_0	0	0
1	τ	α_1	$\alpha_1 \pm \alpha_0$	1	0
2	$[\tau]$	α_2	$\alpha_2 \pm \alpha_1 + \frac{1}{2}\alpha_0$	α_1	α_1
3	$[\tau^2]$	α_3	$\alpha_3 \pm 2\alpha_2 + \alpha_1 \pm \frac{1}{3}\alpha_0$	α_1^2	0
4	$[_2\tau]_2$	α_4	$\alpha_4 \pm \alpha_2 + \frac{1}{2}\alpha_1 \pm \frac{1}{6}\alpha_0$	α_2	α_2
5	$[\tau^3]$	α_5	$\alpha_5 \pm 3\alpha_3 + 3\alpha_2 \pm \alpha_1 + \frac{1}{4}\alpha_0$	α_1^3	0
6	$[\tau[\tau]]$	α_6	$\alpha_6 \pm \alpha_4 \pm \alpha_3 + \frac{3}{2}\alpha_2 \pm \frac{1}{2}\alpha_1 + \frac{1}{8}\alpha_0$	$\alpha_1\alpha_2$	0
7	$[_2\tau^2]_2$	α_7	$\alpha_7 \pm 2\alpha_4 + \alpha_2 \pm \frac{1}{3}\alpha_1 + \frac{1}{12}\alpha_0$	α_3	α_3
8	$[_3\tau]_3$	α_8	$\alpha_8 \pm \alpha_4 + \frac{1}{2}\alpha_2 \pm \frac{1}{6}\alpha_1 + \frac{1}{24}\alpha_0$	α_4	α_4
9	$[\tau^4]$	α_9	$\alpha_9 \pm 4\alpha_5 + 6\alpha_3 \pm 4\alpha_2 + \alpha_1 \pm \frac{1}{5}\alpha_0$	α_1^4	0
10	$[\tau^2[\tau]]$	α_{10}	$\alpha_{10} \pm 2\alpha_6 \pm \alpha_5 + \alpha_4 + \frac{5}{2}\alpha_3 \pm 2\alpha_2 + \frac{1}{2}\alpha_1 \pm \frac{1}{10}\alpha_0$	$\alpha_1^2\alpha_2$	0
11	$[\tau[\tau^2]]$	α_{11}	$\alpha_{11} \pm \alpha_7 \pm 2\alpha_6 + 2\alpha_4 + \alpha_3 \pm \frac{4}{3}\alpha_2 + \frac{1}{3}\alpha_1 \pm \frac{1}{15}\alpha_0$	$\alpha_1\alpha_3$	0
12	$[\tau[_2\tau]_2]$	α_{12}	$\alpha_{12} \pm \alpha_8 \pm \alpha_6 + \alpha_4 + \frac{1}{2}\alpha_3 \pm \frac{2}{3}\alpha_2 + \frac{1}{6}\alpha_1 \pm \frac{1}{30}\alpha_0$	$\alpha_1\alpha_4$	0
13	$[[\tau]^2]$	α_{13}	$\alpha_{13} \pm 2\alpha_6 + \alpha_4 + \alpha_3 \pm \alpha_2 + \frac{1}{4}\alpha_1 \pm \frac{1}{20}\alpha_0$	α_2^2	0
14	$[_2\tau^3]_2$	α_{14}	$\alpha_{14} \pm 3\alpha_7 + 3\alpha_4 \pm \alpha_2 + \frac{1}{4}\alpha_1 \pm \frac{1}{20}\alpha_0$	α_5	α_5
15	$[_2\tau[\tau]]_2$	α_{15}	$\alpha_{15} \pm \alpha_8 \pm \alpha_7 + \frac{3}{2}\alpha_4 \pm \frac{1}{2}\alpha_2 + \frac{1}{8}\alpha_1 \pm \frac{1}{40}\alpha_0$	α_6	α_6
16	$[_3\tau^2]_3$	α_{16}	$\alpha_{16} \pm 2\alpha_8 + \alpha_4 \pm \frac{1}{3}\alpha_2 + \frac{1}{12}\alpha_1 \pm \frac{1}{60}\alpha_0$	α_7	α_7
17	$[_4\tau]_4$	α_{17}	$\alpha_{17} \pm \alpha_8 + \frac{1}{2}\alpha_4 \pm \frac{1}{6}\alpha_2 + \frac{1}{24}\alpha_1 \pm \frac{1}{120}\alpha_0$	α_8	α_8

If we need to operate, not with $hf'(y_0)$ but with $hf'(y_1)$ where y_1 is computed from y_0 by a Runge–Kutta method with image β in G_1, then we can refer the result computed by the method corresponding to α to y_1, operate by J and then preoperate by the method corresponding to β. This gives $\beta J(\beta^{-1}\alpha)$.

Examples of the use of these ideas will be presented in subsection 389.

We conclude the present subsection by giving in table 388 a summary of the effect of carrying out various operations in G/I_5. Note that the αD column is given on the assumption that $\alpha_0 = 1$ and the $J\alpha$ column on the assumption that $\alpha_0 = 0$.

389 Examples of Runge–Kutta space usage.

We will present two simple examples of the use of the Runge-Kutta space to determine the order of accuracy of some methods which fall outside the usual classes of methods. A further example is given in Butcher [1984].

First we consider a method which begins step number n with an approximation y_{n-1} to $y(x_{n-1})$ together with an approximation w_{n-1} to $hy'(x_{n-1} + dh)$. Within the step, $u = hf(y_{n-1} + cw_{n-1})$ is computed and from this are computed in turn $w_n = hf(y_{n-1} + (d - \theta) w_{n-1} + \theta u)$ and $y_n = y_{n-1} + au + bw_n$. Our aim will be to specify values of a, b, c, d, θ such that if y_{n-1} exactly equals

Table 389 Summary of order conditions for a special method.

t	α	$E\alpha$	$1 + c\alpha$	$(1 + c\alpha)D$
$\emptyset$	0	0	1	0
τ	1	1	c	1
$[\tau]$	$d - 1$	d	$c(d - 1)$	c
$[\tau^2]$	$(d - 1)^2$	d^2	$c(d - 1)^2$	c^2
$[_2\tau]_2$	$(d - 1)^2 + \theta(1 + c - d) - \frac{1}{2}$	$d(d - 1) + \theta(1 + c - d)$	$c[(d - 1)^2 + \theta(1 + c - d) - \frac{1}{2}]$	$c(d - 1)$

$1 + (d - \theta)\alpha + \theta(1 + c\alpha)D$	$E^{-1}[1 + (d - \theta)\alpha + \theta(1 + c\alpha)D]D$	$1 + a(1 + c\alpha)D + bE\alpha$	E
1	0	1	1
d	1	$a + b$	1
$(d - \theta)(d - 1) + \theta c$	$d - 1$	$ac + bd$	$\frac{1}{2}$
$(d - \theta)(d - 1)^2 + \theta c^2$	$(d - 1)^2$	$ac^2 + bd^2$	$\frac{1}{3}$
$(d - \theta)[(d - 1)^2 + \theta(1 + c - d) - \frac{1}{2}] + \theta c(d - 1)$	$(d - 1)^2 + \theta(1 + c - d) - \frac{1}{2}$	$ac(d - 1) + bd(d - 1) + b\theta(1 + c - d)$	$\frac{1}{6}$

$y(x_{n-1})$ and w_{n-1} can be computed from y_{n-1} by *some* generalized Runge–Kutta method m then, in each case with an error $O(h^4)$, y_n equals $y(x_n)$ and w_n equals the result computed from y_n using m.

Let the value of $\bar{\Theta}(m)$ be $\alpha \in G$. We will for convenience work in the quotient space G/I_3 and we will use the same notations α, E, D, etc., to represent elements in both G and G/I_3 which correspond under the natural homomorphism. The members of G/I_3 which correspond to the operations which compute y_{n-1}, w_{n-1}, u, w_n and y_n respectively, in each case from y_{n-1}, are given by $1, \alpha, (1 + c\alpha)D$, $[1 + (d - \theta)\alpha + \theta(1 + c\alpha)D]D$ and $1 + a(1 + c\alpha)D + b[1 + (d - \theta)\alpha + \theta(1 + c\alpha)D]D$ respectively, and to achieve the required accuracy we must have

$$\alpha = E^{-1}[1 + (d - \theta)\alpha + \theta(1 + c\alpha)D]D, \tag{389a}$$

$$E = 1 + a(1 + c\alpha)D + bE\alpha. \tag{389b}$$

We now find the values of $\alpha(t)$ for $t = \emptyset, \tau, [\tau], [\tau^2], [{}_2\tau]_2$ by comparing the two sides of (389a) tree by tree. We also find the two sides of (389b). The calculations are summarized in table 389 and make frequent use of table 388.

Comparing the last two columns of table 389, we see that order 3 is achieved if and only if

$$a + b = 1,$$

$$ac + bd = \tfrac{1}{2},$$

$$ac^2 + bd^2 = \tfrac{1}{3},$$

$$ac(d - 1) + bd(d - 1) + b\theta(1 + c - d) = \tfrac{1}{6}.$$

These equations are satisfied for example when

$$a = \tfrac{1}{4}, \qquad b = \tfrac{3}{4}, \qquad c = 0, \qquad d = \tfrac{2}{3}, \qquad \theta = \tfrac{4}{3}$$

or when

$$a = \tfrac{3}{4}, \qquad b = \tfrac{1}{4}, \qquad c = \tfrac{1}{3}, \qquad d = 1, \qquad \theta = 2.$$

It is a simple matter to devise a generalized Runge–Kutta m such that w_0 can be found from y_0 using the method m to begin the integration. Such a method could be chosen in the form

$$w_0 = hf(g(y_0)),$$

where g corresponds to a Runge–Kutta method such that

$$\Phi(\tau) = d - 1,$$

$$\Phi([\tau]) = (d - 1)^2 + \theta(1 + c - d) - \tfrac{1}{2}.$$

Our second example is in the derivation of a Rosenbrock method using the operator J. Specifically we seek to approximate $y(x_n)$ using an approximation y_{n-1} to $y(x_{n-1})$. The approximation y_n will be given by

$$y_n = y_{n-1} + b_1 F_1 + b_2 F_2,$$

Table 389' Order conditions for a special Rosenbrock method.

t	$D(t)$	$\alpha_1(t)$	$(J\alpha_1)(t)$	$(1 - a_{21}\alpha_1)(t)$	$[(1 + a_{21}\alpha_1)D](t)$	α_2	$J\alpha_2$	α
$\emptyset$	0	0	0	1	0	0	0	1
τ	1	1	0	a_{21}	1	1	0	$b_1 + b_2$
$[\tau]$	0	γ	1	$a_{21}\gamma$	a_{21}	$a_{21} + \gamma$	1	$b_1\gamma + b_2(a_{21} + \gamma)$
$[\tau^2]$	0	0	0	0	a_{21}^2	a_{21}^2	0	$b_2 a_{21}^2$
$[_2\tau]_2$	0	γ^2	γ	$a_{21}\gamma^2$	$a_{21}\gamma$	$2a_{21}\gamma + \gamma^2$	$a_{21} + \gamma$	$b_1\gamma^2 + b_2(2a_{21}\gamma + \gamma^2)$

where

$$F_1 = [I - h\gamma f'(y_{n-1})]^{-1} hf(y_{n-1}),$$

$$F_2 = [I - h\gamma f'(y_{n-1})]^{-1} hf(y_{n-1} + a_{21}F_1).$$

We will once more seek order 3 and work in the space G/I_3. Let α_1, α_2 and α correspond to the computation of F_1, F_2 and y_n respectively from y_{n-1}. We then have the equations

$$\alpha_1 = \gamma J\alpha_1 + D,$$

$$\alpha_2 = \gamma J\alpha_2 + (1 + a_{21}\alpha_1)D,$$

$$\alpha = 1 + b_1\alpha_1 + b_2\alpha_2.$$

In table 389', α_1, α_2 and α are computed from these relations. To obtain order 3, equate α to $E \in G/I_3$, so that

$$b_1 + b_2 = 1,$$

$$b_1\gamma + b_2(a_{21} + \gamma) = \tfrac{1}{2},$$

$$b_2 a_{21}^2 = \tfrac{1}{3},$$

$$b_1\gamma^2 + b_2(2a_{21}\gamma + \gamma^2) = \tfrac{1}{6}.$$

For consistency of these equations in b_1, b_2, a_{21} it turns out that $\gamma = (3 \pm \sqrt{3})/6$ and the upper sign is chosen because of its favourable stability. With this choice, the complete solution first given by Calahan (1968) is

$$b_1 = \tfrac{3}{4}, \qquad b_2 = \tfrac{1}{4}, \qquad a_{21} = -\frac{2\sqrt{3}}{3}, \qquad \gamma = \tfrac{1}{2} + \frac{\sqrt{3}}{6}.$$

4

General linear methods

40 INTRODUCTION TO GENERAL LINEAR METHODS

400 General classes of methods.

As we saw in section 21, the most familiar methods, particularly Runge–Kutta methods and linear multistep methods, constitute very special subclasses of a very wide class of methods. Although we will not consider in this chapter methods exhibiting the full generality of the multistep–multistage–multi-derivative scheme, we will attempt to cast both Runge–Kutta methods and linear multistep methods into a common form and use this form as the model for further investigations.

We first consider three examples, the first being the three-stage Runge–Kutta method:

$$Y_1 = y_{n-1},$$

$$Y_2 = y_{n-1} + \tfrac{1}{2} hf(Y_1),$$

$$Y_3 = y_{n-1} + h[-f(Y_1) + 2f(Y_2)],$$

$$y_n = y_{n-1} + h[\tfrac{1}{6}f(Y_1) + \tfrac{2}{3}f(Y_2) + \tfrac{1}{6}f(Y_3)],$$

the second the linear multistep method:

$$y_n = y_{n-2} + h(\tfrac{1}{3}f(y_n) + \tfrac{4}{3}f(y_{n-1}) + \tfrac{1}{3}f(y_{n-2})),$$

and, finally, the (PEC) predictor–corrector method:

$$y_n^{(P)} = y_{n-2} + 2hf(y_{n-1}^{(P)}),$$

$$y_n = y_{n-2} + h[\tfrac{1}{3}f(y_n^{(P)}) + \tfrac{4}{3}f(y_{n-1}^{(P)}) + \tfrac{1}{3}f(y_{n-2}^{(P)})].$$

If we write $y_1^{(n)}, y_2^{(n)}, \ldots, y_N^{(n)}$ for a set of N approximations local to the nth step, then for each of these three methods, each approximation can be written as a linear combination of the corresponding values from the previous step together with linear combinations of $hf(y_1^{(n)}), hf(y_2^{(n)}), \ldots, hf(y_N^{(n)})$, $hf(y_1^{(n-1)}), hf(y_2^{(n-1)}), \ldots, hf(y_N^{(n-1)})$.

In the first example, the three-stage Runge–Kutta method, we can choose $N = 4$ and identify $y_1^{(n)} = Y_1$, $y_2^{(n)} = Y_2$, $y_3^{(n)} = Y_3$, $y_4^{(n)} = y_n$. We have

$$y_1^{(n)} = y_4^{(n-1)},$$

$$y_2^{(n)} = y_4^{(n-1)} + \tfrac{1}{2}hf(y_1^{(n)}),$$

$$y_3^{(n)} = y_4^{(n-1)} - hf(y_1^{(n)}) + 2hf(y_2^{(n)}),$$

$$y_4^{(n)} = y_4^{(n-1)} + \tfrac{1}{6}hf(y_1^{(n)}) + \tfrac{2}{3}hf(y_2^{(n)}) + \tfrac{1}{6}hf(y_3^{(n)}).$$

For the linear multistep method, we can write $N = 2$ and $y_1^{(n)} = y_n$, $y_2^{(n)} = y_{n-1}$, so that

$$y_1^{(n)} = y_2^{(n-1)} + \tfrac{1}{3}hf(y_1^{(n)}) + \tfrac{4}{3}hf(y_1^{(n-1)}) + \tfrac{1}{3}hf(y_2^{(n-1)}).$$

$$y_2^{(n)} = y_1^{(n-1)}.$$

Finally, for the PEC method, with $N = 4$ and $y_1^{(n)} = y_n^{(P)}$, $y_2^{(n)} = y_{n-1}^{(P)}$, $y_3^{(n)} = y_n$, $y_4^{(n)} = y_{n-1}$ we have

$$y_1^{(n)} = y_4^{(n-1)} + 2hf(y_2^{(n)}),$$

$$y_2^{(n)} = y_1^{(n-1)},$$

$$y_3^{(n)} = y_4^{(n-1)} + \tfrac{1}{3}hf(y_1^{(n)}) + \tfrac{4}{3}hf(y_2^{(n)}) + \tfrac{1}{3}hf(y_2^{(n-1)}),$$

$$y_4^{(n)} = y_3^{(n-1)}.$$

Written in this way, the similarity of these methods becomes clear and leads us to study methods of the general class they exemplify. Methods of this type will be referred to as *general linear methods*. They were first studied in their own right by Butcher (1966) and sometimes referred to as $A-B$ methods because of a notation used in that paper.

401 Representation by matrices.

Let A, B and C be $N \times N$ matrices with elements $[a_{ij}]$, $[b_{ij}]$, $[c_{ij}]$ respectively. We shall study methods represented by the triple (A, B, C) such that the formula for $y_i^{(n)} (i = 1, 2, \ldots, N)$ is

$$y_i^{(n)} = \sum_{j=1}^{N} a_{ij} y_j^{(n-1)} + h \sum_{j=1}^{N} b_{ij} f(y_j^{(n)}) + h \sum_{j=1}^{N} c_{ij} f(y_j^{(n-1)}). \qquad (401a)$$

In the special case that $C = 0$, we will identify the method by the pair (A, B).

For convenience in the analysis, we will regard the collection of vectors $y_1^{(n)}$, $y_2^{(n)}, \ldots, y_N^{(n)}$ as being a single point in X^N (X being the vector space on which the original differential equation is defined). If this point is $(y_1^{(n)}, y_2^{(n)}, \ldots, y_N^{(n)}) = Y_n$, say, we can write (401a) in a compact form

$$Y_n = (A \otimes I)Y_{n-1} + h(B \otimes I)F(Y_n) + h(C \otimes I)F(Y_{n-1}), \qquad (401b)$$

where for $V = (v_1, v_2, \ldots, v_N) \in X^N$, the direct product $A \otimes I$ is defined as

$$(A \otimes I)V = W,$$

where $W = (w_1, w_2, \ldots, w_N) \in X^N$ is given by

$$w_i = \sum_{j=1}^{N} a_{ij}v_j, \qquad i = 1, 2, \ldots, N,$$

and

$$F(v_1, v_2, \ldots, v_N) = (f(v_1), f(v_2), \ldots, f(v_N)). \tag{401c}$$

Since an analysis of (401b) will depend on the properties of the matrices $A \otimes I$, $B \otimes I$, $C \otimes I$ and of the function F, and since these properties are inherited from A, B, C and f, we will now consider the relevant relationships between them.

Let $\|\cdot\|$ denote a norm on X. This can be extended to X^N by the definition

$$\| (v_1, v_2, \ldots, v_N) \| = \max_{i=1}^{N} \| v_i \| . \tag{401d}$$

The notation $\|\cdot\|$ will also denote the induced norm on linear operators on X^N as well as on X.

THEOREM 401A.

$$\| A \otimes I \| = \| A \|_\infty.$$

Proof.

We compute

$$\| (A \otimes I)(v_1, v_2, \ldots, v_N) \| = \left\| \left(\sum_{j=1}^{N} a_{1j}v_j, \sum_{j=1}^{N} a_{2j}v_j, \ldots, \sum_{j=1}^{N} a_{Nj}v_j \right) \right\|$$

$$= \max_{i=1}^{N} \left\| \sum_{j=1}^{N} a_{ij}v_j \right\|$$

$$\leq \max_{i=1}^{N} \sum_{j=1}^{N} | a_{ij} | \, \| v_j \|$$

$$\leq \| A \|_\infty \| (v_1, v_2, \ldots, v_N) \| .$$

To verify that this bound can actually be achieved, let i be such that $\Sigma_j | a_{ij} | = \| A \|_\infty$ and, for v an arbitrary non-zero vector in X, let $v_j = v$ (if $a_{ij} > 0$) and $v_j = - v$ (if $a_{ij} \leq 0$). ∎

We now look at the relation between F and f when it is supposed that f satisfies a Lipschitz condition.

THEOREM 401B.

If f satisfies a Lipschitz condition with constant L, then F defined by (401c) also satisfies a Lipschitz condition with the same constant L.

Proof.

If $(u_1, u_2, \ldots, u_N), (v_1, v_2, \ldots, v_N) \in X^N$, then

$$F(u_1, u_2, \ldots, u_N) - F(v_1, v_2, \ldots, v_N)$$

$$= (f(u_1) - f(v_1), f(u_2) - f(v_2), \ldots, f(u_N) - f(v_N))$$

and the norm of this is

$$\max_{i=1}^{N} \| f(u_i) - f(v_i) \| \leq \max_{i=1}^{N} (L \| u_i - v_i \|)$$

$$= L \| (u_1, u_2, \ldots, u_N) - (v_1, v_2, \ldots, v_N) \|. \quad \blacksquare$$

402 Reduction to standard case

Of the methods represented by matrices (A, B, C), those in which $C = 0$ will be referred to as being in the standard case, and we shall represent them by the pair (A, B). As it happens, any method (A, B, C) can be rewritten in an equivalent standard case form with no more than twice the number of approximations.

Consider a method (A, B, C) in which the approximations $y_1^{(n)}$, $y_2^{(n)}, \ldots, y_N^{(n)}$ are given by (401a). If we introduce another set $\bar{y}_1^{(n)}, \bar{y}_2^{(n)}$, $\ldots, \bar{y}_{2N}^{(n)}$ where

$$\bar{y}_i^{(n)} = \begin{cases} y_i^{(n)}, & i \leq N, \\ y_{i-N}^{(n-1)}, & i > N, \end{cases}$$

then we have

$$\bar{y}_i^{(n)} = \sum_{j=1}^{N} \bar{a}_{ij} \bar{y}_j^{(n-1)} + \sum_{j=1}^{N} \bar{b}_{ij} f(\bar{y}_j^{(n)}),$$

where $[\bar{a}_{ij}]$, $[\bar{b}_{ij}]$ constitute the matrices $(\bar{A}, \bar{B})$ given in partitioned form

$$\bar{A} = \begin{bmatrix} A & 0 \\ I & 0 \end{bmatrix}, \tag{402a}$$

$$\bar{B} = \begin{bmatrix} B & C \\ 0 & 0 \end{bmatrix}. \tag{402b}$$

Since this standard form is always available, possibly at the expense of a higher choice of N, we will confine much of our theoretical work to the study of the method (A, B), sure in the knowledge that by a trivial rewriting of any result, its applicability can be extended to (A, B, C).

As we have remarked, the reduction to a standard case may take up to twice the number of approximations, but it can be less than this. For example, the Runge–Kutta example in subsection 400 is already in (A, B) form with $N = 4$, while the linear multistep and PEC methods require an increase of only 1 on the value of N to achieve standard form.

403 Generalizations

In the description of general linear methods that we have considered it is assumed that each of $y_1^{(n)}, y_2^{(n)}, \ldots, y_N^{(n)}$ is computed only from linear combinations of $y_1^{(n-1)}, y_2^{(n-1)}, \ldots, y_N^{(n-1)}$ and derivatives of these $2N$ approximations. However, methods exist in which derivatives of the function f play a role, or where the dependence of $y_i^{(n)}$ on other quantities is non-linear.

While we do not consider these sorts of generalizations in this subsection, it is worth noting that there is a tacit assumption made about the nature of $y_1^{(n)} y_2^{(n)}, \ldots, y_N^{(n)}$. This assumption is that each is an approximation to $y(x_n)$ for small h, where x_n is the nth solution point. In fact, (401a) makes use of $f(y_1^{(n)}), f(y_2^{(n)}), \ldots, f(y_N^{(n)})$ so that, as $h \to 0$, it is natural to suppose that $y_1^{(n)}, y_2^{(n)}, \ldots, y_N^{(n)}$ are close to the y value of a point on the solution trajectory. So that this proximity of the approximations to an exact solution value will be inherited from step to step, it will normally be supposed tht $\sum_{j=1}^{N} a_{ij} = 1$ $(i = 1, 2, \ldots, N)$, and we formally introduce this assumption as the 'preconsistency condition' in section 42.

However, if for a particular j, $b_{ij} = c_{ij} = 0$ for each $i = 1, 2, \ldots, N$, then $f(y_j^{(n)})$ and $f(y_j^{(n-1)})$ are not used in the computation of Y_n. In this case, there is no particular reason why $y_j^{(n)}$ should approximate a point on the solution curve, and it will sometimes be convenient, for example to suppose that $y_j^{(n)} \to 0$ as $h \to 0$ so that it could be an approximation to $hy'(x_n)$, to $h^2 y''(x_n)$ or to any other quantity which enabled us to construct and describe useful methods.

As it happens, the greater generality of this type of widening of the significance of general linear methods is an illusion and there is no essential loss in generality in restricting ourselves to consistent methods. However, we shall briefly discuss the modifications to the theory of section 42 to make it apply to a more general consistency condition introduced in subsection 427.

In section 44, methods of this more general nature fit easily into our formulation and we will deal with them again there.

404 Implementation

If B is strictly lower triangular, or can be converted to this form by permutation of the basis vectors, it is possible to compute $y_1^{(n)}, y_2^{(n)}, \ldots, y_N^{(n)}$ at step

n directly. However, in contrast to these *explicit* methods, *implicit* methods can be found in which the evaluation of $y_1^{(n)}, y_2^{(n)}, \ldots, y_N^{(n)}$ requires the solution of an algebraic system of the form

$$y_i = v_i + \varphi_i(y_1, y_2, \ldots, y_N), \qquad i = 1, 2, \ldots, N,$$

where

$$v_i = \sum_{j=1}^{N} a_{ij} y_j^{(n-1)} + h \sum_{j=1}^{N} c_{ij} f(y_j^{(n-1)})$$

and

$$\varphi_i(y_1, y_2, \ldots, y_N) = h \sum_{j=1}^{N} b_{ij} f(y_j), \qquad i = 1, 2, \ldots, N.$$

Although the solution of this type of system efficiently may be far from simple, we do not discuss it in detail since it is essentially the same as the corresponding problem for implicit Runge–Kutta methods which we discussed in subsection 341.

41   EXAMPLES OF GENERAL LINEAR METHODS

410   Runge–Kutta methods as examples

Consider the (in general, implicit) Runge–Kutta method

$$
\begin{array}{c|cccc}
c_1 & a_{11} & a_{12} & \cdots & a_{1s} \\
c_2 & a_{21} & a_{22} & \cdots & a_{2s} \\
\vdots & \vdots & \vdots & & \vdots \\
c_s & a_{s1} & a_{s2} & \cdots & a_{ss} \\
\hline
 & b_1 & b_2 & \cdots & b_s
\end{array} \; ,
$$

where the internal approximations $Y_1, Y_2, \ldots, Y_s$ and the final result y_n are computed by

$$Y_i = y_{n-1} + h \sum_{j=1}^{s} a_{ij} f(Y_j), \qquad i = 1, 2, \ldots, s,$$

$$y_n = y_{n-1} + h \sum_{j=1}^{s} b_j f(Y_j).$$

If we write $N = s + 1$ and identify $y_1^{(n)}, y_2^{(n)}, \ldots, y_N^{(n)}$ with $Y_1, Y_2, \ldots, Y_s, y_n$ respectively then the method fits quite clearly into the general linear pattern with the matrices (A, B) characterizing it given by

$$A = \begin{bmatrix} 0 & 0 & \cdots & 0 & 1 \\ 0 & 0 & \cdots & 0 & 1 \\ \vdots & \vdots & & \vdots & \vdots \\ 0 & 0 & \cdots & 0 & 1 \end{bmatrix}, \qquad (410a)$$

$$B = \begin{bmatrix} a_{11} & a_{12} & \cdots & a_{1s} & 0 \\ a_{21} & a_{22} & \cdots & a_{2s} & 0 \\ \vdots & \vdots & & \vdots & \vdots \\ b_1 & b_2 & \cdots & b_s & 0 \end{bmatrix}. \qquad (410b)$$

If the method contains an additional approximation $\hat{y}_n$ (for error-estimating purposes), given by

$$\hat{y}_n = y_{n-1} + h \sum_{j=1}^{s} \hat{b}_j f(Y_j),$$

then we write $N = s + 2$ and introduce $y_{s+2}^{(n)}$ with the value of $\hat{y}_n$. The matrices A, B now become

$$A = \begin{bmatrix} 0 & 0 & \cdots & 0 & 1 & 0 \\ 0 & 0 & \cdots & 0 & 1 & 0 \\ \vdots & \vdots & & \vdots & \vdots & \vdots \\ 0 & 0 & \cdots & 0 & 1 & 0 \\ 0 & 0 & \cdots & 0 & 1 & 0 \end{bmatrix},$$

$$B = \begin{bmatrix} a_{11} & a_{12} & \cdots & a_{1s} & 0 & 0 \\ a_{21} & a_{22} & \cdots & a_{2s} & 0 & 0 \\ \vdots & \vdots & & \vdots & \vdots & \vdots \\ b_1 & b_2 & \cdots & b_s & 0 & 0 \\ \hat{b}_1 & \hat{b}_2 & \cdots & \hat{b}_s & 0 & 0 \end{bmatrix}$$

As an example of the more general view of subsection 403, we could define $y_{s+2}^{(n)}$ as $y_n - \hat{y}_n$ rather than as $\hat{y}_n$. In this case, the last rows of A and B would be

$$[0 \; 0 \; \cdots \; 0 \; 0 \; 0] \qquad \text{and} \qquad [b_1 - \hat{b}_1, b_2 - \hat{b}_2, \ldots, b_s - \hat{b}_s, 0, 0].$$

411 Linear multistep methods as examples

For the explicit linear multistep method

$$y_n = \alpha_1 y_{n-1} + \alpha_2 y_{n-2} + \cdots + \alpha_k y_{n-k}$$

$$+ h[\beta_1 f(y_{n-1}) + \beta_2 f(y_{n-2}) + \cdots + \beta_k f(y_{n-k})], \qquad (411a)$$

we write $N = k$ and identify

$$y_1^{(n)} = y_n,$$
$$y_2^{(n)} = y_{n-1},$$
$$\vdots \qquad \vdots$$
$$y_k^{(n)} = y_{n-k+1},$$

so that it is equivalent to the method (A, B, C) with $B = 0$ and

$$A = \begin{bmatrix} \alpha_1 & \alpha_2 & \cdots & \alpha_{k-1} & \alpha_k \\ 1 & 0 & \cdots & 0 & 0 \\ 0 & 1 & \cdots & 0 & 0 \\ \vdots & \vdots & & \vdots & \vdots \\ 0 & 0 & \cdots & 1 & 0 \end{bmatrix},$$

$$C = \begin{bmatrix} \beta_1 & \beta_2 & \cdots & \beta_{k-1} & \beta_k \\ 0 & 0 & \cdots & 0 & 0 \\ 0 & 0 & \cdots & 0 & 0 \\ \vdots & \vdots & & \vdots & \vdots \\ 0 & 0 & \cdots & 0 & 0 \end{bmatrix}.$$

To write this method in (A, B) form, let $N = k + 1$ and $y_{k+1}^{(n)} = y_{n-k}$ so that

$$A = \begin{bmatrix} \alpha_1 & \alpha_2 & \cdots & \alpha_{k-1} & \alpha_k & 0 \\ 1 & 0 & \cdots & 0 & 0 & 0 \\ 0 & 1 & \cdots & 0 & 0 & 0 \\ \vdots & \vdots & & \vdots & \vdots & \vdots \\ 0 & 0 & \cdots & 0 & 1 & 0 \end{bmatrix},$$

$$B = \begin{bmatrix} 0 & \beta_1 & \cdots & \beta_{k-2} & \beta_{k-1} & \beta_k \\ 0 & 0 & \cdots & 0 & 0 & 0 \\ 0 & 0 & \cdots & 0 & 0 & 0 \\ \vdots & \vdots & & \vdots & \vdots & \vdots \\ 0 & 0 & \cdots & 0 & 0 & 0 \end{bmatrix}.$$

If the method is implicit so that a term $h\beta_0 f(y_n)$ appears on the right-hand side of (411a) then the first row of the B matrix is replaced by $[\beta_0 \, \beta_1 \cdots \beta_{k-2} \, \beta_{k-1} \, \beta_k]$.

These formulations of linear multistep methods do not emphasize the fact that only a single derivative is actually carried out in a step (since $f(y_{n-1})$, $f(y_{n-2})$, ... are already known from previous steps). As an alternative that does provide this emphasis, we could write $N = 2k$ and identify

$$y_1^{(n)} = y_n,$$
$$y_2^{(n)} = y_{n-1},$$
$$\vdots \quad \vdots \qquad \vdots$$
$$y_k^{(n)} = y_{n-k+1},$$
$$y_{k+1}^{(n)} = hf(y_n),$$
$$y_{k+2}^{(n)} = hf(y_{n-1}),$$
$$\vdots \quad \vdots \qquad \vdots$$
$$y_{2k}^{(n)} = hf(y_{n-k+1}),$$

and in the method (A, B) we have

$$A = \left[\begin{array}{ccccc|ccccc}
\alpha_1 & \alpha_2 & \cdots & \alpha_{k-1} & \alpha_k & \beta_1 & \beta_2 & \cdots & \beta_{k-1} & \beta_k \\
1 & 0 & \cdots & 0 & 0 & 0 & 0 & \cdots & 0 & 0 \\
\vdots & \vdots & & \vdots & \vdots & \vdots & \vdots & & \vdots & \vdots \\
0 & 0 & \cdots & 1 & 0 & 0 & 0 & \cdots & 0 & 0 \\
\hline
0 & 0 & \cdots & 0 & 0 & 0 & 0 & \cdots & 0 & 0 \\
0 & 0 & \cdots & 0 & 0 & 1 & 0 & \cdots & 0 & 0 \\
\vdots & \vdots & & \vdots & \vdots & \vdots & \vdots & & \vdots & \vdots \\
0 & 0 & \cdots & 0 & 0 & 0 & 0 & \cdots & 1 & 0
\end{array}\right],$$

$$B = \left[\begin{array}{ccccc|ccccc}
\beta_0 & 0 & \cdots & 0 & 0 & 0 & 0 & \cdots & 0 & 0 \\
0 & 0 & \cdots & 0 & 0 & 0 & 0 & \cdots & 0 & 0 \\
\vdots & \vdots & & \vdots & \vdots & \vdots & \vdots & & \vdots & \vdots \\
0 & 0 & \cdots & 0 & 0 & 0 & 0 & \cdots & 0 & 0 \\
\hline
1 & 0 & \cdots & 0 & 0 & 0 & 0 & \cdots & 0 & 0 \\
0 & 0 & \cdots & 0 & 0 & 0 & 0 & \cdots & 0 & 0 \\
\vdots & \vdots & & \vdots & \vdots & \vdots & \vdots & & \vdots & \vdots \\
0 & 0 & \cdots & 0 & 0 & 0 & 0 & \cdots & 0 & 0
\end{array}\right],$$

where broken lines are used to show how these matrices partition into $k \times k$ blocks.

Since only $y_1^{(n)}$ occurs as an argument for f in the method represented in this way, we can, if we wish, apply a linear transformation such that of the new variables $z_1^{(n)}, z_2^{(n)}, \ldots, z_N^{(n)}$, one of these, say $z_1^{(n)}$, is just $y_1^{(n)}$ and the remainder are independent linear combinations of $y_1^{(n)}, y_2^{(n)}, \ldots, y_N^{(n)}$. Writing T as the transformation matrix expressing the dependence of the z system on the y system, this means that T has the first row $[10 \cdots 00]$ and T^{-1} has a similar form.

In the transformed representation of the method it is characterized by the pair $(T^{-1}AT, T^{-1}BT)$ and it is easy to see that $T^{-1}BT$ has only its first column non-zero.

Although this formal change of basis makes no difference to the numerical result, certain choices of T can result in benefits, particularly as regards ease of change of step size in sophisticated algorithms.

412 Predictor–corrector methods as examples.

Consider the predictor–corrector method made up from the pair

$$y_n = \bar{\alpha}_1 y_{n-1} + \bar{\alpha}_2 y_{n-2} + \cdots + \bar{\alpha}_k y_{n-k}$$
$$+ h[\bar{\beta}_1 f(y_{n-1}) + \bar{\beta}_2 f(y_{n-2}) + \cdots + \bar{\beta}_k f(y_{n-k})] \qquad (412a)$$

as predictor and

$$y_n = \alpha_1 y_{n-1} + \alpha_2 y_{n-2} + \cdots + \alpha_k y_{n-k}$$
$$+ h[\beta_0 f(y_n) + \beta_1 f(y_{n-1}) + \cdots + \beta_k f(y_{n-k})] \qquad (412b)$$

as corrector. As a PEC combination, the method can be put in (A, B) form with $N = 2k + 1$ and

$$y_1^{(n)} = y_n,$$
$$y_2^{(n)} = y_{n-1},$$
$$\vdots$$
$$y_k^{(n)} = y_{n-k+1},$$
$$y_{k+1}^{(n)} = \bar{y}_n,$$
$$y_{k+2}^{(n)} = \bar{y}_{n-1},$$
$$\vdots$$
$$y_{2k+1}^{(n)} = \bar{y}_{n-k},$$

$$
A = \left[
\begin{array}{cccc|cccc}
\alpha_1 & \alpha_2 & \cdots & \alpha_k & 0 & 0 & \cdots & 0 \\
1 & 0 & \cdots & 0 & 0 & 0 & \cdots & 0 \\
\vdots & \vdots & & \vdots & \vdots & \vdots & & \vdots \\
0 & 0 & \cdots & 0 & 0 & 0 & \cdots & 0 \\
\hline
\bar{\alpha}_1 & \bar{\alpha}_2 & \cdots & \bar{\alpha}_k & 0 & 0 & \cdots & 0 \\
0 & 0 & \cdots & 0 & 1 & 0 & \cdots & 0 \\
\vdots & \vdots & & \vdots & \vdots & \vdots & & \vdots \\
0 & 0 & \cdots & 0 & 0 & 0 & \cdots & 0
\end{array}
\right],
$$

$$
B = \left[
\begin{array}{cccc|cccc}
0 & 0 & \cdots & 0 & \beta_0 & \beta_1 & \cdots & \beta_k \\
0 & 0 & \cdots & 0 & 0 & 0 & \cdots & 0 \\
\vdots & \vdots & & \vdots & \vdots & \vdots & & \vdots \\
0 & 0 & \cdots & 0 & 0 & 0 & \cdots & 0 \\
\hline
0 & 0 & \cdots & 0 & 0 & \bar{\beta}_1 & \cdots & \bar{\beta}_k \\
0 & 0 & \cdots & 0 & 0 & 0 & \cdots & 0 \\
\vdots & \vdots & & \vdots & \vdots & \vdots & & \vdots \\
0 & 0 & \cdots & 0 & 0 & 0 & \cdots & 0
\end{array}
\right],
$$

where the broken lines partition the matrices into $(k) + (k+1)$ rows and columns.

For the PECE method, N can be chosen as $k + 2$ and in this case we identify $y_1^{(n)}, y_2^{(n)}, \ldots, y_k^{(n)}$ as for the PEC method and $y_{k+1}^{(n)} = y_{n-k}$, $y_{k+2}^{(n)} = \bar{y}_n$, the predicted value. In this case the matrices are

$$
A = \left[
\begin{array}{ccccccc}
\alpha_1 & \alpha_2 & \cdots & \alpha_k & 0 & 0 \\
1 & 0 & \cdots & 0 & 0 & 0 \\
\vdots & \vdots & & \vdots & \vdots & \vdots \\
0 & 0 & \cdots & 0 & 0 & 0 \\
0 & 0 & \cdots & 1 & 0 & 0 \\
\bar{\alpha}_1 & \bar{\alpha}_2 & \cdots & \bar{\alpha}_k & 0 & 0
\end{array}
\right],
\tag{412c}
$$

$$
B = \left[
\begin{array}{cccccc}
0 & \beta_1 & \cdots & \beta_{k-1} & \beta_k & \beta_0 \\
0 & 0 & \cdots & 0 & 0 & 0 \\
\vdots & \vdots & & \vdots & \vdots & \vdots \\
0 & 0 & \cdots & 0 & 0 & 0 \\
0 & 0 & \cdots & 0 & 0 & 0 \\
0 & \bar{\beta}_1 & \cdots & \bar{\beta}_{k-1} & \bar{\beta}_k & 0
\end{array}
\right].
$$

In a similar way, we can write down matrix pairs for arbitrarily complex $P(EC)^m$ and $P(EC)^m E$ methods. We give as our final example the $P(EC)^2 E$ method with $N = k + 3$ and $y_{k+2}^{(n)}$ the predicted value and $y_{k+3}^{(n)}$ the first corrected value at step n. The matrices are

$$
A = \begin{bmatrix}
\alpha_1 & \alpha_2 & \cdots & \alpha_k & 0 & 0 & 0 \\
1 & 0 & \cdots & 0 & 0 & 0 & 0 \\
\vdots & \vdots & & \vdots & \vdots & \vdots & \vdots \\
0 & 0 & \cdots & 0 & 0 & 0 & 0 \\
0 & 0 & \cdots & 1 & 0 & 0 & 0 \\
\bar{\alpha}_1 & \bar{\alpha}_2 & \cdots & \bar{\alpha}_k & 0 & 0 & 0 \\
\alpha_1 & \alpha_2 & \cdots & \alpha_k & 0 & 0 & 0
\end{bmatrix},
$$

$$(412\text{d})$$

$$
B = \begin{bmatrix}
0 & \beta_1 & \cdots & \beta_{k-1} & \beta_k & 0 & \beta_0 \\
0 & 0 & \cdots & 0 & 0 & 0 & 0 \\
\vdots & \vdots & & \vdots & \vdots & \vdots & \vdots \\
0 & 0 & \cdots & 0 & 0 & 0 & 0 \\
0 & 0 & \cdots & 0 & 0 & 0 & 0 \\
0 & \bar{\beta}_1 & \cdots & \bar{\beta}_{k-1} & \bar{\beta}_k & 0 & 0 \\
0 & \beta_1 & \cdots & \beta_{k-1} & \beta_k & \beta_0 & 0
\end{bmatrix}.
$$

413 Modified multistep methods as examples.

Also called generalized multistep or hybrid methods (see Butcher, 1965, Gear, 1965, and Gragg and Stetter, 1964), these can be thought of as extensions of predictor–corrector methods in which approximations are computed not only at step values but at intermediate values as well. Thus, some of the flavour of Runge–Kutta methods is introduced into their structure.

As general linear methods, these take on a form essentially the same as for $P(EC)^m E$ methods, but with some coefficients taking on different values. For example, replacing the last row of each of A and B in (412d) by the coefficients in the second predictor and introducing an extra coefficient (for the derivative at the off-step point) in the $(1, k + 2)$ position of B gives the form of the methods described in Butcher (1965).

We present here the single case with $k = 3$ and with a midway off-step point. The method is given in the form of a matrix pair but with the approximations renumbered so as to give the two predicted results first. The expression for the result at the end of a step is

$$
\begin{bmatrix} \hat{y}_{n-1/2} \\ \hat{y}_n \\ y_n \\ y_{n-1} \\ y_{n-2} \\ y_{n-3} \end{bmatrix}
=
\begin{bmatrix}
0 & 0 & \dfrac{-225}{128} & \dfrac{200}{128} & \dfrac{153}{128} & 0 \\[2mm]
0 & 0 & \dfrac{540}{31} & \dfrac{-297}{31} & \dfrac{-212}{31} & 0 \\[2mm]
0 & 0 & \dfrac{783}{617} & \dfrac{-135}{617} & \dfrac{-31}{617} & 0 \\[2mm]
0 & 0 & 1 & 0 & 0 & 0 \\[1mm]
0 & 0 & 0 & 1 & 0 & 0 \\[1mm]
0 & 0 & 0 & 0 & 1 & 0
\end{bmatrix}
\begin{bmatrix} \hat{y}_{n-3/2} \\ \hat{y}_{n-1} \\ y_{n-1} \\ y_{n-2} \\ y_{n-3} \\ y_{n-4} \end{bmatrix}
$$

$$
+\, h
\begin{bmatrix}
0 & 0 & 0 & \dfrac{225}{128} & \dfrac{300}{128} & \dfrac{45}{128} \\[2mm]
\dfrac{384}{155} & 0 & 0 & \dfrac{-1395}{155} & \dfrac{-2130}{155} & \dfrac{-309}{155} \\[2mm]
\dfrac{2304}{3085} & \dfrac{465}{3085} & 0 & \dfrac{-135}{3085} & \dfrac{-495}{3085} & \dfrac{-39}{3085} \\[2mm]
0 & 0 & 0 & 0 & 0 & 0 \\[1mm]
0 & 0 & 0 & 0 & 0 & 0 \\[1mm]
0 & 0 & 0 & 0 & 0 & 0
\end{bmatrix}
\begin{bmatrix} f(\hat{y}_{n-1/2}) \\ f(\hat{y}_n) \\ f(y_n) \\ f(y_{n-1}) \\ f(y_{n-2}) \\ f(y_{n-3}) \end{bmatrix}.
$$

$$(413a)$$

414 One-leg methods as examples.

Corresponding to every consistent and stable linear multistep method there is a related method characterized by the same k and the same coefficients $\alpha_1, \alpha_2, \ldots, \alpha_k, \beta_0, \beta_1, \beta_2, \ldots, \beta_k$. These methods were given the name 'one-leg' by Dahlquist (1976) to emphasize the dependence of the computed result at each step point on only a single derivative value.

If a linear multistep method is given by

$$
y_n = \alpha_1 y_{n-1} + \alpha_2 y_{n-2} + \cdots + \alpha_k y_{n-k}
$$
$$
+ h[\beta_0 f(y_n) + \beta_1 f(y_{n-1}) + \beta_2 f(y_{n-2}) + \cdots + \beta_k f(y_{n-k})], \qquad (414a)
$$

where it can be assumed that $\beta_0 + \beta_1 + \beta_2 + \cdots + \beta_k \neq 0$ because of the consistency and stability of the method, then the corresponding one-leg method is given by

$$y_n = \alpha_1 y_{n-1} + \alpha_2 y_{n-2} + \cdots + \alpha_k y_{n-k}$$

$$+ h(\beta_0 + \beta_1 + \cdots + \beta_k) f\left(\frac{1}{\beta_0 + \beta_1 + \cdots + \beta_k}(\beta_0 y_n + \beta_1 y_{n-1} + \cdots + \beta_k y_{n-k})\right).$$

$$(414b)$$

To write (414b) in (A, B) form, we choose $N = k + 1$ and identify

$$y_1^{(n)} = \frac{1}{\beta_0 + \beta_1 + \cdots + \beta_k}(\beta_0 y_n + \beta_1 y_{n-1} + \cdots + \beta_k y_{n-k})$$

$$= \frac{\beta_0 \alpha_1 + \beta_1}{\beta_0 + \beta_1 + \cdots + \beta_k} y_{n-1}$$

$$+ \frac{\beta_0 \alpha_2 + \beta_2}{\beta_0 + \beta_1 + \cdots + \beta_k} y_{n-2} + \cdots + \frac{\beta_0 \alpha_k + \beta_k}{\beta_0 + \beta_1 + \cdots + \beta_k} y_{n-k}$$

$$+ h\beta_0 f(y_1^{(n)})$$

$$y_2^{(n)} = y_n,$$

$$\vdots$$

$$y_{k+1}^{(n)} = y_{n-k+1}.$$

Thus we choose the matrices A, B as

$$A = \begin{bmatrix} 0 & \dfrac{\beta_0\alpha_1 + \beta_1}{\beta_0 + \beta_1 + \cdots + \beta_k} & \dfrac{\beta_0\alpha_2 + \beta_2}{\beta_0 + \beta_1 + \cdots + \beta_k} & \cdots & \dfrac{\beta_0\alpha_{k-1} + \beta_{k-1}}{\beta_0 + \beta_1 + \cdots + \beta_k} & \dfrac{\beta_0\alpha_k + \beta_k}{\beta_0 + \beta_1 + \cdots + \beta_k} \\ 0 & \alpha_1 & \alpha_2 & \cdots & \alpha_{k-1} & \alpha_k \\ 0 & 1 & 0 & \cdots & 0 & 0 \\ \vdots & \vdots & \vdots & & \vdots & \vdots \\ 0 & 0 & 0 & \cdots & 1 & 0 \end{bmatrix}$$

$$B = \begin{bmatrix} \beta_0 & 0 & 0 & \cdots & 0 & 0 \\ \beta_0 + \beta_1 + \cdots + \beta_k & 0 & 0 & \cdots & 0 & 0 \\ 0 & 0 & 0 & \cdots & 0 & 0 \\ \vdots & \vdots & \vdots & & \vdots & \vdots \\ 0 & 0 & 0 & \cdots & 0 & 0 \end{bmatrix}$$

We consider one special case for which $k = 1$, $\alpha_1 = 1$ and $\beta_0 = \beta_1 = \frac{1}{2}$. The linear multistep version of this method, given by (414a), is

$$y_n = y_{n-1} + \frac{h}{2}[f(y_n) + f(y_{n-1})],$$

represented by

$$A = \begin{bmatrix} 1 & \frac{1}{2} \\ 0 & 0 \end{bmatrix}, \qquad B = \begin{bmatrix} \frac{1}{2} & 0 \\ 1 & 0 \end{bmatrix},$$

whereas the one-leg method is

$$y_n = y_{n-1} + hf(\tfrac{1}{2}y_n + \tfrac{1}{2}y_{n-1}),$$

and is represented by

$$A = \begin{bmatrix} 0 & 1 \\ 0 & 1 \end{bmatrix}, \qquad B = \begin{bmatrix} \frac{1}{2} & 0 \\ 1 & 0 \end{bmatrix}.$$

415 Miscellaneous examples

We will consider three types of methods all related in some way to Runge–Kutta, or to linear multistep methods. The first type of example is the cyclic composite methods introduced by Donelson and Hansen (1971). In these methods, a collection of l linear k-step methods, say

$$y_n = \alpha_{11}y_{n-1} + \alpha_{12}y_{n-2} + \cdots + \alpha_{1k}y_{n-k}$$
$$+ h[\beta_{10}f(y_n) + \beta_{11}f(y_{n-1}) + \cdots + \beta_{1k}f(y_{n-k})], \qquad (415a)$$

$$y_n = \alpha_{21}y_{n-1} + \alpha_{22}y_{n-2} + \cdots + \alpha_{2k}y_{n-k}$$
$$+ h[\beta_{20}f(y_n) + \beta_{21}f(y_{n-1}) + \cdots + \beta_{2k}f(y_{n-k})], \qquad (415b)$$
$$\vdots \qquad\qquad \vdots$$

$$y_n = \alpha_{l1}y_{n-1} + \alpha_{l2}y_{n-2} + \cdots + \alpha_{lk}y_{n-k}$$
$$+ h[\beta_{l0}f(y_n) + \beta_{l1}f(y_{n-1}) + \cdots + \beta_{lk}f(y_{n-k})], \qquad (415c)$$

is used cyclically. That is, if $y_{m-1}, y_{m-2}, \ldots$ are already computed (in a previous cycle), then y_m is evaluated using (415a) with $n = m$, y_{m+1} is evaluated using (415b) with $n = m+1$ and so on until y_{m+l-1} is evaluated with $n = m+l-1$ in (415c).

To see how this scheme can be written as a general linear method, we first note that as soon as $y_m, y_{m+1}, \ldots$ are evaluated, the formula for these can be substituted into the right-hand sides of (415b), etc., so that the formula for y_{m+i-1} $(i = 1, 2, \ldots l)$ can be written as

$$y_{m+i-1} = \bar{\alpha}_{i1}y_{m-1} + \bar{\alpha}_{i2}y_{m-2} + \cdots + \bar{\alpha}_{ik}y_{m-k}$$
$$+ h[\bar{\beta}_{i0}f(y_{m+i-1}) + \bar{\beta}_{i1}f(y_{m+i-2}) + \cdots + \bar{\beta}_{i,k+i-1}f(y_{m-k})],$$

where $\bar{\alpha}_{ij}$, $\bar{\beta}_{ij}$ are defined recursively for $i = 1, 2, \ldots, l$ by

$$\bar{\alpha}_{1j} = \alpha_{1j}, \qquad\qquad\qquad j = 1, 2, \ldots, k,$$

$$\bar{\beta}_{1j} = \beta_{1j}, \qquad\qquad\qquad j = 0, 1, \ldots, k,$$

$$\bar{\alpha}_{ij} = \alpha_{i,i+j-1} + \sum_{\nu=1}^{\min(i-1,\,k)} \alpha_{i\nu}\bar{\alpha}_{i-\nu,j}, \qquad j = 1, 2, \ldots, k,$$

with the convention that $\alpha_{i,i+j-1} = 0$ if $i + j - 1 > k$, and

$$\bar{\beta}_{ij} = \beta_{ij} + \sum_{\nu=1}^{\min(i-1,\,j,\,k)} \alpha_{i\nu}\bar{\beta}_{i-\nu,\,j-\nu},$$

with the conventions that $\beta_{ij} = 0$ if $j > k$ and that an empty sum is replaced by zero.

Regarding the whole cycle of l steps as a single step of length lh, we see that the whole scheme is equivalent to the general linear method (A, B) with $N = k + l$ and

$$A = \begin{bmatrix}
\bar{\alpha}_{l1} & \bar{\alpha}_{l2} & \cdots & \bar{\alpha}_{lk} & 0 & \cdots & 0 \\
\bar{\alpha}_{l-1,1} & \bar{\alpha}_{l-1,2} & \cdots & \bar{\alpha}_{l-1,k} & 0 & \cdots & 0 \\
\vdots & \vdots & & \vdots & \vdots & & \vdots \\
\bar{\alpha}_{11} & \bar{\alpha}_{12} & \cdots & \bar{\alpha}_{1k} & 0 & \cdots & 0 \\
1 & 0 & \cdots & 0 & 0 & \cdots & 0 \\
0 & 1 & \cdots & 0 & 0 & \cdots & 0 \\
\vdots & \vdots & \ddots & \vdots & \vdots & & \vdots \\
0 & 0 & \cdots & 1 & \cdots & 0 & 0 & \cdots & 0
\end{bmatrix}, \quad (415d)$$

$$B = \frac{1}{l} \begin{bmatrix}
\bar{\beta}_{l0} & \bar{\beta}_{l1} & \cdots & \bar{\beta}_{l,\,k+l-1} \\
0 & \bar{\beta}_{l-1,0} & \cdots & \bar{\beta}_{l-1,\,k+l-2} \\
\vdots & \vdots & \ddots & \vdots \\
0 & 0 & \cdots & \bar{\beta}_{10} & \cdots & \bar{\beta}_{1k} \\
0 & 0 & \cdots & & & 0 \\
\vdots & \vdots & & & & \vdots \\
0 & 0 & \cdots & & & 0
\end{bmatrix}. \quad (415e)$$

The next example is for the pseudo Runge–Kutta methods of Byrne and Lambert (1966) and Byrne (1967). In this type of method, y_n is computed from y_{n-1} using the formula

$$y_n = y_{n-1} + h\sum_{i=1}^{s} b_i f(Y_i^{(n)}) + h\sum_{i=1}^{s} \bar{b}_i f(Y_i^{(n-1)}), \quad (415f)$$

where $Y_i^{(n)}$ (computed within the nth step) is given as for the standard explicit

Runge–Kutta methods by

$$Y_i^{(n)} = y_{n-1} + h \sum_{j < i} a_{ij} f(Y_j^{(n)}), \qquad i = 1, 2, 3, \ldots, s.$$

As an (A, B, C) method this can be written as

$$A = \begin{bmatrix} 0 & 0 & \cdots & 0 & 1 \\ 0 & 0 & \cdots & 0 & 1 \\ \vdots & \vdots & & \vdots & \vdots \\ 0 & 0 & \cdots & 0 & 1 \end{bmatrix},$$

$$B = \begin{bmatrix} 0 & 0 & \cdots & 0 & 0 \\ a_{21} & 0 & \cdots & 0 & 0 \\ \vdots & \vdots & & \vdots & \vdots \\ a_{s1} & a_{s2} & \cdots & 0 & 0 \\ b_1 & b_2 & \cdots & b_s & 0 \end{bmatrix},$$

$$C = \begin{bmatrix} 0 & 0 & \cdots & 0 & 0 \\ 0 & 0 & \cdots & 0 & 0 \\ \vdots & \vdots & & \vdots & \vdots \\ \bar{b}_1 & \bar{b}_2 & \cdots & \bar{b}_s & 0 \end{bmatrix},$$

and it can be rewritten in (A, B) form as described in subsection 402.

Finally, we look at a different type of modification to a Runge–Kutta method which we consider in one specific instance. The following four-stage Runge–Kutta method is easily verified to have order 4:

$$
\begin{array}{c|cccc}
0 \\
-1 & -1 \\
\frac{1}{2} & \frac{5}{8} & -\frac{1}{8} \\
1 & -\frac{3}{2} & \frac{1}{2} & 2 \\
\hline
& \frac{1}{6} & 0 & \frac{2}{3} & \frac{1}{6}
\end{array}
\tag{415g}
$$

and we note that its second stage gives a first-order approximation at the previous step value. To form the modification to this method that we wish to consider, we actually replace the result computed in this second stage by y_{n-2}. The method now becomes a general linear method (A, B) where $N = 5$ and

$$A = \begin{bmatrix} 0 & 0 & 0 & 0 & 1 \\ 1 & 0 & 0 & 0 & 0 \\ 0 & 0 & 0 & 0 & 1 \\ 0 & 0 & 0 & 0 & 1 \\ 0 & 0 & 0 & 0 & 1 \end{bmatrix}, \qquad B = \begin{bmatrix} 0 & 0 & 0 & 0 & 0 \\ 0 & 0 & 0 & 0 & 0 \\ \frac{5}{8} & -\frac{1}{8} & 0 & 0 & 0 \\ -\frac{3}{2} & \frac{1}{2} & 2 & 0 & 0 \\ \frac{1}{6} & 0 & \frac{2}{3} & \frac{1}{6} & 0 \end{bmatrix}. \tag{415h}$$

What is interesting about this method is that it has the order 4, the same as (415g), but can be implemented with only three derivative calculations per step.

42 STABILITY, CONSISTENCY AND CONVERGENCE

420 Stability.

Consider a general linear method (A, B, C). If this is used to solve the trivial differential equation in which f is the zero function, so that the solution is bounded (and, in fact, constant), it is natural to require that all components $y_1^{(n)}, y_2^{(n)}, \ldots, y_N^{(n)}$ of the computed solution Y_n are also bounded as $n \to \infty$. We focus our attention on methods which have this property no matter what initial value Y_0 is used. Thus we give the following definition.

DEFINITION 420A.

A general linear method (A, B, C) is *stable* if A is a stable matrix.

In this definition, we recall definition 105C for stable matrices and theorem 123C which implies that A is stable iff $Y_n = A^n Y_0$ is bounded as $n \to \infty$ for any Y_0.

Necessary and sufficient conditions for stability are given in theorem 105E and we restate part of this theorem for our immediate use.

THEOREM 420B.

A method (A, B, C) is stable iff the minimal polynomial of the matrix A satisfies the condition that it has no zero with magnitude greater than 1 and all zeros with magnitude equal to 1 are simple.

In this subsection, we will not consider further general properties of stability except to remark that the property is independent of any change in basis (because $T^{-1}AT$ has bounded powers iff A has bounded powers) and, furthermore, if we applied the argument given in subsection 402 to construct an (A, B) method from a given (A, B, C) method, then they are both stable or neither is. To see this, we note that the matrix $\bar{A}$ given by (402a) has powers

$$\bar{A}^n = \begin{bmatrix} A^n & 0 \\ A^{n-1} & 0 \end{bmatrix}.$$

We now consider the various example methods discussed in section 41. In the first place we have, for Runge–Kutta methods,

$$A = \begin{bmatrix} 0 & 0 & \cdots & 1 \\ 0 & 0 & \cdots & 1 \\ \vdots & \vdots & & \vdots \\ 0 & 0 & \cdots & 1 \end{bmatrix}, \tag{420a}$$

so that $A = A^2 = A^3 = \cdots$ and, hence, $\| A^n \|$ is bounded. Alternative arguments for the stability of Runge–Kutta methods are (a) the characteristic polynomial of A given by (420a) has a single zero at 1 and the remainder at 0, and (b) $\| A \|_\infty = 1$.

Now turn to linear multistep methods written in (A, B, C) form. For the method given by (411a) we have

$$A = \begin{bmatrix} \alpha_1 & \alpha_2 & \cdots & \alpha_{k-1} & \alpha_k \\ 1 & 0 & \cdots & 0 & 0 \\ 0 & 1 & \cdots & 0 & 0 \\ \vdots & \vdots & & \vdots & \vdots \\ 0 & 0 & \cdots & 1 & 0 \end{bmatrix}, \tag{420b}$$

which could just as easily have been written with rows and columns in the reverse order giving the kind of matrix occurring in theorem 122B.

The polynomial p given by

$$p(z) = z^k - \alpha_1 z^{k-1} - \cdots - \alpha_k \tag{420c}$$

is of central interest in the study of this type of method and, according to definition 232B, the location and multiplicity of its zeros (the so-called root condition) is of particular significance in the study of the stability of linear multistep methods. We formally state the result that establishes this significance.

THEOREM 420C.

A linear multistep is stable iff it satisfies the root condition.

Proof.

According to theorem 420B, it is only necessary to prove that the minimal polynomial of A given by (420b) is the polynomial p given by (420c). We first show that this is in fact the characteristic polynomial. If A_m is the $m \times m$

submatrix

$$A_m = \begin{bmatrix} \alpha_1 & \alpha_2 & \cdots & \alpha_{m-1} & \alpha_m \\ 1 & 0 & \cdots & 0 & 0 \\ 0 & 1 & \cdots & 0 & 0 \\ \vdots & \vdots & & \vdots & \vdots \\ 0 & 0 & \cdots & 1 & 0 \end{bmatrix},$$

where $m = 1, 2, \cdots, k$, and if p_m is its characteristic polynomial given by

$$p_m(z) = \det(zI - A_m),$$

then $p_m(z) = z p_{m-1}(z) - \alpha_m$, as we see by expanding about the last column, and this fact together with $p_1(z) = z - \alpha_1$ leads to a verification that p is the characteristic polynomial. That it is also the minimal polynomial follows by noting that the final row of powers of A from $A^0, A^1, \ldots$ to A^{k-1} each have a single non-zero element but each in a different position so that no minimal polynomial of degree lower than k could exist. ∎

Questions concerning predictor–corrector methods are typically more complicated than the corresponding questions for linear multistep methods. However, stability is a conspicuous exception to this general remark. In fact, we have the following theorem.

THEOREM 420D.

A predictor–corrector method is stable iff the corresponding corrector method is stable.

Proof.

For any predictor–corrector scheme, considering powers of the A matrix is equivalent to considering the corresponding difference scheme but with h replaced by zero. Thus, a method will be stable if predicted and corrected values computed in this way are necessarily bounded. However, with $h = 0$, corrected results satisfy the same difference equation as for the corresponding corrector method (again with $h = 0$). Also, the predicted value (with $h = 0$) is given by

$$\bar{y}_n = \bar{\alpha}_1 y_{n-1} + \bar{\alpha}_2 y_{n-2} + \cdots + \bar{\alpha}_k y_{n-k},$$

and this is bounded if the y sequence is bounded. ∎

By the same argument as this, it is easy to see that a modified multistep method is stable iff its final (corrector) formula is stable.

For example, the particular method given by (413a) will be stable if

$$
\begin{bmatrix}
\dfrac{783}{617} & \dfrac{-135}{617} & \dfrac{-31}{617} \\[2mm]
1 & 0 & 0 \\[1mm]
0 & 1 & 0
\end{bmatrix}
$$

has bounded powers. However, this matrix has the characteristic polynomial

$$617z^3 - 783z^2 + 135z + 31 = (z-1)(617z^2 - 166z - 31),$$

and this clearly satisfies the root condition.

Of the miscellaneous examples given in subsection 415, the stability of the cyclic composite scheme hinges on the boundedness of the powers of (415d) which can be written in partitioned form as

$$
\begin{bmatrix}
\bar{A} & 0 \\
\bar{\bar{A}} & 0
\end{bmatrix},
$$

where the form of the $l \times k$ submatrix $\bar{\bar{A}}$ depends on the relative values of l and k and is, for example, equal to the unit matrix if $k = l$ and where the $k \times k$ submatrix $\bar{A}$ is given by

$$\bar{A} = A_l A_{l-1} \cdots A_1, \tag{420d}$$

in which $A_1, A_2, \ldots, A_l$ are given by

$$
A_i =
\begin{bmatrix}
\alpha_{i1} & \alpha_{i2} & \cdots & \alpha_{i,\,k-1} & \alpha_{ik} \\
1 & 0 & \cdots & 0 & 0 \\
0 & 1 & \cdots & 0 & 0 \\
\vdots & \vdots & & \vdots & \vdots \\
0 & 0 & \cdots & 1 & 0
\end{bmatrix},
\qquad i = 1, 2, \cdots, l.
$$

Since powers of A are bounded iff the same is true for powers of $\bar{A}$, we have a simple criterion for stability of these methods. That is to say, a cyclic composite method is stable if the minimal polynomial of $\bar{A}$ given by (420d) has all its zeros in the closed unit disc and all zeros on the boundary are isolated.

It will turn out to be the case that the characteristic polynomial of $\bar{A}$, because of the consistency requirement to be considered in the next subsection, will always have a zero at 1. It is, therefore, natural to use as a working criterion for stability the sufficient (but not, of course, necessary) condition that the characteristic polynomial of $\bar{A}$ has a single zero at 1 and the remainder in the *open* unit disc.

The stability condition for the pseudo Runge–Kutta type of method is clearly satisfied simply because the A matrix is precisely as for a Runge–Kutta method. Although the special method given by (415h) has a different matrix, it too is stable because $\| A \|_\infty = 1$.

421 Consistency.

If we wish the values computed by a general linear method (A, B, C) to faithfully represent the exact solution of the trivial one-dimensional differential equation $y' = 0$, with exact solution $y(x) = 1$, both at the beginning and end of a step, it is necessary to require that

$$Ae = e, \qquad (421a)$$

where e is the vector in $\mathbb{R}^N$ with 1 in each component. If, furthermore, we wish each of $y_i^{(n)}$ $(i = 1, 2, \ldots, N)$ to represent the solution precisely at a point near x_n, say at $x_n + v_i h = x_0 + (n + v_i)h$, where y_0 satisfies the one-dimensional equation $y' = 1$ with exact solution $y(x) = x - x_0$, we must require that

$$Av + (B + C)e = v + e, \qquad (421b)$$

where $v = [v_1, v_2, \ldots, v_N]^T$. These two requirements make up the condition of consistency which we now formalize.

DEFINITION 421A.

A general linear method (A, B, C) is *preconsistent* if (421a) is satisfied and *consistent* if it is preconsistent and if, in addition, for some $v \in \mathbb{R}^N$, (421b) is also satisfied. The vector v is known as the *consistency* vector.

We note that if $(\bar{A}, \bar{B})$ is given by (402a) and (402b) and (A, B, C) is consistent with the consistency vector v, then $(\bar{A}, \bar{B})$ is consistent in the $2N$-dimensional setting with the consistency vector

$$\bar{v} = \begin{bmatrix} v \\ v - e \end{bmatrix}.$$

Because of this fact, we will usually avoid the slight complication incurred by considering methods of the form (A, B, C), secure in the knowledge that results concerning methods of the form (A, B) generalize appropriately, particularly as regards questions of consistency.

The more general interpretation of (A, B) methods discussed in subsection 403 does give a formally different view of consistency and we will consider this in detail in subsection 427. In the meantime, however, we will consider some basic properties of consistency and see how the concept applies in some of the example situations of section 41. As we have already indicated, we will use a representation of methods in which C is zero so that (421b) is replaced by

$$Av + Be = v + e. \qquad (421c)$$

Given only the condition (421a), it is interesting to ask when v exists such that (421c) also holds. If the characteristic polynomial of A has only a single zero at 1, then the more general equation

$$Av + Be = v + ce \qquad (421d)$$

where c is a scalar, always has a solution because Be can be written as the sum of a member of $\ker(A - I)$ and a member of $\text{im}(A - I)$. Furthermore, if A is stable, then the value of c is unique and completely characterized by A and B because if $\bar{v}$ and $\bar{c}$ give an alternative solution to (421d) then

$$(A - I)(v - \bar{v}) = (c - \bar{c})e$$

implying, if $c \neq \bar{c}$, that $(z - 1)^2$ is a divisor of the minimal polynomial.

On the other hand, if v is a consistency vector then $\bar{v}$ is also a consistency vector iff $\bar{v} - v \in \ker(A - I)$. Thus, for a consistent method, v certainly cannot be unique because $e \in \ker(A - I)$. In the examples that we consider, we will exploit this non-uniqueness by arbitrarily replacing any particular consistency vector $[v_1, v_2, \ldots, v_N]^T$ by $[v_1 + \theta, v_2 + \theta, \ldots, v_N + \theta]^T$, where the number θ is chosen to achieve simplicity. When we state that *the* consistency vector of a particular method is some given vector, we will always intend this to mean that the given vector is *a* consistency vector and that the set of all possible consistency vectors is generated from the given vector by adding the same number to each component.

In the first example, for a general Runge–Kutta method, we note that the consistency condition is the same as the condition that the order of the method, in the sense of chapter 3, is at least one.

THEOREM 421B.

The method (A, B) given by (410a) and (410b) is consistent iff $\sum_{j=1}^{s} b_j = 1$. If it is consistent, the consistency vector is given by

$$v = \begin{bmatrix} c_1 \\ c_2 \\ \vdots \\ c_s \\ 1 \end{bmatrix}, \tag{421e}$$

where

$$c_i = \sum_{j=1}^{s} a_{ij}, \qquad i = 1, 2, \ldots, s. \tag{421f}$$

Proof.

The condition (421a) is trivial in this case. We arbitrarily choose v to have the last component equal to 1 and denote the other components by $c_1, c_2, \ldots, c_s$. It remains to prove that $\Sigma b_j = 1$ and that (421f) holds if and only if (421c) holds. Substitute (421e) into (421c) and we see that this is indeed the case. ∎

We also have, for linear multistep methods, an exact correspondence between consistency as defined here and consistency (in the sense of having order at least 1) introduced in chapter 2.

THEOREM 421C.

The method (A, B) given by

$$
A = \begin{bmatrix} \alpha_1 & \alpha_2 & \cdots & \alpha_k & 0 \\ 1 & 0 & & 0 & 0 \\ \vdots & \vdots & & \vdots & \vdots \\ 0 & 0 & \cdots & 1 & 0 \end{bmatrix},
$$

$$
B = \begin{bmatrix} \beta_0 & \beta_1 & \cdots & \beta_{k-1} & \beta_k \\ 0 & 0 & \cdots & 0 & 0 \\ \vdots & \vdots & & \vdots & \vdots \\ 0 & 0 & \cdots & 0 & 0 \end{bmatrix},
$$

is consistent iff

$$
\sum_{i=1}^{k} \alpha_i = 1 \tag{421g}
$$

and

$$
\sum_{i=1}^{k} i\alpha_i = \sum_{i=0}^{k} \beta_i, \tag{421h}
$$

and, if it is consistent, the consistency vector is $[-1, -2, -3, \ldots, -k-1]$.

Proof.

The condition (421a) is equivalent to (421g).

We write $v = [-m_1, -m_2, \ldots, -m_{k+1}]^T$ with $m_1 = 1$, and substitute into (421c). We find

$$
-\sum_{i=1}^{k} m_i \alpha_i + \sum_{i=0}^{k} \beta_i = -m_1 + 1 \tag{421i}
$$

$$
-m_{i-1} = -m_i + 1, \qquad i = 2, \ldots, k+1, \tag{421j}
$$

so that the values for $m_2, m_3, \ldots, m_{k+1}$ follow and (421h) becomes equivalent to (421i). ∎

Rather than seek a general type of result for predictor–corrector methods, we give a particular but typical statement in the case of PECE methods. This result is stated without its trivial verification.

THEOREM 421D.

If (412a) and (412b) are each consistent linear multistep methods then (A, B) given by (412c) is consistent with the consistency vector $[-1, -2, \ldots, -k-1, -1]$.

Not only is this result typical of predictor–corrector methods, but it applies in an analogous way to modified multistep methods. In (413a), for example, the consistency vector is implicit in the notation for the subscripts on $\hat{y}_{n-3/2}, \hat{y}_{n-1}, y_{n-1} \ldots, y_{n-4}$. In fact, we can verify that the sum for each row of the first matrix occurring in (413a) is just 1 and that

$$
\begin{bmatrix}
0 & 0 & \frac{-225}{128} & \frac{200}{128} & \frac{153}{128} & 0 \\
0 & 0 & \frac{540}{31} & \frac{-297}{31} & \frac{-212}{31} & 0 \\
0 & 0 & \frac{783}{617} & \frac{-135}{617} & \frac{-31}{617} & 0 \\
0 & 0 & 1 & 0 & 0 & 0 \\
0 & 0 & 0 & 1 & 0 & 0 \\
0 & 0 & 0 & 0 & 1 & 0
\end{bmatrix}
\begin{bmatrix}
-\frac{3}{2} \\ -1 \\ -1 \\ -2 \\ -3 \\ -4
\end{bmatrix}
+
\begin{bmatrix}
0 & 0 & 0 & \frac{225}{128} & \frac{300}{128} & \frac{45}{128} \\
\frac{384}{155} & 0 & 0 & \frac{-1395}{155} & \frac{-2130}{155} & \frac{-309}{155} \\
\frac{2304}{3085} & \frac{465}{3085} & 0 & \frac{-135}{3085} & \frac{-495}{3085} & \frac{-39}{3085} \\
0 & 0 & 0 & 0 & 0 & 0 \\
0 & 0 & 0 & 0 & 0 & 0 \\
0 & 0 & 0 & 0 & 0 & 0
\end{bmatrix}
\begin{bmatrix}
1 \\ 1 \\ 1 \\ 1 \\ 1 \\ 1
\end{bmatrix}
=
\begin{bmatrix}
-\frac{3}{2} \\ -1 \\ -1 \\ -2 \\ -3 \\ -4
\end{bmatrix}
+
\begin{bmatrix}
1 \\ 1 \\ 1 \\ 1 \\ 1 \\ 1
\end{bmatrix}.
$$

The miscellaneous examples of subsection 415 do not warrant being dealt with in detail. For the method given by (415d) and (415e) the manipulations are tedious, although it can be shown that if each of the constituent methods (415a), (415b), $\ldots$, (415c) is a consistent linear multistep method then the overall cyclic composite scheme is consistent with consistency vector $[-1/l, -2/l, \ldots, -(l+k)/l]$. For the pseudo Runge–Kutta method given by (415f), it is trivial to show that consistency hinges on whether or not $\Sigma b_j + \Sigma \bar{b}_j = 1$. The method with matrices (A, B) given by (415h) is consistent with the vector $[0, -1, \frac{1}{2}, 1, 1]$.

422 Convergence.

The concept we discuss here expresses the ability of a method to represent the exact solution to certain problems in the limiting case as the number of steps tends to infinity. Let

$$y'(x) = f(y(x)) \tag{422a}$$

be a given differential equation on an interval I. In computing the solution using a particular general linear method, it is first necessary to determine the vector Y_0 made up from the N subvectors $y_1^{(0)}, y_2^{(0)}, \ldots, y_N^{(0)}$ which represents the approximation at step number zero. From Y_0, the method computes in turn $Y_1, Y_2, \ldots$ until sufficient steps have been performed to yield an approximation at a desired point. If h denotes the step size to be used in these computations, it is appropriate to calculate Y_0 from the given initial value to the problem in some way which depends on h. Accordingly, we will write

$Y_n(h)$ (made up from the subvectors $y_1^{(n)}(h), y_2^{(n)}(h), \ldots, y_N^{(n)}(h)$) for the solution computed at the end of n steps each of length h starting from $Y_0(h)$. If I contains x_0 and x, and the initial value $y(x_0)$ is given, then we consider schemes in which each of $y_1^{(0)}(h), \ldots, y_N^{(0)}(h)$ converges to $y(x_0)$ as $h \to 0$. The convergence concept is concerned with the requirement that if $Y_0(h)$ is computed in such a way as this then $y_1^{(n)}(h), y_2^{(n)}(h), \ldots, y_N^{(n)}(h)$ should each converge to $y(x)$ as n tends to infinity and h tends to zero, but in such a way that nh remains equal to the constant value of $x - x_0$. We formalize this as follows.

DEFINITION 422A.

A general linear method is *convergent for a problem* (422a) on I if for any x_0, $x \in I$ and for $Y_0(h)$ such that $\| y_i^{(0)}(h) - y(x_0) \| \to 0$ as $h \to 0$ for $i = 1, 2, \ldots, N$, it holds that $\| y_i^{(n)}((x - x_0)/n) - y(x) \| \to 0$ as $n \to \infty$ for $i = 1, 2, \ldots, N$.

DEFINITION 422B.

A general linear method is *convergent* if it is convergent on any interval for any problem satisfying a Lipschitz condition.

The key result of this section is the fact that convergent methods are just those which are consistent and stable.

THEOREM 422C.

A general linear method is convergent if and only if it is consistent and stable.

We devote the next four subsections to a proof of this result.

423 Necessity of stability.

Given a convergent general linear method we will prove it to be stable. If a convergent method is represented by the matrices (A, B), it is convergent for all Lipschitzian problems and in particular for $y'(x) = 0$ on an interval containing $[0, 1]$ with constant solution $y(x) = 0$. We will suppose that A is not stable and obtain a contradiction.

Using the terminology of subsection 422, we find that

$$Y_n(h) = A^n Y_0(h),$$

so that, taking $x_0 = 0$, $x = 1$, we see that convergence implies that if $\| Y_0(h) \| \to 0$ as $h \to 0$ then $\| A^n Y_0(1/n) \| \to 0$ as $n \to \infty$. Define $Y_0(h)$ to be the zero vector in $\mathbb{R}^N$ unless h is the reciprocal of a positive integer n, in which case its value is to be $(1/\beta_n)v_n$, where v_n is a vector of norm 1 such that $\| A^n v_n \| = \alpha_n = \| A^n \|$ and β_n is the greatest of $\alpha_1, \alpha_2, \ldots, \alpha_n$. The unboundedness of the sequence $(\alpha_1, \alpha_2, \ldots)$ implies that $1/\beta_n \to 0$. We now compute $\| A^n Y_0(1/n) \|$ and find it to be α_n/β_n which must have limit 0 if the method is convergent. However, by the unboundedness of $(\alpha_1, \alpha_2, \ldots)$ this ratio is equal to 1 infinitely often. This contradiction completes the proof of this part of theorem 422C.

424 Necessity of consistency.

Let (A, B) be a convergent general linear method so that, according to the part of the proof of theorem 422C that has just been given, A is stable. One consequence of this is that the image and kernel of $A - I$ are disjoint, since if there were a non-zero vector v in both $\text{im}(A - I)$ and $\ker(A - I)$ then we would have $Av = v$ and there would exist a vector u so that $v = (A - I)u$. It would then follow by induction on $n = 1, 2, \ldots$ that $A^n u = u + nv$, because $A^n u = A(A^{n-1}u) = A[u + (n-1)v] = Au + (n-1)v = u + v + (n-1)v = u + nv$, but this is impossible since the norm of $A^n u$ would be at least $n \| v \| - \| u \|$ which is not bounded.

The vector $Be \in \mathbb{R}^N$ can now be written as the sum of vectors $u \in \ker(A - I)$ and $w \in \text{im}(A - I)$ since these two linear subspaces of $\mathbb{R}^N$ have dimensions totalling N and, as we have just remarked, these are disjoint spaces. Write $w = -(A - I)v$ and we have

$$Av + Be = u + v. \tag{424a}$$

Just as in the proof that stability was necessary, we consider a particular problem; in this case

$$y'(x) = 1, \tag{424b}$$

with solution $y(x) = x$ on the interval $[0, 1]$. We define $Y_0(h)$ to be $0 \in \mathbb{R}^N$ for any h so that $y_i^{(0)}(h)$ certainly converges to $y(0) = y(x_0) = 0$ as $h \to 0$. If we compute the solution after n steps, we find that

$$Y_n(h) = A Y_{n-1}(h) + hBe,$$

so that

$$Y_n(h) = A^n Y_0(h) + h(I + A + \cdots + A^{n-1})Be$$

If we now choose $x = 1$, $h = 1/n$ (as appropriate for an application of definition 422A) and $Y_0(h) = 0$, we have

$$Y_n\left(\frac{1}{n}\right) = \frac{1}{n}(I + A + \cdots + A^{n-1})Be.$$

Substitute Be from (424a) and make use of the fact that $Au = u$ and we find

$$Y_n\left(\frac{1}{n}\right) = \frac{1}{n}(I - A^n)v + u.$$

Since each component of $Y_n(1/n)$ converges to 1 as $n \to \infty$ to satisfy the requirements of convergence for problem (424b) for which $y(1) = 1$, it follows that $\| Y_n(1/n) - e \| \to 0$ so that

$$\| \frac{1}{n}(I - A^n)v + u - e \| \to 0.$$

However, $\| (1/n)(I - A^n)v \| \to 0$ by stability and accordingly $u = e$ implying, because $Au = u$, that

$$Ae = e$$

and, from (424a), that

$$Av + Be = v + e.$$

425 A lemma on stable sequences.

We prove a basic lemma in preparation for a proof that stability and consistency are not only necessary but also sufficient for a convergence of a general linear method.

LEMMA 425A.

Let A be a stable $N \times N$ matrix with α a bound on $\| A^n \|$ for $n = 0, 1, 2, \dots$. Let sequences $(u_0, u_1, \dots)$, $(w_1, w_2, \dots)$ in $\mathbb{R}^N$ be such that

$$u_n = Au_{n-1} + w_n \tag{425a}$$

and

$$\| w_n \| \leq \beta \| u_n \| + \gamma \| u_{n-1} \| + \delta \tag{425b}$$

for $n = 0, 1, 2, \dots$, where β, γ, δ are non-negative numbers and $\alpha\beta < 1$. Then for $n = 0, 1, 2, \dots, u_n$ is bounded by

$$\| u_n \| \leq \alpha \left(\frac{1 + \alpha\gamma}{1 - \alpha\beta} \right)^n \| u_0 \| + \frac{\delta}{\beta + \gamma} \left[\left(\frac{1 + \alpha\gamma}{1 - \alpha\beta} \right)^n - 1 \right] \tag{425c}$$

if $\beta + \gamma > 0$ and by

$$\| u_n \| \leq \alpha \| u_0 \| + n\alpha\delta \tag{425d}$$

if $\beta = \gamma = 0$.

Proof.

From (425a) applied n times, we find

$$u_n = A^n u_0 + A^{n-1} w_1 + A^{n-2} w_2 + \cdots + w_n, \tag{425e}$$

so that bounding the powers of A and using the inequality (425b) to bound

$\| w_1 \|, \| w_2 \|, \ldots, \| w_n \|$, we find that

$$\| u_n \| \lesssim (1 + \gamma)\alpha \| u_0 \| + \alpha(\beta + \gamma)(\| u_1 \| + \cdots + \| u_{n-1} \|) + \alpha\beta \| u_n \| + n\alpha\delta,$$

(425f)

and (425d) follows immediately if $\beta = \gamma = 0$. Assuming now that $\beta + \gamma > 0$, we consider the case $n = 1$ in (425f). It is found that

$$\| u_1 \| \leq \frac{(1 + \gamma)\alpha}{1 - \alpha\beta} \| u_0 \| + \frac{\alpha\delta}{1 - \alpha\beta}$$

and the coefficient of $\| u_0 \|$ is bounded by $\alpha(1 + \alpha\gamma)/(1 - \alpha\beta)$ because $\alpha \geq 1$. The last term can be rearranged as $[\delta/(\beta + \gamma)] [(1 + \alpha\gamma)/ (1 - \alpha\beta) - 1]$ to give the bound (425c) in the case $n = 1$. For $n > 1$, we prove (425c) by induction. Substituting the bound on $\| u_1 \|, \| u_2 \|, \ldots, \| u_{n-1} \|$ from (425c) into (425f), we find

$$\| u_n \| \leq \alpha P \| u_0 \| + \delta Q,$$

(425g)

where

$$P = \frac{1}{1 - \alpha\beta} \left\{ (1 + \gamma) + \alpha(\beta + \gamma)\left[\left(\frac{1 + \alpha\gamma}{1 - \alpha b}\right) + \left(\frac{1 + \alpha\gamma}{1 - \alpha\beta}\right)^2 + \cdots + \left(\frac{1 + \alpha\gamma}{1 - \alpha\beta}\right)^{n-1} \right] \right\}.$$

(425h)

On summation of the geometric series in (425h) we find that

$$P = \left(\frac{1 + \alpha\gamma}{1 - \alpha\beta}\right)^n - \frac{\gamma(\alpha - 1)}{1 - \alpha\beta} \leq \left(\frac{1 + \alpha\gamma}{1 - \alpha\beta}\right)^n.$$

(425i)

Similarly, an evaluation of Q gives

$$Q \leq \frac{1}{\beta + \gamma} \left[\left(\frac{1 + \alpha\gamma}{1 - \alpha\beta}\right)^n - 1 \right].$$

(425j)

Substitution of (425i) and (425j) into (425g) gives the bound (425c). ∎

426 Sufficiency of stability with consistency.

In this subsection, we carry out two tasks. The first is the definition and analysis of the local truncation error. We will find, for a consistent method applied to a problem satisfying a Lipschitz condition, that this can be estimated, just as for the Euler method, by the second power of the step size. Thus the order of a consistent method is at least 1. We then use lemma 425A to obtain a bound on the global error and we show that it tends to zero uniformly as the (constant) step size tends to zero.

In working with an (A, B, C) method, there turns out to be no great simplification in assuming that $C = 0$ so we do not make this assumption. We will, however, always assume that the method is consistent with the consistency vector v. Thus,

$$Ae = e,$$

$$Av + (B + C)e = v + e.$$

For a problem $y'(x) = f(y(x))$, where $f: X \to X$, we shall always assume a Lipschitz condition with constant L and write m as the maximum of $\| f(y(x)) \|$ for x in a closed interval I sufficient to contain the initial point x_0, the point where the solution is to be computed, $\bar{x}$, and all numbers of the form $x_i + hv_j$, where $x_i = x_0 + ih$ is the distance integrated by i steps of size h and $j = 1, 2, \ldots, N$. Thus I contains the interval $[\min(x_0, x_0 + h_0 \min_j v_j), \max(\bar{x}, \bar{x} + h_0 \max_j v_j)]$, where it is assumed by convention that $\bar{x} > x_0$ and h_0 is the greatest of a sequence of step sizes which tends to zero through values of the form $(\bar{x} - x_0)/n$ with n a positive integer.

Let Z_n denote the vector made up from subvectors $z_i^{(n)} = y(x_n + hv_i)$, for $i = 1, 2, \ldots, N$. We recall that in the Euler method, the local truncation error was defined as the inequality between the sides of the equation defining the value at the end of a step when all numerical values are replaced by exact solution values. Similarly, we define the local truncation error for a general linear method as follows.

DEFINITION 426A.

The *local truncation error relative to the exact solution* of (A, B, C) at step n with step size h is

$$E_n = Z_n - (A \otimes I) Z_{n-1} - h(B \otimes I) F(Z_n) - h(C \otimes I) F(Z_{n-1}).$$

With the subvectors of E_n written as $e_1^{(n)}, e_2^{(n)}, \ldots, e_N^{(n)}$, we then have

$$e_i^{(n)} = y(x_n + hv_i) - \sum_{j=1}^{N} a_{ij} y(x_{n-1} + hv_j) - h \sum_{j=1}^{N} b_{ij} y'(x_n + hv_j)$$

$$- h \sum_{j=1}^{N} c_{ij} y'(x_{n-1} + hv_j). \tag{426a}$$

With the assumptions we have made about the method and the problem, we have the following result.

THEOREM 426B.

The local truncation error relative to the exact solution is bounded in such a way that

$$\| e_i^{(n)} \| \leq h^2 L m l_i, \qquad i = 1, 2, \ldots, N,$$

where

$$l_i = \sum_{j=1}^{N} [|a_{ij}(1 + v_i - v_j)| \max(|v_i|, |1 - v_j|) + |b_{ij} v_j| + |c_{ij}(1 - v_j)|].$$

Proof.

If $x_n + h\xi$ and $x_n + h\eta$ are in I, then

$$\| y(x_n + h\xi) - y(x_n + h\eta) \| \le h | \xi - \eta | m, \tag{426b}$$

$$\| y'(x_n + h\xi) - y'(x_n + h\eta) \| \le hL | \xi - \eta | m. \tag{426c}$$

To verify (426b), let $p \in X^*$ so that

$$p[y(x_n + h\xi) - y(x_n + h\eta)] = py'(x_n + h\theta)h(\xi - \eta),$$

by the mean value theorem, where θ lies between ξ and η and may vary with p. We now have

$$\| y(x_n + h\xi) - y(x_n + h\eta) \| = \sup_{\substack{p \in X^* \\ \|p\| \ne 0}} \frac{| p[y(x_n + h\xi) - y(x_n + h\eta)] |}{\| p \|}$$

$$\le h | \xi - \eta | \| y'(x_n + h\theta) \| \le h | \xi - \eta | m.$$

To verify (426c), we use the differential equation and note that

$$\| f(y(x_n + h\xi)) - f(y(x_n + h\eta)) \| \le L \| y(x_n + h\xi) - y(x_n + h\eta) \|.$$

By the consistency condition, it follows easily that for $i = 1, 2, \ldots, N$,

$$\sum_{j=1}^{N} [a_{ij}(1 + v_i - v_j) - b_{ij} - c_{ij}] = 0.$$

Hence, subtracting this quantity multiplied by $hy'(x_n)$ from (426a), we find that $e_i^{(n)}$ can be written in the form

$$e_i^{(n)} = \sum_{j=1}^{N} a_{ij}[y(x_n + hv_i) - y(x_{n-1} + hv_j) - h(1 + v_i - v_j)y'(x_n + \theta h)]$$

$$+ h \sum_{j=1}^{N} a_{ij}(1 + v_i - v_j)[y'(x_n + \theta h) - y'(x_n)]$$

$$+ h \sum_{j=1}^{N} b_{ij}[y'(x_n) - y'(x_n + hv_j)]$$

$$+ h \sum_{j=1}^{N} c_{ij}[y'(x_n) - y'(x_{n-1} + hv_j)]. \tag{426d}$$

Estimate, for $p \in X^*$, the value of $p(e_i^{(n)})$ and for each j and p choose θ so that the coefficient of a_{ij} in the first summation vanishes. This is clearly possible by the mean value theorem. Since θ lies between $v_j - 1$ and v_i, (426c) implies that

$$| p[y'(x_n + \theta h) - y'(x_n)] | \le \| p \| hLm \max(| v_i |, | 1 - v_j |).$$

Similarly, bounding the other terms in $p(e_i^{(n)})$ and combining them gives the estimate

$$| p(e_i^{(n)}) | \le \| p \| h^2 L m l_i$$

and the result follows from $\| e_i^{(n)} \| = \sup_{\| p \| \ne 0} | p(e_i^{(n)}) | / \| p \|.$ ∎

Having obtained a bound on the local truncation error relative to the exact solution which tends to zero as the second power of h, we are in a position to estimate the combined effect of errors at every step as the solution develops.

With the method (A, B, C) applied to a standard problem on the interval $[x_0, \bar{x}]$, we write h for the constant step size and $u_0, u_1, u_2, \ldots$ for the errors at integration points $x_0, x_0 + h, x_0 + 2h, \ldots$. Thus, for $n = 0, 1, 2, \ldots$, u_n is a vector in X^N made up from subvectors $y(x_n + h v_i) - y_i^{(n)} (i = 1, 2, \ldots, N)$. Thus, $u_n = Z_n - Y_n$. We use the norm defined by (401d) on X^N so that $\| (A \otimes I)^n \|$ is bounded by $\sup \| A^n \|_\infty$ for a stable method.

Noting that u_0 represents the error in the starting procedure for the method, we can estimate the global errors $u_1, u_2, \ldots$ in subsequent steps.

THEOREM 426C.

The global truncation error for a stable consistent method with step size $h \le h_0$ satisfying $L \| B \|_\infty h_0 < 1$ is bounded by

$$\| u_n \| \le \alpha P_n \| u_0 \| + h Q (P_n - 1), \tag{426e}$$

where $\alpha = \sup \| A^n \|_\infty$,

$$P_n = \exp(\alpha L (\| B \|_\infty + \| C \|_\infty)(x_n - x_0)/(1 - \alpha L \| B \|_\infty h_0)),$$

$Q = m \max l_i/(\| B \|_\infty + \| C \|_\infty)$ and l_i is given by theorem 426B.

Proof.

Note first of all that, as a consequence of the stability and consistency conditions, $\| B \| + \| C \|$ cannot vanish. We use lemma 425A with A replaced by $A \otimes I$ and

$$w_n = h(B \otimes I) [F(Z_n) - F(Y_n)] + h(C \otimes I) [F(Z_{n-1}) - F(Y_{n-1})] + E_n$$

so that (425b) holds with $\beta = h \| B \| L$, $\gamma = h \| C \| L$ and, because of theorem 426B, $\delta = h^2 L m \max l_i$.

The result now follows from lemma 425A where we have used the inequalities $(1 + x)^n \le \exp(nx)$, $(1 - x)^{-n} \le \exp(nx/(1 - x))$ to approximate the powers of n occurring in (425c). ∎

It is now a trivial matter to complete the work of this subsection.

THEOREM 426D.

A stable consistent method is convergent.

Proof.

Let $\bar{y}_0$ denote the point in X^N made up from N copies of the subvector $y(x_0)$. We then have $u_0 = (Z_0 - \bar{y}_0) + (\bar{y}_0 - Y_0)$ and, for a sequence of step sizes converging to zero, $\| Z_0 - \bar{y}_0 \| \to 0$ by continuity of the solution to a differential equation. Thus, the assumption $\| \bar{y}_0 - Y_0 \| \to 0$ occuring in the definition of convergence is equivalent to $\| u_0 \| \to 0$. In this case, with $h \to 0$ and P_n bounded by $\exp(\alpha L (\| B \| + \| C \|)(\bar{x} - x_0)/(1 - \alpha L \| B \| h_0))$, it is clear that the right-hand side of (426e) tends to zero. ∎

427 Generalizations.

The generalizations we consider are based on the use of a partitioning of the stages into those which are used temporarily to allow the step to be evaluated and those which are passed on as input information for later steps. Consider a method (A, B) in which A and B can be partitioned as

$$A = \begin{bmatrix} A_{11} & 0 \\ A_{21} & 0 \end{bmatrix}, \qquad B = \begin{bmatrix} 0 & B_{12} \\ 0 & B_{22} \end{bmatrix},$$

where the partitioning of the $N \times N$ matrices is in accordance with $N = r + s$.

The quantities $y_1^{(n)}, y_2^{(n)}, \ldots, y_r^{(n)}$ are used to hold information passed on from step to step whereas the remaining stages computed within the step are not needed again once the step is completed. These remaining s stages will now be denoted by $Y_1 (= y_{r+1}^{(n)}), Y_2 (= y_{r+2}^{(n)}), \ldots, Y_s (= y_{r+s}^{(n)})$. If Y is the vector made up from subvectors $Y_1, Y_2, \ldots, Y_s$ and $y^{(n)}$ the vector made up from subvectors $y_1^{(n)}, y_2^{(n)}, \ldots, y_r^{(n)}$ then we see that $Y^{(n)}$ is partitioned as $(y^{(n)}, Y)$. The lack of any step number designation to Y reflects its temporary nature in an actual computation.

We rename the four non-zero matrices $A_{11} = C_{22}, A_{21} = C_{12}$, $B_{12} = C_{21}, B_{22} = C_{11}$ and we now characterize the method by a single partitioned $(s + r) \times (s + r)$ matrix C thus:

$$C = \begin{bmatrix} C_{11} & C_{12} \\ C_{21} & C_{22} \end{bmatrix}.$$

Using this matrix, we can write the definition of the method by the formulae

$$Y = (C_{11} \otimes I) \, hF(Y) + (C_{12} \otimes I) \, y^{(n-1)}, \qquad (427a)$$

$$y^{(n)} = (C_{21} \otimes I) \, hF(Y) + (C_{22} \otimes I) \, y^{(n-1)}. \qquad (427b)$$

It is easy to see that this generalization is purely formal in the sense that we can always write an (A, B) method as a partitioned method with twice the number of stages and with

$$C = \begin{bmatrix} B & A \\ B & A \end{bmatrix}.$$

However, now that we have distinguished between those stages for which f is to be applied and those used purely for passing information between steps, we see that the latter quantities can be replaced by any non-singular set of linear combinations of them. While this new method is equivalent in an obvious sense, it cannot necessarily be put back into the form of a consistent (A, B) method.

Our task now is to see how stability and consistency conditions should be generalized for the more general type of partitioned method produced by such a transformation as this.

Let T^{-1} be the matrix of coefficients representing the new set of information vectors $z_1^{(n)}, z_2^{(n)}, \ldots, z_r^{(n)}$ written in terms of $y_1^{(n)}, y_2^{(n)}, \ldots, y_r^{(n)}$. If T has elements t_{ij}, this means that

$$y_i^{(n)} = \sum_{j=1}^{r} t_{ij} z_j^{(n)}, \qquad i = 1, 2, \ldots, r.$$

Writing this in the form $y^{(n)} = (T \otimes I) z^{(n)}$ and substituting into (427a) and (427b), we find that

$$Y = (C_{11} \otimes I) hF(Y) + [(C_{12}T) \otimes I] z^{(n-1)}, \tag{427c}$$

$$z^{(n)} = [(T^{-1}C_{21}) \otimes I] hF(Y) + [(T^{-1}C_{22}T) \otimes I] z^{(n-1)}. \tag{427d}$$

so that the matrix representing this method is $\bar{C}$, where

$$\bar{C} = \begin{bmatrix} \bar{C}_{11} & \bar{C}_{12} \\ \bar{C}_{21} & \bar{C}_{22} \end{bmatrix} = \begin{bmatrix} C_{11} & C_{12}T \\ T^{-1}C_{21} & T^{-1}C_{22}T \end{bmatrix}.$$

It is easy to see that stability of (A, B) (that is boundedness of $\| A^n \|$) is equivalent to the boundedness of $\| C_{22}^n \|$ which is unaffected by the transformation ($\| \bar{C}_{22}^n \|$ is bounded iff $\| C_{22}^n \|$ is).

The following definition is appropriate.

DEFINITION 427A.

A partitioned general linear method

$$C = \begin{bmatrix} C_{11} & C_{12} \\ C_{21} & C_{22} \end{bmatrix}$$

is *stable* if $\sup\{\| C_{22}^n \| : n = 1, 2, \ldots \} < \infty$.

We now consider the generalized formulation of consistency. For the original method (A, B), suppose the consistency vector to be partitioned as $[\bar{w}, w]$. Also, write $[\bar{e}, e]$ for the vector of ones in the partitioned vector space

$\mathbb{R}^{r+s}$. The consistency conditions are

$$\begin{bmatrix} A_{11} & 0 \\ A_{21} & 0 \end{bmatrix} \begin{bmatrix} \bar{e} \\ e \end{bmatrix} = \begin{bmatrix} \bar{e} \\ e \end{bmatrix},$$

and

$$\begin{bmatrix} A_{11} & 0 \\ A_{21} & 0 \end{bmatrix} \begin{bmatrix} \bar{w} \\ w \end{bmatrix} + \begin{bmatrix} 0 & B_{12} \\ 0 & B_{22} \end{bmatrix} \begin{bmatrix} \bar{e} \\ e \end{bmatrix} = \begin{bmatrix} \bar{w} \\ w \end{bmatrix} + \begin{bmatrix} \bar{e} \\ e \end{bmatrix},$$

which we write as

$$A_{11}\bar{e} = \bar{e}, \tag{427e}$$

$$A_{21}\bar{e} = e, \tag{427f}$$

$$A_{11}\bar{w} + B_{12}e = \bar{w} + \bar{e}, \tag{427g}$$

$$A_{21}\bar{w} + B_{22}e = w + e. \tag{427h}$$

Clearly (427h), the only equation involving w, imposes no constraint on $A_{11}, A_{21}, B_{12}, B_{22}$ so we withdraw it from any further consideration.

We now write $A_{11} = C_{22}$, $A_{21} = C_{12}$, $B_{12} = C_{21}$, transform to the partitioned matrix $\bar{C}$ and write $u = T^{-1}\bar{e}$, $v = T^{-1}\bar{w}$. Equations (427e), (427f) and (427g) now become

$$\bar{C}_{22}u = u, \tag{427i}$$

$$\bar{C}_{12}u = e, \tag{427j}$$

$$\bar{C}_{22}v + \bar{C}_{21}e = v + u. \tag{427k}$$

We shall call u (which generalizes one of the roles played by e in the usual consistency definition) 'the preconsistency vector' and v will, as usual, be called 'the consistency vector'. Their interpretation is that, if

$$z_i^{(n-1)} = u_i y(x_{n-1}) + v_i hy'(x_{n-1}) + O(h^2), \qquad i = 1, 2, \ldots, r, \tag{427l}$$

then after a single step of the method $\bar{C}$,

$$z_i^{(n)} = u_i y(x_n) + v_i hy'(x_n) + O(h^2), \qquad i = 1, 2, \ldots, r. \tag{427m}$$

Formally, we state the following definition for any partitioned method.

DEFINITION 427B.

A partitioned general linear method

$$C = \begin{bmatrix} C_{11} & C_{12} \\ C_{21} & C_{22} \end{bmatrix}$$

is *preconsistent* with the preconsistency vector u if

$$C_{22}u = u,$$
$$C_{12}u = e,$$

and *consistent* if, in addition, for some vector v,

$$C_{22}v + C_{21}e = u + v.$$

43 ORDER OF ACCURACY

430 Natural definitions of order.

Experience in the analysis of traditional types of methods for solving ordinary differential equations leads to more than one way of formulating questions about numerical accuracy.

In the case of Runge–Kutta methods, the approach we have used is equivalent to supposing that the solution at the end of step number $n - 1$ is exact and to estimating the error in the computed value at the end of step number n. A method is said to have order of accuracy p if the local error estimate is $O(h^{p+1})$, where h is the step size.

An alternative approach, which should be regarded as standard for theoretical studies of linear multistep methods, is to replace the various computed quantities that occur in the formulation of the method by the exact quantities they are intended to approximate and to estimate the size of the expression which would be zero if exact agreement were achieved.

Thus for the method

$$y_n - \alpha_1 y_{n-1} - \alpha_2 y_{n-2} - \cdots - \alpha_k y_{n-k} - h[\beta_0 f(y_n) + \beta_1 f(y_{n-1}) + \cdots$$
$$+ \beta_k f(y_{n-k})] = 0$$

the local truncation method would be defined as

$$y(x_n) - \alpha_1 y(x_{n-1}) - \alpha_2 y(x_{n-2}) - \cdots - \alpha_k y(x_{n-k}) - h[\beta_0 y'(x_n)$$
$$+ \beta_1 y'(x_{n-1}) + \cdots + \beta_k y'(x_{n-k})],$$

and if this is $O(h^{p+1})$, the method would have order p.

In attempting to obtain useful generalizations of these order definitions, so as to be applicable to general linear methods, we encounter a number of difficulties.

The first of these is that if we attempt to substitute exact values in place of all or some of the numerically computed quantities occurring in the formulation of a general linear method, it is not completely obvious what values to substitute. If we were to use exact values of the solution evaluated at various points on the solution trajectory, then our definition of order would not be invariant under the type of transformation introduced in subsection 427.

If we consider only the (A, B) type of formulation and restrict ourselves to methods in which B has full rank, so that the question of transformations does not arise, then this difficulty might seem to be overcome. It would then seem to be natural to substitute for $y_i^{(n)}$ the value of $y(x_n + hv_i)$, where v_i is a component of the consistency vector. In this case, the local truncation error could be measured component by component and estimated in terms of powers of h with the lowest power used to determine the order of the method.

There are serious difficulties with this approach as well, since it ignores the part played by different components of the solution vector in possible later steps. In the case of Runge–Kutta methods, for example, even though all the internal stages may be essential ingredients towards finding the final approximation, they will, as approximations to $y(x_{n-1} + hc)$ for various values of c, be generally of rather low order. In terms of accuracy, this apparent loss of order is quite irrelevant since the internal stages are not used in later steps. However, if our definition of order were expressed in terms of a uniform bound on the errors of all the quantities computed in a step, it is these low-order approximations that would determine the value ascribed to the order of a method.

This difficulty cannot be overcome by simply choosing the highest of the orders of various stages, since this would ignore the possible corruption of such accurate approximations through their dependence on less accurately computed quantities in earlier steps.

Although these difficulties lead us to use a fresh approach to the order concept, it should be pointed out that the more natural approaches which we feel we must discard are capable of a rather refined analysis that makes them completely rigorous and acceptable. In this point of view, as developed in the papers of Cooper (1978, 1981) each approximation in a general linear method is given an independent order of accuracy and the overall accuracy of the method is determined by considering the possible effect of low-order approximations in one step on the accuracy of higher-order approximations in later steps.

431 Order relative to a starting procedure.

We will consider general linear methods expressed in partitioned form. For a method C given by

$$C = \begin{bmatrix} C_{11} & C_{12} \\ C_{21} & C_{22} \end{bmatrix},$$

where the partitioning is according to $(s + r) \times (s + r)$, the vectors $y_i^{(n-1)} (i = 1, 2, \ldots, r)$ are supposed to approximate various quantities $z_i^{(n-1)} (i = 1, 2, \ldots, r)$ related to the exact solution. Our approach to local truncation error and to order of accuracy is concerned with how well $y_i^{(n)}$ agrees with $z_i^{(n)} (i = 1, 2, \ldots, r)$, given that $y_i^{(n-1)}$ is precisely equal to $z_i^{(n-1)}$ $(i = 1, 2, \ldots, r)$.

To relate this formal type of definition to actual computational practice, we will interpret $z_i^{(n-1)}$ as being computed from the exact solution $y(x_{n-1})$ at x_{n-1} with $z_i^{(n)}$ computed in just the same way from $y(x_n)$. The computation of $z_i^{(n-1)}$ we will think of as being performed by something like a Runge–Kutta method, the only difference being that of the generalization already discussed in subsection 380. Combining the schemes for each of $z_1^{(n-1)}, z_2^{(n-1)}, \ldots z_r^{(n-1)}$ into one overall starting process leads us to a method of the same general type as a partitioned general linear method except that the matrices corresponding to C_{12} and C_{22} have only a single column. In fact C_{12} must correspond to the vector e and C_{22} to the preconsistency vector, u say, for the method C itself.

Let S denote this starting method:

$$S = \begin{bmatrix} S_{11} & e \\ S_{21} & u \end{bmatrix}.$$

We now consider the effect of applying the method C to the result computed by S. This gives a method in which r results are computed from a single input result using a total of $\bar{s} + s$ stages where $\bar{s}$ is the number of stages in S. The combined method is

$$SC = \begin{bmatrix} S_{11} & 0 & | & e \\ C_{12}S_{21} & C_{11} & | & e \\ \hline C_{22}S_{21} & C_{21} & | & u \end{bmatrix}.$$

In our definition, we compose the r results computed by this method with the r results computed using just S but applied to a point a distance h further along the trajectory. To illustrate the idea better, we present figure 431. In this diagram we have shown the spaces X, X^r in which various vectors lie, the operations S and C and their product SC, the operation E which advances the solution a distance h and ES the product of E with S. Also the values on the exact trajectory $y(x_{n-1})$ and $y(x_n)$ are shown together with the values of $z^{(n-1)}$ and $z^{(n)}$, computed in each case using S from the exact solution values. In addition to the computed solutions, $y^{(n-1)}$ and $y^{(n)}$, related through a single operation of C, the vector $\bar{z}^{(n)}$, found by applying C to $z^{(n-1)}$, is shown.

The vector $z^{(n)} - \bar{z}^{(n)}$ will be thought of as the local truncation error and its asymptotic behaviour as $h \to 0$ will be used to determine order of accuracy. Note that if S were chosen in a different way, the estimate of $z^{(n)} - \bar{z}^{(n)}$ would possibly give a different order. Thus the choice of starting procedure S is an intrinsic part of the definition of order of accuracy and of truncation error.

DEFINITION 431A.

Let C denote a general linear method with preconsistency vector $u \in \mathbb{R}^r$. A *starting method* S for C is a method of the form

$$S = \begin{bmatrix} S_{11} & e \\ S_{21} & u \end{bmatrix}.$$

DEFINITION 431B.

Let $y(x_0)$ denote a point on the solution trajectory of a differential equation

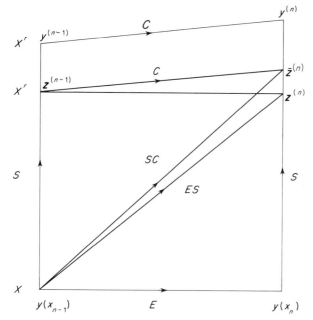

Figure 431 Local truncation error representation.

$y'(x) = f(y(x))$, where $f: X \rightarrow X$ is continuously differentiable arbitrarily often. Let S denote a starting method for a method C and let $S(h, y_0)$ and $C(h, y^{(0)})$ for $y_0 \in X$ and $y^{(0)} \in X^r$ denote the solutions computed after one step using S and C respectively in each case using step size h. Then the *local truncation error at x_0 for C relative to S* is defined by

$$l(C, S)(x_0) = S(h, y(x_0 + h)) - C(h, S(h, y(x_0))) \qquad (431a)$$

DEFINITION 431C.

In definition 431B, the *order of accuracy of C relative to S* is defined as the greatest integer p, such that

$$\| C(h, S(h, y(x_0))) - S(h, y(x_0 + h)) \| \leq Ph^{p+1}$$

for $h < h_0$ where P and h_0 may depend on the problem.

DEFINITION 431D.

The *order of accuracy* of C is the greatest p such that there exists S such that C is of order p relative to S.

In our discussion of global truncation error, it will be convenient to make use of a 'finishing method' of the form

$$F = \begin{bmatrix} F_{11} & F_{12} \\ F_{21} & F_{22} \end{bmatrix},$$

partitioned as $(\bar{s} + 1) \times (\bar{s} + r)$ say, and satisfying the preconsistency condition $F_{12}u = e$, $F_{22}u = 1$.

The only property of F that will interest us is that it will be chosen to approximately undo the work of S in the sense that the method SF formed by applying F to the result computed by S gives the initial value back again, at least to within $O(h^{\bar{p}+1})$, for some integer $\bar{p}$.

This finishing method can always be formed by choosing F_{22} arbitrarily subject to $F_{22}u = 1$ and then choosing F_{12} as eF_{22}. It then remains to choose F_{11} and F_{21} as the coefficients in a Runge–Kutta method

$$\begin{array}{c|c} F_{11}e & F_{11} \\ \hline & F_{21} \end{array}$$

which exactly undoes the work of the Runge–Kutta method

$$\begin{array}{c|c} S_{11}e & S_{11} \\ \hline & F_{22}S_{21} \end{array}.$$

432 Algebraic criteria for order.

Our object now is to reinterpret definitions 431B and 431C using the algebraic tools developed in section 38. In particular, we wish to be able to write $C(h, S(h, y(x_0)))$ in the form

$$C(h, S(h, y(x_0))) = v(\emptyset)\, y(x_0) + \sum_{t \in T} v(t)\, \frac{h^{r(t)}}{\sigma(t)}\, F(t, y(x_0)) \qquad (432a)$$

and $S(h, y(x_0 + h))$ in the form

$$S(h, y(x_0 + h)) = w(\emptyset)\, y(x_0) + \sum_{t \in T} w(t)\, \frac{h^{r(t)}}{\sigma(t)}\, F(t, y(x_0)), \qquad (432b)$$

so that the conditions for C to have order p relative to S become $v(t) = w(t)$ for all $t \in T^{\#}$ such that $r(t) \leq p$.

Let the members of the Runge–Kutta space G corresponding to the r components of the vector of generalized Runge–Kutta methods in the starting procedure S be denoted by $\xi_1, \xi_2, \ldots, \xi_r$ and let ξ denote the vector made up from the components $\xi_1, \xi_2, \ldots, \xi_r$. Thus the formal Taylor series for $S(h, y(x_0))$ is

$$S(h, y(x_0)) = \xi(\emptyset)\, y(x_0) + \sum_{t \in T} \xi(t) \frac{h^{r(t)}}{\sigma(t)}\, F(t, y(x_0)). \qquad (432c)$$

In terms of individual components we can immediately express w occurring in (432b) in terms of ξ. In fact, $w_i = E\xi_i$ and using an obvious extension of our terminology to vectors we write

$$w = E\xi. \tag{432d}$$

Let η_i $(i = 1, 2, \ldots, s)$ denote the members of G corresponding to Y_i computed in the first step of C using S as the starting method. Thus, $hf(Y_i)$ has as its Runge–Kutta space partner the element $\eta_i D$ and the equations relating η_i with $\xi_1, \xi_2, \ldots, \xi_r$ is

$$\eta_i = \sum_{j=1}^{s} c_{ij}^{11}(\eta_j D) + \sum_{j=1}^{r} c_{ij}^{12}\xi_j \tag{432e}$$

Writing ηD as the vector made up from components $\eta_1 D, \eta_2 D, \ldots, \eta_s D$ and adapting the notation for multiplying a matrix by a vector to our purposes, we write (432e) more compactly as

$$\eta = C_{11}(\eta D) + C_{12}\xi. \tag{432f}$$

The parentheses seem to be necessary since the product $C_{11}\eta D$ is not, in general, associative.

In the same way we find a compact formula for v occurring in (432a) as

$$v = C_{21}(\eta D) + C_{22}\xi, \tag{432g}$$

and using the partitioned matrix C we combine (432f) and (432g) as

$$\begin{bmatrix} \eta \\ v \end{bmatrix} = C\begin{bmatrix} \eta D \\ \xi \end{bmatrix}. \tag{432h}$$

If we wish to attain order of accuracy p, it is convenient to reinterpret (432h) with all members of G which occur, replaced by their cosets relative to the ideal I_p. In this case, since $v = w = E\xi$, we have as the order condition

$$\begin{bmatrix} \eta \\ E\xi \end{bmatrix} = C\begin{bmatrix} \eta D \\ \xi \end{bmatrix}, \tag{432i}$$

which can also be interpreted as

$$\begin{bmatrix} \eta(t) \\ (E\xi)(t) \end{bmatrix} = C\begin{bmatrix} (\eta D)(t) \\ \xi(t) \end{bmatrix}, \tag{432j}$$

for all $t \in T^{\#}$ such that $r(t) \le p$.

It $\varepsilon \in K$ is defined by $\varepsilon(t) = w(t) - v(t)$ for all $t \in T^{\#}$, so that the local truncation error is

$$l(C, S)(x_0) = \sum_{r(t) > p} \varepsilon(t) \frac{h^{r(t)}}{\sigma(t)} F(t, y(x_0))$$

for a method of order p, then we can define ε in terms of ξ by the equation

$$\begin{bmatrix} \eta \\ E\xi - \varepsilon \end{bmatrix} = C\begin{bmatrix} \eta D \\ \xi \end{bmatrix}$$

so that, for all t such that $r(t) > p$,

$$\varepsilon(t) = (E\xi)(t) - C_{21}(\eta D)(t) - C_{22}\xi(t).$$

As for Runge–Kutta methods, the terms for which $r(t) = p + 1$ are of special significance, since for h sufficiently small, they dominate the local error behaviour.

Let $d^T = [d_1, d_2, \ldots, d_r]$ denote an eigenvector of C_{22} corresponding to eigenvalue 1, so normalized that $d^T u = 1$. It is convenient to define the principal error function not as the h^p terms of $l(C, S)(x_0)$ but rather as the h^p terms in $(d^T l(C, S)(x_0))u$.

DEFINITION 432A.

In terms of the terminology already introduced, the *principal error coefficient* at x_0 is defined as

$$P(x_0) = \sum_{r(t)=p} \frac{d^T \varepsilon(t)}{\sigma(t)} F(t, y(x_0))$$

and the *principal error function* as $P(x_0)h^{p+1}u$.

In cases in which C_{22} has multiple eigenvalues with magnitude 1, it is also convenient to define 'essential error coefficients' corresponding to projections in the directions of the corresponding eigenvectors. However, for simplicity, we will restrict ourselves from now on to 'strongly stable' methods in which only the eigenvalue 1, corresponding to the preconsistency vector u as eigenvector, lies on the unit circle. It is the opinion of the author that all general linear methods which have prospects of being practically useful are strongly stable.

433 Global error estimates.

As we have indicated, we will restrict ourselves to strongly stable methods. Thus, by selecting a basis in which the preconsistency vector u is e_1, and the projection vector d^T used in the last subsection is e_1^T it is possible to write

$$C_{22} = \begin{bmatrix} 1 & 0 \\ 0 & \bar{C}_{22} \end{bmatrix}, \tag{433a}$$

where $\| \bar{C}_{22} \|_\infty < 1$. Let E_n denote the local truncation error generated in the nth integration step so that

$$E_n = h^{p+1} \begin{bmatrix} P(x_{n-1}) \\ Q(x_{n-1}) \end{bmatrix} + h^{p+2}R(x_{n-1}, h), \tag{433b}$$

where $P(x_{n-1})$ denotes the principal error coefficient, $Q(x_{n-1})$ contains the remaining components of the h^{p+1} terms in the local truncation error and $\| R(x_{n-1}, h)\|$ is bounded as $h \to 0$. In terms of the sequence of approximations $z^{(n)}$ generated by the starting procedure S, that is

$$z^{(n)} = S(h, y(x_n)),$$

we have

$$E_n = z^{(n)} - C(h, z^{(n-1)}).$$

For a given normed space X, we will define a norm on X^s or on X^r by the formulae

$$\| V \| = \max_{i=1}^{s} \| v_i \|$$

or

$$\| V \| = \max_{i=1}^{r} \| v_i \|,$$

where $v_1, v_2, \ldots, v_s$ or $v_1, v_2, \ldots, v_r$ are the subvectors in V. In particular, the accumulated error at the nth integration step is $z^{(n)} - y^{(n)}$, and the norm of this is $\max_{i=1}^{r} \| z_i^{(n)} - y_i^{(n)} \|$.

We will denote by $F^{(n)}$ (respectively $G^{(n)}$) the vector made up from subvectors $f(Y_1^{(n)}), f(Y_2^{(n)}), \ldots, f(Y_s^{(n)})$ (respectively $f(Z_1^{(n)}), f(Z_2^{(n)}), \ldots, f(Z_s^{(n)})$), where the sequence of values of $Y^{(n)}, y^{(n)}, Z^{(n)}, z^{(n)}$ are interrelated by

$$Y^{(n)} = h(C_{11} \otimes I) F^{(n)} + (C_{12} \otimes I) y^{(n-1)}, \tag{433c}$$

$$y^{(n)} = h(C_{21} \otimes I) F^{(n)} + (C_{22} \otimes I) y^{(n-1)}, \tag{433d}$$

$$Z^{(n)} = h(C_{11} \otimes I) G^{(n)} + (C_{12} \otimes I) z^{(n-1)}, \tag{433e}$$

$$z^{(n)} = h(C_{21} \otimes I) G^{(n)} + (C_{22} \otimes I) z^{(n-1)} + E_n. \tag{433f}$$

A Lipschitz condition on the function f implies the bound

$$\| F^{(n)} - G^{(n)} \| \le L \| Y^{(n)} - Z^{(n)} \|.$$

Assuming that h is sufficiently small so that $hL \| C_{11} \|_\infty < 1$, we find from (433c) and (433e) that

$$\| Z^{(n)} - Y^{(n)} \| \le hL \| C_{11} \|_\infty \| Z^{(n)} - Y^{(n)} \| + \| C_{12} \|_\infty \| z^{(n-1)} - y^{(n-1)} \|$$

and hence

$$\| Z^{(n)} - Y^{(n)} \| \le (1 - hL \| C_{11} \|_\infty)^{-1} \| C_{12} \|_\infty \| z^{(n-1)} - y^{(n-1)} \|. \tag{433g}$$

It is convenient to define

$$L^* = L \sum_{i=1}^{s} | c_{1i}^{21} |.$$

Note that $L^* \ge L$ because the consistency condition, in terms of the special basis we have used, becomes

$$\begin{bmatrix} 1 & 0 \\ 0 & \bar{C}_{22} \end{bmatrix} v + C_{21} e = e_1 + v,$$

where v is the consistency vector and premultiplying by e_1^T gives

$$\sum_{i=1}^{s} c_{1i}^{21} = 1.$$

It will turn out that L^* governs the growth of error in a computed solution in the same way that L determines the rate of error growth for a perturbed exact solution.

We now have sufficient background to express the main result of this subsection.

THEOREM 433A.

If f satisfies a Lipschitz condition with constant L and sufficient smoothness conditions, if the method C has order p relative to S and (433a) holds with $\| \bar{C}_{22} \|_\infty < 1$, if the principal error coefficient $P(x)$ is bounded by $\| P(x) \| \le T$ for all $x \in [x_0, \bar{x}]$ and if other notations are in accordance with the discussion in this subsection, then there exist positive constants h_0 and U such that the global truncation error is bounded by

$$\| z_1^{(n)} - y_1^{(n)} \| \le \frac{\exp(L^*(x_n - x_0)) - 1}{L^*} Th^p + Uh^{p+1},$$

$$\| z_k^{(n)} - y_k^{(n)} \| \le Uh^{p+1}, \qquad k = 2, 3, \ldots, r,$$

whenever $h < h_0$ and $x_n = x_0 + nh \in [x_0, \bar{x}]$.

Proof.

We will assume throughout that h_0 is chosen to satisfy $h_0 L \| C_{11} \|_\infty < 1$ so that (433g) necessarily holds.

By repeated use of (433d) and (433f) with n replaced in turn by $n = 1, 2, \ldots, m$ we find that

$$z^{(m)} - y^{(m)} = (C_{22}^m \otimes I)(z^{(0)} - y^{(0)}) + h \sum_{i=0}^{m-1} [(C_{22}^i C_{21}) \otimes I](G^{(m-i)} - F^{(m-i)})$$

$$+ \sum_{i=0}^{m-1} (C_{22}^i \otimes I) E_{m-i}$$

and we note that the first term is zero because $y^{(0)} = z^{(0)}$. We will bound the first subvector in $z^{(m)} - y^{(m)}$ as

$$\| z_1^{(m)} - y_1^{(m)} \| \le h \sum_{i=0}^{m-1} \sum_{j=1}^{s} | c_{1j}^{21} | \| f(Z_j^{(m-i)}) - f(Y_j^{(m-i)}) \|$$

$$+ \sum_{i=0}^{m-1} h^{p+1} [\| P(x_{m-1-i}) \| + h \| R(x_{m-1-i}, h) \|]$$

$$\leq hL^* \sum_{i=0}^{m-1} \| Z^{(m-i)} - Y^{(m-i)} \| + h^p(x_m - x_0)T + U_1 h^{p+1},$$

$$(433\text{h})$$

where U_1 depends on the sequence of R values occurring in (433b).

Let θ denote $\| \bar{C}_{22} \|_\infty$ so that $0 \leq \theta < 1$. Let k be chosen as one of the integers $k = 2, 3, \ldots, r$ so that we may bound the kth subvector in $z^{(m)} - y^{(m)}$:

$$\| z_k^{(m)} - y_k^{(m)} \| \leq h \sum_{i=0}^{m-1} \theta^i \| C_{21} \| \| G^{(m-i)} - F^{(m-i)} \| + \sum_{i=0}^{m-1} \theta^i \| E_{m-i} \|$$

$$\leq (1 - \theta)^{-1} h \| C_{21} \| \max_{i=1}^{m} \| G^{(i)} - F^{(i)} \| + h^{p+1} U_2$$

$$\leq (1 - \theta)^{-1} hL \| C_{21} \| \max_{i=1}^{m} \| Z^{(i)} - Y^{(i)} \| + h^{p+1} U_2, \qquad (433\text{i})$$

where U_2 depends on the sequence of Q and R values and on the value of θ.

By noting that $\theta < 1$ in the first term in the first right-hand side of (433i), an alternative bound is

$$\| z_k^{(m)} - y_k^{(m)} \| \leq hL \| C_{21} \| \sum_{i=0}^{m-1} \| Z^{(m-i)} - Y^{(m-i)} \| + h^{p+1} U_2$$

which can be combined with (433h) to give

$$\| z^{(m)} - y^{(m)} \| \leq hL \| C_{21} \| \sum_{i=0}^{m-1} \| Z^{(m-i)} - Y^{(m-i)} \| + h^{p+1} mT + h^{p+1} U_3$$

so that, also using (433g), we find

$$\| z^{(m)} - y^{(m)} \| \leq hU_4 \sum_{i=0}^{m-1} \| z^{(m-i)} - y^{(m-i)} \| + h^{p+1} mT + h^{p+1} U_3 \qquad (433\text{j})$$

where U_3 and U_4 are further constants. We assume, in addition to other constraints on h_0, that $h_0 U_4 < 1$.

Let $w_0, w_1, w_2, \ldots$ be defined by

$$w_0 = 0,$$

$$w_m = hU_4 \sum_{i=0}^{m-1} w_{m-i} + h^{p+1} mT + h^{p+1} U_3,$$

$$(433\text{k})$$

so that, because of (433j), $\| z^{(m)} - y^{(m)} \| \leq w_m$.

Evaluating w_1 and differencing (433k), we find

$$(1 - hU_4) w_1 = h^{p+1}(T + U_3), \qquad (433\text{l})$$

$$(1 - hU_4) w_m - w_{m-1} = h^{p+1} T, \qquad m = 2, 3, \ldots, \qquad (433\text{m})$$

and we note that (433m) can be written in the form

$$w_m + \frac{h^p T}{U_4} = (1 - hU_4)^{-1} \left(w_{m-1} + \frac{h^p T}{U_4} \right),$$

where we have assumed without loss of generality that $U_4 > 0$. It follows, also using (433l), that

$$\| z^{(m)} - y^{(m)} \| \le w_m \le (1 - hU_4)^{-m} h^{p+1} U_3 + h^p T \frac{(1 - hU_4)^{-m} - 1}{U_4}$$

$$\le U_5 h^p + U_6 h^{p+1}$$

for further constants U_5 and U_6, because hm is bounded by $\bar{x} - x_0$.

From this bound on $\| z^{(m)} - y^{(m)} \|$, together with (433g), we obtain in turn bounds on $\| Z^{(m)} - Y^{(m)} \|$ and on $\| z_k^{(m)} - y_k^{(m)} \|$ making use of (433i). In fact it is easy to see that

$$\| z_k^{(m)} - y_k^{(m)} \| \le h^{p+1} U_7 \tag{433n}$$

for some constant U_7.

Since $\| z_k^{(m)} - y_k^{(m)} \|$ has a tighter bound for $k > 1$ than the bound on $\| z_1^{(m)} - y_1^{(m)} \|$, we can now refine the bound on $\| Y^{(m)} - Z^{(m)} \|$. Subtracting (433c) from (433e) with n replaced by m and selecting component number i, we find

$$\| Z_i^{(m)} - Y_i^{(m)} \| \le hL \| C_{11} \|_\infty \| Z^{(m)} - Y^{(m)} \| + | c_{i1}^{12} | \| z_1^{(m-1)} - y_1^{(m-1)} \|$$

$$+ \sum_{j=2}^{r} | c_{ij}^{12} | \| z_j^{(m-1)} - y_j^{(m-1)} \|.$$

By the preconsistency condition, $c_{i1}^{12} = 1$ for $i = 1, 2, \ldots, s$, so that, using known bounds on various terms in (433h), we find that

$$\| Z_i^{(m)} - Y_i^{(m)} \| \le \| z_1^{(m-1)} - y_1^{(m-1)} \| + U_8 h^{p+1}.$$

Substitute into (433h) and we find that

$$\| z_1^{(m)} - y_1^{(m)} \| \le hL^* \sum_{i=1}^{m} \| z_1^{(m-i)} - y_1^{(m-i)} \| + h^{p+1} mT + U_1 h^{p+1}.$$

Now define $w_0, w_1, w_2, \ldots$ by the recursion

$$w_0 = 0,$$

$$w_m = hL^* \sum_{i=0}^{m-1} w_i + h^{p+1} mT + U_1 h^{p+1}, \qquad m = 1, 2, \ldots,$$

so that $\| z_1^{(m)} - y_1^{(m)} \| \le w_m$. We now find that

$$w_1 = h^{p+1}(T + U_1),$$

$$w_m - (1 + hL^*) w_{m-1} = h^{p+1} T,$$

so that

$$w_m + \frac{h^p T}{L^*} = (1 + hL^*) \left(w_{m-1} + \frac{h^p T}{L^*} \right)$$

and

$$\| z_1^{(n)} - y_1^{(n)} \| \leq w_n$$

$$\leq (1 + hL^*)^{n-1} h^{p+1} U_1 + h^p T \frac{(1 + hL^*)^n - 1}{L^*}$$

$$\leq \frac{\exp(L^*(x_n - x_0)) - 1}{L^*} Th^p + \exp(L^*(\bar{x} - x_0)) U_1 h^{p+1}$$

and the result of the theorem follows by defining U as the maximum of $\exp(L^*(\bar{x} - x_0)) U_1$ and U_7 in (433n). ∎

434 Particular cases.

We discuss three examples of general linear methods from the point of view of their orders of accuracy. The first is the method given by (415h) which can be rewritten in the form

$$Y_1 = y_{n-1} + \tfrac{5}{8} hf(y_{n-1}) - \tfrac{1}{8} hf(y_{n-2}) \tag{434a}$$

$$Y_2 = y_{n-1} - \tfrac{3}{2} hf(y_{n-1}) + \tfrac{1}{2} hf(y_{n-2}) + 2hf(Y_1) \tag{434b}$$

$$y_n = y_{n-1} + \tfrac{1}{6} hf(y_{n-1}) + \tfrac{2}{3} hf(Y_1) + \tfrac{1}{6} hf(Y_2). \tag{434c}$$

We will verify that this has order 4 relative to a starting method which would ensure that $y_0 = y(x_0) + O(h^5)$, $y_1 = y(x_1) + O(h^5)$. Let α, β denote members of G corresponding to Y_1, Y_2. If y_{n-1} and y_{n-2} have corresponding elements 1 and E^{-1} respectively, then in the factor space G/I_4, y_n must have the corresponding element $E + I_4$.

Translating (434a), (434b) and (434c) into corresponding statements in G, we find

$$\alpha = 1 + \tfrac{5}{8} D - \tfrac{1}{8} E^{-1} D$$

$$\beta = 1 - \tfrac{3}{2} D + \tfrac{1}{2} E^{-1} D + 2\alpha D$$

$$E(t) = (1 + \tfrac{1}{6} D + \tfrac{2}{3} \alpha D + \tfrac{1}{6} \beta D)(t), \qquad r(t) \leq 4.$$

For the trees with four or less vertices we find the values in Table 434 of the relevant members of G evaluated at these trees. Each of the columns from $\alpha(t)$ onwards is computed from preceding columns. The fact that the last column agrees exactly with the $E(t)$ column confirms the fourth-order requirements.

We next consider a method in which two quantities y_n, z_n are computed at step n, using the formulae

$$y_n = z_{n-1} + ahf(z_{n-1}) + bhf(y_{n-1}),$$

$$z_n = y_n + chf(y_n) + dhf(z_{n-1}),$$

where a, b, c, d are constants to be determined. Our aim is to achieve order 3. Note that this method may be thought of as a cyclic generalization of the Adams–Bashforth two-step method.

Table 434

t	$1(t)$	$D(t)$	$E(t)$	$E^{-1}(t)$	$(E^{-1}D)(t)$	$\alpha(t)$	$(\alpha D)(t)$	$\beta(t)$	$(\beta D)(t)$	$(1 + \tfrac{1}{6}D + \tfrac{2}{3}\alpha D + \tfrac{1}{6}\beta D)(t)$
$\emptyset$	1	0	1	1	-0	1	0	1	0	1
τ	0	1	1	-1	1	$\frac{1}{2}$	1	1	1	1
$[\tau]$	0	0	$\frac{1}{2}$	$\frac{1}{2}$	-1	$\frac{1}{8}$	$\frac{1}{2}$	$\frac{1}{2}$	1	$\frac{1}{2}$
$[\tau^2]$	0	0	$\frac{1}{3}$	$-\frac{1}{3}$	1	$-\frac{1}{8}$	$\frac{1}{4}$	1	1	$\frac{1}{3}$
$[_2\tau]_2$	0	0	$\frac{1}{6}$	$-\frac{1}{6}$	$\frac{1}{2}$	$-\frac{1}{16}$	$\frac{1}{8}$	$\frac{1}{2}$	$\frac{1}{2}$	$\frac{1}{6}$
$[\tau^3]$	0	0	$\frac{1}{4}$	$\frac{1}{4}$	-1	$\frac{1}{8}$	$\frac{1}{8}$	$-\frac{1}{4}$	1	$\frac{1}{4}$
$[\tau[\tau]]$	0	0	$\frac{1}{8}$	$\frac{1}{8}$	$-\frac{1}{2}$	$\frac{1}{16}$	$\frac{1}{16}$	$-\frac{1}{8}$	$\frac{1}{2}$	$\frac{1}{8}$
$[_2\tau^2]_2$	0	0	$\frac{1}{12}$	$\frac{1}{12}$	$-\frac{1}{3}$	$\frac{1}{24}$	$-\frac{1}{8}$	$-\frac{5}{12}$	1	$\frac{1}{12}$
$[_3\tau]_3$	0	0	$\frac{1}{24}$	$\frac{1}{24}$	$-\frac{1}{6}$	$\frac{1}{48}$	$-\frac{1}{16}$	$-\frac{5}{24}$	$\frac{1}{2}$	$\frac{1}{24}$

Let α, β denote members of G corresponding to y_{n-1}, z_{n-1} respectively. We assume that $\alpha(\emptyset) = \beta(\emptyset) = 1$ because of the requirements of preconsistency. The order requirements written in G/I_3 are

$$E\alpha = \beta + a\beta D + b\alpha D, \tag{434d}$$

$$E\beta = E\alpha + cE\alpha D + d\beta D. \tag{434e}$$

Writing these in terms of $\alpha_i = \alpha(t_i)$, $\beta_i = \beta(t_i)$ for $i = 1, 2, 3, 4$ where t_1, t_2, t_3, t_4 are particular trees, we find that the various quantities associated with (434d) and (434e) are as given in Table 434'. Equate the entries in the $(E\alpha)(t_i)$ and $(\beta + a\beta D + b\alpha D)(t_i)$ columns in the $(E\beta)(t_i)$ and $(E\alpha + cE\alpha D + d\beta D)(t_i)$ columns to arrive at eight equations relating a, b, c, d with $\alpha_1, \beta_1, \ldots, \alpha_4, \beta_4$. Eliminating the α_i, β_i leads to the three constraints on the values of a, b, c, d for third-order accuracy:

$$a + b + c + d = 1,$$

$$ac = \tfrac{2}{3},$$

$$bd = \tfrac{1}{6}.$$

For example, we may choose $a = \tfrac{4}{3}, c = \tfrac{1}{2}, b = -\tfrac{1}{3}, d = -\tfrac{1}{2}$ or $a = \tfrac{4}{3}, c = \tfrac{1}{2}, b = -\tfrac{1}{2}, d = -\tfrac{1}{3}$. If it is required to minimize the difference between $a + b$ and $c + d$, so that the method approximates as closely as possible the traditional type of cyclic composite method, we find that this is achieved by the choice $a = 1 + \sqrt{3}/3, b = -(\tfrac{1}{2} + \sqrt{3}/6), c = 1 - \sqrt{3}/3, d = -(\tfrac{1}{2} - \sqrt{3}/6)$. In any of these cases, appropriate values of α and β to provide a suitable starting method are easy to find.

We next consider a cyclic generalization of the two-step backward difference method which we write in the form

$$(1 + a)\, y_n = z_{n-1} + ay_{n-1} + hbf(y_n),$$

$$(1 + c)\, z_n = y_n + cz_{n-1} + hdf(z_n).$$

Table 434′

i	t_i	$E(t_i)$	$\alpha(t_i)$	$\beta(t_i)$	$(E\alpha)(t_i)$	$(E\beta)(t_i)$	$(\alpha D)(t_i)$	$(\beta D)(t_i)$	$(E\alpha D)(t_i)$	$(\beta + a\beta D + b\alpha D)(t_i)$	$(E\alpha + cE\alpha D + d\beta D)(t_i)$
1	τ	1	α_1	β_1	$\alpha_1 + 1$	$\beta_1 + 1$	1	1	1	$\beta_1 + a + b$	$\alpha_1 + 1 + c + d$
2	$[\tau]$	$\tfrac{1}{2}$	α_2	β_2	$\alpha_2 + \alpha_1 + \tfrac{1}{2}$	$\beta_2 + \beta_1 + \tfrac{1}{2}$	α_1	β_1	$\alpha_1 + 1$	$\beta_2 + a\beta_1 + b\alpha_1$	$\alpha_2 + \alpha_1 + \tfrac{1}{2} + c(\alpha_1 + 1) + d\beta_1$
3	$[\tau^2]$	$\tfrac{1}{3}$	α_3	β_3	$\alpha_3 + 2\alpha_2 + \alpha_1 + \tfrac{1}{3}$	$\beta_3 + 2\beta_2 + \beta_1 + \tfrac{1}{3}$	α_1^2	β_1^2	$(\alpha_1 + 1)^2$	$\beta_3 + a\beta_1^2 + b\alpha_1^2$	$\alpha_3 + 2\alpha_2 + \alpha_1 + \tfrac{1}{3} + c(\alpha_1 + 1)^2 + d\beta_1^2$
4	$[_2\tau]_2$	$\tfrac{1}{6}$	α_4	β_4	$\alpha_4 + \alpha_2 + \tfrac{1}{2}\alpha_1 + \tfrac{1}{6}$	$\beta_4 + \beta_2 + \tfrac{1}{2}\beta_1 + \tfrac{1}{6}$	α_2	β_2	$\alpha_2 + \alpha_1 + \tfrac{1}{2}$	$\beta_4 + a\beta_2 + b\alpha_2$	$\alpha_4 + \alpha_2 + \tfrac{1}{2}\alpha_1 + \tfrac{1}{6} + c(\alpha_2 + \alpha_1 + \tfrac{1}{2}) + d\beta_2$

Table 434″

i	t_i	$\alpha(t_i)$	$(\beta^{-1}\alpha)(t_i)$	$\beta(t_i)$	$(E^{-1}\beta)(t_i)$	$(\alpha D)(t_i)$	$(\beta D)(t_i)$	$(E^{-1}\beta + aE^{-1}\alpha + b\alpha D)(t_i)$	$(\alpha + cE^{-1}\beta + d\beta D)(t_i)$
1	$\tau[\,]$	α_1	$\alpha_1 - 1$	β_1	$\beta_1 - 1$	1	1	$\beta_1 + a\alpha_1 - (a + 1) + b$	$\alpha_1 + c(\beta_1 - 1) + d$
2	$[\tau]$	α_2	$\alpha_2 - \alpha_1 + \tfrac{1}{2}$	β_2	$\beta_2 - \beta_1 + \tfrac{1}{2}$	α_1	β_1	$\beta_2 - \beta_1 + \beta_1 + a(\alpha_2 - \alpha_1) + \tfrac{1}{2}(a + 1) + b\alpha_1$	$\alpha_2 + c(\beta_2 - \beta_1 + \tfrac{1}{2}) + d\beta_1$
3	$[\tau^2]$	α_3	$\alpha_3 - 2\alpha_2 + \alpha_1 - \tfrac{1}{3}$	β_3	$\beta_3 - 2\beta_2 + \beta_1 - \tfrac{1}{3}$	α_1^2	β_1^2	$\beta_3 - 2\beta_2 + \beta_1 + a(\alpha_3 - 2\alpha_2 + \alpha_1) - \tfrac{1}{3}(a + 1) + b\alpha_1^2$	$\alpha_3 + c(\beta_3 - 2\beta_2 + \beta_1 - \tfrac{1}{3}) + d\beta_1^2$
4	$[_2\tau]_2$	α_4	$\alpha_4 - \alpha_2 + \tfrac{1}{2}\alpha_1 - \tfrac{1}{6}$	β_4	$\beta_4 - \beta_2 + \tfrac{1}{2}\beta_1 - \tfrac{1}{6}$	α_2	β_2	$\beta_4 - \beta_2 + \tfrac{1}{2}\beta_1 + a(\alpha_4 - \alpha_2 + \tfrac{1}{2}\alpha_1) - \tfrac{1}{6}(a + 1) + b\alpha_2$	$\alpha_4 + c(\beta_4 - \beta_2 + \tfrac{1}{2}\beta_1 - \tfrac{1}{6}) + d\beta_2$

Our aim will be to select values of a, b, c and d so that this method has order 3. It will be especially interesting if $b/(1 + a)$ and $d/(1 + c)$ can be chosen as equal since this would lead to cheaper implementation costs. Denoting α and β as the members of G corresponding to y_n and z_n respectively, our order conditions in G/I_3 become

$$(1 + a) \alpha = E^{-1}\beta + aE^{-1}\alpha + b\alpha D,$$

$$(1 + c) \beta = \alpha + cE^{-1}\beta + d\beta D.$$

Using the same notation for $\alpha_1, \beta_1, \ldots, \alpha_4, \beta_4$ as in the previous example, we find the table of relevant quantities (Table 434″). Again we equate $(1 + a) \alpha_i$ to $(E^{-1}\beta + aE^{-1}\alpha + b\alpha D)(t_i)$ and $(1 + c)\beta_i$ to $(\alpha + cE^{-1}\beta + d\beta D) (t_i)$ for $i = 1, 2, 3, 4$. We then eliminate $\alpha_i, \beta_i (i = 1, 2, 3, 4)$ from the resulting equations.

It is found that

$$1 + a + c = b + d,$$

$$(a - b)(c - d) = \tfrac{1}{2}(b + d),$$

$$3(a - b)(a - c) = -1 - 4a - c,$$

and these can be solved in a convenient parametric form as

$$a = -\frac{1 + w^2}{4} + \frac{1 + 5w^2}{12w},$$

$$b = \frac{1 - w^2}{4} + \frac{1 + 5w^2}{12w} - \frac{w}{2}$$

$$c = -\frac{1 + w^2}{4} - \frac{1 + 5w^2}{12w}$$

$$d = \frac{1 - w^2}{4} - \frac{1 + 5w^2}{12w} + \frac{w}{2}.$$

The value of w which results in $b/(1 + a) = d/(1 + c)$ is $w = \sqrt{3}/3$, and we note that the alternative choice of sign in $\sqrt{3}$ is equivalent to interchanging the two components in the cyclic method.

435 Effective order of Runge–Kutta methods.

It is perhaps surprising that order in the sense of this section, as applied to Runge–Kutta methods, can sometimes exceed the conventional meaning of order. In other words, there may exist starting methods relative to which the order is higher than it is relative to the identity method. Let α denote the element of G_1 defined from the given Runge–Kutta method according to (387f) and (387g) in theorem 387B. Furthermore, let β denote the element of G_1 corresponding to the Runge–Kutta method constituting the starting method. The element β^{-1} in the group G_1 corresponds to the Runge–Kutta method which *exactly* undoes the work of the starting method. Thus, the method corresponding to $\beta\alpha\beta^{-1}$ which comprises the successive use of the three methods corresponding to β, α and β^{-1} respectively may have an order p, regarded as a Runge–Kutta method in its own right. In this case, the method corresponding to α would have an order p relative to the method corresponding to β.

This generalization of order, introduced in Butcher (1969) may be formalized using the criterion for order given in theorem 387B.

DEFINITION 435A.

Let a, b denote Runge–Kutta methods and α, β the corresponding members of G_1. Then a is of *effective order* p if b exists such that

$$\beta\alpha\beta^{-1} \in E + I_p.$$

In practice, it is more convenient to use the criterion $\beta\alpha = E\beta$ in the factor group $G_1/(1 + I_p)$. Thus we state using the notation already introduced.

THEOREM 435B.

A Runge–Kutta method a is of effective order p iff there exists $\beta \in G_1$ such that

$$(\beta\alpha)(t) = (E\beta)(t), \tag{435a}$$

for all $t \in T$ such that $r(t) \le p$.

While no great practical importance is claimed for this generalization, it is perhaps interesting that the bounds on attainable order derived in section 32 do not apply directly to attainable effective order. For example, Butcher (1969) showed how to derive an explicit method with only five stages for which the effective order is 5.

Consider the three methods:

0					
$\dfrac{1}{5}$	$\dfrac{1}{5}$				
$\dfrac{2}{5}$	0	$\dfrac{2}{5}$			
$\dfrac{3}{4}$	$\dfrac{75}{64}$	$-\dfrac{9}{4}$	$\dfrac{117}{64}$		
1	$-\dfrac{37}{36}$	$\dfrac{7}{3}$	$-\dfrac{3}{4}$	$\dfrac{4}{9}$	
	$\dfrac{19}{144}$	0	$\dfrac{25}{48}$	$\dfrac{2}{9}$	$\dfrac{1}{8}$

$$\tag{435b}$$

$$
\begin{array}{c|ccccc}
0 & & & & & \\
\dfrac{1}{5} & \dfrac{1}{5} & & & & \\
\dfrac{2}{5} & 0 & \dfrac{2}{5} & & & \\
\dfrac{1}{2} & \dfrac{3}{16} & 0 & \dfrac{5}{16} & & \\
1 & \dfrac{1}{4} & 0 & -\dfrac{5}{4} & 2 & \\
\hline
 & \dfrac{1}{6} & 0 & 0 & \dfrac{2}{3} & \dfrac{1}{6}
\end{array}
\qquad (435c)
$$

$$
\begin{array}{c|ccccc}
0 & & & & & \\
\dfrac{1}{5} & \dfrac{1}{5} & & & & \\
\dfrac{2}{5} & 0 & \dfrac{2}{5} & & & \\
\dfrac{3}{4} & \dfrac{161}{192} & -\dfrac{19}{12} & \dfrac{287}{192} & & \\
1 & -\dfrac{27}{28} & \dfrac{19}{7} & -\dfrac{291}{196} & \dfrac{36}{49} & \\
\hline
 & \dfrac{7}{48} & 0 & \dfrac{475}{1008} & \dfrac{2}{7} & \dfrac{7}{72}
\end{array}
\qquad (435d)
$$

If β corresponds to (435b), α to (435c) and γ to (435d), then it can be verified that (435a) is satisfied with $p = 5$. Hence, (435c) has effective order 5 relative to (435b). It can also be verified that $\beta\gamma \in E^2 + I_5$ so that γ not only undoes the work of β, to order 5 accuracy, but at the same time takes the solution two steps further.

The practical interpretation of the relationship between these three methods is that if one integrates between x_0 and $\bar{x}$ using a step size $h = (\bar{x} - x_0)/n$ with the first step given by (435b), the last step given by (435d) and the $n - 2$ steps in between, all given by (435c), then the error will tend to zero at the same asymptotic rate as if a single order 5 method had been used throughout.

We present experimental verification for this behaviour using the trivial example problem introduced for illustrative purposes in chapter 2. For this problem,

$$f(u) = \frac{u(1-u)}{2u-1}, \qquad x_0 = 0, \qquad \bar{x} = 1, \qquad y(x_0) = \frac{5}{6}, \qquad y(\bar{x}) = \frac{1}{2} + \left(\frac{1}{4} - \frac{5}{36e}\right)^{\!\frac{1}{2}}.$$

The following represent the observed errors in the computation of $y(\bar{x})$ using n steps of the methods given here and the method given by (332e), the latter being a six-stage fifth-order method in the conventional sense. The value of n^5 multiplied by the error is given to confirm the fifth-order behaviour:

	Method of this subsection		Conventional fifth-order method	
n	Error	Error $\times n^5$	Error	Error $\times n^5$
4	$4.59_{10} - 7$	$4.70_{10} - 4$	$1.71_{10} - 7$	$1.75_{10} - 4$
8	$1.14_{10} - 8$	$3.73_{10} - 4$	$4.65_{10} - 9$	$1.52_{10} - 4$
16	$3.22_{10} - 10$	$3.37_{10} - 4$	$1.35_{10} - 10$	$1.41_{10} - 4$
32	$9.60_{10} - 12$	$3.22_{10} - 4$	$4.07_{10} - 12$	$1.37_{10} - 4$

Although the conventional method gives greater accuracy, this is at the expense of 20 per cent more function evaluations. If this is taken into account, by performing six steps of the unconventional method for five of the conventional method, then the two methods show almost identical accuracy for this problem.

Method of this subsection		Conventional fifth-order method	
n	Error	n	Error
6	$5.16_{10} - 8$	5	$5.31_{10} - 8$
12	$1.40_{10} - 9$	10	$1.48_{10} - 9$
24	$4.11_{10} - 11$	20	$4.37_{10} - 11$
48	$1.27_{10} - 12$	40	$1.34_{10} - 12$

It has been pointed out by Stetter (1971) that the phenomenon described here can be interpreted as providing an efficient global error estimate.

Finally, we remark that a method of effective order 6 with six stages was derived in the thesis of Gifkins [1972]. It also appears to be the case that methods with eight stages and effective order 7 are possible.

44 STABILITY PROPERTIES OF GENERAL LINEAR METHODS

440 Linear and non-linear stability.

We consider a consistent general linear method

$$C = \begin{bmatrix} C_{11} & C_{12} \\ C_{21} & C_{22} \end{bmatrix}$$

with preconsistency vector u and consistency vector v. As for traditional types of methods, we can define its stability region and determine whether or not it is A-stable. Furthermore, as for Runge–Kutta methods, we can consider a non-autonomous version of the linear test problem. However, for these more general methods, the property of AN-stability is somewhat more complicated and it is convenient to define also a weak form of this property. The stronger type of AN-stability evidently depends on a choice of norm and the special choice where the norm is based on an inner-product space is singled out and given the name Euclidean AN-stability.

A non-linear stability property first proposed by Burrage and Butcher (1980) is introduced. This combines features of G-stability as for one-leg methods and algebraic stability as for Runge–Kutta methods and the latter name is extended to the general linear case. It is shown that Euclidean AN-stability is equivalent to algebraic stability but that the other forms of linear stability are distinct and form a chain of implications. It is not known whether or not Euclidean AN-stability is implied by AN-stability.

441 Linear stability

For the solution of the differential equation

$$y'(x) = q(x)y(x),$$

using a general linear method with step size h, we find that

$$Y_i = h \sum_{j=1}^{s} c_{ij}^{11} q(X_j) Y_j + \sum_{j=1}^{r} c_{ij}^{12} y_j^{(n-1)}, \qquad i = 1, 2, \ldots, s,$$

$$X_i = h \sum_{j=1}^{s} c_{ij}^{11} + \sum_{j=1}^{r} c_{ij}^{12} (u_j x_{n-1} + h v_j), \qquad i = 1, 2, \ldots, s,$$

$$y_i^{(n)} = h \sum_{j=1}^{s} c_{ij}^{21} q(X_j) Y_j + \sum_{j=1}^{r} c_{ij}^{22} y_j^{(n-1)}, \qquad i = 1, 2, \ldots, r,$$

where, for $i = 1, 2, \ldots s$, X_i denotes the points at which the derivative is evaluated in step number n. Using the fact that $C_{12} u = e$ and defining $\xi_i = (X_i - x_{n-1})/h$, we find

$$Y = C_{11} Z Y + C_{12} y^{(n-1)},$$

$$y^{(n)} = C_{21} Z Y + C_{22} y^{(n-1)},$$

where $Z = \text{diag}(z_1, z_2, \ldots, z_s)$ with $z_i = hq(x_{n-1} + h\xi_i)$. Thus, assuming that $I - C_{11}Z$ is non-singular, we find that

$$Y = (I - C_{11}Z)^{-1}C_{12}y^{(n-1)},$$

$$y^{(n)} = [C_{22} + C_{21}Z(I - C_{11}Z)^{-1}C_{12}]y^{(n-1)}. \qquad (441a)$$

In the special case where q is constant we write $hq = z$ and (441a) becomes

$$y^{(n)} = [C_{22} + zC_{21}(I - zC_{11})^{-1}C_{12}]y^{(n-1)}, \qquad (441b)$$

which defines a stable sequence if the matrix $C_{22} + zC_{21}(I - zC_{11})^{-1}C_{12}$ has bounded powers. Criteria for this are given in theorem 105E. The 'stability region' is the set of points in the complex plane for which this situation holds.

Extending definition 350A, we have the following definition.

DEFINITION 441A.

A general linear method is *A-stable* if its stability region includes the left half-plane.

From our foregoing discussion we can state the following result.

THEOREM 441B.

A general linear method C is A-stable iff for all $Z \in \mathbb{C}^-$, $I - zC_{11}$ is non-singular, and $C_{22} + zC_{21}(I - zC_{11})^{-1}C_{12}$ is a stable matrix.

In the case of a non-autonomous problem, the matrix given by

$$L(Z) = C_{22} + C_{21}Z(I - C_{11}Z)^{-1}C_{12} \qquad (441c)$$

plays an analogous role. It is of only passing interest to consider the concept covered by the following definition.

DEFINITION 441C.

A general linear method C is *weakly AN-stable* if for all $z_1, z_2, \ldots, z_s \in \mathbb{C}^-$, the matrix $I - C_{11}Z$ is non-singular, where $Z = \text{diag}(z_1, z_2, \ldots, z_s)$ and $L(Z)$ is a stable matrix.

The reason for the unimportance of weak AN-stability is the unreasonable assumption behind it, that although $q(x)$ may vary quite freely among the stages of a method, it is periodic over successive steps. A much more meaningful concept is given in the following definition.

DEFINITION 441D.

A general linear method C is *AN-stable* if there exists a norm on $\mathbb{C}^r$ such that, for all $z_1, z_2, \ldots, z_s \in \mathbb{C}^-$, $I - C_{11}Z$ is non-singular and, in the subordinate norm, $\|L(Z)\| \le 1$ where $Z = \mathrm{diag}(z_1, z_2, \ldots, z_s)$.

Also, we consider a special case.

DEFINITION 441E.

A general linear method C is *Euclideanly AN-stable* if it satisfies the conditions of definition 441D, where the norm is an inner-product norm.

442 Relationships between linear stability properties.

It is clear that A-stability, weak AN- stability, AN-stability and Euclidean AN-stability are successive strengthenings in the sense of theorem 442A, which we state without proof.

THEOREM 442A.

A Euclideanly AN-stable method is AN-stable, an AN-stable method is weakly AN-stable and a weakly AN-stable method is A-stable.

The remainder of this subsection is devoted to showing that most of these implications do not hold in the reverse directions.

THEOREM 442B.

A-stability does not imply weak AN-stability.

Proof.

A counterexample was given in subsection 355 with $r = 1$. ∎

Note that for the Runge–Kutta case, as used in this proof, weak AN-stability is identical with AN-stability. On the other hand, this is not true when $r > 1$.

THEOREM 442C.

Weak AN-stability does not imply AN-stability.

Proof.

Consider the method

$$C = \begin{bmatrix} C_{11} & C_{12} \\ C_{21} & C_{22} \end{bmatrix},$$

where

$$C_{11} = \begin{bmatrix} \frac{3}{4} & \frac{3}{4} \\ \frac{1}{4} & \frac{1}{2} \end{bmatrix}, \quad C_{12} = \begin{bmatrix} 1 & \frac{1}{2} \\ 1 & \frac{1}{4} \end{bmatrix}, \quad C_{21} = \begin{bmatrix} \frac{1}{2} & \frac{1}{2} \\ -1 & -1 \end{bmatrix}, \quad C_{22} = \begin{bmatrix} 1 & 0 \\ 0 & -1 \end{bmatrix}.$$

For this method, the preconsistency vector is $u = [1 \quad 0]^T$. We will show that this method is weakly AN-stable by showing that $L(Z)$ is power bounded for $Z = \mathrm{diag}(z_1, z_2)$, where $z_1, z_2 \in \mathbb{C}^-$. If either $z_1 = 0$ or $z_2 = 0$, this is trivial to verify. Otherwise, let $z_1^{-1} = iy_1/4$, $z_2^{-1} = iy_2/4$, where y_1, y_2 are real. The characteristic polynomial of $L(Z)$ is

$$c_0 \lambda^2 + c_1 \lambda + c_2, \tag{442a}$$

$$c_0 = 3 - 2iy_1 - 3iy_2 - y_1 y_2, \qquad c_1 = -iy_1, \qquad c_2 = 1 - iy_1 - iy_2 + y_1 y_2.$$

For (442a) to have its zeros in the open unit disc, it is necessary and sufficient that

$$|c_0|^2 - |c_2|^2 > 0$$

and that

$$(|c_0|^2 - |c_2|^2)^2 - |c_1\bar{c}_0 - \bar{c}_1 c_2|^2 > 0.$$

It is found that

$$|c_0|^2 - |c_2|^2 = 8 + 2y_1^2 + (y_1 + y_2)^2 + 7y_2^2$$

and that

$$(|c_0|^2 - |c_2|^2)^2 - |c_1\bar{c}_0 - \bar{c}_1 c_2|^2 = 64 + 8(2y_1 + y_2)^2 + 120y_2^2 + 2(2y_1^2 + y_1 y_2)^2$$
$$+ 14y_1^2 y_2^2 + 8(2y_1 y_2 + y_2^2)^2 + 56y_2^4,$$

and these are evidently always positive.

It is easy to verify that $I - C_{11}Z$ is never singular so that the stability of $L(Z)$ for all $z_1, z_2 \in \mathbb{C}^-$ follows from the maximum modulus principle.

Having shown that the method is weakly AN-stable, we show that it is not AN-stable. If it were, then for Z_1, Z_2 arbitrary diagonal matrices of purely imaginary numbers, $L(Z_1)L(Z_2)$ would be power bounded. Choose $Z_1 = -Z_2 = \mathrm{diag}(2i/3, -i)$. It is found that

$$L(Z_1) = \frac{1}{27} \begin{bmatrix} 25 - 4i & -1 + i \\ 4 + 8i & -25 - 2i \end{bmatrix}$$

with $L(Z_2)$ the same except for the sign of i. It is further found that

$$L(Z_1)L(Z_2) = \frac{1}{729} \begin{bmatrix} 645 + 12i & -6 - 48i \\ -48 + 408i & 633 - 12i \end{bmatrix}$$

with the characteristic polynomial $729\lambda^2 - 1278\lambda + 533$. This polynomial has a real zero greater than unity because its value is negative when $\lambda = 1$. ∎

443 Monotonic problems and methods.

To avoid the complication of dealing with the difference of two solutions to the same differential equation, we use a device introduced by Burrage and Butcher (1980) of using as a test problem

$$y'(x) = f(y(x)), \qquad f : \mathbb{R}^N \to \mathbb{R}^N, \tag{443a}$$

where

$$\langle v, f(v) \rangle \le 0, \tag{443b}$$

for a pseudo inner product $\langle \cdot, \cdot \rangle$. If $\|\cdot\|$ denotes the corresponding pseudo norm, an immediate consequence of (443b) is that $\| y(x) \|$ is a non-decreasing function, because $(d/dx) \| y(x) \|^2 = 2\langle y(x), f(y(x)) \rangle$. We note that a pseudo inner product satisfies all the requirements of an inner product except that $\langle v, v \rangle = 0$ does not necessary imply that $v = 0$.

DEFINITION 443A.

The differential equation (443a) is said to be *monotonic* with respect to $\langle \cdot, \rangle$ if (443b) holds for all $v \in \mathbb{R}^N$.

Given a symmetric positive semidefinite $r \times r$ matrix G it is possible to construct an extension $\langle \cdot, \cdot \rangle_G$ of $\langle \cdot, \cdot \rangle$, with corresponding extension $\|\cdot\|_G$ of $\|\cdot\|$, in each case to deal with the vector space $\mathbb{R}^{Nr}$. If $v, w \in \mathbb{R}^{Nr}$ are made up from subvectors $v_1, v_2, \ldots, v_r$ and $w_1, w_2, \ldots, w_r$ then we define

$$\langle v, w \rangle_G = \sum_{i,j=1}^{r} g_{ij} \langle v_i, w_j \rangle,$$

so that

$$\| v \|_G = \left(\sum_{i,j=1}^{r} g_{ij} \langle v_i, v_j \rangle \right)^{1/2},$$

where $g_{11}, g_{12}, \ldots, g_{rr}$ are the elements of G. We note that $\langle \cdot, \cdot \rangle_G$ is also a pseudo inner product.

We are interested in general linear methods for which a G exists such that monotonicity in the sense that $\| y^{(n)} \|_G \le \| y^{(n-1)} \|_G$ is a consequence of monotonicity of the underlying problem.

DEFINITION 443B.

A general linear method C is said to be *monotonic* with respect to G if for any problem which is monotonic with respect to $\langle \cdot, \cdot \rangle$ it holds that

$$\| y^{(n)} \|_G \le \| y^{(n-1)} \|_G.$$

For a method with this property it is easy to see that A-stability is a consequence because we may choose $N = 2$ and

$$f(u) = \begin{bmatrix} \mathrm{Re}(q) & -\mathrm{Im}(q) \\ \mathrm{Im}(q) & \mathrm{Re}(q) \end{bmatrix} u$$

with $\langle u, v \rangle = u^\mathrm{T} v$. The equation $y'(x) = f(y(x))$ is then equivalent to

$$y_1'(x) + iy_2'(x) = q(y_1(x) + iy_2(x))$$

for a complex number q, where y_1 and y_2 denote the components of y. If $\mathrm{Re}(q) \le 0$, then $\langle u, f(u) \rangle \le 0$. In the same way, it is easy to see that Euclidean algebraic stability is a consequence of monotonicity if G happens to be positive definite.

We now consider the test problem introduced in definition 356D (*BN*-stability). We write this problem here as

$$y'(x) = g(x, y(x))$$

and consider also a second solution z to the same equation. The two solutions together satisfy the single system corresponding to the function f given by

$$f\left(\begin{bmatrix} u \\ v \\ w \end{bmatrix}\right) = \begin{bmatrix} g(w, u) \\ g(w, v) \\ 1 \end{bmatrix}$$

and it can be seen that if $[g(w, u) - g(w, v)]^\mathrm{T}(u - v) \le 0$ then

$$\left\langle \begin{bmatrix} u \\ v \\ w \end{bmatrix}, \ f\left(\begin{bmatrix} u \\ v \\ w \end{bmatrix}\right) \right\rangle \le 0,$$

where the pseudo inner product is given by

$$\left\langle \begin{bmatrix} u_1 \\ v_1 \\ w_1 \end{bmatrix}, \ \begin{bmatrix} u_2 \\ v_2 \\ w_2 \end{bmatrix} \right\rangle = u_1^\mathrm{T} u_2 - v_1^\mathrm{T} u_2 - u_1^\mathrm{T} v_2 + v_1^\mathrm{T} v_2.$$

444 Algebraic stability for general linear methods.

Let M denote the partitioned matrix

$$M = \begin{bmatrix} G - C_{22}^\mathrm{T} G C_{22} & C_{12}^\mathrm{T} D - C_{22}^\mathrm{T} G C_{21} \\ D C_{12} - C_{21}^\mathrm{T} G C_{22} & D C_{11} + C_{11}^\mathrm{T} D - C_{21}^\mathrm{T} G C_{21} \end{bmatrix}, \tag{444a}$$

where D is a diagonal $s \times s$ matrix of non-negative numbers and G is a symmetric positive semidefinite $r \times r$ matrix. It can be seen that M is a symmetric matrix. Even though the generality of allowing G to be singular was built into the definition of algebraic stability, given by Burrage and Butcher (1980) it is perhaps more convenient to exclude this possibility since non-zero but non-positive definite matrices occurring in the roles of G and D are always associated with the use of some type of redundancy or reducibility. As a compromise we express the concept through a pair of definitions.

DEFINITION 444A.

A general linear method C is said to be *algebraically stable relative to a symmetric positive semidefinite $r \times r$ matrix G* if there exists a positive semidefinite $s \times s$ diagonal matrix D such that M given by (444a) is positive semidefinite.

DEFINITION 444B.

A general linear method C is said to be *algebraically stable* if there exists a symmetric positive definite $r \times r$ matrix G such that C is algebraically stable relative to G.

For convenience in the proof of theorem 444C, the main theorem of Burrage and Butcher (1980), we write the submatrices in M as

$$M = \begin{bmatrix} U & V^{\mathrm{T}} \\ V & W \end{bmatrix},$$

where u_{ij}, v_{ij}, w_{ij} denote elements of U, V, W.

THEOREM 444C.

An algebraically stable general linear method is monotonic.

Proof.

Let $D = \mathrm{diag}(d_1, d_2, \ldots, d_s)$ and let g_{ij} denote an element of G. If C is the method, then

$$\| y^{(n)} \|_G^2 - \| y^{(n-1)} \|_G^2 - 2 \sum_{i=1}^{s} d_i \langle Y_i, hf(Y_i) \rangle$$

$$= \sum_{i,j=1}^{r} g_{ij} \langle y_i^{(n)}, y_j^{(n)} \rangle - \sum_{i,j=1}^{r} g_{ij} \langle y_i^{(n-1)}, y_j^{(n-1)} \rangle - 2 \sum_{i=1}^{s} d_i \langle Y_i, hf(Y_i) \rangle$$

$$= \sum_{i,j=1}^{r} g_{ij} \left\langle \sum_{k=1}^{r} c_{ik}^{22} y_k^{(n-1)} + h \sum_{k=1}^{s} c_{ik}^{21} f(Y_k), \sum_{k=1}^{r} c_{jk}^{22} y_k^{(n-1)} + h \sum_{k=1}^{s} c_{jk}^{21} f(Y_k) \right\rangle$$

$$- \sum_{i,j=1}^{r} g_{ij} \langle y_i^{(n-1)}, y_j^{(n-1)} \rangle - 2 \sum_{i=1}^{s} d_i \left\langle \sum_{j=1}^{r} c_{ij}^{12} y_j^{(n-1)} + h \sum_{j=1}^{s} c_{ij}^{11} f(Y_j), hf(Y_i) \right\rangle$$

$$= - \sum_{i,j=1}^{r} u_{ij} \langle y_i^{(n-1)}, y_j^{(n-1)} \rangle - 2 \sum_{i=1}^{s} \sum_{j=1}^{r} v_{ij} \langle y_j^{(n-1)}, hf(Y_i) \rangle - \sum_{i,j=1}^{s} w_{ij} \langle hf(Y_i) hf(Y_j) \rangle$$

$$= - \sum_{i,j=1}^{r+s} m_{ij} \langle \xi_i, \xi_j \rangle \leq 0,$$

where $\xi_i = y_i^{(n-1)} (i = 1, 2, \ldots, r)$, $\xi_{r+j} = hf(Y_j)$ $(j = 1, 2, \ldots, s)$. Since $\Sigma_{i=1}^s d_i \langle Y_i, hf(Y_i) \rangle \le 0$ by the monotonicity condition it follows that $\| y^{(n)} \|_G^2 \le \| y^{(n-1)} \|_G^2$. ∎

In definition 444A, it might seem that both D and G are to some extent at our disposal. We now find a condition on G and show how D is completely determined once G is known. We will confine our attention to the special case in which C has a strengthened version of the condition of stability given in definition 427A.

THEOREM 444D.

If C is algebraically stable relative to G and $I - C_{22}$ has rank $r - 1$, then for any vector v orthogonal to the preconsistency vector u, $u^T G v = 0$. Furthermore, the diagonal matrix D occurring in definition 444A is $\mathrm{diag}(d_1, d_2, \ldots, d_s)$, where $d = (d_1, d_2, \ldots, d_s)^T$ is given by $d = C_{21}^T G u$.

Proof.

For a real number ε compute the quadratic form

$$[u^T + \varepsilon v^T, 0] M [u^T + \varepsilon v^T, 0]^T$$
$$= u^T(G - C_{22}^T G C_{22})u + 2\varepsilon u^T(G - C_{22}^T G C_{22})v + \varepsilon^2 v^T(G - C_{22}^T G C_{22})v$$
$$= 2\varepsilon(u^T G v - u^T G C_{22} v) + \varepsilon^2 v^T(G - C_{22}^T G C_{22})v.$$

Since ε can be made arbitrarily small but of either sign, it follows that $u^T G v = u^T G C_{22} v$. Repeating this argument with v replaced in turn by $C_{22}^2 v, \ldots, C_{22}^{r-2} v$ and forming a linear combination of the resulting equations, we find that

$$u^T G v = u^T G \varphi(C_{22})v,$$

where φ is a polynomial of degree $r - 1$ such that $\varphi(1) = 1$. Choose φ such that $\varphi(z)(z - 1)$ is a constant times the characteristic polynomial of C_{22} and it follows that $\varphi(C_{22})v = 0$ so that $u^T G v = 0$.

To derive the result relating D and G, choose ε as a real number and x as an arbitrary member of $\mathbb{R}^s$ and form the quadratic form

$$[u^T, \varepsilon x^T] M [u^T, \varepsilon x^T]^T = 2\varepsilon x^T(DC_{12} - C_{21}^T G C_{22})u$$
$$+ \varepsilon^2 x^T(DC_{11} + C_{11}^T D - C_{21}^T G C_{21})x$$

and again the ε coefficient is zero. Since x is arbitrary, it follows that $(DC_{12} - C_{21}^T G C_{22})u = 0$ and we note that $DC_{12}u = De = d$ and that $C_{22}u = u$. ∎

445   An example of an algebraically stable method.

The following method constructed by Dekker [1981] is of order 4 and is diagonally implicit:

$$
C = \left[
\begin{array}{cc:cc}
\frac{2}{3} & 0 & 1 & -\frac{7}{6} \\
\frac{2}{3} & \frac{2}{3} & 1 & \frac{1}{6} \\
\hdashline
\frac{1}{2} & \frac{1}{2} & 1 & 0 \\
-\frac{1}{11} & \frac{7}{11} & 0 & \frac{5}{11}
\end{array}
\right].
$$

Since the preconsistency vector is $u = [1 \quad 0]^T$, we see from theorem 444D that G and D may be assumed to have the forms

$$
G = \begin{bmatrix} 1 & 0 \\ 0 & g \end{bmatrix}, \qquad D = \begin{bmatrix} \frac{1}{2} & 0 \\ 0 & \frac{1}{2} \end{bmatrix},
$$

where $g > 0$. The matrix M given by (444a) is found to be

$$
M = \left[
\begin{array}{cccc}
0 & 0 & 0 & 0 \\
0 & \dfrac{96g}{121} & -\dfrac{7}{12} + \dfrac{5g}{121} & \dfrac{1}{12} - \dfrac{35g}{121} \\
0 & -\dfrac{7}{12} + \dfrac{5g}{121} & \dfrac{5}{12} - \dfrac{g}{121} & \dfrac{1}{12} + \dfrac{7g}{121} \\
0 & \dfrac{1}{12} - \dfrac{35g}{121} & \dfrac{1}{12} + \dfrac{7g}{121} & \dfrac{5}{12} - \dfrac{49g}{121}
\end{array}
\right].
$$

It is easily found that $e^T M e = 0$ and, accordingly, if M is to be non-negative definite, then $Me = 0$. This determines the value of $g = 11/12$ and a reevaluation of M gives the result

$$
M = \frac{1}{22} \begin{bmatrix} 0 \\ 4 \\ -3 \\ -1 \end{bmatrix} [0 \quad 4 \quad -3 \quad -1].
$$

446   The equivalence of Euclidean AN- and algebraic stabilities.

If C is Euclideanly AN-stable with an inner product defined by an Hermitian matrix H, then we will prove that this implies the same property for G, the real part of H. Also we will show that a non-negative diagonal matrix D exists satisfying the requirements of theorem 444D and finally that using the G and D thus defined, C satisfies the requirements of definition 444A.

If H has elements $h_{ij}(i, j = 1, 2, \ldots, r)$, where $h_{ij} = \bar{h}_{ji}$ then an inner product on $\mathbb{R}^r$ can be defined by

$$\langle v, w \rangle = \sum_{i,j=1}^{r} \bar{v}_i h_{ij} w_j,$$

where $v_1, v_2, \ldots, v_r$ (respectively $w_1, w_2, \ldots, w_r$) are the components of v (respectively w). It is well known that *every* inner product norm on $\mathbb{C}^r$ has this form. If C is AN-stable with respect to the norm $\| v \| = \langle v, v \rangle^{\frac{1}{2}}$, then it follows that $K(Z) = H - L(Z)^* H L(Z)$ is a positive semidefinite Hermitian matrix, where $L(Z)$ is given by (441c). Since $K(Z^*)^* = H^* - L(Z)^* H^* L(Z)$ will also be positive semidefinite, the same will also be true of $\frac{1}{2}[K(Z) + K(Z^*)^*] = G - L(Z)^* G L(Z)$, where $G = \frac{1}{2}(H + H^*)$ is the real part of H. We thus state the following lemma.

LEMMA 446A.

If C is Euclideanly AN-stable, then it is AN-stable with respect to an inner product norm based on a real symmetric matrix G.

For the remainder of this subsection, we will assume that the inner product in a Euclideanly AN-stable method is always chosen as this matrix G. We now proceed to the construction of D.

LEMMA 446B.

If C is Euclideanly AN-stable and a diagonal matrix D is defined by $De = d = C_{21}^{\mathrm{T}} G u$, where u is the preconsistency vector of C, then all the elements of D are non-negative.

Proof.

Let X be a diagonal matrix of non-positive real numbers, $x = Xe$, ε a positive real number and write $Z = \varepsilon X$.

Because $G - L(Z)^* G L(Z) = G - L(\varepsilon X)^{\mathrm{T}} G L(\varepsilon X)$,

$$0 \leq u^{\mathrm{T}} [G - L(\varepsilon X)^{\mathrm{T}} G L(\varepsilon X)] u$$

$$= u^{\mathrm{T}} G u - [C_{22} u + \varepsilon C_{21} X (I - \varepsilon C_{11} X)^{-1} C_{12} u]^{\mathrm{T}} G [C_{22} u + \varepsilon C_{21} X (I - \varepsilon C_{11} X)^{-1} C_{12} u]$$

$$= u^{\mathrm{T}} G u - [u + \varepsilon C_{21} X (I - \varepsilon C_{11} X)^{-1} e]^{\mathrm{T}} G (u + \varepsilon C_{21} X (I - \varepsilon C_{11} X)^{-1} e)$$

$$= -2\varepsilon (e^{\mathrm{T}} X) C_{21}^{\mathrm{T}} G u + \varepsilon^2 P(\varepsilon),$$

where $|P(\varepsilon)|$ is bounded as $\varepsilon \to 0$. Since $e^{\mathrm{T}} X = x^{\mathrm{T}}$, $C_{21}^{\mathrm{T}} G u = d$ and ε can be made arbitrarily small, it follows that $-x^{\mathrm{T}} d \geq 0$. Since $-x$ is an arbitrary vector of non-negative numbers, it follows that all elements in d are non-negative. ∎

Finally, we come to the main result of this subsection which generalizes theorem 355B.

THEOREM 446C.

If C is Euclideanly AN-stable, then C is algebraically stable.

Proof.

Let Y be a diagonal matrix of s real numbers, $y = Ye$ and $x \in \mathbb{R}^r$. Write $v = u + i\varepsilon x$ and $Z = i\varepsilon Y$. We compute

$$L(Z)v = [C_{22} + i\varepsilon C_{21} Y(I - i\varepsilon C_{11} Y)^{-1} C_{12}] (u + i\varepsilon x)$$

$$= u + i\varepsilon(C_{22}x + C_{21}y) - \varepsilon^2(C_{21} YC_{12}x + C_{21} YC_{11}y) + \varepsilon^3 q(\varepsilon),$$

where $\| q(\varepsilon) \|$ is bounded as $\varepsilon \to 0$. Because $G - L(Z)^* GL(Z)$ is positive semidefinite, we have

$$0 \le v^* Gv - [L(Z)v]^* G[L(Z)v]$$

$$= \varepsilon^2 \bigg[x^\mathrm{T} Gx - (C_{22}x + C_{21}y)^\mathrm{T} G(C_{22}x + C_{21}y)$$

$$+ u^\mathrm{T} G(C_{21} YC_{12}x + C_{21} YC_{11}y) + (C_{21} YC_{12}x + C_{21} YC_{11}y)^\mathrm{T} Gu \bigg] + \varepsilon^3 Q(\varepsilon),$$

where $|Q(\varepsilon)|$ is bounded as $\varepsilon \to 0$. Substitute $C_{21}^\mathrm{T} Gu = d$ and note that, because ε can be made arbitrarily small, the coefficient of ε^2 is non-negative. Hence,

$$0 \le x^\mathrm{T} Gx - (C_{22}x + C_{21}y)^\mathrm{t} G(C_{22}x + C_{21}y)$$

$$+ d^\mathrm{T} Y(C_{12}x + C_{11}y) + (C_{12}x + C_{11}y)^\mathrm{T} Yd$$

and, because $Yd = Dy$, it follows that

$$[x^\mathrm{T} \quad y^\mathrm{T}] M \begin{bmatrix} x \\ y \end{bmatrix} \ge 0,$$

where M is the matrix given by (444a). ■

447 Generalizations of algebraic stability.

If the test equation (443a) satisfies the inequality

$$\langle v, f(v) \rangle \le \lambda \| v \|^2, \tag{447a}$$

rather than the special case given by (443b) in which the real number λ equals zero, then we find that $\| y(x)\exp(-x\lambda) \|$, rather than $\| y(x) \|$, is a non-decreasing function. Thus we are able to study the error growth for equations satisfying a one-sided Lipschitz condition with constant λ, as an extension of the result given for Runge–Kutta methods in subsection 357. Following the notation we employed there, we write $l = h\lambda$ so that l becomes the one-sided Lipschitz constant normalized for h the unit for distance in the x direction. Also following subsection 357, we will consider the possible existence of a number k such that the growth of the squared norm of the solution grows by no more than a factor k in any step. Furthermore, as for the Runge–Kutta case, we approach questions of non-linear stability directly, rather than through the corresponding linear questions.

For a given positive semidefinite symmetric $r \times r$ matrix G, a non-negative diagonal matrix D, a real number l and a positive number k, we will consider a matrix M given in partitioned form as

$$M = \begin{bmatrix} kG - C_{22}^T G C_{22} - 2lC_{12}^T G C_{12} & C_{12}^T D - C_{22}^T G C_{21} - 2lC_{12}^T D C_{11} \\ DC_{12} - C_{21}^T G C_{22} - 2lC_{11}^T D C_{12} & DC_{11} + C_{11}^T D - C_{21}^T G C_{21} - 2lC_{11}^T D C_{11} \end{bmatrix}.$$

(447b)

We note that if $k = 1$, $l = 0$, M simplifies to the form of (444a).

DEFINITION 447A.

A method C is (k, l)-*algebraically stable* if G and D exist with G positive definite such that M given by (447b) is positive semidefinite.

Generalizing definitions in subsection 443 we have the following definition.

DEFINITION 447B.

The differential equation (443a) is said to be λ-*monotonic with respect to* $\langle \cdot, \cdot \rangle$ if for all $v \in \mathbb{R}^N$, (447a) is satisfied.

DEFINITION 447C.

A method C is said to be (k, l)-*monotonic* with respect to a positive semidefinite symmetric $r \times r$ matrix G if, for any problem which is λ-monotonic with respect to $\langle \cdot, \cdot \rangle$, it holds that

$$\| y^{(n)} \|_G^2 \le k \| y^{(n-1)} \|_G^2,$$

with $l = h\lambda$.

The main result on generalized non-linear stability is the following theorem.

THEOREM 447D.

A (k, l)-algebraically stable method is (k, l)-monotonic.

Proof.

We evaluate $\| y^{(n)} \|_G^2 - k \| y^{(n-1)} \|_G^2 - 2\Sigma_{i=1}^2 d_i(\langle Y_i, hf(Y_i) \rangle - l \| Y_i \|^2)$ as in the proof of theorem 444C. It is again found to equal $-\Sigma_{i,j=1}^{r+s} m_{ij} \langle \xi_i, \xi_j \rangle \le 0$, where $\xi_1, \xi_2, \ldots, \xi_{r+s}$ have their former meanings but $m_{ij}(i, j = 1, 2, \ldots, r+s)$ are now the elements of M given by (447b). Since $\langle Y_i, hf(Y_i) \rangle - l \| Y_i \|^2 \le 0$, it follows that $\| y^{(n)} \|_G^2 \le k \| y^{(n-1)} \|_G^2$. ∎

Although this result has the potential for providing sharp error growth estimates for general linear methods, the theoretical details have not yet been fully worked through. Furthermore, it is rather difficult to find the best k as a function of l. Some preliminary results for some well-known methods have been given by Burrage and Butcher (1980) where the results presented in this subsection were first published. Further examples are given by Butcher (1981).

45 PRACTICAL PROSPECTS FOR GENERAL LINEAR METHODS

450 Survey of possible methods.

The specific ideas which led to the development of these methods were aimed at enhancing the behaviour of existing methods. The so-called Dahlquist barrier which limits the order of convergent linear k-step methods to $k + 2$ (k even) or $k + 1$ (k odd) is an irresistible challenge. Attempts to break this barrier have led, on the one hand, to cyclic composite methods and, on the other hand, to hybrid methods (generalized multistep methods, modified multistep methods).

The literature on cyclic methods starts with the work of Donelson and Hansen (1971) and has been pursued more recently by Albrecht (1979, 1980) and others. These methods can achieve genuine improvements over classical linear multistep methods and have been implemented in at least one published program for stiff problems by Tendler, Bickart and Picel (1978, 1978a).

The idea of modifying linear multistep methods by adding one or more off-step points has a literature dating from Gragg and Stetter (1964), Gear (1965), Butcher (1965) and Kohfeld and Thompson (1967). These methods also achieve improvements over classical linear multistep methods and, although some test results have been published, they have not been incorporated into published programs.

We will consider some methods based on simple generalizations of Runge–Kutta methods in subsections 451 and 452. It is hoped that these methods might offer efficient alternatives to some known explicit Runge–Kutta methods. In subsection 453 progress in the search for general linear methods suitable for stiff problems will be reported.

451 A fourth-order Runge–Kutta–Nordsieck method.

We consider the construction of a method in which each step begins with approximations to the solution value as well as to appropriately scaled values of its first and second derivatives. Three stages are allowed in the step and it is required to produce approximations at the end of the step to quantities corresponding to those available at the beginning, to fourth-order accuracy. Modifying the usual notation for general linear methods to avoid superscipts and to emphasize the similarity to Runge–Kutta methods, we write

$$
C = \begin{bmatrix}
0 & 0 & 0 & \vdots & 1 & c_{12} & c_{13} \\
a_{21} & 0 & 0 & \vdots & 1 & c_{22} & c_{23} \\
b_{11} & b_{12} & 0 & \vdots & 1 & d_{12} & 0 \\
\cdots & \cdots & \cdots & \cdots & \cdots & \cdots & \cdots \\
b_{11} & b_{12} & 0 & \vdots & 1 & d_{12} & 0 \\
b_{21} & b_{22} & b_{23} & \vdots & 0 & d_{22} & 0 \\
b_{31} & b_{32} & b_{33} & \vdots & 0 & d_{32} & 0
\end{bmatrix} .
$$

Note that the C_{11} matrix is strictly lower triangular for explicitness of the method and that the first column of C_{12} is the vector e so that the method is preconsistent with the preconsistency vector $[1, 0, 0]^T$. The choice of the first column of C_{22} is also for reasons of preconsistency whereas the last column of C_{22} is chosen as the zero vector as a design decision.

We will assume that $y_1^{(n-1)} \approx y(x_{n-1})$, $y_2^{(n-1)} \approx hy'(x_{n-1})$, $y_3^{(n-1)} \approx (h^2/2)y''(x_{n-1})$ and we wish to impose conditions that will guarantee that $y_1^{(n)} \approx y(x_n)$, $y_2^{(n)} \approx hy'(x_n)$ to fourth-order accuracy. It is, however, only necessary to require that $y_3^{(n)} \approx (h^2/2)y''(x_n)$ to third order because of the choice of zeros in the last column of C_{22}.

By analogy with fourth-order four-stage Runge–Kutta methods, we will assume that each of Y_2 and Y_3 approximates $y(x_n)$ and that Y_3 is identical with $y_1^{(n)}$. This is the reason for the zero in each of the b_{13} and c_{33} positions and for the values $a_{31} = b_{11}$, $a_{32} = b_{12}, c_{32} = d_{12}$. Finally, we assume that Y_1 approximates, to second-order accuracy, the value of $y(x_{n-1} + h\xi)$, where ξ is a number to be chosen.

Let α and β denote the elements of the Runge–Kutta space G corresponding to Y_1 and Y_2 respectively. The elements corresponding to $y_1^{(n-1)}, y_2^{(n-1)}, y_3^{(n-1)}$ will be 1, D and T_2 and to $y_1^{(n)}, y_2^{(n)}, y_3^{(n)}$ the elements are E, ED and ET_2. Thus we have the order conditions

$$\alpha(t) = 1 + c_{12}D(t) + c_{13}T_2(t), \qquad\qquad r(t) \leq 3, \tag{451a}$$

$$\beta(t) = 1 + c_{22}D(t) + c_{23}T_2(t) + a_{21}(\alpha D)(t), \qquad\qquad r(t) \leq 3, \tag{451b}$$

$$E(t) = 1 + d_{12}D(t) + b_{11}(\alpha D)(t) + b_{12}(\beta D)(t), \qquad\qquad r(t) \leq 4, \tag{451c}$$

$$(ED)(t) = d_{22}D(t) + b_{21}(\alpha D)(t) + b_{22}(\beta D)(t) + b_{23}(ED)(t), \quad r(t) \leq 4, \tag{451d}$$

$$(ET_2)(t) = d_{32}D(t) + b_{31}(\alpha D)(t) + b_{32}(\beta D)(t) + b_{33}(ED)(t), \quad r(t) \leq 3, \tag{451e}$$

with the additional assumption that

$$\alpha(t) = \xi^{r(t)} E(t), \qquad r(t) \le 2. \tag{451f}$$

Evaluate $E([\tau^2]) - 2E([_2\tau]_2) = 0$ using (451c) and it follows that $\beta([\tau]) = \frac{1}{2}\beta(\tau)^2$ or else $b_{12} = 0$. A similar argument applied to (451d) and (451e) leads to the conclusion that either $\beta([\tau]) = \frac{1}{2}$ or else $b_{22} = b_{32} = 0$. We reject the possibility that $b_{12} = b_{22} = b_{32} = 0$ since in this case no subsequent use would ever be made of Y_2, and instead assume that

$$\beta(t) = E(t), \qquad r(t) \le 2. \tag{451g}$$

A similar argument leads to the conclusion that (451g) cannot hold also for $r(t) = 3$ since (451f) does not. It is then easy to see from (451d) that $d_{22} = b_{21} = b_{22} = 0, \ b_{23} = 1$.

We now substitute into (451c) the trees $t = \tau, \ [\tau], [\tau^2], \ [\tau^3]$, and it follows that

$$d_{12} + b_{11} + b_{12} = 1,$$
$$b_{11}\xi + b_{12} = \tfrac{1}{2},$$
$$b_{11}\xi^2 + b_{12} = \tfrac{1}{3},$$
$$b_{11}\xi^3 + b_{12} = \tfrac{1}{4},$$

leading to the unique solution, related to Simpson's rule, $\xi = \frac{1}{2}$, $d_{12} = b_{12} = \frac{1}{6}$, $b_{11} = \frac{2}{3}$. We now evaluate $\alpha(\tau)^3 - 3\alpha([\tau^2]) = \frac{1}{8}$ and $\beta(\tau)^3 - 3\beta([\tau^2]) = 1 - \frac{3}{4}a_{21}$, which we substitute into $E([\tau^3]) - 3E([_2\tau^2]_2) = 0$, from (451c). It is found that $\frac{1}{8}b_{11} + b_{12}(1 - \frac{3}{4}a_{21}) = 0$, from which it follows that $a_{21} = 2$. Substitute $t = \tau, \ [\tau]$ into (451a), (451b) and it follows that $c_{12} = \frac{1}{2}$, $c_{13} = \frac{1}{4}$, $c_{22} = c_{23} = -1$. Substitute $t = \tau, \ [\tau], [\tau^2]$ into (451e) and we find $d_{32} = \frac{1}{2}$, $b_{31} = -2, \ b_{32} + b_{33} = \frac{3}{2}$.

It is now a simple matter to verify that (451c) and (451d) are satisfied for the only trees that have not been involved in the derivation, $t = [\tau[\tau]]$ and $t = [_3\tau]_3$. Writing $b_{33} = w$, we now have a one-parameter family of fourth-order methods:

$$
C = \left[
\begin{array}{ccc:ccc}
0 & 0 & 0 & 1 & \frac{1}{2} & \frac{1}{4} \\
2 & 0 & 0 & 1 & -1 & -1 \\
\frac{2}{3} & \frac{1}{6} & 0 & 1 & \frac{1}{6} & 0 \\
\hdashline
\frac{2}{3} & \frac{1}{6} & 0 & 1 & \frac{1}{6} & 0 \\
0 & 0 & 1 & 0 & 0 & 0 \\
-2 & \frac{3}{2} - w & w & 0 & \frac{1}{2} & 0
\end{array}
\right].
$$

The most natural choice of w is $w = 1.5$ since this makes the method most similar to a PECE method. In figure 451 we show the boundaries of the stability regions for the method for $w = 1.3, \ 1.5, \ 1.7$ and it will be noted that these do not suggest an overwhelming reason for a different choice than $w = 1.5$.

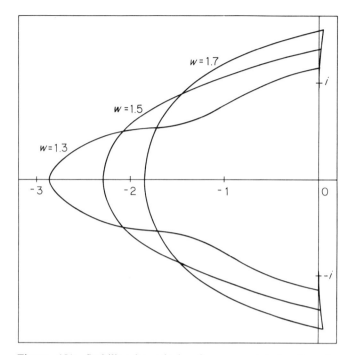

Figure 451: Stability boundaries for a special fourth-order method for three values of w.

The method we have described can easily be developed into a convenient adaptive procedure with local error estimation and variable step size (using rescaling of the Nordsieck vector), and it is typical of methods of various orders with some use of the Nordsieck idea and some of the Runge–Kutta idea. By way of contrast, we introduce in the next subsection a fifth-order method which also balances multistep aspects against Runge–Kutta aspects, but in this case information from previous steps is more directly related to the quantities actually computed in the step.

452 A fifth-order Runge–Kutta multistep method.

Consider a general linear method with $r = s = 3$ of the form

$$
C = \begin{bmatrix}
0 & 0 & 0 & \vdots & 1 & 0 & 0 \\
a_{21} & 0 & 0 & \vdots & 1 & c_{22} & c_{23} \\
a_{31} & a_{32} & 0 & \vdots & 1 & c_{32} & c_{33} \\
\cdots & \cdots & \cdots & \vdots & \cdots & \cdots & \cdots \\
b_1 & b_2 & b_3 & \vdots & 1 & 0 & d \\
1 & 0 & 0 & \vdots & 0 & 0 & 0 \\
0 & 1 & 0 & \vdots & 0 & 0 & 0
\end{bmatrix} ,
$$

in which $y_1^{(n-1)} = Y_1 \approx y(x_{n-1})$, $y_2^{(n-1)} \approx hy'(x_{n-2})$, $y_3^{(n-1)} \approx hy'(x_{n-2} + h\xi)$, $Y_2 \approx y(x_{n-1} + h\xi)$, $Y_3 \approx y(x_n)$. We will seek values of $a_{21}, a_{31}, a_{32}, c_{22}, c_{23}$, $c_{32}, c_{33}, b_1, b_2, b_3, d$ such that Y_2 and Y_3 agree with the quantities they are supposed to approximate to third-order accuracy and where the error constants, ε_2 and ε_3 say, are such that

$$(b_2 + d)\varepsilon_2 + b_3\varepsilon_3 = 0 \tag{452a}$$

and such that fifth-order accuracy is achieved for $y_1^{(n)}$. Note that $y_2^{(n)} = hf(y_1^{(n-1)}) \approx hy'(x_{n-1})$ and that $y_3^{(n)} = hf(Y_2) \approx hy'(x_{n-1} + h\xi)$.

Define rational functions ϕ, ϕ_2, ϕ_3 of a complex variable by

$$\phi(z) = \frac{Az + B}{z^2(z-1)^2(z+1-\xi)^2(z-\xi)^2},$$

$$\phi_2(z) = \frac{Cz^2 + Dz + E}{(z+1)^2z^2(z+1-\xi)^2(z-\xi)},$$

$$\phi_3(z) = \frac{Fz^4 + Gz^3 + Hz^2 + Iz + J}{(z+1)^2z^2(z-1)(z+1-\xi)^2(z-\xi)^2},$$

where the residues of these functions at some of its poles are given by

Function \ Pole	-1	1	$\xi - 1$	ξ	
ϕ		-1	0	0	
ϕ_2	0		0	-1	(452b)
ϕ_3	0	-1	0	0	

The three residues specified for ϕ are sufficient to determine A and B once ξ is known and to determine ξ itself as either $(5 - \sqrt{5})/5$ or $(5 + \sqrt{5})/5$. We will, for convenience, continue the analysis using only the former choice of ξ. It is found that

$$A = \frac{\sqrt{5}}{50}, \qquad B = \frac{11 - 3\sqrt{5}}{300}.$$

We write $\phi(z)$ in partial fractions:

$$\phi(z) = -\frac{1}{z-1} + \frac{1}{z} + \frac{d}{(z+1-\xi)^2} + \frac{b_1}{z^2} + \frac{b_2}{(z-\xi)^2} + \frac{b_3}{(z-1)^2},$$

where the various numerators are

$$d = \frac{15 - 7\sqrt{5}}{48}, \qquad b_1 = \frac{9 + \sqrt{5}}{48}, \qquad b_2 = \frac{15 + 7\sqrt{5}}{48}, \qquad b_3 = \frac{9 - \sqrt{5}}{48}.$$

Using lemma 132E and the discussion following it, we see that this construction does indeed produce an approximation to fifth-order accuracy.

In a similar way, we determine the C, D, E in ϕ_2 as

$$C = \frac{-204 + 68\sqrt{5}}{75}, \qquad D = \frac{-550 + 126\sqrt{5}}{375}, \qquad E = \frac{90 - 62\sqrt{5}}{375}$$

so that

$$\phi_2(z) = -\frac{1}{z - \xi} + \frac{1}{z} + \frac{c_{22}}{(z + 1)^2} + \frac{c_{23}}{(z + 1 - \xi)^2} + \frac{a_{21}}{z^2},$$

where

$$c_{22} = \frac{9 - \sqrt{5}}{30}, \qquad c_{23} = \frac{35 - 27\sqrt{5}}{30}, \qquad a_{21} = \frac{-7 + 11\sqrt{5}}{15}.$$

We specify F, G, H, I, J, occurring in $\phi_3(z)$, to satisfy the requirements in (452b) and also to satisfy (452a). This last condition is equivalent to

$$(b_2 + d)C + b_3 F = 0$$

and we find

$$F = \frac{748 - 204\sqrt{5}}{95}.$$

Rather than evaluate G, H, I and J directly, we now observe that $\phi_3(z)$ can be written in the form

$$\phi_3(z) = \phi(z) - \frac{F}{4} \frac{d}{dz} [z^{-1}(z - 1)^{-1}(z - \xi)^{-1}(z - \xi + 1)^{-1}]$$

$$+ \left(\frac{F}{2} + 2(1 - \xi)(2 - \xi)b_3\right) \frac{d}{dz} [(z + 1)^{-1}z^{-1}(z - 1)^{-1}(z - \xi)^{-1}(z - \xi + 1)^{-1}]$$

since $z^5\phi_3(z) \to F$ as $|z| \to \infty$, $(z - 1)^2\phi_3(z) \to 0$ as $z \to 1$ and the correct values of the residues are found at -1, 1, $\xi - 1$ and ξ. The outcome of this is that

$$\phi_3(z) = -\frac{1}{z - 1} + \frac{1}{z} + \frac{c_{32}}{(z + 1)^2} + \frac{c_{33}}{(z + 1 - \xi)^2} + \frac{a_{31}}{z^2} + \frac{a_{32}}{(z - \xi)^2},$$

where

$$c_{32} = \frac{-857 - 25\sqrt{5}}{912}, \qquad c_{33} = \frac{-151 + 1537\sqrt{5}}{912},$$

$$a_{31} = \frac{569 - 1727\sqrt{5}}{912}, \qquad a_{32} = \frac{1351 + 215\sqrt{5}}{912}.$$

To see why the method constructed in the way we have described has fifth-order accuracy we refer to the construction of related methods in Butcher (1965, 1967). The distinctive feature of the new method described here is that the off-step approximation is used in each of two successive steps. This is the reason for the factor $(b_2 + d)$, rather than b_2, in (452a).

In summary we write the full matrix C in the form involving surds as well as in the form of real numbers. Finally, we give the conjugate $\tilde{C}$ of the matrix (with $\sqrt{5}$ replaced by $-\sqrt{5}$), to give an alternative method:

$$
C = \left[
\begin{array}{ccc:ccc}
0 & 0 & 0 & 1 & 0 & 0 \\[2mm]
\dfrac{-7+11\sqrt{5}}{15} & 0 & 0 & 1 & \dfrac{9-\sqrt{5}}{30} & \dfrac{35-27\sqrt{5}}{30} \\[3mm]
\dfrac{569-1727\sqrt{5}}{912} & \dfrac{1351+215\sqrt{5}}{912} & 0 & 1 & \dfrac{-857-25\sqrt{5}}{912} & \dfrac{-151+1537\sqrt{5}}{912} \\[3mm]
\hdashline
\dfrac{9+\sqrt{5}}{48} & \dfrac{15+7\sqrt{5}}{48} & \dfrac{9-\sqrt{5}}{48} & 1 & 0 & \dfrac{15-7\sqrt{5}}{48} \\[3mm]
1 & 0 & 0 & 0 & 0 & 0 \\[2mm]
0 & 1 & 0 & 0 & 0 & 0
\end{array}
\right],
$$

$$
C = \left[
\begin{array}{ccc:ccc}
0 & 0 & 0 & 1 & 0 & 0 \\
1.1731165168 & 0 & 0 & 1 & 0.2254644008 & -0.8457945131 \\
-3.6104050407 & 2.0085028675 & 0 & 1 & -1.0009887055 & 3.6028908787 \\
\hdashline
0.2340847495 & 0.6385932467 & 0.1409152505 & 1 & 0 & -0.0135932467 \\
1 & 0 & 0 & 0 & 0 & 0 \\
0 & 1 & 0 & 0 & 0 & 0
\end{array}
\right].
$$

$$
\tilde{C} = \left[
\begin{array}{ccc:ccc}
0 & 0 & 0 & 1 & 0 & 0 \\
-2.1064498502 & 0 & 0 & 1 & 0.3745355992 & 3.1791278464 \\
4.8582120583 & 0.9542164307 & 0 & 1 & -0.8783972594 & -3.9340312296 \\
\hdashline
0.1409152505 & -0.0135932467 & 0.2340847495 & 1 & 0 & 0.6385932467 \\
1 & 0 & 0 & 0 & 0 & 0 \\
0 & 1 & 0 & 0 & 0 & 0
\end{array}
\right].
$$

453 General linear methods for stiff problems.

While methods suitable for non-stiff problems, as exemplified in the last two subsections, are easy to find and to analyse, methods appropriate for stiff problems seem to be able to hide their presence much more successfully.

However, there are such methods and we will discuss just a single family of them. These methods, which generalize singly implicit Runge–Kutta methods, were introduced by Butcher (1981a). They are characterized by the values of

r and s and by a positive real parameter θ. If H is defined as h/θ, then the quantities $y_1^{(n)}, y_2^{(n)}, \ldots, y_r^{(n)}$ are required to approximate scaled derivatives

$$y_i^{(n)} = \frac{H^{i-1}}{(i-1)!} \, y^{(i-1)}(x_n), \qquad 1, 2, \ldots, r,$$

$Y_1, Y_2, \ldots, Y_s$ are required to approximate the solution value at off-step points

$$Y_i = y(x_{n-1} + H\xi_i), \qquad i = 1, 2, \ldots, s,$$

in each case to order of accuracy $r + s - 1$. It can be shown that $\sigma(\theta C_{11}) = \{1\}$ iff $\xi_1, \xi_2, \ldots, \xi_s$ are the zeros of the generalized Laguerre polynomial $L_s^{(r-1)}$.

For $r = s = 2$, the range of θ values which yield A-stability can be identified as $(0, \theta_1]$, where $\theta_1 = 1.405$ is a zero of a known polynomial. Further information on the stability properties of methods in this family has been obtained using computer search techniques by Burrage and Chipman [1985]. They find A-stable methods as high as $r = 6$, $s = 9$ giving order 14. If s is restricted to $s = r$, $s = r + 1$ or $s = r + 2$, then a complete sequence up to order 10 can be found from methods which appear to be completely satisfactory from the point of view of stability. They are all either A-stable or deviate from strict A-stability by a negligible amount.

The difficulty of resolving various other questions concerned with incorporating general linear methods into practical software should not be underestimated. Nevertheless, there seem to be quite good prospects that in the foreseeable future a useful implementation of these methods for stiff and for non-stiff problems can be constructed.

Bibliography

Abbott, N. E., see Calahan, D. A.

Abd-el-Nabby, M. A., see Greig, D. M.

Adams, J. C., see Bashforth, F.

Ageev, V. N. (1979): A method for the numerical integration of a system of ordinary differential equations (Russian), *Trudy Moskov. Orden. Lenin. Ènerget. Inst. No. 438*, pp. 48–50.

Aguado, P. M. and Correas, J. M. (1981): An A-stable formula that overrides the Dahlquist ≤ 2 condition (Spanish), in *Proc. Eighth Portuguese–Spanish Conf. on Math. Coimbra, 1981*, vol. III, University of Coimbra, Coimbra, pp. 7–14.

Aiken, R. C. and Lapidus, L. (1974): An effective numerical integration method for typical stiff systems, *AIChE J.*, **20**, 368–375.

Aiken, R. C. and Lapidus, L. (1975): Pseudo steady state approximation for the numerical integration of stiff kinetic systems, *AIChE J.*, **21**, 817–820.

Aiken, R. C. and Lapidus, L. (1975a): Problem approximation for stiff ordinary differential equations, *AIChE J.*, **21**, 1227–1230.

Albrecht, J. (1955): Beiträge zum Runge–Kutta-Verfahren, *Z. Angew. Math. Mech.*, **35**, 100–110.

Albrecht, P. (1978): Zum Stabilitätsbegriff von Diskretisierungsverfahren. Neuere Ergebnisse zu einem alten Thema, *Jahrbuch Überblicke Mathematik*, **1978**, 107–130.

Albrecht, P. (1978a): On the order of composite multistep methods for ordinary differential equations, *Numer. Math.*, **29**, 381–396.

Albrecht, P. (1978b): Explicit, optimal stablity functionals and their application to cyclic discretization methods, *Computing*, **19**, 233–249.

Albrecht, P. (1979): *Die numerische Behandlung gewöhnlicher Differentialgleichungen*, Hanser, Munich; Akademie Verlag, Berlin.

Albrecht, P. (1980): Survey of recent results on composite integration methods for ordinary differential equations, especially cyclic methods, in *Information Processing 80: Proc. IFIP Congress, 1980* (Ed. S. H. Lavington), North Holland, Amsterdam, pp. 711–716.

Alemayehu, A. and März, R. (1982): A report on development, analysis and implementation of parametric Adams-*PC*-methods, in *Proc. Third Conf. on Numerical Treatment of Ordinary Differential Equations* (Ed. R. März), Seminarberichte No. 46, Humboldt University, Berlin, pp. 3–18.

Alexander, R. (1977): Diagonally implicit Runge–Kutta methods for stiff ODEs, *SIAM J. Numer. Anal.*, **14**, 1006–1021.

Alfeld, P. (1979): Inverse linear multistep methods for the numerical solution of initial value problems of ordinary differential equations, *Math. Comp.*, **33**, 111–124.

Alfeld, P. (1979a): An improved version of the reduction to scalar CDS method for the numerical solution of separably stiff initial value problems, *Math. Comp.*, **33**, 535–539.

Alfeld, P. (1979b): A special class of explicit linear multistep methods as basic methods for the correction in the dominant space technique, *Math. Comp.*, **33**, 1195–1212.

Alfeld, P. (1980): A method of skipping the transient phase in the solution of separably stiff ordinary initial value problems, *Math. Comp.*, **35**, 1173–1176.

Alfeld, P. and Lambert, J. D. (1977): Correction in the dominant space: a numerical technique for a certain class of stiff initial value problems, *Math. Comp.*, **31**, 922–938.

Allen, B. T. (1966): A new method of solving second-order differential equations when the first derivative is present, *Comput. J.*, **8**, 392–394.

Allen, R. H. and Pottle, C. (1968): Stable integration methods for electronic circuit analysis with widely separated time constants, in *Proc. Sixth Annual Allerton Conf. on Circuit and System Theory*, (Eds. T. N. Trick and R. T. Chien), University of Illinois, Urbana, Ill., pp. 311–320.

Al-Mutib, A., see Fawzy, T.

Alonso, R. (1960): A starting method for the three-point Adams predictor–corrector method, *J. Assoc. Comput. Mach.*, **7**, 176–180.

Alt, R. (1972): Deux théorèmes sur la *A*-stabilité des schémas de Runge–Kutta simplement implicites, *Rev. Française d'Automat. Informat. Recherche Opérationelle*, **6**, sér. R-3, 99–104.

Alt, R. (1978): *A*-stable one-step methods with step-size control for stiff systems of ordinary differential equations, *J. Comput. Appl. Math.*, **4**, 29–35.

Alt, R. (1982): Evaluation de l'erreur de discrétisation des méthodes à pas séparés à l'aide d'interpolation rationelle, in *Les Mathématiques de l'Informatique, AFCET Colloq., Paris, 1982* (Ed. B. Robinet), Centre National d'Études des Télécommunications, Thomson CSF, Paris, pp. 515–524.

Alt, R. and Ceschino, F. (1972): Mise en oeuvre de schémas *A*-stables du type de Runge–Kutta pour l'intégration des systèmes différentiels, *C. R. Acad. Sci. Paris sér. A*, **274**, 846–849.

Aluffi, F., Incerti, S. and Zirilli, F. (1980): Systems of equations and *A*-stable integration of second order ODEs, in *Numerical Optimisation of Dynamic Systems* (Eds. L. C. W. Dixon and G. P. Szegö), North-Holland, Amsterdam, pp. 289–307.

Amdursky, V., see Ziv, A.

Amel'chenko, V. V. see Vysotskiĭ, L. I.

Ananda, M., see Dyer, J.

Anatha Krishnaiah, U. = Anantha Krishnaiah, U. = Krishnaiah, U. A., see Jain, M. K.

Anderson, W. H. (1960): The solution of simultaneous ordinary differential equations using a general purpose digital computer, *Comm. ACM*, **3**, 355–360.

Andreassen, D. (1973): On k-step methods with almost constant coefficients, *BIT*, **13**, 265–271.

Andreev, A. S., Popov, V. A. and Sendov, B. (1981): Error estimates for the numerical solution of ordinary differential equations (Russian), *Zh. Vychisl. Mat. i. Mat. Fiz.*, **21**, 635–650; English translation: *USSR Comput. Math. and Math. Phys.*, **21**, No. 3, 104–120.

Andres, T. H., see Usmani, R. A.

Andria, G. D., Byrne, G. D. and Hill, D. R. (1973): Natural spline block implicit methods, *BIT*, **13**, 131–144.

Andria, G. D., Byrne, G. D. and Hill, D. R. (1973a): Integration formulas and schemes based on g-splines, *Math. Comp.*, **27**, 831–838.

Andrus, J. F. (1979): Numerical solution of systems of ordinary differential equations separated into subsystems, *SIAM J. Numer. Anal.*, **16**, 605–611.

410

Anīsīmova, Z. P. (1981): Limit theorems for recurrence sequences and their applications (Russian, English summary), *Dokl. Akad. Nauk Ukrain. SSR*, ser. A, **1981**, No. 1, 69–72.

Anīsīmova, Z. P. (1981a): The asymptotic behaviour of recurrent algorithms involving random errors (Russian), *Vychisl. Prikl. Mat. (Kiev)*, **45**, 42–46.

Ansorge, R. and Törnig, W. (1960): Zur Stabilität des Nyströmschen Verfahrens, *Z. Angew. Math. Mech.*, **40**, 568–570.

Ansorge, R. and Törnig, W. (Eds.) (1974): *Numerische Behandlung nichtlinearer Integrodifferential- und Differentialgleichungen*, Lecture notes in Mathematics No. 395, Springer, Berlin (see Micula, G.; Spijker, M. N.; Taubert, K.).

Antosiewicz, H. A. and Gautschi, W. (1962): Numerical methods in ordinary differential equations, in *A Survey of Numerical Analysis* (Ed. J. Todd), McGraw-Hill, New York, pp. 314–346.

Arndt, H. (1979): Lösung von gewöhnlichen Differentialgleichungen mit nichtlinearen Splines, *Numer. Math.*, **33**, 323–338.

Arndt, H. (1980): Über die Verwendung von nichlinearen Splines zur Integration gewöhnlicher Differentialgleichungen, *Z. Angew. Math. Mech.*, **60**, T280–T281.

Artem'ev, S. S. (1976): The construction of semi-implicit Runge–Kutta methods (Russian), *Dokl. Akad. Nauk SSSR*, **228**, 776–778; English translation: *Soviet Math. Dokl.*, **17**, 802–805.

Artem'ev, S. S. and Demidov, G. V. (1975): A-stable method for the solution of the Cauchy problem for stiff systems of ordinary differential equations, in *Optimization Techniques, IFIP Technical Conf. Novosibirsk, 1974* (Ed. G. I. Marchuk), Lecture Notes in Computer Science No. 27, Springer, New York, pp. 270–274.

Artem'ev, S. S. and Demidov, G. V. (1975a): A-stable method of Rosenbrock type of fourth-order accuracy for the solution of the Cauchy problem for stiff systems of ordinary differential equations (Russian), in *Some Problems of Computational and Applied Mathematics*, Nauka, Novosibirsk, pp. 214–220.

Artem'ev, S. S. and Demidov, G. V. (1978): An algorithm of variable order and step for the numerical solution of stiff systems of ordinary differential equations (Russian), *Dokl. Akad. Nauk SSSR*, **238**, 517–520; English translation: *Soviet Math. Dokl.*, **19**, 53–56.

Artemov, G. A. (1955): On a variant of Chaplygin's method for systems of ordinary differential equations of first order (Russian), *Dokl. Akad. Nauk SSSR*, **101**, 197–200.

Atanacković, T. M., see Djukić, Dj. S.

Avenhaus, J. (1971): Ein Verfahren zur Einschließung der Lösung des Anfangswertproblems, *Computing*, **8**, 182–190.

Axelsson, O (1964): Global integration of differential equations through Lobatto quadrature, *BIT*, **4**, 69–86.

Axelsson, O. (1969): A class of A-stable methods, *BIT*, **9**, 185–199.

Axelsson, O. (1972): A note on a class of strongly A-stable methods, *BIT*, **12**, 1–4.

Axelsson, O. (1974): On the efficiency of a class of A-stable methods, *BIT*, **14**, 279–287.

Babinkostov, V. (1979): A way of application of quadrature formulas for solving ordinary linear differential equations (Macedonian), *Mat. Fak. Univ. Kiril Metodij Skopje Godishen Zb.*, **30**, 57–63.

Babinkostov, V. (1980): A way of applying quadrature formulas for the numerical solution of systems of ordinary linear differential equations (Macedonian), *Mat. Bilten*, **29(30)**, No. 3–4, 67–72.

Babkin, B. N. (1948): Approximate solution of ordinary differential equations of any order by a method of successive approximations based on a theorem by S. A. Chaplygin on differential inequalities (Russian), *Dokl. Akad. Nauk SSSR*, **59**, 419–422.

Babuška, I., Práger, M. and Vitásek, E. (1966): *Numerical Processes in Differential Equations*, Interscience, London.

Bachmann, K. H. (1968): Genäherte Integration von Differentialgleichungssystemen mit Schrittweitensteuerung, *Z. Angew. Math. Mech.*, **48**, 210–212.

Bachmann, K. H. (1982): Explizite Runge–Kutta-Verfahren mit erweitertem Stabilitätsgebiet, in *Mathematische Verfahrenstechnik, Bielatal, 1981*, Weiterbildungszentrum Math. Kybernet. Rechentech. Informationsverarbeitung no. 58, Technical University of Dresden, Dresden, pp. 3–8.

Bader, G., see Hairer, E.

Baggott, E. A., see Levy, H.

Bakhvalov, N. S. (1955): On the estimation of the error in the numerical integration of differential equations by the Adams extrapolation method (Russian), *Dokl. Akad. Nauk SSSR*, **104**, 683–686.

Bakhvalov, N. S. (1955a): Some remarks concerning the numerical integration of differential equations by the method of finite differences (Russian), *Dokl. Akad. Nauk SSSR*, **104**, 805–808.

Bakiev, R. R. (1982): A method of approximate solution of the Cauchy problem by means of cubic splines (Russian), *Dokl. Akad. Nauk. UzSSR.*, **1982**, No. 7, 9–11.

Ball, W. E., see Papian, L. E.

Banchev, V. T. (1979): GERUN—a program for integration of ordinary differential equations by means of the generalized Runge–Kutta method (Russian, English summary), *Godishnik Vissh. Uchebn. Zaved. Tekhn. Fiz.*, **16**, No. 1, 125–140.

Bard, A., Ceschino, F., Kuntzmann, J. and Laurent, P. J. (1961): Formules de base de la méthode de Runge–Kutta, *Chiffres*, **4**, 31–37.

Barney, J. R., see Johnson, A. I.

Barozzi, G. C. and Bellomo, B. R. (1975): Un algoritmo per la soluzione numerica delle equazioni differenziali lineari del secondo ordine, *Boll. Un. Mat. Ital.*, ser. 4, **11**, 578–589.

Barton, D. (1980): On Taylor series and stiff equations, *ACM Trans. Math. Software*, **6**, 280–294.

Barton, D., Willers, I. M. and Zahar, R. V. M. (1971): Taylor series methods for ordinary differential equations—an evaluation, in *Mathematical Software* (Ed. I. R. Rice), Academic Press, New York, pp. 369–390.

Barton, D., Willers, I. M. and Zahar, R. V. M. (1971a): The automatic solution of systems of ordinary differential equations by the method of Taylor series, *Comput. J.*, **14**, 243–248.

Bashforth, F. and Adams, J. C. (1883): *An Attempt to Test the Theories of Capillary Action by Comparing the Theoretical and Measured Forms of Drops of Fluid, with an Explanation of the Method of Integration Employed in Constructing the Tables which Give the Theoretical Forms of Such Drops*, Cambridge University Press, Cambridge.

Bateman, H., see Bennett, A. A.

Battin, R. H. (1976): Resolution of Runge–Kutta–Nyström condition equations through eighth order, *AIAA J.*, **14**, 1012–1021.

Battye, D. J. and Parsaei, M. (1980): The effect of computer dependent factors when solving ordinary differential equations, in *Production and Assessment of Numerical Software* (Eds. M. A. Hennel and L. M. Delves), Academic Press, London, pp. 83–96.

Bauch, H. (1977): Zur Lösungseinschließung bei Anfangswertaufgaben gewöhnlicher Differentialgleichungen nach der Defektmethode, *Z. Angew. Math. Mech*, **57**, 387–396.

Bauch, H. (1980): Zur iterativen Lösungseinschließung bei Anfangswertproblemen mittels Intervallmethoden, *Z. Angew. Math. Mech.*, **60**, 137–145.

Bauch, H. (1982): Zur iterativen Lösung von Anfangswertaufgaben bei gewöhnlichen

412

Differentialgleichungen, *Zh. Vychisl. Mat. i Mat. Fiz.*, **22**, 309–321; English translation: *USSR Comput. Math. and Math. Phys.*, **22**, No. 2, 58–72.

Baumgarte, J. (1973): Asymptotische Stabilisierung von Integralen bei gewöhnlichen Differentialgleichungen 1. Ordnung, *Z. Angew. Math. Mech.*, **53**, 701–704.

Baumgarte, J. and Stiefel, E. (1974): Examples of transformations improving the numerical accuracy of the integration of differential equations in *Proc. Conf. on the Numerical Solution of Ordinary Differential Equations, University of Texas at Austin, 1972* (Ed. D. G. Bettis), Lecture Notes in Mathematics No. 362, Springer, Berlin, pp. 207–236.

Bayer, G. (1966): Remark on algorithm 218, Kutta Merson (P. M. Lukehart, *Comm. ACM*, **6**, 737, 1963), *Comm. ACM*, **9**, 273.

Beaudet, P. R. (1974): Multi-off-grid methods in multi-step integration of ordinary differential equations, in *Proc. Conf. on the Numerical Solution of Ordinary Differential Equations, University of Texas at Austin, 1972* (Ed. D. G. Bettis), Lecture Notes in Mathematics No. 362, Springer, Berlin, pp. 128–148.

Beaudet, P. R. and Feagin, T. (1975): The use of back corrections in multistep methods of numerical integration, *SIAM. J. Numer. Anal.*, **12**, 895–918.

Becker, R. I. (1981): Two-sided stability and convergence of multistage and multistep methods for ordinary differential equations. *J. Math. Anal. Appl.*, **81**, 453–473.

Békéssy, A., Kis, O. and Tarnay, Gy. (1969): On the method of R. H. Merson (Hungarian), *Közl. –MTA Számitástech. Automat. Kutató Int. Budapest*, **4**, 3–31.

Bellman, R. (1979): Selective computation—VI: stiff differential equations, *Nonlinear Anal.*, **3**, 909–911.

Bellomo, B. R., see Barozzi, G. C.

Belykh, V. M., see Gavurin, M. K.

Bennett, A. A., Milne, W. E. and Bateman, H. (1956): *Numerical Integration of Differential Equations*, Dover, New York.

Bennett, A. W. and Vichnevetsky, R. (Eds.)(1978): *Numerical Methods for Differential Equations and Simulation: Proc. IMACS (AICA) Internat. Sympos. Blacksburg, Virginia, 1977*, North-Holland, Amsterdam (see Blum, H.; Brown, R. L.; Byrne, G. D.; Carver, M. B.; Irby, T. C.; Rosenbaum, J. S.; Shampine, L. F.).

Bennett, M. M., see Brown, R. R.

Berbente, C. and Blumenfeld, M. (1979): A numerical method for ordinary differential equations using the derivative of the slope-function, *Bul. Inst. Politehn. Bucureşti Ser. Mec.*, **41**, No. 3, 3–12.

Berecová, A. (1972): An error estimation of a Huťa formula of the Runge–Kutta type of the sixth order, *Acta. Math. Univ. Comenian*, **28**, 67–90.

Beregovenko, G. Ya and Saukh, S. E. (1981): On the organization of parallel calculations for the solution of nonlinear differential equations (Russian, English summary), *Dokl. Akad. Nauk Ukrain. SSR*, ser A, **1981**, No. 9, 64–67.

Berman, M., see Chu, S. C.

Bettis, D. G. (1970): Stablization of finite difference methods of numerical integration, *Celestial Mech.*, **2**, 282–295.

Bettis, D. G. (1970a): Numerical integration of products of Fourier and ordinary polynomials, *Numer. Math.*, **14**, 421–434.

Bettis, D. G. (1973): A Runge–Kutta Nyström algorithm, *Celestial Mech.*, **8**, 229–233.

Bettis, D. G. (Ed.) (1974): *Proc. Conf. on the Numerical Solution of Ordinary Differential Equations, University of Texas at Austin 1972*, Lecture Notes in Mathematics No. 362, Springer, Berlin (see Baumgarte, J. and Stiefel, E.; Beaudet, P. R.; Bettis, D. G.; Burcher, J. C.; Danchick, R.; Howard, B. E.; Krogh, F. T.; Moore, H.; Stoer, J.).

Bettis, D. G. (1974a): Equations of condition for high order Runge–Kutta–Nyström formulae, in *Proc. Conf. on the Numerical Solution of Ordinary Differential Equa-*

tions, University of Texas at Austin, 1972 (Ed. D. G. Bettis), Lecture Notes in Mathematics No. 362, Springer, Berlin, pp. 76–91.

Bettis, D. G. (1978): Efficient embedded Runge–Kutta methods, in *Numerical Treatment of Differential Equations: Proc. Oberwolfach, 1976*, (Eds. R. Bulirsch, R. D. Grigorieff and J. Schröder) Lecture Notes in Mathematics No. 631, Springer, Berlin, pp. 9–18.

Bettis, D. G. (1979): Runge–Kutta algorithms for oscillatory problems, *Z. Angew. Math. Phys.*, **30**, 699–704.

Bettis, D. G., see also Stiefel, E.

Bettis, D. G. and Horn, M. K. (1979): An optimal $(m+3)[m+4]$ Runge–Kutta algorithm, *Celestial Mech.*, **14**, 133–140.

Bezvershenko, I. I. (1979): On a method for the solution of differential equations on the basis of the approximation of function by surface of parabolic form (Russian), in *Investigations on Modern Problems of Summation and Approximation of Functions and Their Applications. Collection of Scientific Works* (Ed. V. P. Motornyĭ), Dnepropetrovskĭ Gosudarstvennyĭ University, Dnepropetrovsk.

Bhattacharya, D. K., see Palanisamy, K. R.

Bickart, T. A. (1977): An efficient solution process for implicit Runge–Kutta methods, *SIAM J. Numer. Anal.*, **14**, 1022–1027.

Bickart, T. A. (1979): Transformed methods for the solution of stiff system equations, in *Internat. Symposium on Circuits and Systems, Tokyo, 1979*, IEEE Computer Society, New York, pp. 56–59.

Bickart, T. A., see also Rubin, W. B.; Sloate, H. M.; Tendler, J. M.

Bickart, T. A., Burgess, D. A. and Sloate, H. M. (1971): High order A-stable composite multistep methods for numerical integration of stiff differential equations, in *Proc. Ninth Annual Allerton Conference on Circuit and System Theory*, University of Illinois, Urbana, Ill, pp. 465–473.

Bickart, T. A. and Jury, E. I. (1978): Arithmetic tests for A-stability, $A[\alpha]$-stability, and stiff-stability, *BIT*, **18**, 9–21.

Bickart, T. A. and Picel, Z. (1973): High order stiffly stable composite multistep methods for numerical integration of stiff differential equations, *BIT*, **13**, 272–286.

Bickart, T. A. and Rubin, W. B. (1974): Composite multistep methods and stiff stability, in *Stiff Differential Systems* (Ed. R. A. Willoughby), Plenum Press, New York, pp. 21–36.

Bieberbach, L. (1930): *Theorie der Differentialgleichungen*, Springer, Berlin (reprinted 1944, Dover, New York).

Bieberbach, L. (1951): On the remainder of the Runge–Kutta formula in the theory of ordinary differential equations, *Z. Angew. Math. Phys.*, **2**, 233–248.

Birnbaum, I., see Chan, Y. N. I.

Birta, L. G. (1980): A quasiparallel method for the simulation of loosely coupled continuous subsystems, *Math. Comput. Simulation*, **22**, 189–199.

Bjurel, G. (1972): Modified linear multistep methods for a class of stiff ordinary differential equations, *BIT*, **12**, 142–160.

Blanc, C. (1954): Sur les formules d'intégration approchée d'équations différentielles, *Arch. Math. (Basel)*, **5**, 301–308.

Blanch, G. (1952): On the numerical solution of equations involving differential operators with constant coefficients, *Math. Comp.*, **6**, 219–223.

Blue, J. L. and Gummel, H. K. (1970): Rational approximations to matrix exponential for systems of stiff differential equations. *J. Comput. Phys.*, **5**, 70–83.

Blum, E. K. (1962): A modification of the Runge–Kutta fourth-order method, *Math. Comp.*, **16**, 176–187.

Blum, E. K. (1972): Numerical solution of ordinary differential equations, in *Numerical Analysis and Computation: Theory and Practice*, Addison-Wesley, Reading, Massachusetts, pp. 431–488.

414

Blum, H. (1978): Numerical integration of large scale systems using separate step sizes, in *Numerical Methods for Differential Equations and Simulation: Proc. IMACS (AICA) Internat. Symposium, Blacksburg, Virginia, 1977* (Eds. A. W. Bennett and R. Vichnevetsky), North-Holland, Amsterdam, pp. 19–23.

Blumenfeld, M., see Berbente, C.

Bobkov, V. V. (1973): A particular method of construction of rules for the approximate solution of differential equations (Russian), *Vestsi Akad. Navuk BSSR Ser. Fiz.-Mat. Navuk*, **1973**, No. 2, 52–58.

Bobkov, V. V. (1975): Methods of two-sided type for the numerical solution of differential equations (Russian), *Vestsi Akad. Navuk BSSR Ser. Fiz. - Mat. Navuk*, **1975**, No. 1, 25–28.

Bobkov, V. V. (1977): Explicit *A*-stable methods for the numerical integration of differential equations (Russian), *Dokl. Akad. Nauk BSSR*, **21**, 395–397.

Bobkov, V. V. (1977a): New one-step rules for solving differential equations (Russian), *Vestnik Beloruss. Gos. Univ.*, ser. I, **1977**, No. 1, 9–12.

Bobkov, V. V. (1978): A family of three-layer differences schemes (Russian, English summary), *Dokl. Akad. Nauk BSSR*, **22**, 21–24.

Bobkov, V. V. (1978a): New explicit *A*-stable methods for the numerical solution of differential equations (Russian), *Differentsial'nye Uravneniya*, **14**, 2249–2251; English translation: *Differential Equations*, **14**, 1595–1597.

Bobkov, V. V. (1978b): Some explicit stable methods for the numerical integration of differential equations (Russian, English summary), *Differentsial'nye Uravneniya i Primenen—Trudy Sem. Protsessy Optimal, Upravleniya I Sektsiya*, No. 21, pp. 9–19.

Bobkov, V. V. (1978c): A method for the construction of one-step methods for the numerical integration of differential equations (Russian), *Vestnik Beloruss. Gos. Univ.*, ser. I, **1978**, No. 1, 70–72.

Bobkov, V. V. (1980): Some nonlinear methods of numerical solution of differential equations (Russian), *Vestnik Beloruss. Gov. Univ.*, ser. I, **1980**, No. 2. 55–56.

Bobkov, V. V. (1981): Numerical methods with improved properties of agreement of differential and difference problems (Russian), *Vestnik Beloruss. Gos. Univ.*, ser. I, **1981**, No. 3, 61–65.

Bobkov, V. V. (1982): A family of numerical methods with improved stability properties (Russian), *Vestnik Beloruss. Gos. Univ.*, ser. I, **1982**, No. 2, 63–65.

Bobkov, V. V., see also Krylov, V. I.

Bobkov, V. V. and Din', Kh. Kh. (1970): A certain method of approximate solution of the Cauchy problems for the equation $y'' = f(x, y)$, *Vestsi Akad. Navuk BSSR Ser. Fiz. -Mat. Navuk*, **1970**, No. 5, 43–49.

Bobkov, V. V. and Gorodetskiĭ, L. M. (1980): Quasi-Newtonian methods of numerical integration of differential equations (Russian, English summary), *Dokl. Akad. Nauk BSSR*, **24**, No. 1, 11–14.

Bobkov, V. V. and Shalima, V. (1978): Explicit *A*-stable methods for second-order differential equations (Russian) *Differentsial'nye Uraveniya i Primenen—Trudy Sem. Protsessy Optimal, Upravleniya I Sektsiya*, No. 21, 21–30.

Bobkov, V. V. and Shkel', V. A. (1970): One-step rules for the approximate solution to the Cauchy problem for the equation $y' = f(x, y)$ in terms of the values of the right hand side of the equation and its derivatives, *Vestsi Akad. Navuk BSSR Ser. Fiz. -Mat. Navuk*, **1970**, No. 3, 60–67.

Bodnarchuk, P. Ĭ. and Maksimĭv, Ė. M. (1978): Nonlinear multipoint methods of solution of differential equations (Ukranian), in *Studies in the Qualitative Theory of Differential Equations and Its Applications* (Ed. V. Ya. Skorobogat'ko), Inst. Mat., Akad. Nauk Ukrain. SSR, Kiev, pp. 9–10.

Bodnarchuk, P. Ĭ. and Pelekh, Ya. M. (1978): Some methods of solution of differential equations (Ukranian), in *Studies in the Qualitative Theory of Differential Equations and Its Applications* (Ed. V. Ya. Skorobogat'ko), Inst. Mat., Akad. Nauk Ukrain. SSR, Kiev, pp. 11–12.

Böhmer, K. and Fleischmann, H. J. (1980): Self-adaptive discrete Newton methods for Runge–Kutta methods, in *Numerical Methods of Approximation Theory. (Conf. Math. Res. Inst., Oberwolfach, 1979)* (Eds. L. Collatz, G. Meinardus and H. Werner), Internat. Ser. Numer. Math. No. 52, Birkhäuser, Basel, vol. 5, 28–48.

Bokhoven, W. M. G. van = van Bokhoven, W. M. G.

Bond, J. E. and Cash, J. R. (1979): A block method for the numerical integration of stiff systems of ordinary differential equations, *BIT*, **19**, 429–447.

Bonnemoy, C. (1965): Comparaison stochastique des méthodes de Runge–Kutta de rang deux, trois, quatre, *Chiffres*, **8**, 161–177.

Bozzini, M. and Lenarduzzi, L. (1981): Formula a un passo basata sui polinomi di ottima approssimazione secondo Sard., *Riv. Mat. Univ. Parma*, ser. 4, **7**, 237–244.

Brandon, D. M. (1974): A new single-step implicit integration algorithm with *A*-stability and improved accuracy, *Simulation*, **23**, 17–29.

Bräuer, K., see Scholz, S.

Brayton, R. K. (1974): Numerical *A*-stability for difference-differential systems, in *Stiff Differential Systems* (Ed. R. A. Willoughby), Plenum Press, New York, pp. 37–48.

Brayton, R. K., see also Hatchel, G. D.

Brayton, R. K. and Conley, C. C. (1973): Some results on the stability and instability of the backward differentiation methods with non-uniform time steps, in *Topics in Numerical Analysis* (Ed. J. J. H. Miller), Academic Press, London, pp. 13–33.

Brayton, R. K. Gustavson, F. G. and Hachtel, G. D. (1972): A new efficient algorithm for solving differential-algebraic systems using implicit backward differentiation formulas, *Proc. IEEE*, **60**, 98–108.

Brayton, R. K., Gustavson, F. G. and Liniger, W. (1966): A numerical analysis of the transient behavior of a transistor circuit, *IBM J. Res. Develop.*, **10**, 292–299.

Brennan, P. A., see Ruehli, A. E.

Brezinski, C. (1974): Intégration des systèmes différentiels à l'aide du ρ-algorithme, *C. R. Acad. Sci. Paris*, sér. A, **278**, 875–878.

Brianzi, P. and Rebolia, L. (1982): Caratterisiche numeriche delle forma integrale delle equazioni differentiali di ordine *n, Calcolo*, **19**, 71–86.

Brock, P. and Murray, F. J. (1952): The use of exponential sums in step by step integration, *Math. Comp.*, **6**, 63–78.

Brodskiï, M. L. (1953): Asymptotic estimates of the errors in the numerical integration of systems of ordinary differential equations by difference methods (Russian), *Dokl. Akad. Nauk SSSR*, **93**, 599–602.

Brosilow, C. see Merluzzi, P.

Brouwer, D. (1937): On the accumulation of errors in numerical integration, *Astronom. J.*, **46**, No. 16, 149–153.

Brown, H. G., see Byrne, G. D.

Brown, R., see Sarafyan, D.

Brown, R. L. (1976): Numerical integration of linearized stiff ordinary differential equations, in *Numerical Methods for Differential Systems* (Eds. L. Lapidus and W. E. Schiesser), Academic Press, New York, pp. 39–44.

Brown, R. L. (1977): Some characteristics of implicit multistep multi-derivative integration formulas, *SIAM J. Numer. Anal.*, **14**, 982–993.

Brown, R. L. (1978): Evaluation of ordinary differential equation software, *BIT*, **18**, 103–105.

Brown, R. L. (1978a): Investigation of ODE integrators using interactive graphics, *Numerical Methods for Differential Equations and Simulation: Proc. IMACS (AICA) Internat. Symposium Blacksburg, Virginia, 1977* (Eds. A. W. Bennett and R. Vichnevetsky), North-Holland, Amsterdam, pp. 33–38.

Brown, R. L. (1979): Stability of sequences generated by nonlinear differential systems, *Math. Comp.*, **33**, 637–645.

Brown, R. L. (1979a): Stability analysis of nonlinear differential sequences generated numerically, *Comput. Math. Appl.*, **5**, 187–192.

Brown, R. R., Riley, J. D. and Bennett, M. M. (1965): Stability properties of Adams–Moulton type methods, *Math. Comp.*, **19**, 90–96.

Brunner, H. (1967): Stabilization of optimal difference operators, *Z. Angew. Math. Phys.*, **18**, 438–444.

Brunner, H. (1970): Marginal stability and stabilization in the numerical integration of ordinary differential equations, *Math. Comp.*, **24**, 635–646.

Brunner, H. (1971): Optimale lineare Mehrschrittverfahren: Elimination von schwacher Stabilität im Falle eines Systems nichtlinearer Differentialgleichungen, *B. I. Hochschultaschenbücher*, **4**, 139–163.

Brunner, H. (1972): A class of *A*-stable two-step methods based on Schur polynomials, *BIT*, **12**, 468–474.

Brunner, H. (1972a): A note on modified optimal linear multistep methods, *Math. Comp.*, **26**, 625–631.

Brunner, H. (1974): Recursive collocation for the numerical solution of stiff ordinary differential equations, *Math. Comp.*, **28**, 475–481.

Brunner, H. (1974a): Über Klassen von *A*-stabilen linearen Mehrschrittverfahren maximaler Ordnung, *Internat. Ser. Numer. Math.*, **19**, 67–75.

Brunner, H. (1975): The solution of systems of stiff nonlinear differential equations by recursive collocation using exponential functions, *Internat. Ser. Numer. Math.*, **27**, 29–44.

Brush, D. G., Kohfeld, J. J. and Thompson, G. T. (1967): Solution of ordinary differential equations using two 'off-step' points, *J. Assoc. Comput. Mach.*, **14**, 769–784.

Budak, B. M., see Gorbunov, A. D.

Budak, B. M. and Gorbunov, A. D. (1959): Stability of calculation processes in the solution of the Cauchy problem for the equation $dy/dx = f(x, y)$ by multipoint difference methods (Russian), *Dokl. Akad. Nauk SSSR*, **124**, 1191–1194.

Bui, T. D. (1977): On an *L*-stable method for stiff differential equations, *Inform. Process. Lett.*, **6**, 158–161.

Bui, T. D. (1977a): Errata and comments on a paper by J. R. Cash: 'Semi-implicit Runge–Kutta procedures with error estimates for the numerical integration of stiff systems of ordinary differential equations' (see Cash, J. R. (1976): *J. Assoc. Comput. Mach*, **23**, 455–460), *J. Assoc. Comput. Mach.*, **24**, 623.

Bui, T. D. (1979): A note on the Rosenbrock procedure, *Math. Comp.*, **33**, 971–975.

Bui, T. D. (1979a): Some *A*-stable and *L*-stable methods for the numerical integration of stiff ordinary differential equations, *J. Assoc. Comput. Mach.*, **26**, 483–493.

Bui, T. D. (1981): Solving stiff differential equations for simulation. *Math. Comput. Simulation*, **23**, No. 2, 149–156.

Bui, T. D. (1981a): Solving stiff differential equations in the simulation of physical systems, *Simulation*, **37**, No. 2, 37–46.

Bui, T. D. and Bui, T. R. (1979): Numerical methods for extremely stiff systems of ordinary differential equations, *Appl. Math. Modelling*, **3**, 355–358.

Bui, T. D. and Bui, T. R. (1980): Solving stiff differential equations for simulation: applications to the simulation of a new molecular laser system, in *Simulation of Systems '79. Proc. Ninth IMACS Congress, Sorrento, 1979* (Eds. L. Dekker, G. Savastano and G. C. Vansteenkiste), North-Holland, Amsterdam, pp. 89–94.

Bui, T. D. and Ghaderpanah, S. S. (1978): Modified Richardson extrapolation scheme for error estimate in implicit Runge–Kutta procedures for stiff systems of ordinary differential equations, in *Proc. Seventh Manitoba Conf. on Numerical Mathematics and Computing, Winnipeg, Man., 1977* (Eds. D. McCarthy and H. C. Williams), Congress Numer. No. 20, Utilitas Mathematica, Winnipeg, Man., pp. 251–268.

Bui, T. D. and Poon, S. W. H. (1981): On the computational aspects of Rosenbrock procedures with built-in error estimates for stiff systems, *BIT*, **21**, 168–174.

Bui, T. R., see Bui, T. D.

Bukovics, E. (1950): Eine Verbesserung und Verallgemeinerung des Verfahrens von Blaess zur numerischen Integration gewöhnlicher Differentialgleichungen, *Österreich. Ing. -Archiv*, **4**, 338–349.

Bukovics, E. (1953): Beiträge zur numerischen Integration I, *Monatsh. Math.*, **57**, 217–245.

Bukovics, E. (1953a): Beiträge zur numerischen Integration II, *Monatsh. Math.*, **57**, 333–350.

Bukovics, E. (1954): Beiträge zur numerischen Integration III, *Monatsh. Math.*, **58**, 258–265.

Bulirsch, R. (1964): Bemerkungen zur Romberg-Integration, *Numer. Math.*, **6**, 6–16.

Bulirsch, R., Grigorieff, R. D. and Schröder, J. (Eds.) (1978): *Numerical Treatment of Differential Equations: Proc. Oberwolfach, 1976*, Lecture Notes in Mathematics No. 631, Springer, Berlin (see Bettis, D. G.; Frank, R. and Ueberhuber, C. W.; Friedli, A.; Jeltsch, R.; Mannshardt, R.; Scherer, R.; and Stetter, H. J.)

Bulirsch, R. and Stoer, J. (1964): Fehlerabschätzungen und Extrapolation mit rationalen Funktionen bei Verfahren vom Richardson-Typus, *Numer. Math.*, **6**, 413–427.

Bulirsch, R. and Stoer, J. (1966): Numerical treatment of ordinary differential equations by extrapolation methods, *Numer. Math.*, **8**, 1–13.

Bulirsch, R. and Stoer, J. (1966a): Asymptotic upper and lower bounds for results of extrapolation methods, *Numer. Math*, **8**, 93–104.

Bulirsch, R. and Stoer, J. (1967): Asymptotische Fehlerschranken bei Extrapolationsverfahren, *Internat. Ser. Numer. Math.*, **7**, 165–170.

Burgess, D. A., see Bickart, T. A.

Burka, M. K. (1982): Solution of stiff ordinary differential equations by decomposition and orthogonal collocation, *AIChE J.*, **28**, 11–20.

Burrage, K. (1978): A special family of Runge–Kutta methods for solving stiff differential equations, *BIT*, **18**, 22–41.

Burrage, K. (1978a): High order algebraically stable Runge–Kutta methods, *BIT*, **18**, 373–383.

Burrage, K. (1980): Non-linear stability of multivalue multiderivative methods, *BIT*, **20**, 316–325.

Burrage, K. (1982): Efficiently implementable algebraically stable Runge–Kutta methods, *SIAM J. Numer. Anal.*, **19**, 245–258.

Burrage, K. and Butcher, J. C. (1979): Stability criteria for implicit Runge–Kutta methods, *SIAM J. Numer. Anal.*, **16**, 46–57.

Burrage, K. and Butcher, J. C. (1980): Non-linear stability of a general class of differential equation methods, *BIT*, **20**, 185–203.

Burrage, K., Butcher, J. C. and Chipman, F. H. (1980): An implementation of singly-implicit Runge–Kutta methods, *BIT*, **20**, 326–340.

Burrage, K. and Moss, P. (1980): Simplifying assumptions for the order of partitioned multivalue methods, *BIT*, **20**, 452–465.

Butcher, J. C. (1963): Coefficients for the study of Runge–Kutta integration processes, *J. Austral. Math. Soc.*, **3**, 185–201.

Butcher, J. C. (1963a): On the integration processes of A. Huťa, *J. Austral. Math. Soc.*, **3**, 202–206.

Butcher, J. C. (1964): Implicit Runge–Kutta processes, *Math. Comp.*, **18**, 50–64.

Butcher, J. C. (1964a): Integration processes based on Radau quadrature formulas, *Math. Comp.*, **18**, 233–244.

Butcher, J. C. (1964b): On Runge–Kutta processes of high order, *J. Austral. Math. Soc.*, **4**, 179–194.

Butcher, J. C. (1965): A modified multistep method for the numerical integration of ordinary differential equations, *J. Assoc. Comput. Mach.* **12**, 124–135.

Butcher, J. C. (1965a): On the attainable order of Runge–Kutta methods, *Math. Comp.*, **19**, 408–417.

Butcher, J. C. (1966): Some developments in the theory of Runge–Kutta methods, in *Information Processing 65: Proc. IFIP Congress 1965* (Ed. W. A. Kalenich), vol. 2, Spartan Books, Washington, D. C., pp. 561–562.

Butcher, J. C. (1966a): On the convergence of numerical solutions to ordinary differential equations, *Math. Comp.*, **20**, 1–10.

Butcher, J. C. (1967): A multistep generalization of Runge–Kutta methods with four or five stages, *J. Assoc. Comput. Mach.*, **14**, 84–99.

Butcher, J. C. (1969): The effective order of Runge–Kutta methods, in *Conf. on the Numerical Solution of Differential Equations, Dundee, 1969* (Ed. J. L. Morris), Lecture Notes in Mathematics No. 109, Springer, Berlin, pp. 133–139.

Butcher, J. C. (1971): An approximation theorem in numerical analysis, in *A Spectrum of Mathematics* (Ed. J. C. Butcher), Auckland and Oxford University Presses, pp. 121–125.

Butcher, J. C. (1972): An algebraic theory of integration methods, *Math. Comp.*, **26** 79–106.

Butcher, J. C. (1972a): A convergence criterion for a class of integration methods, *Math. Comp.*, **26**, 107–117.

Butcher, J. C. (1973): The order of numerical methods for ordinary differential equations, *Math. Comp.*, **27**, 793–806.

Butcher, J. C. (1973a): Order conditions for a general class of numerical methods for ordinary differential equations, in *Topics in Numerical Analysis* (Ed. J. J. H. Miller), Academic Press, London, pp. 35–40.

Butcher, J. C. (1974): The order of differential equation methods, in Bettis, D. G. (Ed.): *Proc. Conf. on the Numerical Solution of Ordinary Differential Equations, University of Texas at Austin, 1972* (Ed. D. G. Bettis), Lecture Notes in Mathematics No. 362, Springer, Berlin, pp. 72–75.

Butcher, J. C. (1974a): Order conditions for general linear methods for ordinary differential equations, *Internat. Ser. Numer. Math.*, **19**, 77–81.

Butcher, J. C. (1974b): Computation and theory in ordinary differential equations, *Math. Chronicle*, **3**, 63–69.

Butcher, J. C. (1975): An order bound for Runge–Kutta methods, *SIAM J. Numer. Anal.*, **12**, 304–315.

Butcher, J. C. (1975a): A stability property of implicit Runge–Kutta methods, *BIT*, **15**. 358–361.

Butcher, J. C. (1976): A class of implicit methods for ordinary differential equations, in *Numerical Analysis: Proc. Dundee Conf., 1975* (Ed. G. A. Watson), Lecture Notes in Mathematics No. 506, Springer, Berlin, pp. 28–37.

Butcher, J. C. (1976a): On the implementation of implicit Runge–Kutta methods, *BIT*, **16**, 237–240.

Butcher, J. C. (1977): On *A*-stable implicit Runge–Kutta methods, *BIT*, **17**, 375–378.

Butcher, J. C. (1979): A transformed implicit Runge–Kutta method, *J. Assoc. Comput. Mach.*, **26**, 731–738.

Butcher, J. C. (1980): Some implementation schemes for implicit Runge–Kutta methods in *Numerical Analysis: Proc. Dundee Conference, 1979* (Ed. G. A. Watson), Lecture Notes in Mathematics No. 773, Springer, Berlin, pp. 12–24.

Butcher, J. C. (1981): Stability properties for a general class of methods for ordinary differential equations, *SIAM J. Numer. Anal.*, **18**, 37–44.

Butcher, J. C. (1981a): A generalization of singly-implicit methods, *BIT*, **21**, 175–189.

Butcher, J. C. (1982): A short proof concerning B-stability, *BIT*, **22**, 528–529.

Butcher, J. C., see also Burrage, K.

Butorin, N. N. (1974): Improvement of the accuracy of the Runge–Kutta method (Russian), *Zh. Vychisl. Mat. i Mat. Fiz.*, **14**, 493–495; English translation: *USSR Comput. Math. and Math. Phys.*, **14**, No. 2, 221–222.

Byrne, G. D. (1967): Parameters for pseudo Runge–Kutta methods, *Comm. ACM*, **10**, 102–104.

Byrne, G. D. (1978): Some software for stiff systems of differential equations, in *Numerical Methods for Differential Equations and Simulation: Proc. IMACS (AICA) Internat. Sympos. Blacksburg, Virginia, 1977* (Eds. A. W. Bennett and R. Vichnevetsky), North-Holland, Amsterdam, pp. 45–50.

Byrne, G. D., see also Andria, G. D.; Hindmarsh, A. C.

Byrne, G. D. and Chi, D. N. H. (1972): Linear multistep formulas based on g-splines, *SIAM J. Numer. Anal.*, **9**, 316–324.

Byrne, G. D., Gear, C. W., Hindmarsh, A. C., Hull, T. E., Krogh, F. T. and Shampine, L. F. (1976): Panel discussion of quality software for ODEs, in *Numerical Methods for Differential Systems*, (Eds. L. Lapidus and W. E. Schiesser), Academic Press, New York, pp. 267–285.

Byrne, G. D. and Hindmarsh, A. C. (1975): A polyalgorithm for the numerical solution of ordinary differential equations, *ACM Trans. Math. Software*, **1**, 71–96.

Byrne, G. D., Hindmarsh, A. C., Jackson, K. R. and Brown, H. G. (1977): A comparison of two ODE codes: GEAR and EPISODE, *Computers and Chem. Engrg.*, **1**, 133–147.

Byrne, G. D. and Lambert, R. J. (1966): Pseudo-Runge–Kutta methods involving two points, *J. Assoc. Comput. Mach.*, **13**, 114–123.

Cahill, L. W. (1974): A stable numerical integration scheme for nonlinear systems, in *Proc. Seventh Hawaii Internat. Conf. on System Sciences*, Western Periodicals, California, pp. 183–185.

Calahan, D. A. (1967): Numerical solution of linear systems with widely separated time constants, *Proc. IEEE*, **55**, 2016–2017.

Calahan, D. A. (1968): A stable, accurate method of numerical integration for nonlinear systems, *Proc. IEEE*, **56**, 744.

Calahan, D. A. and Abbott, N. E. (1970): Stability analysis of numerical integration, in *Proc. Tenth Midwest Symposium on Circuit Theory*, **1970**, 1-2-1 to 1-2-20.

Calistru, N. (1981): On the convergence of some multistep methods, *An. Ştiinţ. Univ. 'Al. I. Cuza' Iaşi, Secţ. I a Mat. (new ser.)*, **27**, 393–402.

Callender, E. D. (1971): Single step methods and low order splines for solutions of ordinary differential equations, *SIAM J. Numer. Anal.*, **8**, 61–66.

Calligani, I., see Casulli, V.

Calvo, M. and Lisbona, F. J. (1979): On the concept of stability in step-by-step methods (Spanish), in *Contributions in Probability and Mathematical Statistics, Teaching of Mathematics and Analysis* (Spanish), Grindley, Granada, pp. 296–307.

Calvo, M. and Quemada, M. M. (1982): On the stability of rational Runge–Kutta methods, *J. Comput. Appl. Math.*, **8**, 289–292.

Capozza, M. and Costabile, F. (1974): Metodi pseudo Runge–Kutta del quinto ordine, *Bulzoni Editore*, **1974**, 1–12.

Capra, V. (1956): Nuove formule per l'integrazione numerica delle equazioni differenziali ordinarie del $1°$ e del $2°$ ordine, *Rend. Sem. Mat. Univ. Politec. Torino*, **16**, 301–359.

Capra, V. (1957): Valutazione degli errori nella integrazione numerica dei systemi di equazione differenziali ordinarie, *Atti Accad. Sci. Torino Cl. Sci. Fis. Mat. Natur.*, **91**, 188–203.

Carr, J. W. (1958): Error bounds for the Runge–Kutta single-step integration process, *J. Assoc. Comput. Mach.*, **5**, 39–44.

420

Carroll, J. (1982): On the implementation of exponentially fitted one-step methods for the numerical integration of stiff linear initial value problems, in *Computational and Asymptotic Methods for Boundary and Interior Layers: Proc. BAIL II Conf., Dublin, 1982*, (Ed. J. J. H. Miller), Boole Press Conf. Ser, No. 4, Boole Press, Dublin, pp. 165–170.

Carver, M. B. (1977): Efficient handling of discontinuities and time delays in ordinary differential equations, in *Proc. Simulation '77 Symposium, Montreux, Switzerland*, (Ed. M. Hamza), Acta Press, Zurich.

Carver, M. B. (1978): Efficient integration over discontinuities in ordinary differential equations, in *Numerical Methods for Differential Equations and Simulation: Proc. IMACS (AICA) Internat. Symposium, Blacksburg, Virginia, 1977* (Eds. A. W. Bennett and R. Vichnevetsky), North-Holland, Amsterdam, pp. 51–56.

Carver, M. B. and MacEwen, S. R. (1981): On the use of sparse matrix approximation to the Jacobian in integrating large sets of ordinary differential equations, *SIAM J. Sci. Statist. Comput.*, **2**, 51–64.

Carver, M. B. and MacEwen, S. R. (1982): Automatic partitioning in ordinary differential equation integration, in *Progress in Modelling and Simulation*, Academic Press, London, pp. 265–280.

Case, J. (1969): A note on the stability of predictor–corrector techniques, *Math. Comp.*, **23**, 741–749.

Cash, J. R. (1972): The numerical solution of linear stiff differential equations, *Proc. Cambridge Philos. Soc.*, **71**, 505–515.

Cash, J. R. (1974): The numerical solution of systems of stiff ordinary differential equations, *Proc. Cambridge Philos. Soc.*, **76**, 443–456.

Cash, J. R. (1975): A class of implicit Runge–Kutta methods for the numerical integration of stiff ordinary differential equations, *J. Assoc. Comput. Mach.*, **22**, 504–511.

Cash, J. R. (1976): Semi-implicit Runge–Kutta procedures with error estimates for the numerical integration of stiff systems of ordinary differential equations, *J. Assoc. Comput. Mach.*, **23**, 455–460. For comments on this paper see Bui, T. D. (1977a).

Cash, J. R. (1977): A class of iterative algorithms for the integration of stiff systems of ordinary differential equations. *J. Inst. Math. Appl.*, **19**, 325–335.

Cash, J. R. (1977a): On a class of cyclic methods for the numerical integration of stiff systems of ODEs, *BIT*, **17**, 270–280.

Cash, J. R. (1977b): On a class of implicit Runge–Kutta procedures, *J. Inst. Math. Appl.*, **19**, 455–470.

Cash, J. R. (1977c): A note on the computational aspects of a class of implicit Runge–Kutta procedures, *J. Inst. Math. Appl.*, **20**, 425–441.

Cash, J. R. (1978): A note on a class of modified backward differentiation schemes. *J. Inst. Math. Appl.*, **21**, 301–313.

Cash, J. R. (1978a): High order methods for the numerical integration of ordinary differential equations, *Numer. Math.*, **30**, 385–409.

Cash, J. R. (1979): Diagonally implicit Runge–Kutta formulae with error estimates, *J. Inst. Math. Appl.*, **24**, 293–301.

Cash, J. R. (1979a): *Stable Recursions with Applications to the Numerical Solution of Stiff Systems*, Academic Press, London.

Cash, J. R. (1980): On the integration of stiff systems of ODEs using extended backward differentiation formulae, *Numer. Math.* **34**, 235–246.

Cash, J. R. (1980a): A semi-implicit Runge–Kutta formula for the integration of stiff systems of ordinary differential equations, *Chem. Engrg. J.*, **20**, 219–224.

Cash, J. R. (1981): On the design of high order exponentially fitted formulae for the numerical integration of stiff systems, *Numer. Math.*, **36**, 253–266.

Cash, J. R. (1981a): Second derivative extended backward differentiation formulas for the numerical integration of stiff systems, *SIAM J. Numer. Anal.*, **18**, 21–36.

Cash, J. R. (1981b): On the exponential fitting of composite, multiderivative linear multistep methods, *SIAM J. Numer. Anal.*, **18**, 808–821.

Cash, J. R. (1981c): High order *P*-stable formulae for the numerical integration of periodic initial value problems, *Numer. Math.*, **37**, 355–370.

Cash, J. R. (1981d): A note on the exponential fitting of blended, extended linear multistep methods, *BIT*, **21**, 450–454.

Cash, J. R. (1982): On improving the absolute stability of local extrapolation, *Numer. Math.*, **40**, 329–337.

Cash, J. R. (1982a): On the solution of block tridiagonal systems of linear algebraic equations having a special structure, *SIAM J. Numer. Anal.*, **19**, 1220–1232.

Cash, J. R., see Bond, J. E.

Cash, J. R. and Liem, C. B. (1978): On the computational aspects of semi-implicit Runge–Kutta methods, *Comput. J.*, **21**, 363–365.

Cash, J. R. and Liem, C. B. (1980): On the design of a variable order, variable step diagonally implicit Runge–Kutta algorithm, *J. Inst. Math. Appl.*, **26**, 87–91.

Cash, J. R. and Miller, J. C. P. (1978): On an iterative approach to the numerical solution of difference schemes, *Comput. J.*, **22**, 184–187.

Cash, J. R. and Singhal, A. (1982): Mono-implicit Runge–Kutta formulae for the numerical integration of stiff differential systems, *IMA J. Numer. Anal.*, **2**, 211–227.

Cassity, C. R. (1966): Solutions of the fifth-order Runge–Kutta equations, *SIAM J. Numer. Anal.*, **3**, 598–606.

Cassity, C. R. (1969): The complete solution of the fifth-order Runge–Kutta equations, *SIAM J. Numer. Anal.*, **6**, 432–436.

Casulli, V., Galligani, I. and Trigiante, D. (1978): *Alcune applicazioni del calcolo dell'errore globale nell'integrazione numerica di equazioni differenziali ordinarie*, Pubbl. ser. III no. 123, Istituto per le Applicazioni del Calcolo 'Mauro Picone', Rome, 14pp.

Cauchy, A. (1840): Mémoire sur l'intégration des équations différentielles, in *Oeuvres complètes d'Augustin Cauchy*, 2nd series, vol 11, Gauthier-Villars, Paris (1913), pp. 399–465.

Céa, J. (1965): Equations différentielles méthode d'approximation discrète *p*-implicite, *Chiffres*, **8**, 179–194.

Céa, J. (1966): *p*-implicit methods for ordinary differential equations, in *Information Processing 65: Proc. IFIP Congress 1965* (Ed. W. A. Kalenich), vol. 2, Spartan Books, Washington, D. C., pp. 563–564.

Certaine, J. (1960): The solution of ordinary differential equations with large time constants, in *Mathematical Methods for Digital Computers*, (Eds. A. Ralston and H. S. Wilf), Wiley, New York, pp. 128–132.

Ceschino, F. (1954): Critère d'utilisation du procédé de Runge–Kutta, *C. R. Acad. Sci. Paris*, **238**, 986–988.

Ceschino, F. (1954a): Critère d'utilisation du procédé de Runge–Kutta, *C.R. Acad. Sci. Paris*, **238**, 1553–1555.

Ceschino, F. (1956): L'intégration approchée des équations différentielles, *C. R. Acad. Sci. Paris*, **243**, 1478–1479.

Ceschino, F. (1959): Sur une formule de Runge–Kutta de rang cinq, *Chiffres*, **2**, 39–42.

Ceschino, F. (1960): Sur certaines applications de l'intégration approchée, in *Rome Symposium: Symposium on the Numerical Treatment of Ordinary Differential Equations—Integral and Integro-differential Equations: Proc.*, Birkhauser, Basel, pp. 42–47.

Ceschino, F. (1961): Une méthode de mise en oeuvre des formules d'Obrechkoff pour l'intégration des équations différentielles, *Chiffres*, **4**, 49–54.

Ceschino, F. (1961a): Modification de la longueur du pas dans l'intégration numérique par les méthodes à pas liés, *Chiffres*, **4**, 101–106.

Ceschino, F. (1962): Evaluation de l'erreur par pas dans les problèmes différentiels, *Chiffres*, **5**, 223–229.

Ceschino, F., see also Alt, R.; Bard, A.

Ceschino, F. and Kuntzmann, J. (1958): Impossibilité d'un certain type de formule d'intégration approchée à pas liés, *Chiffres*, **1**, 95–101.

Ceschino, F. and Kuntzmann, J. (1959): Remarques sur l'erreur dans la résolution approchée des problèmes de conditions initiales, *Chiffres*, **2**, 249–252.

Ceschino, F and Kuntzmann, J (1960): Faut-il passer à la forme canonique dans les problèmes différentiels de conditions initiales?, in *Information Processing*, UNESCO, Paris, pp. 33–36.

Ceschino, F. and Kuntzmann, J. (1963): *Problèmes Différentiels de Conditions Initiales*, Dunod, Paris: English translation: *Numerical Solution of Initial Value Problems*, Prentice Hall, Englewood Cliffs, N. J., 1966.

Chai, A. S. (1968): Error estimate of a fourth-order Runge–Kutta method with only one initial derivative evaluation, *AFIPS Conf. Proc.*, **32**, 467–471.

Chai, A. S. (1968a): A modified Runge–Kutta method, *Simulation*, **10**, 221–223.

Chai, A. S. (1972): A fifth-order modified Runge–Kutta integration algorithm, *Simulation*, **18**, 21–27.

Chakravarti, P. C. and Worland, P. B. (1971): A class of self-starting methods for the numerical solution of $y'' = f(x, y)$, *BIT*, **11**, 368–383.

Chakravarti, P. C. and Worland, P. B. (1973): Automatic stepsize control in the numerical solution of $y'' = f(x, y)$, in *Proc. Second Manitoba Conf. Numerical Math Winnipeg, 1972*. (Eds. R. S. D. Thomas and H. C. Williams), Congress, Numer. No. 7, Utilitas Mathematica, Winnipeg, pp. 417–432.

Chan, Y. N. I., Birnbaum, I. and Lapidus, L. (1978): Solution of stiff differential equations and the use of imbedding techniques. *Indusr. and Engin. Chemistry Fundamentals*, **17**, 133–148.

Chang, K. T. (1959): On the numerical solution of simultaneous linear differential equations, *J. Soc. Indust. Appl. Math.*, **7**, 468–472.

Chang, Y. F., see Corliss, G. F.

Charnyĭ, I. A., see Vlasov, I. O.

Chartres, B. A. and Stepleman, R. S. (1971): Convergence of difference methods for initial and boundary value problems with discontinuous data *Math. Comp.*, **25**, 729–732.

Chartres, B. A. and Stepleman, R. S. (1972): A general theory of convergence for numerical methods, *SIAM J. Numer. Anal.*, **9**, 476–492.

Chartres, B. A. and Stepleman, R. S. (1974): Actual order of convergence of Runge–Kutta methods on differential equations with discontinuities, *SIAM J. Numer. Anal.*, **11**, 1193–1206.

Chartres, B. A. and Stepleman, R. S. (1976): Convergence of linear multistep methods for differential equations with discontinuities, *Numer. Math.*, **27**, 1–10.

Chase, P. E. (1962): Stability properties of predictor–corrector methods for ordinary differential equations, *J. Assoc. Comput. Mach.*, **9**, 457–468.

Chawla, M .M. (1981): Two-step fourth order P-stable methods for second order differential equations, *BIT*, **21**, 190–193.

Chawla, M. M. and Sharma, S. R. (1980): Families of direct fourth-order methods for the numerical solution of general second-order initial-value-problems, *Z. Angew. Math. Mech*, **60**, 469–478.

Chawla, M. M. and Sharma, S. R. (1981): Intervals of periodicity and absolute stability of explicit Nyström methods for $y'' = f(x, y)$, *BIT*, **21**, 455–464.

Chawla, M. M. and Sharma, S. R. (1981a): Families of fifth order Nyström methods for $y'' = f(x, y)$ and intervals of periodicity, *Computing*, **26**, 247–256.

Chen, Y. M. and Lee, D. T. S. (1981): Numerical methods for solving differential equations with inadequate data, *J. Comput. Phys.*, **42**, 238–256.

Cherepennikov, V. B. (1976): Approximate integration of a system of linear differential equations in a small time space (Russian), in *Asymptotic Methods in Systems Theory* (Ed. A. N. Panchenkov), No. 9, Irkutsk. Gos. Univ., Irkutsk, pp. 163–173.

Chern, I. L., see Miranker, W. L.

Chernorutskiĭ, I. G., see Rakitskiĭ, Yu. V.

Chi, D. N. H., see Byrne, G. D.

Chipman, F. H. (1971): *A*-stable Runge–Kutta processes, *BIT*, **11**, 384–388.

Chipman, F. H. (1973): The implementation of Runge–Kutta implicit processes, *BIT*, **13**, 391–393.

Chipman, F. H. (1976): A note on implicit *A*-stable *R-K* methods with parameters, *BIT*, **16**, 223–225.

Chipman, F. H., see Burrage, K.

Chistyakova, O. G., see Dombrovskaya, L. M.

Christensen, B., see Skelboe, S.

Christiansen, J. (1970): Numerical solution of ordinary simultaneous differential equations of the 1st order using a method for automatic step change, *Numer. Math.*, **14**, 317–324; Corrigendum: *Numer. Math.*, **15**, before p. 1.

Christiansen, J. (1973): Algorithm 77, solving a system of simultaneous ordinary differential equations of the first order using a method for automatic step change, *Comput. J.*, **16**, 187–188.

Chu, S. C. and Berman, M. (1974): An exponential method for the solution of systems of ordinary differential equations, *Comm. ACM*, **17**, 699–702.

Clasen, R. J., Garfinkel, D., Shapiro, N. Z. and Roman, G. C. (1978): A method for solving certain stiff differential equations, *SIAM J. Appl. Math.*, **34**, 732–742.

Clenshaw, C. W. (1957): The numerical solution of linear differential equations in Chebyshev series, *Proc. Cambridge Philos. Soc.*, **53**, 134–149.

Clough, D. E., see Weimer, A. W.

Cohen, C. J. and Hubbard, E. C. (1960): An algorithm applicable to numerical integration of orbits in multirevolution steps, *Astronom. J.*, **65**, 454–456.

Coleman, J. P. (1980): A new fourth-order method for $y'' = g(x)y + r(x)$, *Comput. Phys. Comm.*, **19**, 185–195.

Coleman, J. P. and Mohamed, J. (1978): On de Vogelaere's method for $y'' = f(x, y)$, *Math. Comp.*, **32**, 751–762.

Coleman, J. P. and Mohamed, J. (1979): De Vogelaere's method with automatic error control, *Comput. Phys. Comm.*, **17**, 283–300.

Collatz, L. (1942): Natürliche Schrittweite bei numerischer Integration von Differentialgleichungssystemen, *Z. Angew. Math. Mech.*, **22**, 216–225.

Collatz, L. (1949): Differenzenverfahren zur numerischen Integration von gewöhnlichen Differentialgleichungen *n*-ter Ordnung, *Z. Angew. Math. Mech.*, **29**, 199–209; Corrigendum: *Z. Angew. Math. Mech.*, **29**, 319.

Collatz, L. (1953): Über die Instablität beim Verfahren der zentralen Differenzen für Differentialgleichungen zweiter Ordnung, *Z. Angew. Math. Phys.*, **4**, 153–154.

Collatz, L. (1955): *Numerische Behandlung von Differentialgleichungen*, 2nd ed., Springer, Berlin.

Collatz, L. (1960): *The Numerical Treatment of Differential Equations*, 3rd ed., Springer, Berlin.

Collatz, L. and Zurmühl, R. (1942): Zur Genauigkeit verschiedener Integrationsverfahren bei gewöhnlichen Differentialgleichungen, *Ingenieur-Archiv*, **13**, 34–36.

Comincioli, V., Monti, G. P., Pierini, G. and Dalle Rive, L. (1972): *La risoluzione numerica delle equazioni differenziali ordinarie di tipo 'stiff'* (Ed. E. Magenes), Publication No. 36, Laboratorio di Analisi Numerica del Consiglio Nazionale delle Ricerche, Pavia.

Conley, C. C., see Brayton, R. K.

Conte, S. D. and Reeves R. F. (1956): A Kutta third-order procedure for solving differential equations requiring minimum storage, *J. Assoc. Comput. Mach*, **3**, 22–25.

Conti, R. (1956): Sulla prolungabilità delle soluzioni di un sistema di equazioni differenziali ordinarie, *Boll. Un. Mat. Ital*, ser. 3, **11**, 510–514.

Cooke, C. H. (1972): On stiffly stable implicit linear multistep methods, *SIAM J. Numer. Anal*, **9**, 29–34.

Cooke, C. H. (1973): A characterization of stiffly stable linear multistep methods, *Internat. J. Numer. Methods Engrg*, **7**, 117–124.

Cooper, G. J. (1967): A class of single-step methods for systems of nonlinear differential equations, *Math. Comp.*, **21**, 597–610.

Cooper, G. J. (1968): Interpolation and quadrature methods for ordinary differential equations, *Math. Comp.*, **22**, 69–76.

Cooper, G. J. (1969): Error bounds for some single step methods, in *Conf. on the Numerical Solution of Differential Equations, Dundee, 1969* (Ed. J. L. Morris), Lecture Notes in Mathematics No. 109, Springer, Berlin, pp. 140–147.

Cooper, G. J. (1969a): The numerical solution of stiff differential equations, *FEBS Letters*, **2** (suppl.), S22–S29.

Cooper, G. J. (1971): Bounds for the error in approximate solutions of ordinary differential equations, in *Conf. on Applications of Numerical Analysis, Dundee, 1971* (Ed J. L. Morris), Lecture Notes in Mathematics No. 228, Springer, Berlin, 270–276.

Cooper, G. J. (1971a): Error bounds for numerical solutions of ordinary differential equations, *Numer. Math.*, **18**, 162–170.

Cooper, G. J. (1978): The order of convergence of general linear methods for ordinary differential equations, *SIAM J. Numer. Anal.* **15**, 643–661.

Cooper, G. J. (1981): Error estimates for general linear methods for ordinary differential equations, *SIAM J. Numer. Anal.* **18**, 65–82.

Cooper, G. J. and Gal, E. (1967): Single step methods for linear differential equations, *Numer. Math.*, **10**, 307–315.

Cooper, G. J. and Sayfy, A. (1979): Semiexplicit A-stable Runge–Kutta methods, *Math. Comp.*, **33**, 541–556.

Cooper, G. J. and Sayfy, A. (1980): Additive methods for the numerical solution of ordinary differential equations, *Math. Comp.*, **35**, 1159–1172.

Cooper, G. J. and Verner, J. H. (1972): Some explicit Runge–Kutta methods of high order, *SIAM J. Numer. Anal.*, **9**, 389–405.

Cooper, G. J. and Whitworth, F. C. P. (1978): Liapunov functions and error bounds for approximate solutions of ordinary differential equations, *Numer. Math.*, **30**, 411–414.

Cordón, J. A. and Gasca González, M. (1978): Generation of one-step methods for the approximate solution of ordinary differential equations (Spanish, English summary), *Rev. Acad. Cienc. Zaragoza*, **33**(2), 23–32.

Corliss, G. F. (1978): Integrating ODEs along paths on Riemann surfaces, in *Proc. Seventh Manitoba Conf. on Numerical Math. and Computing, Winnipeg, 1977* (Eds. D. McCarthy and H. C. Williams), Congress numer. No. 20, Utilitas Mathematica, Winnipeg, pp. 279–295.

Corliss, G. F. (1980): Integrating ODEs in the complex plane—pole vaulting, *Math. Comp.*, **35**, 1181–1189.

Corliss, G. F. and Chang, Y. F. (1982): Solving ordinary differential equations using Taylor series, *ACM Trans. Math. Software*, **8**, 114–144.

Corliss, G. F. and Lowery, D. (1977): Choosing a stepsize for Taylor series methods for solving ODEs, *J. Comput. Appl. Math.*, **3**, 251–256.

Coroian, I. (1979): A fifth order family Runge–Kutta type methods, *Studia Univ. Babeş–Bolyai Math*, **24**, 57–63.

Coroian, I. (1980): An error estimation formula for numerical methods of high order, *Mathematica (Cluj)*, **22**(45), 241–245.

Correas, J. M., see Aguado, P. M.

Costabile, C., see Costabile, F.

Costabile, F. (1969): *Un Metodo Pseudo Runge–Kutta del Quarto Ordine*, Pubbl. Ser. III No. 20, Istit. Appl. Calcolo (IAC), Rome, 12 pp.

Costabile, F. (1970): Metodi pseudo Runge–Kutta di seconda specie, *Calcolo*, **7**, 305–313.

Costabile, F. (1971): Un metodo del terzo ordine per l'integrazione numerica dell'equazione differenziale ordinaria, *Calcolo*, **8**, 61–75.

Costabile, F. (1971a): Sulla stabilità dei metodi pseudo Runge–Kutta, *Calcolo*, **8**, 293–300.

Costabile, F. (1973): Un metodo con ascisse gaussiane per l'integrazione numerica dell'equazione differenziale ordinaria, *Rend. Mat.*, ser. 6, **6**, 733–748.

Costabile, F. (1973a): Metodi pseudo Runge–Kutta ottimali, *Calcolo*, **10**, 101–116.

Costabile, F. (1975): Un metodo per l'integrazione numerica della equazione differenziale ordinaria $y'' = f(x, y)$ con condizioni iniziali, *Calcolo*, **12**, 249–258.

Costabile, F., see also Capozza, M.

Costablie, F. and Costabile, C. (1982): Two-step fourth order P-stable methods for second order differential equations, *BIT*, **22**, 384–386.

Costabile, F. and Varano, A. (1981): Convergence, stability and truncation error estimation of a method for the numerical integration of the initial value problem $y'' = f(x, y)$, *Calcolo*, **18**, 371–384.

Coţiu, A. (1959): Some formulae for the numerical integration of differential equations of the first order making use of differences (Romanian, French summary)., *Lucrări, Ştiinţ. Inst. Politehn. Cluj.*, **1959**, (suppl.), 49–59.

Coţiu, A. (1961): A process of eight-order accuracy for the numerical integration of differential equations of the first order (Romanian, French summary), *Stud. Cerc. Mat.*, **12**, 29–40.

Coţiu, A. (1962): A Runge–Kutta process of order $n + 4$ with three nodes for the numerical integration of first order differential equations (Romanian, French summary), *Bul. Ştiinţ. Inst. Politehn. Cluj*, **5**, 39–49.

Coţiu, A. (1962a): A method of sixth-order precision on the two modes for the numerical integration of first-order differential equations (Romanian, French summary), *Stud. Cerc. Mat.*, **13**, 27–33.

Coţiu, A. (1962b): A process of Runge–Kutta type order of accuracy $n + 4$ ($n \geq 2$) on two nodes for the numerical integration of differential equations of the first order (Romanian, French summary), *Stud. Cerc. Mat.*, **13**, 253–262. •

Coţiu, A (1964): Des procédés de type Runge–Kutta, sur deux noeuds, pour l'intégration numérique des équations différentielles du deuxième ordre, *Mathematica (Cluj)*, **6**(29), 189–206.

Coţiu, A. (1964a): On the limitation of error in the process of Runge and Heun for the numerical integration of first order differential equations (Romanian, French summary), *Bul. Ştiinţ. Inst. Politehn. Cluj*, **7**, 33–41.

Coţiu, A. (1968): Procedures of Runge–Kutta type for the numerical integration of systems of first order differential equations (Romanian, French summary), *Bul. Ştiinţ. Inst. Politehn. Cluj*, **11**(2), 19–24.

Coţiu, A. (1968a): On certain quadrature formulas from which can be obtained processes of numerical integration of differential equations (Romanian, French summary), *Stud. Cerc. Mat.*, **20**, 819–831.

Coţiu, A. (1970): Runge–Kutta type procedures of high order of exactness for the numerical integration of differential equations of order n (Romanian, French summary), *Bul. Ştiinţ. Inst. Politehn. Cluj*, **13**, 19–24.

Coţiu, A. (1971): Procedures of Runge–Kutta–Fehlberg type of order of exactness 5 and 6 on 2 and 3 nodes, for the numerical integration of systems of first order ordinary differential equations (Romanian), in *Scientific Conference of the Teaching Staff* (Romanian, French summary), Ministry of Education, Cluj Politehn. Inst., Cluj, pp. 113–118.

Coţiu, A. (1981): New procedures of numerical integration of first-order differential equations (Romanian, French summary), *Studia Univ. Babeş–Bolyai Math.*, **26**, No. 2, 32–36.

Coţiu, A. see also Coţiu, F.

Coţiu, A. and Indoleanu, I. (1965): Runge–Kutta type processes of order $p + 5$ ($p \geq 1$) of exactness, on two nodes, for numerical integration of second order differential equations (Romanian, English summary), *Bul. Ştiinţ. Inst. Politehn. Cluj*, **8**, 57–71.

Coţiu, A. and Micula, M. (1968): High order of accuracy Runge–Kutta procedures, on two nodes, for the numerical integration of nth order differential equations (Romanian, French summary), *Bul. Ştiinţ Inst. Politehn. Cluj*, **11**, 25–32.

Coţiu, F. and Coţiu, A. (1960): Reduction of error in the Kutta procedure of fifth order for the numerical integration of differential equations (Romanian, French summary), *Studia Univ. Babeş-Bolyai Ser. Math.-Phys.*, **1960**, No. 1, 193–198.

Craggs, J. W., see Mitchell, A. R.

Crane, R. L. and Klopfenstein, R. W. (1965): A predictor–corrector algorithm with an increased range of absolute stability, *J. Assoc. Comput. Mach.*, **12**, 227–241.

Crane, R. L. and Lambert, R. J. (1962): Stability of a generalized corrector formula, *J. Assoc. Comput. Mach.*, **9**, 104–117.

Creedon, D. M. and Miller, J. J. H. (1975): The stability properties of q-step backward difference schemes, *BIT*, **15**, 244–249.

Creemer, A. L., see Hull, T. E.

Crisci, M. R. and Russo, E. (1982): A-stability of a class of methods for the numerical integration of certain linear systems of ordinary differential equations, *Math. Comp.*, **38**, 431–435.

Crouzeix, M. (1976): Sur les méthodes de Runge Kutta pour l'approximation des problèmes d'évolution, in *Computing Methods in Applied Sciences and Engineering: Second Internat. Symposium, 1975* (Eds. R. Glowinski and J. L. Lions), Lecture Notes in Econom. and Math. Systems No. 134, Springer, Berlin, pp. 206–223.

Crouzeix, M. (1979): Sur la B-stabilité des méthodes de Runge–Kutta, *Numer. Math.*, **32**, 75–82.

Crouzeix, M. and Ruamps, F. (1977): On rational approximations to the exponential, *RAIRO Anal. Numér.*, **11**, No. 3, 241–243.

Crown, J. C. (1980): A new PEC algorithm for the numerical solution of ordinary differential equations, *Appl. Math. Comput.*, **6**, 189–209.

Cryer, C. W. (1972): On the instability of high order backward-difference multistep methods, *BIT*, **12**, 17–25.

Cryer, C. W. (1973): A new class of highly-stable methods: A_0-stable methods, *BIT*, **13**, 153–159.

Cui, K. F., see Han, T. M.

(*Note*: The following four entries are the work of A. R. Curtis of the National Physical Laboratory, Teddington, England)

Curtis, A. R. (1965): Estimation of the truncation error in Runge–Kutta and allied processes, *Comput. J.*, **8**, 52.

Curtis, A. R. (1970): An eighth order Runge–Kutta process with eleven function evaluations per step, *Numer. Math.*, **16**, 268–277.

Curtis, A. R. (1975): High order explicit Runge–Kutta formulae, their uses, and limitations, *J. Inst. Math. Appl.*, **16**, 35–55.

Curtis, A. R. (1979): The implementation of implicit Runge–Kutta formulae for the solution of initial value problems, *J. Inst. Math. Appl.*, **23**, 339–353.

(*Note*: The following two entries are the work of A. R. Curtis of the Atomic Energy Research Establishment, Harwell, England)

Curtis, A. R. (1978): Solution of large, stiff initial value problems—the state of the art, in *Numerical Software—Needs and Availability: Proc. Conf., University of Sussex, 1977* (Ed. D. A. H. Jacobs), Academic Press, London, pp. 257–278.

Curtis, A. R. (1980): The FACSIMILE numerical integrator for stiff initial value problems, in *Computational Techniques for Ordinary Differential Equations: Proc. Conf., University of Manchester, 1978* (Eds. I. Gladwell and D. K. Sayers), Academic Press, London, pp. 47–82.

Curtiss, C. F. and Hirschfelder, J. O. (1952): Integration of stiff equations, *Proc. Nat. Acad. Sci.*, **38**, 235–243.

Cusimano, M., Genco, A. and Tortorici, M. (1977): Costruzione automatica di algoritmi di Runge–Kutta dei vari ordini, *Atti Accad. Sci. Lett. Arti Palermo*, ser. 4, pt. 1, **36**, 117–147.

Dahlquist, G. (1951): Fehlerabschätzungen bei Differenzenmethoden zur numerischen Integration gewöhnlicher Differentialgleichungen, *Z. Angew. Math. Mech.*, **31**, 239–240.

Dahlquist, G. (1956): Convergence and stability in the numerical integration of ordinary differential equations, *Math. Scand.*, **4**, 33–53.

Dahlquist, G. (1959): Stability and error bounds in the numerical integration of ordinary differential equations, *Kungl. Tekn. Högsk. Handl. Stockholm (Trans. Royal Inst. Tech. Stockholm)*, No. 130, 87 pp.

Dahlquist, G. (1963): A special stability problem for linear multistep methods, *BIT*, **3**, 27–43.

Dahlquist, G. (1963a): Stability questions for some numerical methods for ordinary differential equations, *Proc. Symposia in Applied Math.*, **15**, 147–158.

Dahlquist, G. (1966): On rigorous error bounds in the numerical solution of ordinary differential equations, in *Numerical Solutions of Nonlinear Differential Equations* (Ed. D. Greenspan), Wiley, New York, pp. 89–96.

Dahlquist, G. (1969): A numerical method for some ordinary differential equations with large Lipschitz constants, in *Information Processing 68: Proc. IFIP Congress, 1968, I-Math. Software* (Ed. A. J. H. Morrell), North Holland, Amsterdam, pp. 183–186.

Dahlquist, G. (1974): Problems related to the numerical treatment of stiff differential equations, in *International Computing Symposium, 1973* (Eds. A. Günther *et al.*), North-Holland, Amsterdam, pp. 307–314.

Dahlquist, G. (1974a): The sets of smooth solutions of differential and difference equations, in *Stiff Differential Systems* (Ed. R. A. Willoughby), Plenum Press, New York, pp. 67–80.

Dahlquist, G. (1976): Error analysis for a class of methods for stiff non-linear initial value problems, in *Numerical Analysis: Proc. Dundee Conf., 1975* (Ed. G. A. Watson), Lecture Notes in Mathematics, No. 506, Springer, Berlin, pp. 60–72.

Dahlquist, G. (1977): On the relation of G-stability to other stability concepts for linear multistep methods, in *Topics in Numerical Analysis III* (Ed. J. J. H. Miller), Academic Press, London pp. 67–80.

Dahlquist, G. (1978): On accuracy and unconditional stability of linear multistep methods for second order differential equations, *BIT*, **18**, 133–136.

Dahlquist, G. (1978a): G-stability is equivalent to A-stability, *BIT*, **18**, 384–401.

Dahlquist, G. (1978b): Positive functions and some applications to stability questions for numerical methods, in *Recent Advances in Numerical Analysis* (Eds. C. de Boor and G. H. Golub), Academic Press, New York, pp. 1–29.

Dahlquist, G. (1982): On the control of the global error in stiff initial value problems, in *Numerical Analysis: Proc. Dundee, 1981* (Ed. G. A. Watson), Lecture Notes in Mathematics No. 912, Springer, Berlin, pp. 38–49.

Dahlquist, G., Edsberg, L., Sköllermo, G. and Söderlind, G. (1982): Are the numerical methods and software satisfactory for chemical kinetics? in *Numerical Integration of Differential Equations and Large Linear Systems: Proc. Bielefeld, 1980* (Ed. J. Hinze), Lecture Notes in Mathematics No. 968, Springer, Berlin, pp. 149–164.

Dahlquist, G. and Söderlind, G. (1982): Some problems related to stiff nonlinear differential systems in *Computing Methods in Applied Sciences and Engineering* (Eds. R. Glowinski and J. L. Lions), vol. V, North-Holland, Amsterdam, pp. 57–74.

Dalle Rive, L., see Comincioli, V.

Dalle Rive, L. and Merli, C. (1974): FALCON—a conversational polyalgorithm for ordinary differential equation problems, in *Information Processing 74: Proc. IFIP Congress, 1974* (Ed. J. L. Rosenfeld), North Holland, Amsterdam, pp. 537–541.

Dalle Rive, L. and Pasciutti, F. (1975): Runge–Kutta methods with global error estimates, *J. Inst. Math. Appl.*, **16**, 381–386.

Danchick, R. (1968): Further results on generalized predictor–corrector methods, *J. Comput. System Sci.*, **2**, 203–218.

Danchick, R. (1974): On the non-equivalence of maximum polynomial degree Nordsieck–Gear and classical methods, in *Proc. Conf. on the Numerical Solution of Ordinary Differential Equations, University of Texas at Austin, 1972* (Ed. D. G. Bettis), Lecture Notes in Mathematics No. 362, Springer, Berlin, pp. 92–106.

Danchick, R. and Pope, D. A. (1977): Starting in maximum-polynomial-degree Nordsieck–Gear methods, *Appl. Math. Comput.*, **3**, 317–329.

d'Angelo, H., see Windeknecht, T. G.

Daniel, J. W. and Moore, R. E. (1970): *Computation and Theory in Ordinary Differential Equations*, W. H. Freeman, San Francisco.

Davenport, S. M., see Shampine, L. F.

Davey, D. P. and Stewart, N. F. (1976): Guaranteed error bounds for the initial value problem using polytope arithmetic, *BIT*, **16**, 257–268.

Davis, C. B., see Klopfenstein, R. W.

Davison, E. J. (1967): A high-order Crank–Nicholson technique for solving differential equations, *Comput. J.*, **10**, 195–197.

Day, J. D. (1980): On the internal S-stability of Rosenbrock methods, *Inform. Process. Lett.*, **11**, 27–30.

Day, J. D., (1980a): Comments on: T. D. Bui, 'On an L-stable method for stiff differential equations', (*Inform. Process. Lett.*, **6**, 158–161, 1977) *Inform. Process. Lett.*, **11**, 31–32.

Day, J. D., see also Murthy, D. N. P.

Day, J. D. and Murthy, D. N. P. (1981): Two classes of explicit generalized Runge–Kutta processes for non-stiff systems of ordinary differential equations, *Internat. J. Comput. Math.*, **9**, 249–268. Corrigendum (1982): *Internat. J. Comput. Math.*, **12**, 91–92.

Day, J. D. and Murthy, D. N. P. (1982): An $L(\alpha)$–stable fourth order Rosenbrock method with error estimator, *J. Comput. Appl. Math.*, **8**, 21–27.

Day, J. D. and Murthy, D. N. P. (1982a): Two classes of internally S-stable generalized Runge–Kutta processes which remain consistent with an inaccurate Jacobian, *Math. Comp.*, **39**, 491–509.

Day, J. T. (1964): A one-step method for the numerical solution of second order linear ordinary differential equations. *Math. Comp.*, **18**, 664–668.

Day, J. T. (1965): A Runge–Kutta method for the numerical integration of the differential equation $y'' = f(x, y)$, *Z. Angew. Math. Mech.*, **45**, 354–356.

Day, J. T. (1965a): A one-step method for the numerical integration of the differential equation $y'' = f(x)y + g(x)$, *Comput. J.*, **7**, 314–317.

Day, J. T. (1966): Quadrature methods of arbitrary order for solving linear ordinary differential equations, *BIT*, **6**, 181–190.

Decell, H. P., Guseman, L. F. and Lea, R. N. (1966): Concerning the numerical solution of differential equations, *Math. Comp.*, **20**, 431–434.

de Hoog, F., see Williams, J.

de Hoog, F. and Weiss, R. (1977): The application of linear multistep methods to singular initial value problems, *Math. Comp.*, **31**, 676–690.

Dejon, B. (1966): Stronger than uniform convergence of multistep difference methods, *Numer. Math.*, **8**, 29–41; Addendum: *Numer. Math.*, **9**, 268–270.

Dejon, B. (1967): Numerical stability of difference equations with matrix coefficients, *SIAM J. Numer. Anal.*, **4**, 119–128.

Dejon, B. and Kelz, W. (1979): Zur Konsistenz linearer Mehrschrittverfahren für gewöhnliche Differentialgleichungen in allgemeiner Input-Outputform, *Z. Angew. Math. Mech.*, **59**, T51–T53.

Dekker, K. (1977): Generalized Runge–Kutta methods for coupled systems of hyperbolic differential equations, *J. Comput. Appl. Math.*, **3**, 221–233.

Dekker, K. (1981): Stability of linear multistep methods on the imaginary axis, *BIT*, **21**, 66–79.

Delchambre, M. (1973): De l'avantage des méthodes aux différences modifiées de Stiefel–Bettis pour la résolution d'équations différentielles du second ordre perturbées, *Numer. Math.*, **21**, 33–36.

Delfour, M., Hager, W. and Trochu, F. (1981): Discontinuous Galerkin methods for ordinary differential equations, *Math. Comp.*, **36**, 455–473.

Delsarte, P., Genin, Y. and Kamp, Y. (1981): A proof of the Daniel–Moore conjectures for *A*-stable multistep two-derivative integration formulas, *Philips J. Res.*, **36**, 77–86.

Delves, L. M., see Hennel, M. A.

Demidov, G. V. (1970): About one method of constructing stable high order schemes (Russian), *Chisl. Metody Mekh. Sploshnoĭ. Sredy.*, **1**, No. 6, 60–69.

Demidov, G. V., see also Artem'ev, S. S.; Novikov, V. A.

Dennis, J. E. and Sweet, R. A. (1972): Some minimum properties of the trapezoidal rule, *SIAM J. Numer. Anal.*, **9**, 230–236.

Dennis, S. C. R. (1960): The numerical integration of ordinary differential equations possessing exponential type solutions, *Proc. Cambridge Philos. Soc.*, **56**, 240–246.

Dennis, S. C. R. (1960a): Finite differences associated with second-order differential equations, *Quart. J. Mech. Appl. Math.*, **13**, 487–507.

Dennis, S. C. R. (1962): Step-by-step integration of ordinary differential equations, *Quart. Appl. Math.*, **20**, 359–372.

de Ridder, R., see van Dyck, D.

de Sitter, J., see van Dyck, D.

der Houwen, P. J. van = van der Houwen, P. J.

Deuflhard, P. (1979): A study of extrapolation methods based on multistep schemes without parasitic solutions, *Z. Angew. Math. Phys.*, **30**, 177–189.

de Vogelaere, R. (1955): A method for the numerical integration of differential equations of second order without explicit first derivatives, *J. Res. Nat. Bur. Standards*, **54**, 119–125.

de Vogelaere, R. (1957): On a paper of Gaunt concerned with the start of numerical solutions of differential equations, *Z. Angew. Math. Phys.*, **8**, 151–156.

Devyatko, V. I. (1963): On a two-sided approximation for the numerical integration of ordinary differential equations (Russian), *Zh. Vychisl. Mat. i. Mat. Fiz.*, **3**, 254–265; English translation: *USSR Comput. Math. and Math. Phys.*, **3**, 336–350.

Dew, P. M. and West, M. R. (1979): Estimating and controlling the global error in Gear's method, *BIT*, **19**, 135–137.

Diekhoff, H. J., Lory, P., and Oberle, H. J., Pesch, H. J., Rentrop, P. and Seydel, R. (1977): Comparing routines for the numerical solution of initial value problems of ordinary differential equations in multiple shooting, *Numer. Math.*, **27**, 449–469.

di Lena, G. (1979): Formule Runge–Kutta per equazioni stiff con un intervallo invariante ed attrattivo, *Istituto per le Applicazioni del Calcolo 'Mauro Picone' (IAC), Rome*, Pubbl. Ser. III, No. 194, 12 pp.

di Lena, G. and Piazza, G. (1981): Una classe di metodi Runge–Kutta con esistenza dei parametri implicitamente definiti, *Istituto per le Applicazioni del Calcolo 'Mauro Picone' (IAC), Rome*, Pubbl. Ser. III, No. 201, 22 pp.

Dill, C. and Gear, C. W. (1971): A graphical search for stiffly stable methods for ordinary differential equations, *J. Assoc. Comput. Mach.*, **18**, 75–79.

Din', Kh. Kh., see Bobkov, V. V.

Ding, T. R. (1980): A necessary and sufficient condition for the convergence of Gear's difference methods in ordinary differential equations (Chinese), *Acta Math. Appl. Sinica*, **3**, 293–300.

Distefano, G. P. (1968): Stability of numerical integration techniques, *AIChE J.*, **14**, 946–955.

Djukić, Dj. S. and Atanacković, T. M. (1982): Contribution to error estimate, *J. Math. Anal. Appl.*, **88**, 183–195.

Doedel, E. J. (1978): The construction of finite difference approximations to ordinary differential equations, *SIAM J. Numer. Anal.*, **15**, 450–465.

Dolinnyĭ, O. B. (1978): On two-sided approximations of Runge–Kutta type with increased accuracy (Russian), *Differentsial'nye Uravneniya*, **14**, 1131–1132.

Dolinnyĭ, O. B., see also Lyashchenko, N. Ya.

Dombrovskaya, L. M., Nazarenko, T. I. and Chistyakova, O. G. (1979): Interval variants of Adams' type methods for the solution of the Cauchy problem for ordinary differential equations (Russian), in *Optimization Methods and Their Applications. No. 9. Work Collection* (Ed. B. A. Bel'tyukov), Akademiya Nauk SSSR, Sibirskoe Otdelenie, Sibirskiĭ Ehnergeticheskiĭ, Institut, Irkutsk, pp. 130–139.

Domingo, C. (1963): Algorithm 194, ZERSOL, *Comm. ACM*, **8**, 441.

Donelson, J. and Hansen, E. (1971): Cyclic composite multistep predictor-corrector methods, *SIAM J. Numer. Anal.*, **8**, 137–157.

Doolan, E. P. and Schilders, W. H. A. (1980): Uniformly convergent difference schemes for stiff initial value problems in *Boundary and Interior Layers— Computational and Asymptomic Methods: Proc. Conf., Dublin, 1980* (Ed. J. J. H. Miller), Boole Press, Dublin, pp. 256–259.

Dooren, R. van = van Dooren, R.

Dörband, W. (1974): Ein numerisches Verfahren zur näherungsweisen Lösung des Anfangswertproblems $y' = f(x, y)$ mit $y(x_0) = y_0$ durch iterative Berechnung eines Interpolationspolynoms n-ten Grades für die Ableitung $y'(x)$ in einem Intervall $[x_0, x_1]$, *Wiss. Z. Ernst-Moritz-Arndt-Univ. Greifswald, Math-Natur. Reihe*, **23**, 133–139.

Dormand, J. R., see Prince, P. J.

Dormand, J. R. and Prince P. J. (1980): A family of embedded Runge–Kutta formulae, *J. Comput. Appl. Math.*, **6**, 19–26.

Duffin, R. J. (1969): Algorithms for classical stability problems, *SIAM Rev.*, **11**, 196–213.

Duncan, W. J. (1948): Technique of the step-by-step integration of ordinary differential equations, *Philos. Mag.*, ser. 7, **39**, 493–509.

Durieu, J. and Genin, Y. (1973): On the existence of linear multistep formulas enjoying an 'h^2-process' property, *Philips Research Reports*, **28**, 120–129.

Dyck, D. van = van Dyck, D.

Dyer, J. (1968): Generalized multistep methods in satellite orbit computation. *J. Assoc. Comput. Mach.*, **15**, 712–719.

Dyer, J. and Ananda, M. (1971): Generalized multistep methods for second-order equations and applications to orbit computation, *AIAA J*, **9**, 614–620.

Dzhalilov, D., see Nikol'skiĭ, È. V.

Edsberg, L., see Dahlquist, G.

Ehle, B. L. (1968): High order A-stable methods for the numerical solution of systems of DEs, *BIT*, **8**, 276–278.

Ehle, B. L. (1973): A-stable methods and Padé approximations to the exponential, *SIAM J. Math. Anal.*, **4**, 671–680.

Ehle, B. L. (1975): On the S-stability and stiff order of generalized Runge–Kutta, in *Proc. Fourth Manitoba Conf. Numerical Math., Winnipeg, 1974* (Eds. B. L. Hartnell and H. C. Williams), Congress, Numer. No. 12, Utilitas Mathematica, Winnipeg, pp. 223–242.

Ehle, B. L. (1976): On certain order constrained Chebyshev rational approximations, *J. Approx. Theory*, **17**, 297–306.

Ehle, B. L., see also Lawson, J. D.

Ehle, B. L. and Lawson, J. D. (1975): Generalized Runge–Kutta processes for stiff initial-value problems, *J. Inst. Math. Appl.*, **16**, 11–21.

Ehle, B. L. and Picel, Z. (1975): Two-parameter, arbitrary order, exponential approximations for stiff equations, *Math. Comp.*, **29**, 501–511.

Eitelberg, E. (1979): Numerical simulation of stiff systems with a diagonal splitting method, *Math. Comput. Simulation*, **21**, 109–115.

Ellison, D. (1981): Efficient automatic integration of ordinary differential equations with discontinuities, *Math. Comput. Simulation*, **23**, 12–20.

Eltermann, H. (1955): Fehlerabschätzung bei näherungsweiser Lösung von Systemen von Differentialgleichungen erster Ordnung, *Math. Z.*, **62**, 469–501.

Emanuel, G. (1963): The Wilf stability criterion for numerical integration, *J. Assoc. Comput. Mach.*, **10**, 557–561.

Emel'yanov, K. V. (1980): A difference scheme for a differential equation with a small parameter multiplying the highest derivative (Russian), *Chisl. Metody Mekh. Sploshn. Sredy*, **11**, No. 5, *Mat. Modelirovanie*, 54–74.

Engels, H. (1974): Runge–Kutta–Verfahren auf der Basis von Quadraturformeln, *Internat. Ser. Numer. Math.*, **19**, 83–102.

Engels, H. (1975): Allgemeine Einschrittverfahren, *Internat. Ser. Numer. Math.*, **26**, 47–62.

England, R. (1969): Error estimates for Runge–Kutta type solutions to systems of ordinary differential equations, *Comput. J.*, **12**, 166–170.

England, R. (1982): Some hybrid implicit stiffly stable methods for ordinary differential equations, in *Numerical Analysis: Proc. Third IIMAS Workshop, Cocoyoc, Mexico, 1981* (Ed. J. P. Hennart), Lecture Notes in Mathematics No. 909, Springer, Berlin, pp. 147–158.

Enright, W. H. (1974): Second derivative multistep methods for stiff ordinary differential equations, *SIAM J. Numer. Anal.*, **11**, 321–331.

Enright, W. H. (1974a): Optimal second derivative methods for stiff systems, in *Stiff Differential Systems* (Ed. R. A. Willoughby), Plenum Press, New York, 95–109.

Enright, W. H. (1978): The efficient solution of linear constant-coefficient systems of differential equations, *Simulation*, **28**, 129–133.

Enright, W. H. (1978a): Improving the efficiency of matrix operations in the numerical solution of stiff ordinary differential equations, *ACM Trans. Math. Software*, **4**, 127–136.

Enright, W. H. (1979): Using a testing package for the automatic assessment of numerical methods for ODEs, in *Performance Evaluation of Numerical Software: Proc., IFIP TC2.5 Working Conf., Baden, Austria, 1978* (Ed. L. D. Fosdick), North-Holland, Amsterdam, pp. 199–213.

Enright, W. H. (1979a): On the efficient and reliable numerical solution of large linear systems of ODEs, in: *Proc. 1978 IEEE conf. on decision and control San Diego, Calif., 1979*, IEEE Control Systems Soc., New York, 54–59.

Enright, W. H. (1980): On the efficient time integration of systems of second–order equations arising in structural dynamics, *Internat. J. Numer. Methods Engrg.*, **16**, 13–18.

Enright, W. H. (1982): Developing effective multistep methods for the numerical solution of systems of second order initial value problems, in *Numerical Analysis: Proc. Third IIMAS Workshop. Cocoyoc, Mexico, 1981* (Ed. J. O. Hennart), Lecture Notes in Mathematics No. 909, Springer, Berlin, pp. 159–165.

Enright, W. H., see also Hull, T. E.; Jackson, K. R.

Enright, W. H. and Hull, T. E. (1976): Comparing numerical methods for stiff systems of ODEs arising in chemistry, in *Numerical Methods for Differential Systems* (Eds. L. Lapidus and W. E. Schiesser), Academic Press, New York, pp. 45–63.

Enright, W. H. and Hull, T. E. (1976a): Test results on initial value methods for non-stiff ordinary differential equations, *SIAM J. Numer. Anal.*, **13**, 944–961.

Enright, W. H., Hull, T. E. and Lindberg, B. (1975): Comparing numerical methods for stiff systems of ODEs, *BIT*, **15**, 10–48.

Enright, W. H. and Kamel, M. S. (1979): Automatic partitioning of stiff systems and exploiting the resulting structure, *ACM Trans. Math. Software*, **5.**, 374–385.

Èphstein, B. S., see Shuster, A. R.

Epton, M. A. (1980): Methods for the solution of $AXD - BXC = E$ and its application in the numerical solution of implicit ordinary differential equations, *BIT*, **20**, 341–345.

Esser, H. and Scherer, K. (1974): Konvergenzordnungen von Ein- und Mehrschritt-verfahren bei gewöhnlichen Differentialgleichungen, *Computing*, **12**, 127–143.

Euler, L. (1913); De integratione aequationum differentialium per approximationem, in *Opera Omnia*, 1st series, Vol. 11, Institutiones Calculi Integralis, Teubner, Leipzig and Berlin, pp. 424–434.

Evans, D. J., see Murphy, C. P.

Evans, D. J. and Fatunla, S. O. (1975): Accurate numerical determination of the intersection point of the solution of a differential equation with a given algebraic relation, *J. Inst. Math. Appl.*, **16**, 355–359.

Evans, D. J. and Fatunla, S. O. (1977): A linear multistep numerical integration scheme for solving systems of ordinary differential equations with oscillatory solutions, *J. Comput. Appl. Math*, **3**, 235–241.

Everhart, E. (1974): Implicit single-sequence methods for integrating orbits, *Celestial Mech.*, **10**, 35–55.

Fagaraş, S. (1975): An algorithm for determining *p*-stability of methods of numerical integration (Romanian, English summary), *Studia Univ. Babeş-Bolyai Math.*, **20**, 65–69.

Fair, W., see Luke, Y. L.

Falkner, V. M. (1936): A method of numerical solution of differential equations, *Phil. Mag.*, ser. 7, **21**, 624–640.

Fatunla, S. O. (1976): A new algorithm for numerical solution of ordinary differential equations, *Comput. Math. Appl.*, **2**, 247–253.

Fatunla, S. O. (1978): A variable order one-step scheme for numerical solution of ordinary differential equations, *Comput. Math. Appl.*, **4**, 33–41.

Fatunla, S. O. (1978a): An implicit two-point numerical integration formula for linear and nonlinear stiff systems of ordinary differential equations, *Math. Comp.*, **32**, 1–11.

Fatunla, S. O. (1980): Numerical integrators for stiff and highly oscillatory differential equations, *Math. Comp.*, **34**, 373–390.

Fatunla, S. O. (1982): Nonlinear multistep methods for initial value problems, *Comput. Math. Appl.*, **8**, 231–239.

Fatunla, S. O. (1982a): Numerical treatment of special initial value problems, in *Computational and Asymptotic Methods for Boundary and Interior Layers: Proc.*

BAIL II Conf., Dublin, 1982 (Ed. J. J. H. Miller), Boole Press Conf. Ser. 4, Boole Press, Dublin, pp 28–45.

Fatunla, S. O., see also Evans, D. J.

Fawzy, T. (1977): Spline functions and the Cauchy problems II. Approximate solution of the differential equation $y'' = f(x, y, y')$ with spline functions, *Acta Math. Acad. Sci. Hungar.*, **29**, 259–271.

Fawzy, T. (1977a): Spline functions and the Cauchy problems. IV. On the stability of the method, *Acta Math. Acad. Sci. Hungar.*, **30**, 219–226.

Fawzy, T. (1978): Spline functions and the Cauchy problems. I. Approximate solution of the differential equation $y'' = f(x, y, y')$ with spline functions, *Ann. Univ. Sci. Budapest. Sect. Comput.*, **1**, 81–98.

Fawzy, T. (1978a): Spline functions and the Cauchy problems. III. Approximate solution of the differential equation $y' = f(x, y)$ with spline functions. *Ann. Univ. Sci. Budapest. Sect. Comput.*, **1**, 35–45.

Fawzy, T. and Al-Mutib, A. (1980): Spline functions and Cauchy problems. XII. Error of an arbitrary order for the approximate solution of the differential equation $y' = f(x, y)$ with spline functions, in *Boundary and Interior Layers—Computational and Asymptotic Methods, Proc. Trinity College Dublin BAIL I Conf.* (Ed. J. J. H. Miller), Boole, Dublin, pp. 281–285.

Fawzy, T., Kőhegyi, J. and Fekete, I. (1978): Spline functions and the Cauchy problems. V. Application with programs to the method, *Ann. Univ. Sci. Budapest. Sect. Comput.*, **1**, 109–127.

Feagin, T., see Beaudet, P. R.

Fehlberg, E. (1958): Eine Methode zur Fehlerverkleinerung beim Runge–Kutta-Verfahren, *Z. Angew. Math. Mech*, **38**, 421–426.

Fehlberg, E. (1960): Neue genauere Runge–Kutta-Formeln für Differentialgleichungen zweiter Ordnung, *Z. Angew. Math. Mech.*, **40**, 252–259.

Fehlberg, E. (1960a): Neue genauere Runge–Kutta-Formeln für Differentialgleichungen n-ter Ordnung, *Z. Angew. Math. Mech.*, **40**, 449–455; Corrigendum: *Z. Angew. Math. Mech.*, **42**, 424.

Fehlberg, E. (1961): Numerisch stabile Interpolationsformeln mit günstiger Fehlerfortpflanzung für Differentialgleichungen erster und zweiter Ordnung, *Z. Angew. Math. Mech.*, **41**, 101–110.

Fehlberg, E. (1963): Zur Fehlerfortpflanzung einiger Interpolationsformeln für Differentialgleichungen zweiter Ordnung, *Z. Angew. Math. Mech.*, **43**, 199–210.

Fehlberg, E. (1963a): Runge–Kutta type formulas of high-order accuracy and their application to the numerical integration of the restricted problem of three bodies, in *Actes du Colloque Internat. des Tech. Calcul. Analogique et Numérique en Aéronautique*, Liège, pp. 351–359.

Fehlberg, E. (1964): Zur numerischen Integration von Differentialgleichungen durch Potenzreihenansätze, dargestellt an Hand physikalischer Beispiele, *Z. Angew. Math. Mech.*, **44**, 83–88.

Fehlberg, E. (1964a): New high-order Runge–Kutta formulas with step size control for systems of first- and second-order differential equations, *Z. Angew. Math. Mech.*, **44**, T17–T29.

Fehlberg, E. (1966): New high-order Runge–Kutta formulas with an arbitrarily small truncation error, *Z. Angew. Math. Mech.*, **46**, 1–16.

Fehlberg, E. (1969): Klassische Runge–Kutta-Formeln fünfter und siebenter Ordnung mit Schrittweiten-Kontrolle, *Computing*, **4**, 93–106; Corrigendum: *Computing*, **5**, 184.

Fehlberg, E. (1970): Klassische Runge–Kutta-Formeln vierter und niedrigerer Ordnung mit Schrittweiten-Kontrolle und ihre Anwendung auf Wärmeleitungsprobleme, *Computing*, **6**, 61–71.

434

Fehlberg, E. (1972): Klassische Runge–Kutta–Nyström-Formeln mit Schrittweiten-Kontrolle für Differentialgleichungen $\ddot{x} = f(t,x)$, *Computing*, **10**, 305–315.

Fehlberg, E. (1975): Klassische Runge–Kutta–Nyström-Formeln mit Schrittweiten-Kontrolle für Differentialgleichungen $\ddot{x} = f(t, x, \dot{x})$, *Computing*, **14**, 371–387.

Fehlberg, E. (1980): Some new RKT-formulas for first-order differential equations requiring a minimum number of differentiations, *Z. Angew. Math. Mech.*, **60**, 185–193.

Fehlberg, E. (1981): Eine Runge–Kutta–Nyström-Formel 9-ter Ordnung mit Schrittweitenkontrolle für Differentialgleichungen $\ddot{x} = f(t, x)$, *Z. Angew. Math. Mech.*, **61**, 477–485.

Fehlberg, E. and Filippi, S. (1966): Some new high-accuracy one-step methods for the numerical integration of ordinary differential equations, in *Information Processing 65: Proc. IFIP Congress 1965* (Ed. W. A. Kalenich), vol. 2, Spartan Books, Washington, D. C., pp. 562–563.

Feinberg, R. B. (1982): A_0-stable formulas of Adams type, *SIAM J. Numer. Anal.*, **19**, 259–262.

Fekete, I., see Fawzy, T.

Feldstein, A. and Goodman, R. (1973): Numerical solution of ordinary and retarded differential equations with discontinuous derivatives, *Numer. Math.*, **21**, 1–13.

Feldstein, A. and Stetter, H. J. (1963): Simplified predictor–corrector methods, *Proc. Eighteenth ACM National Conference 1963*, Denver, Colarado.

Fellen, B. M., see Hull, T. E.

Feng, B. P. (1982): The Gear program for solving initial value problems in general or stiff ordinary differential equations (Chinese, English summary), *J. Numer. Methods Comput. Appl.*, **3**, 12–23.

Ficker, V. (1961): Une amélioration de la méthode de Runge–Kutta–Nyström pour la résolution numérique des équations différentielles du deuxième ordre, *Acta Math. Univ. Comenian.*, **5**, 503–551.

Filippi, S. (1974): Spezielle verallgemeinerte k-Schrittverfahren der Ordnung $p = 2k$ für gewöhnliche Differentialgleichungen erster Ordnung, *Mitt. Math. Sem. Giessen*, **108**, 1–24.

Filippi, S. (1977): Verallgemeinerte k-Schrittverfahren der Ordnung $p = 3k-m + 2$ und der Ordnung $p = 2k-m + 1$ zur numerischen Lösung von Anfangswertaufgaben bei Differentialgleichungen m-ter Ordnung der Form $y^{(m)} = f(x, y)$, *Computing*, **17**, 361–372.

Filippi, S. (1977a): Über direkte, verallgemeinerte k-Schrittverfahren der Ordnung $p = 3k$ für Differentialgleichungen der Form $y'' = f(x, y)$, *Mitt. Math. Sem. Giessen*, **123**, 129–142.

Filippi, S. (1981): Konstruktion von expliziten Runge–Kutta-Formelpaaren nach dem Prinzip von Stimberg, *Mitt. Math. Sem. Giessen*, **148**, 39–55.

Filippi, S., see also Fehlberg, E.

Filippi, S. and Kraska, E. (1973): Stabile $2k$-Schritt-Verfahren der Ordnung $p = 3k + 1$ zur numerischen Lösung von Anfangswertaufgaben bei gewöhnlichen Differentialgleichungen, *Z. Angew. Math. Mech.*, **53**, 527–539.

Filippi, S. and Krüger, S. (1971): Verallgemeinerte Mehrschrittverfahren—eine Klasse effizienter Methoden zur numerischen Integration gewöhnlicher Differentialgleichungen, *Mitt. Math. Sem. Giessen*, **93**, 32–63.

Filippi, S. and Ostermann, A. (1971): Konvergenz-und Stabilitätssätze, die sich durch Anwendung von Diskretisierungsalgorithmen auf lineare Differentialgleichungen n-ter Ordnung ergeben, *Mitt. Math. Sem. Giessen*, **94**, 1–33.

Filippov, S. S. (1982): On the numerical solution of large stiff systems of ordinary differential equations, *Proc. Third Conf. on Numerical Treatment of Ordinary Differential Equations, Berlin, 1982* (Ed. R. März), Seminarberichte No. 46, Humboldt University, Berlin, pp. 37–51.

Finden, W. F. (1974): An interpolation scheme to solve systems of ordinary differential equations in which a small parameter multiplies some of the derivatives, in *Proc. Third Manitoba Conf. Numerical Math., Winnipeg, 1973* (Eds. R. S. D. Thomas and H. C. Williams), Congress Numer. No. 9, Utilitas Mathematica, Winnipeg, pp. 163–181.

Finden, W. F. (1982): An interpolation procedure for solving systems of ordinary differential equations containing a small parameter, *Utilitas Math.*, **21**, C, 5–29.

Flaherty, J. E., see O'Malley, R. E.

Fleischmann, H. J., see Böhmer, K.

Fleming, C. G., see Snider, A. D.

Forrington, C. V. D. (1961): Extensions of the predictor–corrector method for the solution of systems of ordinary differential equations, *Comput. J.*, **4**, 80–84.

Forsythe, G. E. (1959): Reprint of a note on rounding-off errors, *SIAM Rev.*, **1**, 66–67.

Fosdick, L. D. (Ed.) (1979): *Performance Evaluation of Numerical Software: Proc. IFIP TC 2.5 Working Conf., Baden, Austria, 1978*, North-Holland, Amsterdam, (see Enright, W. H.; Hull, T. E.; Stetter, H. J.).

Fowler, M. E. and Warten, R. M. (1967): A numerical integration technique for ordinary differential equations with widely separated eigenvalues, *IBM J. Res. Develop.*, **11**, 537–543.

Fox, L. (Ed.) (1962): *Numerical Solution of Ordinary and Partial Differential Equations*, Pergamon, Oxford; Addison-Wesley, Reading, Massachusetts.

Fox, L. (1962a): Chebyshev methods for ordinary differential equations, *Comput. J.*, **4**, 318–331.

Fox, L. and Goodwin, E. T. (1949): Some new methods for the numerical integration of ordinary differential equations, *Proc. Cambridge Philos. Soc.*, **45**, 373–388.

Fox, L. and Mayers, D. F. (1981): On the numerical solution of implicit ordinary differential equations, *IMA J. Numer. Anal.*, **1**, 377–401.

Fox, L. and Mitchell, A. R. (1957): Boundary-value techniques for the numerical solution of initial value problems in ordinary differential equations, *Quart. J. Mech. Appl. Math.*, **10**, 232–243.

Fox, P. A. (1971): DESUB: Integration of a first-order system of ordinary differential equations, in *Mathematical Software* (Ed. J. R. Rice), Academic Press, New York, pp. 477–507.

Fox, P. A. (1972): A comparative study of computer programs for integrating differential equations, *Comm. ACM,* **15**, 941–948.

Fraboul, F. (1962): Un critère de stabilité pour l'intégration numérique des équations différentielles, *Chiffres*, **5**, 55–63.

Frâncu, S. (1974): An algorithm for the calculation of the coefficients of formulae of Adams type used in the integration of first order differential equations (Romanian, French summary), *Stud. Cerc. Mat.*, **26**, 153–158.

Frank, R. (1975): Schätzungen des globalen Diskretisierungsfehlers bei Runge–Kutta-Methoden, *Internat. Ser. Numer. Math.*, **27**, 45–70.

Frank R., Schneid, J. and Ueberhuber, C. W. (1981): The concept of *B*-convewrgence, *SIAM J. Numer. Anal.*, **18**, 753–780.

Frank, R. and Ueberhuber, C. W. (1977): Iterated defect correction for the efficient solution of stiff systems of ordinary differential equations, *BIT*, **17**, 146–159.

Frank, R. and Ueberhuber, C. W. (1978): Collocation and iterated defect correction, in *Numerical Treatment of Differential Equations, Proc. Oberwolfach, 1976* (Eds. R. Bulirsch, R. D. Grigorieff and J. Schröder, Lecture Notes in Mathematics No. 631, Springer, Berlin, pp. 19–34.

Frank, R. and Ueberhuber, C. W. (1978a): Iterated defect correction for differential equations, I: Theoretical results, *Computing*, **20**, 207–228.

Franklin, M. A., see Katz, I. N.; Mattione, R. P.

Frei, T. (1954): Application of the moments of integral curves for the numerical

436

solution of differential equations (Hungarian, German summary), *Magyar Tud. Akad. Alkalm, Mat. Int. Közl.*, **2**, 395–414.

Freilich, J. H. and Ortiz, E. L. (1982): Numerical solution of systems of ordinary differential equations with the Tau method: an error analysis, *Math. Comp.*, **39**, 467–479.

Frey, T. (1958): On improvement of the Runge–Kutta–Nyström method I, *Periodica Polytechnica, Electrical Engineering—Electrotechnik*, **2**, 141–165.

Fricke, A. (1949): Über die Fehlerabschätzung des Adamsschen Verfahrens zur Integration gewöhnlicher Differentialgleichungen 1. Ordnung, *Z. Angew. Math. Mech.*, **29**, 165–178.

Friedli, A. (1978): Verallgemeinerte Runge–Kutta-Verfahren zur Lösung steifer Differentialgleichungssysteme, in *Numerical Treatment of Differential Equations, Proc. Oberwolfach, 1976* (Eds. R. Bulirsch, R. D. Grigorieff and J. Schröder), Lecture Notes in Mathematics No. 631, Springer, Berlin, pp. 35–50.

Friedli, A. and Jeltsch, R. (1978): An algebraic test for A_0-stability, *BIT*, **18**, 402–414.

Friedrich, V. and Müller, D. (1975): Untersuchungen zum asymptotischen Stabilitätsbegriff bei numerischen Verfahren für Anfangswertaufgaben zu gewöhnlichen Differentialgleichungen, *Beiträge Numer. Math.*, **3**, 21–35.

Frivaldszky, S. (1972): A numerical method for a first order differential equation having a solution with a pole (Hungarian, German summary), *Magyar Tud. Akad. Mat. Fiz. Oszt. Közl.*, **20**, 361–378.

Frivaldszky, S. (1975): Ein Verfahren zur Berechnung der Lösung mit singulärem Verhalten bei Differentialgleichungen erster Ordnung, *Studia Sci. Math. Hungar.*, **10**, Nos. 1–2, 1–18.

Frivaldszky, S. (1977): Lösung gewöhnlicher Anfangswertaufgaben singulären Typs und singulärer nichtlinearer Gleichungssysteme, *Studia Sci. Math. Hungar.*, **12**, Nos. 1–2, 267–280.

Frivaldszky, S. (1982): A predictor–corrector method with variable coefficients, in *Proc. Third Conf. on Numerical Treatment of Ordinary Differential Equations, Berlin, 1982* (Ed. R. März), Seminarberichte No. 46, Humboldt University, Berlin, pp. 53–63.

Froese, C. (1961): An evaluation of Runge–Kutta type methods for higher order differential equations, *J. Assoc. Comput. Mach.*, **8**, 637–644.

Fu, H. Y. (1982): A class of A-stable or $A(\alpha)$-stable explicit schemes, in *Computational and Asymptotic Methods for Boundary and Interior Layers: Proc. BAIL II Conf., Dublin, 1982* (Ed. J. J. H. Miller), Boole Press, Dublin, pp. 236–241.

Fuchs, F. (1976): A-stability of Runge–Kutta methods with single and multiple nodes, *Computing*, **16**, 39–48.

Fujii, M. (1970): Optimal choice of mesh points for Adams-type-methods with variable step-size, *Bull. Fukuoka Univ. Ed. III*, **20**, 31–46.

Fujii, M. (1973): An a posteriori, error estimation of the numerical solution by step-by-step methods for systems of ordinary differential equations, *Bull. Fukuoka Univ. Ed. III*, **23**, 35–44.

Fujii, M. (1980): Some properties of continuous approximate solutions by discrete variable methods, *Bull. Fukuoka Univ. Ed. III*, **30**, 29–35.

Fyfe, D. J. (1966): Economical evaluation of Runge–Kutta formulae, *Math. Comp.*, **20**, 392–398.

Gaffney, P. W. (1981): Some experiments in solving oscillatory differential equations, in *Elliptic Problem Solvers: Proc. Conf., Santa Fe, N. M., 1980* (Ed. M. H. Schultz), Academic Press, New York, pp. 301–305.

Gagnebin, T., see Liniger, W.

Gaier, D. (1956): Über die Konvergenz des Adamsschen Extrapolationsverfahrens, *Z. Angew. Math. Mech.*, **36**, 230.

Gal, E., see Cooper, G. J.

Galántai, A. (1975): On error estimates for one-step solution methods of ordinary differential equations (Hungarian, English summary), *Alkalmaz. Mat. Lapok*, 1, Nos. 3–4, 265–274.

Galántai, A. (1976): Local error estimates for one-step methods, *Tanulmányok MTA Számitástechn. Automat. Kutató Int. Budapest*, No. 46, 57 pp.

Galántai, A. (1976a): Local error estimate of one-step methods, *Közl.—MTA Számitástech. Automat. Kutató Int. Budapest*, 17, 63–66.

Galántai, A. (1976b): On the automatic error estimates of the Runge–Kutta methods, *Beiträge Numer. Math.*, 5, 43–49.

Galántai, A. (1976c): Convergence theorems and error analysis for one-step methods, *Ann. Univ. Sci. Budapest, Eötvös Sect. Math.*, 19, 69–78.

Galántai, A. (1979): Discrete convergence to generalized solution of Cauchy problems, *Colloq. Math. Soc. János Bolyai*, 30, 257–265.

Galántai, A. (1980): New stability property concerning stiff methods, in *Numerical Methods: Third Colloq., Keszthely, 1977* (Ed. P. Rózsa), Colloq. Math. Soc. János Bolyai No. 22, North-Holland, Amsterdam, pp. 203–212.

Galántai, A. (1980a): On the families of one-step methods, *Bull. Appl. Math.*, 1980, 76–82, 93–107.

Galántai, A. (1981): Discrete convergence to generalized solution of Cauchy problems, in *Qualitative Theory of Differential Equations, Proc. Colloquium Szeged, 1979* (Ed. M. Farkas), vol. 1, North Holland, Amsterdam, pp. 257–265.

Galántai, A. (1982): The study of the families of one-step methods, *Közl.—MTA Számitástech. Automat. Kutató Int. Budapest*, 1982, No. 26, 61–69.

Galler, B. A. and Rozenberg, D. P. (1960): A generalization of a theorem of Carr on error bounds for Runge–Kutta procedures, *J. Assoc. Comput. Mach*, 7, 57–60.

Galligani, I., see Casulli, V.

Gallivan, K. A., see Gear, C. W.

Ganev, K. Z. (1979): Algorithmic estimates of the error of numerical integration of systems of differential equations by the Adams method (Bulgarian, French summary), *Godishnik Vissh. Uchebn. Zaved. Prilozhna Mat.*, 15, No. 4, 65–74.

Ganev, K. Z. (1979a): Linearized error estimates in numerical integration of systems of differential equations by the Adams method (Bulgarian, English summary), *Godishnik Vissh. Uchebn. Zaved. Prilozhna Mat.*, 15, No. 4, 75–82.

Garfinkel, D., see Clasen, R. J.

Garg, D. P., see Geisler, E. G.

Gasca, M. = Gasca González, M., see Cordón, J. A.

Gates, L. D. (1964): Numerical solution of differential equations by repeated quadratures, *SIAM Rev.*, 6, 134–147.

Gaul, H. (1979): Eine 5-stufige Runge–Kutta-Formel für Hand- und Maschinenrechnung, *Z. Angew. Math. Mech.*, 59, 661–663.

Gaunt, J. A. (1927): The deferred approach to the limit, Part II—Interpenetrating lattices, *Philos. Trans. Roy. Soc. London*, ser A, 226, 350–361.

Gautschi, W. (1955): Über den Fehler des Runge–Kutta-Verfahrens für die numerische Integration gewöhnlicher Differentialgleichungen *n*-ter Ordnung, *Z. Angew. Math. Phys*, 6, 456–461.

Gautschi, W. (1961): Numerical integration of ordinary differential equations based on trigonometric polynomials, *Numer. Math*, 3, 381–397.

Gautschi, W. (1975): Stime dell'errore globale nei metodi 'one-step' per equazioni differenziali ordinarie, *Rend. Mat.*, ser. 2, 8, 601–617.

Gautschi, W., see also Antosiewicz, H. A.

Gautschi, W. and Montrone, M. (1980): Metodi multistep con minimo coefficiente dell'errore globale, *Calcolo*, 17, 67–75.

Gavurin, M. K. (1949): On a method of numerical integration of homogeneous linear differential equations convenient for mechanization of the computation (Russian), *Trudy Mat. Inst. Steklov.*, **28**, 152–156.

Gavurin, M. K. and Belykh, V. M. (1971): Certain methods of the numerical integration of ordinary differential equations (Russian), *Metody Vychisl.*, **7**, 3–15.

Gavurin, M. K., Belykh, V. M. and Shidlovskaya, N. A. (1971): Certain special methods of numerical inegration of ordinary differential equations (Russian), *Metody Vychisl.* **7**, 15–23.

Gear, C. W. (1965): Hybrid methods for initial value problems in ordinary differential equations, *SIAM J. Numer. Anal.*, **2**, 69–86.

Gear, C. W. (1967): The numerical integration of ordinary differential equations, *Math. Comp.*, **21**, 146–156.

Gear, C. W. (1969): The automatic integration of stiff ordinary differential equations, in *Information Processing 68: Proc. IFIP Congress, Edinburgh, 1968* (Ed. A. J. H. Morrell), North-Holland, Amsterdam, pp. 187–193.

Gear, C. W. (1969a): The automatic integration of large systems of ordinary differential equations, in *Digest Record of the Joint Conference on Mathematical Aids to Design*, Anaheim, California, pp. 27–58.

Gear, C. W. (1970): Rational approximations by implicit Runge–Kutta schemes, *BIT*, **10**, 20–22.

Gear, C. W. (1971): *Numerical Intitial Value Problems in Ordinary Differential Equations*, Prentice-Hall, Englewood Cliffs, N. J.

Gear, C. W. (1971a): Simultaneous numerical solution of differential-algebraic equations, *IEEE Trans. Circuit Theory*, **18**, 89–95.

Gear, C. W. (1971b): The automatic integration of ordinary differential equations, *Comm. ACM*, **14**, 176–179.

Gear, C. W. (1971c): Algorithm 407, DIFSUB for solution of ordinary differential equations, *Comm. ACM*, **14**, 185–190.

Gear, C. W. (1971d): Experience and problems with the software for the automatic solution of ordinary differential equations, in *Mathematical Software* (Ed. J. R. Rice), Academic Press, New York, pp. 211–227.

Gear, C. W. (1975): Estimation of errors and derivatives in ordinary differential equations, in: *Information Processing 74: Proc. IFIP Congress 1974*, (Ed. J. L. Rosenfeld), North-Holland, Amsterdam, pp 447–451.

Gear, C. W. (1978): The stability of numerical methods for second-order ordinary differential equations, *SIAM J. Numer. Anal.*, **15**, 188–197.

Gear, C. W. (1980): Runge–Kutta starters for multistep methods, *ACM Trans. Math. Software*, **6**, 263–279.

Gear, C. W. (1980a): Automatic multirate methods for ordinary differential equations, in *Information Processing 80: Proc. IFIP Congress 1980* (Ed. S. H. Lavington), North-Holland, Amsterdam, pp. 717–722.

Gear, C. W. (1980b): Initial value problems: practical theoretical developments, in *Computational Techniques for Ordinary Differential Equations: Proc. Conf., University of Manchester, 1978* (Eds. I. Gladwell and D. K. Sayers), Academic Press, London, pp. 143–162.

Gear, C. W. (1981): Numerical solution of ordinary differential equations: is there anything left to do? *SIAM Rev.*, **23**, 10–24.

Gear, C. W. (1982): Automatic detection and treatment of oscillatory and/or stiff ordinary differential equations, in *Numerical Integration of Differential Equations and Large Linear Systems: Proc. Bielefeld, 1980* (Ed. J. Hinze), Lecture Notes in Mathematics No. 968, Springer, Berlin, pp. 190–206.

Gear, C. W., see also Byrne, G. D.; Dill, C.; Shampine, L. F.

Gear, C. W. and Gallivan, K. A. (1982): Automatic methods for highly oscillatory ordinary differential equations in *Numerical Analysis: Proc. Dundee, 1981* (Ed.

G. A. Watson), Lecture Notes in Mathematics No. 912, Springer, Berlin, pp. 115–124.

Gear, C. W. and Tu, K. W. (1974): The effect of variable mesh size on the stability of multistep methods, *SIAM J. Numer. Anal.*, **11**, 1025–1043.

Gear, C. W., Tu, K. W. and Watanabe, D. S. (1974): The stability of automatic programs for numerical problems, in *Stiff Differential Systems* (Ed. R. A. Willoughby), Plenum Press, New York, pp. 111–121.

Gear, C. W. and Watanabe, D. S. (1974): Stability and convergence of variable order multistep methods, *SIAM J. Numer. Anal.*, **11**, 1044–1058.

Geisler, E. G., Tal. A. A. and Garg, D. P. (1975): On the *a posteriori* error bounds for the solution of ordinary nonlinear differential equations, *Comput. Math. Appl.*, **1**, 407–416.

Gekeler, E. (1982): On the stability of backward differentiation methods, *Numer. Math.*, **38**, 467–471.

Gekeler, E. (1982a): Linear multistep methods for stable differential equations $y'' = Ay + B(t)y' + c(t)$, *Math. Comp.*, **39**, 481–490.

Gelinas, R. J. (1972): Stiff systems of kinetic equations—a practitioner's view, *J. Comput. Phys.*, **9**, 222–236.

Genco, A., see Cusimano, M.

Genin, Y. (1973): A new approach to the synthesis of stiffly stable linear multistep formulas, *IEEE Trans. Circuit Theory*, **20**, 352–360.

Genin, Y. (1974): An algebraic approach to A-stable linear multistep-multiderivative integration formulas, *BIT*, **14**, 382–406.

Genin, Y., see also Delsarte, P.; Durieu, J.

Genuys, F. (1960): Rapport général sur le traitement numérique des équations différentielles, in *Rome Symposium: Symposium on the Numerical Treatment of Ordinary Differential Equations—Integral and Integro-differential Equations: Proc.*, Birkhauser, Basel, pp. 89–103.

Gerasimov, B. P. and Kul'chitskaya, I. A. (1982): A family of linear multistep methods for ordinary differential equations (Russian, English summary), *Akad. Nauk SSSR Inst. Prikl. Mat. Preprint*, **1982**, No. 51, 31pp.

Gez', I. F. see Lyashchenko, N. Ya.

Ghaderpanah, S. S., see Bui, T. D.

Gheri, G. and Marzulli, P. (1982): Un metodi di approssimazione bilaterale per equazioni differenziali ordinarie, *Calcolo*, **19**, 301–320.

Gibbons, A. (1960): A program for the automatic integration of differential equations using the method of Taylor series, *Comput. J.*, **3**, 108–111.

Giese, C. (1967): State variable difference methods for digital simulation, *Simulation*, **8**, 263–271.

Gill, S. (1951): A process for the step-by-step integration of differential equations in an automatic digital computing machine, *Proc. Cambridge Philos. Soc.*, **47**, 96–108.

Giloi, W. and Grebe, H. (1968): Construction of multistep integration formulas for simulation purposes, *IEEE Trans. Comput.*, **17**, 1121–1131.

Gladwell, I. (1979): Initial value routines in the NAG Library, *ACM Trans. Math. Software*, **5**, 386–400.

Gladwell, I. and Sayers, D. K. (Eds). (1980): *Computational Techniques for Ordinary Differential Equations: Proc. Conf., University of Manchester, 1978*, Academic Press, London (see: Curtis, A. R.; Gear, C. W.; Hull, T. E.; Lambert, J. D.; Prothero, A.; Shampine, L. F.).

Glasmacher, W. and Sommer, D. (1966): *Implizite Runge–Kutta-Formeln*, Westdeutscher Verlag, Cologne and Opladen, pp. 1–178.

Glinskiĭ, Ya.N. (1981): Explicit methods for solving stiff systems of ordinary differential equations (Russian, English summary), *Dokl. Akad. Nauk Ukrain. SSR*, ser. A, **1981**, No. 2, 74–78.

Glover, K. and Willems, J. C. (1973): On the stability of numerical integration routines for ordinary differential equations, *J. Inst. Math. Appl.*, **11**, 171–180.

Gofen, A. M. (1982): Fast Taylor-series expansion and the solution of the Cauchy problem (Russian), *Zh. Vychisl. Mat. i Mat. Fiz.*, **22**, 1094–1108; English translation: *USSR Comput. Math. and Math. Phys.*, **22**, No. 5, 74–88.

Goldshteĭn, Yu.B. (1978): A form of numerical solution of ordinary differential equations (Russian), *Chisl. Metody Mekh. Sploshn. Sredy*, **9**, No. 4, *Mat. Modelirovanie*, 18–29.

Golovan', N. S. (1979): A two-sided Runge–Kutta type process for second-order differential equations (Russian, English summary), *Vychisl. Prikl. Mat. (Kiev)*, **1979**, No. 39, 60–68.

Gomm, W. (1981): Stability analysis of explicit multirate methods, *Math. Comput. Simulation*, **23**, 34–50.

Goncharova, I. F. and Martynov, A. V. (1962): On a practical method for the automatic selection of scaling factors in the solution of systems of ordinary differential equations (Russian), *Zh. Vychisl. Mat. i Mat. Fiz.*, **2**, 921–924; English translation: *USSR Comput. Math. and Math. Phys.*, **2**, 1074–1079 (1963).

Goodman, R., see Feldstein, A.

Goodwin, E. T., see Fox, L.

Gorbunov, A. D. (1969): The convergence of horizontal successive approximations (Russian), in *Computing Methods and Programming*, (Russian), vol. XII, Izdat. Moskov, Univ., Moscow, pp. 151–156.

Gorbunov, A. D. (1971): New aspects of the construction and proof of Adams' methods, *Dokl. Acad. Nauk SSSR*, **198**, 23–26.

Gorbunov, A. D. (1972): On Adams' methods of maximal accuracy (Russian), *Zh. Vychisl. Mat. i Mat. Fiz.*, **12**, 503–509; English translation: *USSR Comput. Math. and Math. Phys.*, **12**, No. 2, 284–293.

Gorbunov, A. D., see also Budak, B. M.; Tikhonov, A. N.

Gorbunov, A. D. and Budak, B. M. (1958): On the convergence of certain finite difference processes for the equations $y' = f(x, y)$ and $y'(x) = f(x, y(x), y(x - \tau(x)))$ (Russian), *Dokl. Akad. Nauk SSSR*, **119**, 644–647

Gorbunov, A. D. and Pospelov, V. V. (1976): On the solvability and convergence of Adams methods with an unstable operator (Russian), *Zh. Vychisl. Mat. i. Mat. Fiz.*, **16**, 359–371; English translation: *USSR Comput. Math. and Math. Phys*, **16**, No. 2, 75–89.

Gorbunov, A. D. and Shakhov, Yu.A. (1963): On the approximate solutions of Cauchy's problem for ordinary differential equations to a given number of correct figures (Russian), *Zh. Vychisl. Mat. i Mat. Fiz.*, **3**, 239–253; English translation: *USSR Comput. Math. and Math. Phys.*, **3**, No. 2, 316–335.

Gorbunov, A. D. and Shakhov, Yu.A. (1964): On the approximate solution of Cauchy's problem for ordinary differential equations to a given number of correct figures II (Russian), *Zh. Vychisl. Mat. i. Mat. Fiz.*, **4**, 426–433; English translation: *USSR Comput. Math. and Math. Phys.*, **4**, No. 3, 37–47.

Gorbunov, A. D., Zverkina, T. S. and Sokolikhin, V. N. (1974): Two economical methods for the numerical integration of ordinary differential equations, using a displaced mesh (Russian), *Zh. Vychisl. Mat. i Mat. Fiz.*, **14**, 99–112; English translation: *USSR Comput. Math. and Math. Phys.*, **14**, No. 1, 100–113.

Gordon, M. K., see Shampine, L. F.

Gordon, M. K. and Shampine, L. F. (1974): Interpolating numerical solutions of ordinary differential equations, *Proc. ACM*, **74**, 46–53.

Gordon, M. K. and Shampine, L. F. (1975): Typical problems for stiff differential equations, *SIGNUM Newsletter*, **10**, parts 2 and 3, 41.

Gorodetskiĭ, L. M., see Bobkov, K. K.

Gottwald, B. A. and Wanner, G. (1981): A reliable Rosenbrock integrator for stiff differential equations, *Computing*, **26**, 355–360.

Gottwald, B. A. and Wanner, G. (1982): Comparison of numerical methods for stiff differential equations in biology and chemistry, *Simulation*, **38**, 61–66.

Gourgeon, H. and Hennart, J. P. (1982): A class of exponentially fitted piecewise continuous methods for initial value problems, in *Numerical Analysis: Proc. Third IIMAS Workshop, Cocoyoc, Mexico, 1981* (Ed. J. P. Hennart), Lecture Notes in Mathematics No. 909, Springer, Berlin, pp. 200–207.

Gourlay, A. R. (1970): A note on trapezoidal methods for the solution of initial value problems, *Math. Comp.*, **24**, 629–633.

Gourlay, A. R., see also Watson, H. D. D.

Gragg, W. B. (1965): On extrapolation algorithms for ordinary initial value problems, *SIAM J. Numer. Anal.*, **2**, 384–403.

Gragg, W. B. and Stetter, H. J. (1964): Generalized multistep predictor–corrector methods, *J. Assoc. Comput. Mach.*, **11**, 188–209.

Gray, H. J. (1955): Propagation of truncation errors in the numerical solution of ordinary differential equations by repeated closures, *J. Assoc. Comput. Mach.*, **2**, 5–17.

Grebe, H., see Giloi, W.

Greco, A., Pierini, G. and Tagliabue, G. (1972): Valutazione di algoritmi per la risoluzione di sistemi stiff di equazioni differenziali ordinarie, *Atti del Convegno AICA su Techniche di Simulazione e Algoritmi*, 137.

Greenspan, H., Hafner, W. and Ribaric, M. (1965): On varying stepsize in numerical integration of first order differential equations, *Numer. Math.*, **7**, 286–291.

Greig, D. M. and Abd-el-Naby, M. A. (1980), Iterative solutions of nonlinear initial value differential equations in Chebychev series using Lie series, *Numer. Math.*, **34**, 1–13.

Griepentrog, E. (1970): Mehrschrittverfahren zur numerischen Integration von gewöhnlichen Differentialgleichungssystemen und asymptotische Exaktheit, *Wiss. Z. Humboldt-Univ. Berlin Math.-Natur. Reihe*, **19**, 637–653.

Griepentrog, E. (1980): Numerische Integration steifer Differentialgleichungssysteme mit Einschrittverfahren, *Beiträge Numer. Math.*, **8**, 59–74.

Griepentrog, E. (1982): Numerik steifer Anfangswertprobleme, *Mitt. Math. Ges. DDR*, **2**, 49–61.

Griepentrog, E. and Möbius, A. (1982): Effektive Realisierung numerischer Integrationsverfahren für steife Probleme, in *Proc. Third Conf. on Numerical Treatment of Ordinary Differential Equations, Berlin, 1980* (Ed. R. März), Seminarberichte No. 46, Humboldt University, Berlin, pp. 75–80.

Griga, A. P. (1977): Integration of linear systems of ordinary differential equations (Russian), *Latv. Mat. Ezhegodnik*, **21**, 187–190.

Grigorieff, R. D. (1972): *Numerik gewöhnlicher Differentialgleichungen 1: Einschrittverfahren*, Teubner, Stuttgart.

Grigorieff, R. D. (1977): *Numerik gewöhnlicher Differentialgleichungen 2: Mehrschrittverfahren*, Teubner, Stuttgart.

Grigorieff, R. D., see also Bulirsch, R.

Grigorieff, R. D. and Schroll, J. (1978): Über $A(\alpha)$-stabile Verfahren hoher Konsistenzordnung, *Computing*, **20**, 343–350.

Gröbner, W. (1967): *Die Lie-Reihen und ihre Anwendungen*, 2nd ed., VEB Deutscher Verlag der Wissenschaften, Berlin.

Gronau, D. (1980): Numerische Integration konstant–steifer Differentialgleichungssysteme mit Mehrschrittverfahren, in *Proc. Second Conf. on Numerical Treatment of Ordinary Differential Equations, Berlin, 1980*, (Ed. R. März), Seminarberichte No. 32, Humboldt University, Berlin, pp. 14–22.

Gross, W., and Urbani, A. M. (1982): Economized Runge–Kutta methods, *Rend. Mat.*, ser. 7, **2**, 227–235.

Gruttke, W. B. (1970): Pseudo–Runge–Kutta methods of the fifth order, *J. Assoc. Comput. Mach.*, **17**, 613–628.

Guan, S. R. (1981): The linear multistep methods for the systems of differential equations of neutral type (Chinese, English summary), *Math. Numer. Sinica*, **3**, 365–371.

Gubareva, E. B., see Nazarenko, T. I.

Guderley, K. G. and Hsu, C. C. (1972): A predictor–corrector method for a certain class of stiff differential equations, *Math. Comp.*, **26**, 51–69.

Gudovich, N. N. (1980): Difference methods of arbitrary order of approximation for ordinary differential equations (Russian), in *Theory of Cubature Formulas and Numerical Mathematics. Proc. Conf., Novosibirsk, 1978* (Ed. S. L. Sobolev), 'Nauka' Sibirsk. Otdel, pp. 28–31.

Guercia, L. and Paracelli, C. (1980): Classi de metodi multistep A-stabili tipo Rosenbrock per la risoluzione di sistemi di equazioni differenziali ordinarie, *Rend. Sem. Mat. Univ. Politec. Torino*, **38**, 73–86.

Guerra, S. (1965): Qualche contributo alla teoria della formule di Runge–Kutta, *Calcolo*, **2**, 273–294.

Guerra, S. (1966): Formule di Runge–Kutta di ordine elevato per i sistemi differenziali lineari del $1°$ ordine con applicazione al calcolo delle radici delle equazioni algebriche, *Calcolo*, **3**, 407–440.

Guillou, A. and Soulé, J. C. (1969): La résolution numérique des problèmes différentiels aux conditions initiales par des méthodes de collocation, *RIRO*, **R–3**, 17–44.

Gumen, N. B., see Petrenko, A. I.

Gummel, H. K., see Blue, J. L.

Gupta, G. K. (1976): Some new high-order multistep formulae for solving stiff equations, *Math. Comp.*, **30**, 417–432.

Gupta, G. K. (1978): Implementing second-derivative multistep methods using the Nordsieck polynomial representation, *Math. Comp.*, **32**, 13–18.

Gupta, G. K. (1979): A polynomial representation of hybrid methods for solving ordinary differential equations, *Math. Comp.*, **33**, 1251–1256.

Gupta, G. K. (1980): A note about overhead costs in ODE solvers, *ACM Trans. Math. Software*, **6**, 319–326.

Gupta, G. K., see also Kovvali, S.; Wallace, C. S.

Gupta, G. K. and Wallace, C. S. (1975): Some new multistep methods for solving ordinary differential equations, *Math. Comp.*, **29**, 489–500.

Gupta, G. K. and Wallace, C. S. (1979): A new step-size changing technique for multistep methods, *Math. Comp.*, **33**, 125–138.

Gupta, R. G. (1975): A direct numerical integration method for second-order differential equations, *Z. Angew. Math. Mech.*, **55**, 709–714.

Gupta, R. G., see also Sharma, K. D.

Gur'yanov, A. E. (1980): Recursive calculation of the solutions of a linear homogeneous stationary system (Russian): *Differentsial'nye Uravneniya*, **16**, 1517–1519.

Gur'yanov, V. M. (1964): On the connection between the Runge–Kutta method and Picard's method (Russian), *Prikl. Mat. Mekh.*, **28**, 783–786; English translation: *J. Appl. Math. Mech.*, **28**, 952–956.

Guseman, L. F., see Decell, H. P.

Gustavson, F. G., see Brayton, R. K.; Hachtel, G. D.

Hachtel, G. D., see Brayton, R. K.

Hachtel, G. D., Brayton, R. K. and Gustavson, F. G. (1971): The sparse tableau approach to network analysis and design, *IEEE Trans. Circuit Theory*, **18**, 101–113.

Hadnagy, A. (1978): The stability of systems of finite difference equations (Romanian, French summary), *An. Univ. Timişoara Ser. Ştiinţ. Mat.*, **16**, 41–48.

Hadnagy, A. (1978a): Sur les méthodes des différences finies implicites, *An. Univ. Timişoara Ser. Ştiinţ. Mat.*, **16**, 157–164.

Hafner, W. see Greenspan, H.

Hager, W., see Delfour, M.

Haimovici, A. (1981): On a method for numerical integration of linear differential equations, *Rev. Roumaine Math. Pures Appl.*, **26**, 1067–1073.

Hain, K. and Hertweck, F. (1960): Numerical integration of ordinary differential equations by difference methods with automatic determination of steplength, in *Rome Symposium: Symposium on the Numerical Treatment of Ordinary Differential Equations—Integral and Integro-differential Equations: Proc.*, Birkhauser, Basel, pp. 122-128.

Haines, C. F. (1969): Implicit integration processes with error estimate for the numerical solution of differential equations, *Comput. J.*, **12**, 183–187.

Hairer, E. (1977): Méthodes de Nyström pour l'équation différentielle $y'' = f(x, y)$, *Numer. Math.*, **27**, 283–300.

Hairer, E. (1978): A Runge–Kutta method of order 10, *J. Inst. Math. Appl.*, **21**, 47–59.

Hairer, E. (1978a): On the order of iterated defect correction, *Numer. Math.*, **29**, 409–424.

Hairer, E. (1979): Nonlinear stability of *RAT*, an explicit rational Runge–Kutta method, *BIT*, **19**, 540–542.

Hairer, E. (1979a): Unconditionally stable methods for second order differential equations, *Numer. Math.*, **32**, 373–379.

Hairer, E. (1980): Highest possible order of algebraically stable diagonally implicit Runge–Kutta methods, *BIT*, **20**, 254–256.

Hairer, E. (1980a): Unconditionally stable explicit methods for parabolic equations, *Numer. Math.*, **35**, 57–68.

Hairer, E. (1981): Order conditions for numerical methods for partitioned ordinary differential equations, *Numer. Math.*, **36**, 431–445.

Hairer, E. (1982): A one-step method of order 10 for $y'' = f(x, y)$, *IMA J. Numer. Anal.*, **2**, 83–94.

Hairer, E. (1982a): Constructive characterization of A-stable approximations to exp(z) and its connection with algebraically stable Runge–Kutta methods, *Numer. Math.*, **39**, 247–258.

Hairer, E., see also Wanner, G.

Hairer, E., Bader, G. and Lubich, C. (1982): On the stability of semi-implicit methods for ordinary differential equations, *BIT*, **22**, 211–232.

Hairer, E. and Wanner, G. (1973): Multistep–multistage–multiderivative methods for ordinary differential equations, *Computing*, **11**, 287–303.

Hairer, E. and Wanner, G. (1974): On the Butcher group and general multi-value methods, *Computing*, **13**, 1–15.

Hairer, E. and Wanner, G. (1976): A theory for Nyström methods, *Numer. Math.*, **25**, 383–400.

Hairer, E. and Wanner, G. (1981): Algebraically stable and implementable Runge–Kutta methods of high order, *SIAM J. Numer. Anal.*, **18**, 1098–1108.

Hairer, E and Wanner, G. (1982): Characterization of non-linearly stable implicit Runge–Kutta methods, in *Numerical Integration of Differential Equations and Large Linear Systems, Proc. Bielefeld, 1980* (Ed. J. Hinze), Lecture Notes in Mathematics No. 968, Springer, Berlin, pp. 207–219.

Halin, H. J. (1979): Integration across discontinuities in ordinary differential equations using power saeries, *Simulation*, **29**, 33–45.

Hall, G. (1967): The stability of predictor–corrector methods, *Comput. J.*, **9**, 410–412.

Hall, G. (1974): Stability analysis of predictor–corrector algorithms of Adams type, *SIAM J. Numer. Anal.*, **11**, 494–505.

Hall, G. (1975): Implementation of linear multistep methods, in *Modern Numerical Methods for Ordinary Differential Equations* (Eds. G. Hall and J. M. Watt), Oxford University Press, Oxford, pp. 86–104.

Hall, G. (1982): Numerical solution of stiff systems of ordinary differential equations, in *Differential Equations and Applications, I, II: Proc. Second Conf., Ruse, 1981* (Eds. I. Dimovski and I. Stoyanov), Technical University, Ruse, pp. 775–786.

Hall, G. and Suleiman, M. B. (1981): Stability of Adams-type formulae for second-order ordinary differential equations, *IMA J. Numer. Anal.*, **1**, 427–438.

Hall, G. and Watt, J. M. (Eds.) (1976): *Modern Numerical Methods for Ordinary Differential Equations*, Oxford University Press.

Hamel, G. (1949): Zur Fehlerschätzung bei gewöhnlichen Differentialgleichungen erster Ordnung, *Z. Angew. Math. Mech.*, **29**, 337–341.

Hammer, P. C. and Hollingsworth, J. W. (1955): Trapezoidal methods of approximating solutions of differential equations, *Math. Comp.*, **9**, 92–96.

Hamming, R. W. (1959): Stable predictor–corrector methods for ordinary differential equations, *J. Assoc. Comput. Mach.*, **6**, 37–47.

Han, T. M. (1976): A numerical method for solving the initial value problems of stiff ordinary differential equations, *Sci. Sinica*, **19**, 180–198.

Han, T. M. and Cui, K. F. (1979): The dangerous property of numerical methods for stiff differential equations (Chinese, English summary), *Math. Numer. Sinica*, **1**, 331–335.

Hansen, E. (1969): Cyclic composite multistep predictor–corrector methods, *Proc. Twenty-fourth ACM Conf.*, pp. 135–139.

Hansen, E., see also Donelson, J.

Harris, R. P. (1969): Runge–Kutta processes, *Proc. Fourth Australian Computer Conf., Adelaide*, pp. 429–433.

Harris, R. P. (1976): Algorithms for computing the coefficients for Runge–Kutta methods, *Proc. Seventh Australian Computer Conf., Perth*, pp. 1023–1037.

Hartnell, B. L. and Williams, H. C. (Eds.) (1975): *Proc. Fourth Manitoba Conf. Numerical Math., Winnipeg, 1974*, Congress. Numer. No. 12, Utilitas Mathematica, Winnipeg (see Ehle, B. L.; Nevanlinna, O. and Sipilä, A. H.).

Hartnell, B. L. and Williams, H. (Eds.) (1976): *Proc. Fifth Manitoba Conf. Numerical Math. Winnipeg, 1975*, Congress Numer. No. 16, Utilitas Mathematica, Winnipeg (see Jeltsch, R.; Lambert, J. D. : Usmani, R. A.).

Häussler, W. M. (1981): Zum Diskretisierungsfehler von Einschrittverfahren bei nichtdifferenzierbarer rechter Seite, *Z. Angew. Math. Mech.*, **61**, T287–T289.

Hayashi, K. (1967): On instability in the numerical integration of $y'' = -xy$, by multistep methods, *TRU Math.*, **3**, 30–40.

Hayashi, K. (1968): On orbital stability of numerical solutions of a non-linear autonomous differential system by one-step methods, *TRU Math.*, **4**, 66–79.

Hayashi, K. (1969): On stability of numerical solutions of a differential system by one-step methods, *TRU Math.*, **5**, 67–83.

Hayashi, K. (1979): On *a posteriori* error estimation in the numerical solution of systems of ordinary differential equations, *Hiroshima Math. J.*, **9**, 201–243.

Hayashi, Y, see Shintani, H.

Hebsaker, H. M. (1982): Conditions for the coefficients of Runge–Kutta methods for systems of nth order differential equations, *J. Comput. Appl. Math.*, **8**, 3–14.

Heinrich, H. (1949): Genauigkeitsvergleich für die Halbschrittverfahren der graphischen Integration, *Z. Angew. Math. Mech.*, **29**, 51–52.

Heinze, K. and Hollatz, H. (1981): Zur Anwendung der Grenzwertextrapolation bei der

Lösung von Anfangswertaufgaben, *Wiss. Z. Tech. Hochsch. Otto von Guericke*, **25**, No. 1, 103–108.

Heinze, K., Hollatz, H. and Strümke, M. (1981): Mathematische Modellierung von Prozessen, die auf parameterbhängige lineare Differentialgleichungssysteme führen und ihre numerisch-rechentechnische Behandlung, *Wiss. Z. Tech. Hochsch. Otto von Guericke*, **25**, No. 1, 97–101.

Hellemans, P. (1982): A numerical solution of the Cauchy problem based on trigometric interpolation, *J. Comput. Appl. Math.*, **8**, 305–306.

Hennart, J. P. (1976): Piecewise polynomial multiple collocation methods for initial value problems, in *Differential Equations, 3rd Symposium of United Mexican States, Collección Matemática Superior, Notas de Matemática y Simposia* (Ed. C. Imaz), Fondo de Cultura Económica, Mexico, pp. 253–254.

Hennart, J. P. (1977): One-step piecewise polynomial multiple collocation methods for initial value problems, *Math. Comp.*, **31**, 24–36.

Hennart, J. P. (ed.) (1982): *Numerical Analysis: Proc. Third IIMAS Workshop, Cocoyoc, Mexico, 1981*, Lecture Notes in Mathematics No. 909, Springer, Berlin (see England, R.; Enright, W. H.; Gourgeon, H. and Hennart, J. P.).

Hennart, J. P., also Gourgeon, H.

Hennel, M. A. and Delves, L. M. (Eds.) (1980): *Production and Assessment of Numerical Software*, Academic Press, London.

Henrici, P. (1960): The propagation of round-off error in the numerical solution of initial value problems involving ordinary differential equations of the second order, in *Rome Symposium: Symposium on the Numerical Treatment of Ordinary Differential Equations—Integral and Integro-differential Equations: Proc.*, Birkhauser, Basel, pp. 275–291.

Henrici, P. (1960a): Theoretical and experimental studies on the accumulation of error in the numerical solution of initial value problems for systems of ordinary differential equations, in *Information Processing*, UNESCO, Paris, pp. 36–44.

Henrici, P. (1962): *Discrete Variable Methods in Ordinary Differential Equations*, Wiley, New York.

Henrici, P. (1963): *Error Propagation for Difference Methods*, Wiley, New York.

Henrici, P. (1964): The propagation of error in the digital integration of ordinary differential equations in *Error in Digital Computation* (Ed. L. B. Rall), vol. 1, Wiley, New York, pp. 185–205.

Hermite, C. (1878): Sur la formula d'interpolation de Lagrange, *J. de Crelle*, **84**, 70–79.

Herrick, S. (1951): Step-by-step integration of $\ddot{x} = f(x, y, z, t)$ without a 'corrector', *Math. Comp.*, **5**, 61–67.

Hersh, R. and Kato, T. (1979): High-accuracy stable difference schemes for well-posed initial-value problems, *SIAM J. Numer. Anal*, **16**, 670–682.

Hertweck, F., see Hain, K.

Heun, K. (1900): Neue Methoden zur approximativen Integration der Differentialgleichungen einer unabhängigen Veränderlichen, *Z. Math. Phys.*, **45**, 23–38.

Hiebert, K. L., see Shampine, L. F.

Hill, D. R. (1975): On comparing Adams and natural spline multistep formulas, *Math. Comp.*, **29**, 741–745.

Hill, D. R. (1976): Second derivative multistep formulas based on *g*-splines, in *Numerical Methods for Differential Systems*, (Eds. L. Lapidus and W. E. Schiesser), Academic Press, New York, pp. 25–38.

Hill, D. R., see also Andria, G. D.

Hillion, P. (1979): A new stability criterion for linear discrete systems, *BIT*, **19**, 186–195.

Himmelblau, D. M., see Watanabe, K.

446

Hindmarsh, A. C., see Byrne, G. D.; Sherman, A. H.

Hindmarsh, A. C. and Byrne, G. D. (1976): Applications of EPISODE: an experimental package for the integration of systems of ordinary differential equations, in *Numerical Methods for Differential Systems* (Eds. L. Lapidus and W. E. Schiesser), Academic Press, New York, pp. 147–166.

Hindmarsh, A. C. and Byrne, G. D. (1976a): On the use of rank-one updates in the solution of stiff systems of ordinary differential equations, *SIGNUM Newsletter*, **11**, 23–27.

Hinze, J. (Ed.) (1982): *Numerical Integration of Differential Equations and Large Linear Systems: Proc. Bielefeld, 1980* Lecture Notes in Mathematics No. 968, Springer, Berlin (see Dahlquist, G.; Gear, C. W.; Hairer, E. and Wanner, G.; Jeltsch, R. and Nevanlinna, O.; Shampine, L. F.; Stetter, H. J.; Thomsen, P. G.).

Hirschfelder, J. O., see Curtiss, C. F.

Hobot, G. (1974): Some fourth-order formulae of a certain method of Zurmühl, *Zastos. Mat.*, **14**, 449–460.

Hodgkins, W. R. (1969): A method for the numerical integration of non-linear ordinary differential equations with greatly different time constants, in *Conf. on the Numerical Solution of Differential Equations, Dundee, 1969* (Ed. J. L. Morris) Lecture Notes in Mathematics No. 109, Springer, Berlin, pp. 172–177.

Hofer, E. (1976): A partially implicit method for large stiff systems of ODEs with only few equations introducing small time-constants, *SIAM J. Numer. Anal.*, **13**, 645–663.

Hoffman, V., see Strehmel, K.

Hoog, F. de = de Hoog, F.

Hollatz, H., see Heinze, K.

Hollingsworth, J. W., see Hammer, P. C.

Holmes, W. M., see Rowland, J. R.

Hoppensteadt, F. C., see Miranker, W. L.

Hoppensteadt, F. C. and Miranker, W. L. (1981): Computation by extrapolation of solutions of singular perturbation problems, in *Analytical and Numerical Approaches to Asymptotic Problems in Analysis: Proc. Conf., University of Nijmegen, 1980* (Eds. O. Axelsson, L. S. Frank and A. van der Sluis), North-Holland Math. Stud. No. 47, pp. 73–85.

Horn, M. K., see also Bettis, D. G.

Horner, T. S. (1980): Recurrence relations for the coefficients in Chebyshev series solutions of ordinary differential equations, *Math. Comp.*, **35**, 893–905.

Houwen, P. J. van der = van der Houwen, P. J.

Howard, B. E. (1974): Phase space analysis in numerical integration of ordinary differential equations, in *Proc. Conf. on the Numerical Solution of Ordinary Differential Equations, University of Texas at Austin, 1972* (Ed. D. G. Bettis), Lecture Notes in Mathematics No. 362, Springer, Berlin, pp. 107–127.

Howells, P. B. and Marshall, R. H. (1981): A fast and accurate integration scheme for coupled stiff differential equations, in *Numerical Methods for Coupled Problems: Proc. Conf., Swansea, 1981* (Eds. E. Hinton, P. Bettess and R. W. Lewis), Pineridge Press, Swansea, pp. 49–59.

Hsu, C. C., see Guderley, K. G.

Hubbard, E. C., see Cohen, C. J.

Huddleston, R. E. (1971): Some relations between the values of a function and its first derivative at *n* abscissa points, *Math. Comp.*, **25**, 553–558.

Huddleston, R. E. (1972): Selection of stepsize in the variable-step predictor–corrector method of van Wyk, *J. Comput. Phys.*, **9**, 528–537.

Huddleston, R. E. (1972a): Variable-step truncation error estimates for Runge–Kutta methods of order 4 or less, *J. Math. Anal. Appl.*, **39**, 636–646.

Hull, D. G. (1977): Fourth-order Runge–Kutta integration with stepsize control, *AIAA J.*, **15**, 1505–1507.

Hull, T. E. (1967): A search for optimum methods for the numerical integration of ordinary differential equations, *SIAM Rev.*, **9**, 647–654.

Hull, T. E. (1969): The numerical integration of ordinary differential equations, in *Information Processing 68: Proc. IFIP Congress, Edinburgh, 1968* (Ed. A. J. H. Morrell), North-Holland, Amsterdam, pp. 40–53.

Hull, T. E. (1970): The effectiveness of numerical methods of ordinary differential equations, *Studies in Numerical Analysis 2: Numerical Solutions of Non-linear Problems: Symposium, SIAM, Philadelphia, Pa., 1968*, pp. 114–121.

Hull, T. E. (1974): The development of software for solving ordinary differential equations, in *Conf. on the Numerical Solution of Differential Equations, Dundee, 1973* (Ed. G. A. Watson), Lecture Notes in Mathematics No. 363, Springer, Berlin, pp. 55–63.

Hull, T. E. (1974a): The validation and comparison of programs for stiff systems, in *Stiff Differential Systems* (Ed. R. A. Willoughby), Plenum Press, New York, pp. 151–164.

Hull, T. E. (1975): Numerical solution of initial value problems for ordinary differential equations, in *Numerical Solutions of Boundary Value Problems for Ordinary Differential Equations: Proc. Sympos., University of Maryland, Baltimore, Md., 1974*, Academic Press, New York, pp. 3–26.

Hull, T. E. (1979): Correctness of numerical software, in *Performance Evaluation of Numerical Software* (Ed. L. D. Fosdick), North-Holland, Amsterdam, pp. 3–15.

Hull, T. E. (1980): Comparison of algorithms for initial value problems, in *Computational Techniques for Ordinary Differential Equations: Proc. Conf., University of Manchester, 1978* (Eds. I. Gladwell and D. K. Sayers), Academic Press, London, pp. 129–142.

Hull, T. E., see also Byrne, G. D.; Enright, W. H.; Jackson, K. R.

Hull, T. E. and Creemer, A. L. (1963): Efficiency of predictor–corrector procedures, *J. Assoc. Comput. Mach.*, **10**, 291–301.

Hull, T. E., Enright, W. H., Fellen, B. M. and Sedgwick, A. E. (1972): Comparing numerical methods for ordinary differential equations, *SIAM J. Numer. Anal.* **9**, 603–637; Corrigendum: *SIAM J. Numer. Anal*, **11**, 681.

Hull, T. E. and Johnston, R. L. (1964): Optimum Runge–Kutta methods, *Math. Comp.*, **18**, 306–310.

Hull, T. E. and Luxemburg, W. A. J. (1960): Numerical methods and existence theorems for ordinary differential equations, *Numer. Math.*, **2**, 30–41.

Hull, T. E. and Newbery, A. C. R. (1959): Error bounds for a family of three-point integration procedures, *J. Soc. Indust. Appl. Math.*, **7**, 402–412.

Hull, T. E. and Newbery, A. C. R. (1961): Integration procedures which minimize propagated errors, *J. Soc. Indust. Appl. Math.*, **9**, 31–47.

Hull, T. E. and Newbery, A. C. R. (1962): Corrector formulas for multi-step integration methods, *J. Soc. Indust. Appl. Math.*, **10**, 351–369.

Hulme, B. L. (1971): Piecewise polynomial Taylor methods for initial value problems, *Numer. Math.*, **17**, 367–381.

Hulme, B. L. (1972): One-step piecewise polynomial Galerkin methods for initial value problems, *Math. Comp.*, **26**, 415–426.

Hulme, B. L. (1972a): Discrete Galerkin and related one-step methods for ordinary differential equations, *Math. Comp.*, **26**, 881–891.

Hundsdorfer, W. H. and Spijker, M. N. (1981): A note on *B*-stability of Runge–Kutta methods, *Numer. Math.*, **36**, 319–331.

Hurewicz, W. (1958): *Lectures on Ordinary Differential Equations*, Technology Press (MIT), Cambridge, Massachusetts.

448

Huskey, H. D. (1949): On the precision of a certain procedure of numerical integration (with an appendix by D. R. Hartree), *J. Res. Nat. Bur. Standards*, **42**, 57–62.

Huston, R. L., see Krinke, D. C.

Huťa, A. (1956): Une amélioration de la méthode de Runge–Kutta–Nyström pour la résolution numérique des équations différentielles du premier ordre, *Acta Math. Univ. Comenian*, **1**, 201–224.

Huťa, A. (1957): Contribution à la formule de sixième ordre dans la méthode de Runge–Kutta–Nyström, *Acta Math. Univ. Comenian*, **2**, 21–24.

Huťa, A. (1972): Contribution to the numerical solution of differential equations by means of Runge–Kutta formulas with Newton–Cotes numbers weights, *Acta Math. Univ. Comenian.*, **28**, 51–65.

Huťa, A. (1974): Eine Verallgemeinerung des Runge–Kutta-Verfahrens zur numerischen Lösung der Gleichung $y' = f(x, y)$, *Z. Angew. Math. Mech.*, **54**, T221.

Huťa, A. (1978): An algorithm for the computation of an nth order formula for the numerical solution of initial value problems for ordinary differential equations, in *Proc. Fourth Symposium on Basic Problems of Numerical Mathematics, Plzeň, 1978* (Ed. I. Marek), Charles University, Prague, pp. 87–101.

Huťa, A. (1979): An *a priori* bound of the round-off error in the integration by multistep difference method for the differential equation $y^{(s)} = f(x, y)$, *Acta Math. Univ. Comenian.*, **34**, 51–56.

Huťa, A. (1979a): An *a priori* bound of the discretization error in the integration by multistep difference method for the differential equation $y^{(s)} = f(x, y)$, *Acta Math. Univ. Comenian.*, **34**, 157–163.

Huťa, A. (1982): Algorithm for the construction of explicit n-order Runge–Kutta formulas for the systems of differential equations of the first order, in *Differential Equations and Their Applications: Equadiff 5, Proc. Fifth Czech. Conf., Bratislava, 1981* (Ed. M. Greguš), Teubner-Texte zur Mathematics No. 47, Teubner, Leipzig, pp. 140–144.

Huťa, A. and Strehmel, K. (1982): Construction of explicit and generalized Runge–Kutta formulas of arbitrary order with rational parameters, *Apl. Mat.*, **27**, 259–276.

Hwang, M., see Seinfeld, J. H.

Ibiejugba, M. A. (1978): Stability of difference schemes of linear ordinary differential equations (nth order) with defining polynomial possessing repeated roots on the unit circle, *Math. Student*, **46**, 351–358.

Ibragimov, Sh.I. and Novruzov, G. M. (1982): An approximate method for solving the Cauchy problem for differential equations with a singularity (Russian), in *Approximate Methods of Analysis* (Ed. Ya.D. Mamedov), Azerbaĭdzhan. Gos. University, Baku, pp. 66–72.

Ibragimov, V. R. (1969): A certain numerical method for the solution of higher order differential equations (Russian), *Azerbaĭdhzan. Gos. Univ. Uchen. Zap. Ser. Fiz.-Mat. Nauk*, **1969**, No. 1, 74–77.

Ibragimov, V. R. (1971): Application of a certain numerical method to the solution of higher order differential equations (Russian), *Akad. Nauk Azerbaĭdzhan. SSR Dokl.*, **27**, No. 5, 9–12.

Ibragimov, V. R. (1978): The connection between order and degree of a stable difference formula (Russian), *Azerbaĭdzhan. Gos. Univ. Uchen. Zap.*, **1978**, No. 3, 40–49

Ibragimov, V. R. (1982): A nonlinear method of numerical solution of the Cauchy problem for ordinary differential equations (Russian), in *Differential Equations and Applications, I, II: Proc. Second Conf., Ruse, 1981* (Eds. I. Dimovski and J. Stoyanov), Technical University, Ruse, pp. 310–319.

Ibragimov, V. R. (1982a): A representation of local error of the K-step method (Russian), in *Approximate Methods of Analysis* (Ed. Ya.D. Mamedov), Azerbaĭdzhan, Gos. University Baku, pp. 47–65.

Incerti, S., see Aluffi, F.

Indoleanu, I., see Coţiu, A.

Inerbaev, M. S. (1979): On applications of numerical methods to the Cauchy problem for systems of ordinary differential equations (Russian), *Differential Equations and Their Applications, Work Collect., Alma–Ata,* **1979**, 34–39.

Inerbaev, M. S. (1980): Some applications of computational methods to the Cauchy problem for ordinary differential equations, in *Mathematical Modelling and Optimal Control* (Russian) (Ed. A. T. Lukyanov), Kazah. Gos. University, Alma-Ata, pp. 124–128.

Information Processing 62: (1962): see Kuntzmann, J.

Information Processing 65: Kalenich, W. A. (Ed.) (1966); see Butcher, J. C.; Cea, J.; Fehlberg, E. and Filippi, S.

Information Processing 68: Morrell, A. J. H. (Ed.) (1969); see Dahlquist, G.; Gear, C. W.; Hull, T. E.; Krogh, F. T.; Osborne, M. R.

Information Processing 71: Griffith, J. E. and Rosenfeld, J. L. (Eds.) (1972); see Thacher, H. C.

Information Processing 74: Rosenfeld, J. L. (Ed.) (1974); see Dalle Rive, L. and Merli, C.

Information Processing 80: Lavington, S. H. (Ed.)(1980); see Albrecht, P., Gear, C. W.

Ionescu, D. V. (1954): A generalization of a property which has applications to Runge–Kutta methods for the numerical integration of differential equations (Russian), *Acad. Republ. Pop. Romîne Bul. Şti. Secţ. Şti. Mat. Fiz.*, **6**, 229–241.

Ionescu, D. V. (1956): A generalization of a property which has applications to Runge–Kutta methods for the numerical integration of differential equations (Romanian, French summary), *Acad. Republ. Pop. Romîne Bul. Şti. Secţ. Şti. Mat. Fiz.*, **8**, 67–100.

Ionescu, D. V. (1957): The integration of a differential equation (Romanian, French summary), *Stud. Cerc. Mat.*, **8**, 275–289.

Ionescu, D. V. (1959): The application of numerical differentiation formulas to the numerical integration of differential equations (Romanian, French summary) *Stud. Cerc. Mat.*, **10**, 259–315.

Ionescu, D. V. (1959a): L'application de la méthode des approximations successives à l'intégration numérique des équations différentielles, *Bull. Math. Soc. Sci. Math. Phys. Rép. Pop. Roumaine* (N.S.) 3, (51), 423–431.

Ionescu, D. V. (1960): New Adams type formulae for the numerical integration of first order differential equations (Romanian, French summary), *Stud. Cerc. Mat.*, **11**, 101–116.

Ionescu, D. V. (1960a): The application of the method of successive approximations to the numerical integration of differential equations (Romanian, French summary), *Stud. Cerc. Mat.*, **11**, 273–286.

Ionescu, D. V. (1961): Practical methods of numerical integration of differential equations (Romanian, French summary), *Stud. Cerc. Mat.*, **12**, 257–280.

Ionescu, D. V. (1962): The remainder in the Adams formula for numerical integration (Russian), *Zh. Vychisl. Mat. i Mat. Fiz.*, **2**, 154–157. English translation: *USSR Comput. Math. and Math. Phys*, **2**, 163–165 (1963).

Ionescu, D. V. (1962a): Numerical integration (Romanian, French summary), *Stud. Cerc. Mat.*, **13**, 243–286.

Ionescu, D. V. (1963): Méthodes pratiquées pour l'intégration numérique des équations différentielles du premier ordre, *C. R. Acad. Bulgare Sci.*, **16**, 469–479.

Ionescu, D. V. (1963a): The remainder in the Nyström formula of numerical integration (Romanian, French summary), *Stud. Cerc. Mat.*, **14**, 43–48.

Ionescu, D. V. (1963b): The remainder in the Störmer formula of numerical integration (Romanian, French summary), *Stud. Cerc. Mat.*, **14**, 49–56.

Ionescu, D. V. (1964): L'intégration numérique des équations différentielles du second ordre, *Mathematica (Cluj)*, **6**(29), 217–232.

Ionescu, D. V. (1964a): Quelques formules pratiques d'intégration numérique des équations différentielles, *Rev. Roumaine Math. Pures Appl.*, **9**, 237–243.

Irby, T. C. (1978): A numerical technique for solving a class of n-th order ordinary differential equations, in *Numerical Methods for Differential Equations and Simulation: Proc. IMACS (AICA) Internat. Sympos., Blacksburg, Virginia, 1977* (Eds. A. W. Bennett and R. Vichnevetsky), North-Holland, Amsterdam, pp. 115–118.

Isaĭko, G. M. (1980): Numerical solution of the Cauchy problem for equations of a certain type (Russian), in *Approximate Solution of Some Problems for Differential and Integral Equations of Applied Character*, (Ed. M. N. Zakharova), Preprint No. 17, Akad. Nauk. Ukrain SSR. Inst. Kibernet, Kiev, pp. 12–15.

Iserles, A. (1977): Functional fitting—new family of schemes for integration of stiff ODE, *Math. Comp.*, **31**, 112–123.

Iserles, A. (1978): On the A-stability of implicit Runge–Kutta processes, *BIT*, **18**, 157–169.

Iserles, A. (1978a): A-stability and dominating pairs, *Math. Comp.*, **32**, 19–33.

Iserles, A. (1978b): A-acceptable exponentially fitted combinations of three Padé approximations, *J. Comput. Appl. Math.*, **4**, 143–146.

Iserles, A. (1979): On the A-acceptability of Padé approximations, *SIAM J. Math. Anal.*, **10**, 1002–1007.

Iserles, A. (1979a): On the generalized Padé approximations to the exponential function, *SIAM J. Numer. Anal.*, **16**, 631–636.

Iserles, A. (1980): Nonexponential fitting techniques for numerical solution of stiff equations, *Utilitas Math.*, **17**, 276–302.

Iserles, A. (1981): Quadrature methods for stiff ordinary differential systems, *Math. Comp.*, **36**, 171–182.

Iserles, A. (1981a): Rational interpolation to exp(-x) with application to certain stiff systems, *SIAM J. Numer. Anal.*, **18**, 1–12.

Iserles, A. (1981b): On multivalued exponential approximations, *SIAM J. Numer. Anal.*, **18**, 480–499.

Iserles, A. (1981c): Two-step numerical methods for parabolic differential equations, *BIT*, **21**, 80–96.

Iserles, A. (1982): Composite exponential approximations, *Math. Comp.*, **38**, 99–112.

Iserles, A. and Powell, M. J. D. (1981): On the A-acceptability of rational approximations that interpolate the exponential function, *IMA J. Numer. Anal.*, **1**, 241–251.

Ismail, H. N. A. (1979): Generalized periodic overimplicit multistep methods (GPOM methods), *Apl. Mat.*, **24**, 250–272.

Ismail, H. N. A. (1979a): Necessary conditions for the convergence of the generalized periodic overimplicit multistep methods, *Apl. Mat.*, **24**, 273–283.

Iyengar, S. R. K. (1975): Stability of generalized predictor corrector methods, *J. Math. Phys. Sci.*, **9**, 106–110.

Iyengar, S. R. K., see also Jain, M. K.; Rao, C. P.

Iyengar, S. R. K. and Jain, M. K. (1974): The stability of modified predictor–corrector methods, *J. Math. Phys. Sci.*, **8**, 319–325.

Iyengar, S. R. K. and Rao, C. P. (1976): A note on generalized multistep methods for special second order ordinary differential equations, *J. Math. Phys. Sci.*, **10**, 457–460.

Jackiewicz, Z. and Kwapisz, M. (1978): On the convergence of multistep methods for the Cauchy problem for ordinary differential equations, *Computing*, **20**, 351–361.

Jackiewicz, Z. and Kwapisz, M. (1981): On numerical integration of implicit ordinary differential equations, *Apl. Mat.*, **26**, 97–109.

Jackson, J. (1924): Note on the numerical integration of $d^2x/dt^2 = f(x, t)$, *Monthly Notices Roy. Astronom. Soc.*, **84**, 602–606.

Jackson, K. R., see Byrne, G. D.

Jackson, K. R., Enright, W. H. and Hull, T. E. (1978): A theoretical criterion for comparing Runge–Kutta formulas, *SIAM J. Numer. Anal.*, **15**, 618–641.

Jackson, K. R. and Sacks-Davis, R. (1980): An alternative implementation of variable step-size multistep formulas for stiff ODEs, *ACM Trans. Math. Software*, **6**, 295–318.

Jackson, L. W. (1975): Interval arithmetic error-bounding algorithms, *SIAM J. Numer. Anal.*, **12**, 223–238.

Jackson, L. W. (1976): The *A*-stability of a family of fourth order methods, *BIT*, **16**, 383–387.

Jackson, L. W. and Kenue, S. K. (1974): A fourth order exponentially fitted method, *SIAM J. Numer. Anal.*, **11**, 965–978.

Jain, M. K., see Iyengar, S. R. K.; Jain, R. K.

Jain, M. K., Iyengar, S. R. K. and Saldanha, J. S. V. (1977): Numerical solution of a fourth-order ordinary differential equation, *J. Engrg. Math.*, **11**, 373–380.

Jain, M. K., Jain, R. K. and Anantha Krishnaiah, U. (1979): *P*-stable methods for periodic initial value problems of second order differential equations, *BIT*, **19**, 347–355.

Jain, M. K., Jain, R. K. and Anantha Krishnaiah, U. (1979a): *P*-stable singlestep methods for periodic initial-value problems involving second-order differential equations, *J. Engrg. Math.*, **13**, 317–326.

Jain, M. K., Jain, R. K. and Anantha Krishnaiah, U. (1981): Obrechkoff methods for periodic initial value problems of second order differential equations, *J. Math. Phys. Sci.*, **15**, 239–250.

Jain, M. K., Jain, R. K. and Anantha Krishnaiah, U. (1981a): Hybrid numerical methods for periodic initial value problems involving second-order differential equations, *Appl. Math. Modelling*, **5**, 53–56.

Jain, R. K. (1971): Higher order Runge–Kutta methods with extended region of stability for a system of first order differential equations, in *Proc. [first] Manitoba Conf. Numerical Math., Winnipeg, 1971* (Eds. R. S. D. Thomas and H. C. Williams), Congress Numer. No. 5, Utilitas Mathematica, Winnipeg, pp. 385–400.

Jain, R. K. (1971a): *A*-stable multi-step methods for stiff ordinary differential equations, in *Proc. [first] Manitoba Conf. Numerical Math., Winnipeg, 1971* (Eds. R. S. D. Thomas and H. C. Williams), Congress Numer, No. 5, Utilitas Mathematica, Winnipeg, pp. 401–416.

Jain, R. K. (1972): Some *A*-stable methods for stiff ordinary differential equations, *Math. Comp.*, **26**, 71–77.

Jain, R. K., see also Jain, M. K.

Jain, R. K. and Jain, M. K. (1971): Optimum Runge–Kutta Fehlberg methods for first order differential equations, *J. Inst. Math. Appl.*, **8**, 386–396.

Jain, R. K., Jain, M. K. (1972): Optimum Runge–Kutta–Fehlberg methods for second-order differential equations, *J. Inst. Math. Appl.*, **10**, 202–210.

Jain, R. K., Jain, M. K. and Saldanha, J. S. V. (1976): Some multistep methods for stiff ordinary differential equations, *J. Math. Phys. Sci.*, **10**, 449–456.

Janković, V. (1965): Formulas for numerical solution of $y' = f(t, y)$, containing higher derivatives (Slovak, English summary), *Apl. Mat.*, **10**, 469–482.

Janković, V. (1974): A parallel method of solution of the differential equations by the formulas containing higher derivatives, *Acta Univ. Carolin.—Math. Phys.*, **15**, Nos. 1–2, 55–57.

452

Jankowski, T. (1979): Some remarks on numerical solution of initial problems for systems of differential equations, *Apl. Mat.*, **24**, 421–426.

Jankowski, T. (1981): On the conditions for convergence of multistep methods for ordinary differential equations, *Anal. Numér. Théor. Approx.*, **10**, 49–55.

Jansen, J. K. M. (1978): The numerical solution of differential equations (Dutch), *Nieuw Tijdschr. Wisk.*, **66**, 240–250.

Jeltsch, R. (1973): Integration of iterated integrals by multistep methods, *Numer. Math.*, **21**, 303–316.

Jeltsch, R. (1976): Note on A-stability of multistep multiderivative methods, *BIT*, **16**, 74–78.

Jeltsch, R. (1976a): Stiff stability and its relation to A_0- and $A(0)$-stability, *SIAM J. Numer. Anal.*, **13**, 8–17.

Jeltsch, R. (1976b): A necessary condition for A-stability of multistep multiderivative methods, *Math. Comp.*, **30**, 739–746.

Jeltsch, R. (1976c): Multistep multiderivative methods and Hermite–Birkhoff interpolation, in *Proc. Fifth Manitoba Conf. Numerical Math., Winnipeg, 1975* (Eds. B. L. Hartnell and H. C. Williams), Congress. Numer. No. 16, Utilitas Mathematica, Winnipeg, pp. 417–428.

Jeltsch, R. (1977): Multistep methods using higher derivatives and damping at infinity, *Math. Comp.*, **31**, 124–138.

Jeltsch, R. (1977a): Stiff stability of multistep multiderivative methods, *SIAM J. Numer. Anal.*, **14**, 760–771; Corrigendum: *SIAM J. Numer. Anal.*, **16**, 339–345.

Jeltsch, R. (1977b): A_0-stability of Brown's multistep multiderivative methods, *Proc. of the 1977 Army Numer. Anal. and Computer Conference*, pp. 565–581.

Jeltsch, R. (1978): On the stability regions of multistep multiderivative methods, in *Numerical Treatment of Differential Equations: Proc. Oberwolfach, 1976* (Eds. R. Bulirsch, R. D. Grigorieff and J. Schröder), Lecture Notes in Mathematics No. 631, Springer, Berlin, pp. 63–80.

Jeltsch, R. (1978a): Stability on the imaginary axis and A-stability of linear multistep methods, *BIT*, **18**, 170–174.

Jeltsch, R. (1978b): Complete characterization of multistep methods with an interval of periodicity for solving $y'' = f(x, y)$, *Math. Comp.*, **32**, 1108–1114.

Jeltsch, R. (1979): A_0-stability and stiff stability of Brown's multistep multiderivative methods, *Numer. Math.*, **32**, 167–181.

Jeltsch, R., see also Friedli, A.

Jeltsch, R. and Kratz, L. (1977): On the stability properties of Brown's multistep multiderivative methods, *Numer. Math.*, **30**, 25–38.

Jeltsch, R. and Nevanlinna, O. (1978): Largest disk of stability of explicit Runge–Kutta methods, *BIT*, **18**, 500–502.

Jeltsch, R. and Nevanlinna, O. (1981): Stability of explicit time discretizations for solving initial value problems, *Numer. Math*, **37**, 61–91; Corrigendum: *Numer. Math.*, **39**, 155.

Jeltsch, R. and Nevanlinna, O. (1982): Über das Erhöhen der Fehlerordnung von Mehrschrittverfahren zum Lösen von Anfangswertproblemen, *Z. Angew. Math. Mech.*, **62**, T332–T334.

Jeltsch, R. and Nevanlinna, O. (1982a): Stability and accuracy of time discretizations for initial value problems, *Numer. Math.*, **40**, 245–296.

Jeltsch, R. and Nevanlinna, O. (1982b): Lower bounds for the accuracy of linear multistep methods, in *Numerical Integration of Differential Equations and Large Linear Systems: Proc. Bielefeld, 1980* (Ed. J. Hinze), Lecture Notes in Mathematics No. 968, Springer, Berlin, pp. 280–291.

Jensen, P. S. (1974): Transient analysis of structures by stiffly stable methods, *Comput. & Structures*, **4**, 615–626.

Jensen, P. S. (1976): Stiffly stable methods for undamped second order equations of motion *SIAM J. Numer. Anal.*, **13**, 549–563.

Johnson, A. I. and Barney, J. R. (1976): Numerical solution of large systems of stiff ordinary differential equations in a modular simulation framework, in *Numerical Methods for Differential Systems* (Eds. L. Lapidus and W. E. Schiesser), Academic Press, New York, pp. 97–124.

Johnston, R. L., see Hull, T. E.

Joyce, D. C. (1971): Survey of extrapolation processes in numerical analysis, *SIAM Rev.*, **13**, 435–490.

Jukl, V. (1970): Fehlerabschätzung der Nyström'schen Formel, *Acta Math. Univ. Comenian.*, **24**, 81–100.

Jury, E. I., see Bickart, T. A.

Kacewicz, B. (1982): On the optimal error of algorithms for solving a scalar autonomous ODE, *BIT*, **22**, 503–518.

Kaizuka, T. (1977): A remark on the numerical solution of a single first-order ordinary differential equation, *Yokohama Math. J.*, **25**, 183–189.

Kalitkin, N. N. and Kuz'mina, L. V. (1981): Integration of stiff systems of differential equations (Russian, English summary), *Akad. Nauk SSSR Inst. Prikl. Mat. Preprint*, **1981**, No. 80, 23 pp.

Kalitkin, N. N. and Kuz'mina, L. V. (1981a): Numerical examples of the integration of stiff systems (Russian, English summary), *Akad. Nauk SSSR Inst. Prikl. Mat. Preprint*, **1981**, No. 90, 18 pp.

Kalmykov, S. A. (1980): A two-sided method of solution of the equation $y' = f(y)$ with initial value in the form of an interval (Russian), *Chisl. Metody Mekh. Sploshn. Sredy*, **11**, 111–126.

Kalmykov, S. A., see also Shokin, Yu.I.

Kalmykov, S. A., Shokin, Yu.I. and Yuldashev, Z.Kh. (1976): On the solution of ordinary differential equations by integral methods (Russian), *Dokl. Akad. Nauk SSSR*, **230**, 1267–1270.

Kamel, M. S. (1980): High-order multistep multiderivative formulas for the integration of stiff ordinary differential equations in *Proc. Ninth Conf. Numerical Math. and Computing, Winnipeg, 1979* (Eds. G. H. J. van Rees and H. C. Williams), Congress Numer. No. 27, Utilitas Mathematica, Winnipeg, pp. 271–284.

Kamel, M. S., see also Enright, W. H.

Kamp, Y., see Delsarte, P.

Kaps, P. and Rentrop, P. (1979): Generalized Runge–Kutta methods of order four with stepsize control for stiff ordinary differential equations, *Numer. Math.*, **33**, 55–68.

Kaps, P. and Wanner, G. (1981): A study of Rosenbrock-type methods of high order, *Numer. Math.*, **38**, 279–298.

Karim, A. I. A. (1966): Stability of the fourth order Runge–Kutta method for the solution of systems of differential equations, *Comput. J.*, **9**, 308–311.

Karim, A. I. A. (1966a): The stability of the fourth order Runge–Kutta method for the solution of systems of differential equations, *Comm. ACM*, **9**, 113–116.

Karim, A. I. A. (1968): A theorem for the stability of general predictor–corrector methods for the solution of systems of differential equations, *J. Assoc. Comput. Mach.*, **15**, 706–711.

Karpilovskaya, È. B. (1953): On convergence of an interpolation method for ordinary differential equations (Russian), *Uspekhi Mat. Nauk (N. S.)*, **8**, No. 3 (55), 111–118.

Kastlunger, K. H. and Wanner, G. (1972): Runge–Kutta processes with multiple nodes, *Computing*, **9**, 9–24.

Kastlunger, K. H. and Wanner, G. (1972a): On Turan type implicit Runge–Kutta methods, *Computing*, **9**, 317–325.

454

Kato, T., see Hersh, R.

Katz, I. N., see Mattione, R. P.

Katz, I. N., Franklin, M. A. and Sen. A. (1977): Optimally stable parallel predictors for Adams–Moulton correctors, *Comput. Math. Appl.*, **3**, 217–233.

Keener, H. M. and Meyer, G. E. (1982): Solving differential equations by first-order explicit integration, *Simulation*, **38**, 122–130.

Keitel, G. H. (1956): An extension of Milne's three-point method, *J. Assoc. Comput. Mach.*, **3**, 212–222.

Keller, G. (1982): Numerical solution of initial-value problems by collocation methods using generalized piecewise functions, *Computing*, **28**, 199–211.

Kelz, W., see Dejon, B.

Kenue, S. K., see Jackson, L. W.

Khaliq, A. Q. M., see Twizell, E. H.

Khesina, I.Ya., see Zavorin, A. N.

King, R. (1966): Runge–Kutta methods with constrained minimum error bounds, *Math. Comp.*, **20**, 386–391.

Kirchgraber, U. (1978): Sur une méthode de stabilisation en analyse numérique, *C. R. Acad. Sci. Paris*, ser. A, **286**, 421–422.

Kirk, J. van = van Kirk, J., see Leung, K. V.

Kis, O. (1970): The error estimate for the Runge–Kutta method (Russian), *Studia Sci. Math. Hungar.*, **5**, 427–432.

Kis, O. (1970a): The Runge–Kutta method (Russian), *Studia Sci. Math. Hungar.*, **5**, 433–435.

Kis, O., see also Békéssy, A.

Kish, O = Kis, O.

Klimenko, R. K. (1971): A use of the Lagrange constant for the approximate solution of the Cauchy problem for ordinary differential equations (Russian), *Software of Electronic Digital Computers, Proc. Sem. Akad. Nauk Ukrain SSR Inst. Kibernet., Kiev, 1971*, pp. 317–327.

Klokov, Yu. A. = Klokovs, J. = Klokovs, Yu. A.

Klokovs, Yu. A. and Shkerstena, A. (1982): A matrix exponent and the solution of linear differential equations with constant coefficients (Russian), *Latv. Mat. Ezhegodnik*, **26**, 237–244.

Klopfenstein, R. W. (1965): Applications of differential equations in general problem solving, *Comm. ACM*, **8**, 575–578.

Klopfenstein, R. W. (1971): Numerical differentiation formulas for stiff systems of ordinary differential equations, *RCA Rev.*, **32**, 447–462.

Klopfenstein, R. W., see also Crane, R. L.; Pelios, A.

Klopfenstein, R. W. and Davis, C. B. (1971): PECE algorithms for the solution of stiff systems of ordinary differential equations, *Math. Comp.*, **25**, 457–473.

Klopfenstein, R. W. and Millman, R. S. (1968): Numerical stability of a one-evaluation predictor–corrector algorithm for numerical solution of ordinary differential equations, *Math. Comp.*, **22**, 557–564.

Knapp, H. (1964): Über eine Verallgemeinerung des Verfahrens der sukzessiven Approximation zur Lösung von Differentialgleichungssystemen, *Monatsh. Math.*, **68**, 33–45.

Knapp, H. (1966): Ein Spektrum von Iterationsvorschriften zur numerischen Behandlung von Differentialgleichungen, *Bul. Inst. Politehn. Iaşi Secţ. I*, **12**, 55–63.

Knapp, H., see also Reutter, F.

Knapp, H. and Wanner, G. (1967): On the numerical treatment of ordinary differential equations, in *Contribution to the Method of Lie Series* (Eds. W. Gröbner and H. Knapp), Chap. 2, pp. 43–97.

Knapp, H. and Wanner, G. (1968): Numerische Integration gewöhnlicher Differentialgleichungen: Einschrittverfahren, in *Überblicke Mathematik*. (Ed. D. Laugwitz), vol. I, B. I. Hochschultaschenbücher 161/161a, Bibliographisches Institut, Mannheim, pp. 87–114.

Kobza, I. (1975): Methods of Adams type with second derivatives (Russian), *Apl. Mat.*, **20**, 389–405.

Kobza, J. = Kobza, I.

Kőhegyi, J., see Fawzy, T.

Kohfeld, J. J., see Brush, D. G.

Kohfeld, J. J. and Thompson, G. T. (1967): Multistep methods with modified predictors and correctors, *J. Assoc. Comput. Mach.*, **14**, 155–166.

Kohfeld, J. J. and Thompson, G. T. (1968): A modification of Nordsieck's method using an 'off-step' point *J. Assoc. Comput. Mach.*, **15**, 390–401.

Köhler, J., see Strehmel, K.

Konen, H. P., see Luther, H. A.

Konen, H. P. and Luther, H. A. (1967): Some singular explicit fifth order Runge–Kutta solutions, *SIAM J. Numer. Anal.* **4**, 607–619.

Kong, A. K., see Skeel, R. D.

Kopal, Z. (1958): Operational methods in numerical analysis based on rational approximations, in *On Numerical Approximation: Proc. of a Symposium* (Ed. R. E. Langer), University of Wisconsin Press, Madison, pp. 25–43.

Korganoff, A. (1958): Sur des formules d'intégration numérique des équations différentielles donnant une approximation d'ordre élevé, *Chiffres*, **1**, 171–180.

Korolev, V. K. (1975): A comparison of certain methods for the numerical integration of ordinary differential equations (Russian), *Vychisl. Sistemy*, **64**, 108–127.

Korolev, V. K. (1975a): Quadratic-fractional approximation in numerical differentiation and integration (Russian), *Vychisl. Sistemy*, **65**, 130–142.

Korelev, V. K. (1981): The Butcher method with a variable step (Russian), *Vychisl. Sistemy*, **87**, 114–119.

Kovvali, S. and Gupta, G. K. (1982): Polynomial formulation of second derivative multistep methods, *Math. Comp.*, **38**, 447–458.

Kramarz, L. (1977): The collocation solution of nonlinear differential equations by spline functions, *Z. Angew. Math. Mech.*, **57**, 163.

Kramarz, L. (1978): Global approximations to solutions of initial value problems, *Math. Comp.*, **32**, 35–59.

Kramarz, L. (1979): Hermite methods for the numerical solution of ordinary initial value problems, in *Functional Analysis Methods in Numerical Analysis: Proc. St. Louis, 1977* (Ed. M. Z. Nashed), Lecture Notes in Mathematics No. 701, Springer, Berlin, pp. 134–148.

Kramarz, L. (1980): Stability of collocation methods for the numerical solution of $y'' = f(x, y)$, *BIT*, **20**, 215–222.

Kraska, E, see Filippi, S.

Kratz, L., see Jeltsch, R.

Kreiss, H. O. (1978): Difference methods for stiff ordinary differential equations, *SIAM J. Numer. Anal.*, **15**, 21–58.

Kreiss, H. O. (1979): Problems with different time scales for ordinary differential equations, *SIAM J. Numer. Anal.*, **16**, 980–998.

Krestinin, A. V. and Pavlov, B. V. (1981): The single pole rational approximation of the exponential function in the complex plane (Russian), *Zh. Vychisl. Mat. i Mat. Fiz.*, **21**, 1318–1322; English translation: *USSR Comput. Math. and Math. Phys.*, **21**, No. 5, 244–249.

Krinke, D. C. and Huston, R. L. (1980): An analysis of algorithms for solving differential equations, *Comput Structures*, **11**, 69–74.

Krishnaiah, U. A. (1981): Inverse linear multistep methods for the numerical solution of initial value problems of second order differential equations, *J. Comput. Appl. Math.*, **7**, 111–114.

456

Krishnaiah, U. A. (1982): Adaptive methods for periodic initial value problems of second order differential equations, *J. Comput. Appl. Math*, **8**, 101–104.

Krishnaiah, U. A. = Anantha Krishnaiah, U., see also Jain, M. K.

Krochuk, V. V. (1982): On the question of convergence of an approximation method of solution of differential equations, constructed on the basis of Bernstein polynomials (Russian), in *Methods of Approximation Theory and Their Applications* (Ed. V. K. Dzyadyk), Akad. Nauk Ukrain. SSR, Inst. Mat., Kiev, pp. 66–71.

Krogh, F. T. (1966): Predictor–corrector methods of high order with improved stability characteristics, *J. Assoc. Comput. Mach.*, **13**, 374–385.

Krogh, F. T. (1967): A note on the effect of conditionally stable correctors, *Math. Comp.*, **21**, 717–719.

Krogh, F. T. (1967a): A test for instability in the numerical solution of ordinary differential equations, *J. Assoc. Comput. Mach.*, **14**, 351–354.

Krogh, F. T. (1969): A variable step variable order multistep method for the numerical solution of ordinary differential equations, in *Information Processing 68: Proc. IFIP Congress 1968* (Ed. A. J. H. Morrell), North-Holland, Amsterdam, pp. 194–199.

Krogh, F. T. (1972): Opinions on matters connected with the evaluation of programs and methods for integrating ordinary differential equations, *SIGNUM Newsletter*, **7**, No. 3, 27–48.

Krogh, F. T. (1973): Algorithms for changing the step size, *SIAM J. Numer. Anal.*, **10**, 949–965.

Krogh, F. T. (1973a): On testing a subroutine for the numerical integration of ordinary differential equations, *J. Assoc. Comput. Mach.*, **20**, 545–562.

Krogh, F. T. (1974): Changing stepsize in the integration of differential equations using modified divided differences, in *Proc. Conf. on the Numerical Solution of Ordinary Differential Equations, University of Texas at Austin, 1972* (Ed. D. G. Bettis), Lecture Notes in Mathematics No. 362, Springer, Berlin, pp. 22–71.

Krogh, F. T. (1979): Recurrence relations for computing with modified divided differences, *Math. Comp.*, **33**, 1265–1271; Corrigendum: *Math. Comp.*, **35**, 1445.

Krogh, F. T., see also Byrne, G. D.

Krückeberg, F. (1961): Zur numerischen Integration und Fehlererfassung bei Anfangswertaufgaben gewöhnlicher Differentialgleichungen, *Schriften des Rheinisch-Westfälischen Institutes für instrumentelle Mathematik an der Universität Bonn*, No. 1, Bonn, viii + 118 pp.

Krückeberg, F. (1969): Ordinary differential equations, in *Topics in Interval Analysis 1969* (Ed. E. Hansen), Oxford University Press, pp. 91–97.

Krückeberg, F. and Unger, H. (1960): On the numerical integration of ordinary differential equations and the determination of error bounds in *Proc. Symposium on the Numerical Treatment of Ordinary Differential Equations—Integral and Integrodifferential Equations, Rome, 1960*, Birkhauser, Basel, pp. 369–379.

Krüger, S. see Filippi, S.

Krylov, V. K., Bobkov, V. V. and Monastyrnyĭ, P. I. (1982): *Introduction to the Theory of Numerical Methods. Differential Equations*, (Russian), Institut Matematiki AN BSSR. Belorusskiĭ Gosundarstvennyĭ Univ. im. V. I. Lenina, Minsk., 288 pp. (*Nauka i Tekhnika*).

Kubiček, M., see Višňak, K.

Kubiček, M. and Višňak, K. (1974): Two classes of numerical methods for stiff problems, *Acta Univ. Carolin.—Math. Phys.*, **1974**, Nos. 1–2, 63–66.

Kul'chitskaya, I. A., see Gerasimov, B. P.

Kumar, S. = Kenue, S. K., see Jackson, L. W.

Kuntzmann, J. (1956): Remarques sur la méthode de Runge–Kutta, *C. R. Acad. Sci. Paris*, **242**, 2221–2223.

Kuntzmann, J. (1959): Evaluation de l'erreur sur un pas dans les méthodes à pas séparés, *Chiffres*, **2**, 97–102.

Kuntzmann, J. (1959a): Deux formules optimales du type de Runge–Kutta, *Chiffres*, **2**, 21–26.

Kuntzmann, J. (1961): Neuere Entwicklungen der Methode von Runge und Kutta, *Z. Angew. Math. Mech.*, **41**, T28–T31.

Kuntzmann, J. (1962): Nouvelles méthodes pour l'intégration approchée des équations différentielles, in *Information Processing 62: Proc. IFIP Congress 1962*, pp. 157–162.

Kuntzmann, J., see also Bard, A.; Ceschino, F.

Kunz, K. S. (1957): Numerical solution of ordinary differential equations: methods of starting the solution, in *Numerical Analysis*, McGraw-Hill, pp. 167–191.

Kurchatov, V. A. (1978): A method for the approximate solution of systems of nonlinear differential equations (Russian), *Izv. Vyssh. Uchebn. Zaved. Mat.*, **5**(192), 79–83.

Kurmit, A. (1963): Adaption of an error estimate for difference methods of numerical integration to Runge–Kutta's method, *Latvijas PSR Zinatņ Akad. Véstis*, **192**, 75–84.

Kurmit, A. (1963a): A theory of errors for the Runge–Kutta method (Russian, English summary), *Vestnik Leningrad. Univ. Mat. Mekh. Astronom.*, **1963**, No. 19 (Math. section 4), 35–48.

Kurpel', N. S. and Tivonchuk, V. Ĭ. (1975): A certain two-sided method for the approximate solution of the Cauchy problem for ordinary differential equations (Russian), *Ukrain. Mat. Zh.*, **27**, 528–534.

Kutniv, M. V., see Maksymīv, E. M.

Kutta, W. (1901): Beitrag zur näherungsweisen Integration totaler Differentialgleichungen, *Z. Math. Phys.*, **46**, 435–453.

Kuz'mina, L. V., see Kalitkin, N. N.

Kuznetsov, N. N. (1971): Weak stability and asymptotics for solutions of finite–difference approximations of differential equations (Russian), *Dokl. Akad. Nauk SSSR*, **200**, 1026–1029; English translation: *Soviet Math. Dokl.*, **12**, 1529–1533.

Kwapisz, M. see Jackiewicz, Z.

Lafferriere, G., see Zadunaisky, P. E.

Lambert, J. D. (1970): Linear multistep methods with mildly varying coefficients, *Math. Comp.*, **24**, 81–93.

Lambert, J. D. (1971): Predictor–corrector algorithms with identical regions of stability, *SIAM J. Numer. Anal.*, **8**, 337–344.

Lambert, J. D. (1972): Numerical methods for resolution of stiff systems, *ODEs Proc. Equadiff 3, Brno, Sett.*

Lambert, J. D. (1973): *Computational Methods in Ordinary Differential Equations*, Wiley, New York.

Lambert, J. D. (1974): Nonlinear methods for stiff systems of ordinary differential equations, in *Conf. on the Numerical Solution of Differential Equations, Dundee, 1973* (Ed. G. A. Watson), Lecture Notes in Mathematics, No. 363, Springer, Berlin, 75–88.

Lambert, J. D. (1974a): Two unconventional classes of methods for stiff systems, in *Stiff Differential Systems*, (Ed. R. A. Willoughby), Plenum Press, New York, pp. 171–186.

Lambert, J. D. (1976): The numerical integration of a special class of stiff differential systems, in *Proc. Fifth Manitoba Conf. Numerical Math., Winnipeg, 1975* (Eds B. L. Hartnell and H. C. Williams, Congress Numer. No. 16, Utilitas Mathematica, Winnipeg, pp. 91–108.

Lambert, J. D. (1977): The initial value problem for ordinary differential equations, in *The State of the Art in Numerical Analysis*, (Ed. D. A. H. Jacobs), Academic Press, London, pp. 451–500.

Lambert, J. D. (1978): Frequency fitting in the numerical solution of ordinary differential equations, in *Numerical Treatment of Differential Equations in Applications: Proc. Oberwolfach, Germany*, (Eds. R. Ansorge and W. Törnig), Lecture Notes in Mathematics No. 679, Springer, Berlin, pp. 65–72.

Lambert, J. D. (1980): Stiffness, in *Computational Techniques for Ordinary Differential Equations: Proc. Conf., University of Manchester, 1978* (Eds. I. Gladwell and D. K. Sayers), Academic Press, London, pp. 19–46.

Lambert, J. D. (1981): Safe point methods for separably stiff systems of ordinary differential equations, *SIAM J. Numer. Anal.*, **18**, 83–101.

Lambert, J. D., see also Alfeld, P.

Lambert, J. D. and McLeod, R. J. Y. (1979): Numerical methods for phase-plane problems in ordinary differential equations, in *Numerical Analysis: Proc. Dundee, 1979* (Ed. G. A. Watson), Lecture Notes in Mathematics No. 773, Springer, Berlin, pp. 83–97.

Lambert, J. D. and Mitchell, A. R. (1962): On the solution of $y' = f(x, y)$ by a class of high accuracy, difference formulae of low order, *Z. Angew. Math. Phys.*, **13**, 223–232.

Lambert, J. D. and Shaw, B. (1965): On the numerical solution of $y' = f(x, y)$ by a class of formulae based on rational approximation, *Math. Comp.*, **19**, 456–462.

Lambert, J. D. and Shaw, B. (1966): A method for the numerical solution of $y' = f(x, y)$ based on a self-adjusting non-polynomial interpolant, *Math. Comp.*, **20**, 11–20.

Lambert, J. D. and Shaw, B. (1966a): A generalisation of multistep methods for ordinary differential equations, *Numer. Math.*, **8**, 250–263.

Lambert, J. D. and Sigurdsson, S. T. (1972): Multistep methods with variable matrix coefficients, *SIAM J. Numer. Anal.*, **9**, 715–733.

Lambert, J. D. and Watson, I. A. (1976): Symmetric multistep methods for periodic initial value problems, *J. Inst. Math. Appl.*, **18**, 189–202.

Lambert, R. J. (1967): An analysis of the numerical stability of predictor–corrector solutions of nonlinear ordinary differential equations, *SIAM J. Numer. Anal.*, **4**, 597–606.

Lambert, R. J., see also Byrne, G. D.; Crane, R. L.

Lanczos, C. (1960): Solution of ordinary differential equations by trigonometric interpolation, in *Rome Symposium: Symposium on the Numerical Treatment of Ordinary Differential Equations—Integral and Integro-differential Equations: Proc.*, Birkhauser, Basel, pp. 22–32.

Lanning, W. D., see Richards, P. I.

Lapidus, L., see Aiken, R. C.; Chan, Y. N. I.; Seinfeld, J. H.

Lapidus, L. and Schiesser, W. E. (Eds.) (1976): *Numerical Methods for Differential Systems*, Academic Press, New York.

Lapidus, L. and Seinfeld, J. H. (1971): *Numerical Solution of Ordinary Differential Equations*, Academic Press, New York.

Latyshev, A. V. (1980): Adaptive algorithms for checking the numerical solution of dynamics equations (Russian), *Gibridnye Vychisl. Mashiny i Kompleksy,* **1980**, No. 3, 84–88.

Lau, T. C. Y. (1977): Rational exponential approximation with real poles, *BIT*, **17**, 191–199.

Laurent, P. J. (1961): Méthodes spéciales du type de Runge–Kutta, in *Premier congrès AFCAL*, Gauthier-Villars, pp. 27–36.

Laurent, P. J., see also Bard, A.

Lawson, J. D. (1966): An order five Runge–Kutta process with extended region of stability, *SIAM J. Numer. Anal.*, **3**, 593–597.

Lawson, J. D. (1967): Generalized Runge–Kutta processes for stable systems with large Lipschitz constants, *SIAM J. Numer. Anal.*, **4**, 372–380.

Lawson, J. D. (1967a): An order six Runge–Kutta process with extended region of stability, *SIAM J. Numer. Anal.*, **4**, 620–625.

Lawson, J. D. (1970): A note on Runge–Kutta processes for quadratic derivative functions, in *Proc. Combinatorics, Graph Theory and Computing, Louisiana State University*, pp. 189–198.

Lawson, J. D. (1972): On the exactness of implicit Runge–Kutta processes for particular integrals, *BIT*, **12**, 586–588.

Lawson, J. D., see also Ehle, B. L.

Lawson, J. D. and Ehle, B. L. (1970): Asymptotic error estimation for one-step methods based on quadrature, *Aequationes Math.*, **5**, 236–246.

Lea, R. N., see Decell, H. P.

Lecture Notes in Mathematics No. 109: see Morris, J. L. (Ed.) (1969).

Lecture Notes in Mathematics No. 228: see Morris, J. L. (Ed.) (1971).

Lecture Notes in Mathematics No. 333: see van der Houwen, P. J. (1973).

Lecture Notes in Mathematics No. 362: see Bettis, D. G. (Ed.) (1974).

Lecture Notes in Mathematics No. 363: see Watson, G. A. (Ed.) (1974).

Lecture Notes in Mathematics No. 395: see Ansorge, R. and Törnig, W. (Eds.) (1974).

Lecture Notes in Mathematics No. 501: see Micula, G. (1976).

Lecture Notes in Mathematics No. 506: see Watson, G. A. (Ed.) (1976).

Lecture Notes in Mathematics No. 630: see Stetter, H. J. (1978).

Lecture Notes in Mathematics No. 631: see Bulirsch, R., Grigorieff, R. D. and Schröder, J. (Eds.) (1978).

Lecture Notes in Mathematics No. 679: see Lambert, J. D. (1978).

Lecture Notes in Mathematics No. 701: see Kramarz, L. (1979).

Lecture Notes in Mathematics No. 765: see Wambecq, A. (1979).

Lecture Notes in Mathematics No. 773: see Watson, G. A. (Ed.) (1980).

Lecture Notes in Mathematics No. 888: see Werner, H. (1981).

Lecture Notes in Mathematics No. 909: see Hennart, J. P. (Ed.) (1982).

Lecture Notes in Mathematics No. 912: see Dahlquist, G. (1982); Gear, C. W. and Gallivan, K. A. (1982).

Lecture Notes in Mathematics No. 957: see Schumaker, L. L. (1982).

Lecture Notes in Mathematics No. 968: see Hinze, J. (Ed.) (1982).

Lee, D. and Preiser, S. (1978): A class of nonlinear multistep *A*-stable numerical methods for solving stiff differential equations, *Comput. Math. Appl.*, **4**, 43–51.

Lee, D. T. S., see Chen. Y. M.

Lee, H. B. (1967): Matrix filtering as an aid to numerical integration, *Proc. IEEE*, **55**, 1826–1831.

Lee, J. S. and Ni, W. C. (1980): On the (14,9) explicit Runge–Kutta method, *J. Nat. Chiao Tung Univ.*, **8**, 31–55.

Leech, J. W., see Morino, L.

Legras, J. (1966): Résolution numérique des grands systèmes différentiels linéaires, *Numer. Math.*, **8**, 14–28.

Lemaréchal, C. (1971): La stabilité absolue des schémas de Runge–Kutta semi-implicites, *C. R. Acad. Sci.*, ser A, **272**, 397–400.

Lena, G. di = di Lena, G.

Lenarduzzi, L., see Bozzini, M.

Lensing. J. (1976): Verallgemeinertes Runge–Kutta-Verfahren, *Z. Angew. Math. Mech.*, **55**, 503–504.

Lensing, J. (1977): Die numerische Integration linearer inhomogener Differential-gleichungssysteme zweiter Ordnung mit nichtkonstanten Koeffizienten, *Z. Angew. Math. Mech.*, **57**, 559.

Lether, F. G. (1966): The use of Richardson extrapolation in one-step methods with variable step size, *Math Comp.*, **20**, 379–385.

Leung, K. V. and van Kirk, J. (1974): Numerical solution of ordinary differential equations by the method of undetermined parameters, *Bull. Inst. Politehn. Iaşi.*, **20**, Nos. 3–4, 47–52.

Levy, H. and Baggott, E. A. (1934): *Numerical Studies in Differential Equations*, vol. 1, Watts, London. (Reprinted under title *Numerical Solutions of Differential Equations*, Dover, New York, 1950.

Lewanowicz, S. (1976): Algorithm 47, solution of a first-order non-linear differential equation in Chebyshev series, *Zastos. Mat.*, **15**, 251–269.

Lewis, H. R. and Stovall, E. J. (1967): Comments on a floating-point version of Nordsieck's scheme for the numerical integration of differential equations, *Math. Comp.*, **21**, 157–161.

Li, C. S. = Lee, J. S.

Li, W. Y. (1980): A discussion on a class of explicit linear multistep methods with difference perturbation terms (Chinese, English summary), *Math. Numer. Sinica*, **2**, 203–208.

Li, W. Y. (1982): Some ways of constructing explicit methods with large stability regions, and their interrelation (Chinese, English summary), *J. Numer. Methods Comput. Appl.*, **3**, 125–128.

Liem, C. B., see Cash, J. R.

Lindberg, B. (1971): On smoothing and extrapolation for the trapezoidal rule, *BIT*, **11**, 29–52.

Lindberg, B. (1972): A simple interpolation algorithm for improvement of the numerical solution of a differential equation, *SIAM J. Numer. Anal.*, **9**, 662–668.

Lindberg, B. (1974): A stiff system package based on the implicit midpoint method with smoothing and extrapolation, in *Stiff Differential Systems* (Ed. R. A. Willoughby), Plenum Press, New York, pp. 201–215.

Lindberg, B. (1974a): On a dangerous property of methods for stiff differential equations, *BIT*, **14**, 430–436.

Lindberg, B. (1977): Characterization of optimal stepsize sequences for methods for stiff differential equations, *SIAM J. Numer. Anal.*, **14**, 859–887.

Lindberg, B, see also Enright, W. H.

Lindelöf, E. (1938): Remarques sur l'intégration numérique des équations différentielles ordinaires, *Acta Soc. Sci. Fennicae Nova ser A*, **2**, No. 13, 1–21.

Liniger, W. (1968): A criterion for *A*-stability of linear multistep integration formulae, *Computing*, **3**, 280–285.

Liniger, W. (1969): Global accuracy and *A*-stability of one- and two-step integration formulae for stiff ordinary differential equations, in *Conf. on the Numerical Solution of Differential Equations, Dundee, 1969* (Ed. J. L. Morris), Lecture Notes in Mathematics No. 109, Springer, Berlin, pp. 188–193.

Liniger, W. (1971): A stopping criterion for the Newton–Raphson method in implicit multistep integration algorithms for nonlinear systems of ordinary differential equations, *Comm. ACM*, **14**, 600–601.

Liniger, W. (1975): Connections between accuracy and stability properties of linear multistep formulas, *Comm. ACM*, **18**, 53–56.

Liniger, W. (1976): High-order *A*-stable averaging algorithms for stiff differential systems, in *Numerical Methods for Differential Systems* (Ed. L. Lapidus), Academic Press, New York, pp. 1–23.

Liniger, W. (1977): Stability and error bounds for multistep solutions of nonlinear differential equations, in *Proc. IEEE International Symposium on Circuits and Systems*, pp. 277–280.

Liniger, W. (1979): Multistep and one-leg methods for implicit mixed differential algebraic systems, *IEEE Trans. Circuits and Systems*, **26**, 755–762.

Liniger, W., see also Brayton, R. K.; Miranker, W. L.; Nevanlinna, O.; Odeh, F.; Ruehli, A. E.; Sarkany, E. F.

Liniger, W. and Gagnebin, T. (1974): Construction of a family of second order, *A*-

stable k-step formulas depending on the maximum number, $2k - 2$, of parameters, in *Stiff Differential Systems* (Ed. R. A. Willoughby), Plenum Press, New York, pp. 217–227.

Liniger, W. and Odeh, F. (1972): A-stable, accurate averaging of multistep methods for stiff differential equations, *IBM J. Res. Develop.*, **16**, 335–348.

Liniger, W. and Odeh, F. (1982): Accurate multistep methods for smooth stiff problems, in *Computational and Asymptotic Methods for Boundary and Interior Layers: Proc. BAIL II Conf., Dublin, 1982* (Ed. J. J. H. Miller), Boole Press Conf. Ser. 4, Boole Press, Dublin, pp. 53–67.

Liniger, W. and Willoughby R. A. (1970): Efficient integration methods for stiff systems of ordinary differential equations, *SIAM J. Numer. Anal.*, **7**, 47–66.

Lisbona, F. J., see Calvo, M.

Liseĭkin, V. D. (1978): The optimal grid for numerical solution of an ordinary differential equation (Russian), *Chisl. Metody Mekh. Sploshn. Sredy*, **9**, No. 6, *Mat. Modelirovanie*, 115–118.

Liseĭkin, V. D. (1981): Optimal meshes for second-order schemes approximating an ordinary differential equation (Russian), *Chisl. Metody Mekh. Sploshn. Sredy*, **12**, No. 1, *Mat. Modelirovanie*, 78–81.

Liseĭkin, V. A. (1982): Numerical solution of a second-order ordinary differential equation with a small parameter multiplying the highest derivative (Russian), *Chisl. Metody Mekh. Sploshn. Sredy*, **13**, No. 3, 71–80.

Litewska, K. and Muszyński, J. (1981): On the convergence of the finite elements method for some initial problem, in *Nonlinear Analysis: Theory and Applications: Proc. Seventh Summer School, Berlin, 1979* (Ed. R. Kluge), Abh. Akad. Wiss. DDR, Abt. Math. Naturwiss. Tech. 2, Adakemie, Berlin, pp. 367–370.

Litinsky, E. G. (1978): Best approximation and numerical methods in solution of differential equations, *J. Comput. Appl. Math.*, **4**, 113–116.

Liu, S. T. (1976): On the study of Curtis' (11,8) explicit Runge–Kutta method, *J. Nat. Chiao Tung Univ.*, **1**, 185–205.

Locher, F. (1975): Numerische Lösung linearer Differentialgleichungen mit Hilfe von Chebyshev-Entwicklung, *Internat. Ser. Numer. Math.*, **27**, 155–163.

Lombardi, G. and Marzulli, P. (1980), Metodi di predizione e correzione A-stabili per la risoluzione numerica di sistemi differenziali stiff, *Calcolo*, **14**, 369–384.

López, M. S. (1975): The numerical solution of ordinary differential equations, I (Spanish), *Gaceta. Mat.*, ser 1, **27**, 32–44.

López, M. S. (1975a): The numerical solution of ordinary differential equations, II (Spanish), *Gaceta Mat.*, ser. 1, **27**, 136–157.

Lory, P., see Diekhoff, H. J.

Loscalzo, F. R. (1969): An introduction to the application of spline functions to initial value problems, in *Theory and Applications of Spline Functions* (Ed. T. N. E. Greville), Academic Press, New York, pp. 37–64.

Loscalzo, F. R. and Talbot, T. D. (1967): Spline function approximations for solutions of ordinary differential equations, *SIAM J. Numer. Anal.*, **4**, 433–445 (also *Bull. Amer. Math. Soc.*, **73**, 438–442).

Lotkin, M. (1951): On the accuracy of Runge–Kutta's method, *Math. Comp.*, **5**, 128–133; Corrigendum: *Math. Comp.*, **6**, 61.

Lotkin, M. (1952): A new integrating procedure of high accuracy, *J. Math. Physics*, **31**, 29–34.

Lotkin, M. (1954): The propagation of error in numerical integrations, *Proc. Amer. Math. Soc.*, **5**, 869–887.

Lotkin, M. (1955): On the improvement of accuracy in integration, *Quart. Appl. Math.*, **13**, 47–54.

Lotkin, M. (1956): A note on the midpoint method of integration, *J. Assoc. Comput. Mach.*, **3**, 208–211.

462

Loud, W. S. (1949): On the long-run error in the numerical solution of certain differential equations, *J. Math. Phys.*, **28**, 45–49.

Löwdin, P. O. (1952): On the numerical integration of ordinary differential equations of the first order, *Quart. Appl. Math.*, **10**, 97–111.

Lowery, D., see Corliss, G. F.

Lozinskiĭ, S. M. (1953); Estimate of the error of an approximate solution of a system of differential equations (Russian), *Dokl. Akad. Nauk SSSR*, **92**, 225–228.

Lozinskiĭ, S. M. (1954): On the approximate solution of a system of ordinary differential equations (Russian), *Dokl. Akad, Nauk SSSR*, **97**, 29–32.

Lozinskiĭ, S. M. (1958): Error estimate for numerical integration of ordinary differential equations, I (Russian), *Izv. Vyssh. Uchebn Zaved. Mat.*, **5**(6), 52–90; Corrigendum: *Izv. Vyssh. Uchebn Zaved. Mat.*, **5**(12), 222 (1959).

Lubich, C., see Hairer, E.

Luke, Y. L., Fair, W. and Wimp, J. (1975): Predictor–corrector formulas based on rational interpolants, *Comput. Math. Appl.*, **1**, 3–12.

Lukehart, P. M. (1963): Algorithm 218, Kutta Merson, *Comm. ACM*, **6**, 737 (see Priebe, K. B.; Bayer, G.).

Luther, H. A. (1966): Further explicit fifth-order Runge–Kutta formulas, *SIAM Rev.*, **8**, 374–380.

Luther, H. A. (1968): An explicit sixth-order Runge–Kutta formula, *Math. Comp.*, **22**, 434–436.

Luther, H. A., see also Konen, H. P.

Luther, H. A. and Konen, H. P. (1965): Some fifth-order classical Runge–Kutta formulas, *SIAM Rev.*, **7**, 551–558.

Luther, H. A. and Sierra, H. G. (1970): On the optimal choice of fourth–order Runge–Kutta formulas, *Numer. Math.*, **15**, 354–358.

Luxemburg, W. A. J., see Hull, T. E.

Lyapunov, A. M. (1907): Problème général de la stabilité du mouvement, *Ann. Fac. Sci. Univ. Toulouse*, Ser. II, **9**. 203–474. (Reprinted in Ann. of Math. Studies No. 17, Princeton, 1947).

Lyashchenko, N. Ya. and Dolinnyĭ, O. B. (1977): The construction of a class of two-sided formulas of Runge–Kutta type of sixth order accuracy (Russian, English summary), *Vychisl. Prikl. Mat. (Kiev)*, **33**, 98–109.

Lyashchenko, N. Ya. and Gez', I. F. (1976): A two-sided process of Runge–Kutta type (Russian, English summary), *Vychisl. Prikl. Mat. (Kiev)*, **28**, 48–62.

Lyashchenko, N. Ya., Gez', I. F. and Oleĭnik, A. G. (1976): On two–sided approximations of Runge–Kutta type (Russian), *Dokl. Akad. Nauk SSSR*, **226**, 265–268; English translation: *Soviet Math. Dokl.*, **17**, 94–98.

Lyashchenko, N. Ya. and Oleĭnik, A. G. (1976): Two-sided methods of Runge–Kutta–Fehlberg type (Russian, English summary), *Vychisl. Prikl. Mat. (Kiev)*, **29**, 17–26.

Lyashchenko, N. Ya. and Oleĭnik. A. G., (1978): On a two-sided process of Runge–Kutta type with three double nodes (Russian), *Differentsial'nye Uravneniya*, **14**, 369–370.

Lyche, T. (1969): Optimal order multistep methods with an arbitrary number of nonstep points, in *Conf. on the Numerical Solution of Differential Equations, Dundee, 1969*, (Ed. J. L. Morris), Lecture Notes in Mathematics No. 109, Springer, Berlin, pp. 194–199.

Lyche, T. (1969a): A note on correctors with an arbitrary number of nonstep points, *BIT*, **9**, 239–249.

Lyche, T. (1972): Chebyshevian multistep methods for ordinary differential equations, *Numer. Math.*, **19**, 65–75.

Lyubchenko, I. N. (1978): An iteration method for the solution of the Cauchy problem for a system of ordinary differential equations (Russian), *Ukrain. Mat. Zh.*, **30**, 816–817: English translation: *Ukranian Math. J.*, **30**, 615–616.

MacEwen, S. R., see Carver. M. B.

Maess, G. (1965): Zur Bestimmung des Restgliedes von Lie-Reihen, *Wiss. Z. Friedrich-Schiller-Univ. Jenal/Thüringen*, **14**, 423–425.

Magenes, E. (Ed.) (1972): *Seminari sulla Risoluzione Numerica delle Equazioni Differeziali di Tipo 'Stiff'*, by V. Comincioli, G. P. Monti, G. Pierini and L. Dalle Rive, Laboratorio di Analisi Numerica Consiglio Nazionale della Ricerche, vol. 36.

Mahony, J. J. and Shepherd, J. J. (1981): Stiff systems of ordinary differential equations, I. Completely stiff, homogeneous systems, *J. Austral. Math. Soc.*, Ser. B, **23**, 17–51.

Mäkelä, M. (1971): On a generalized interpolation approach to the numerical integration of ordinary differential equations, *Ann. Acad. Sci. Fenn. ser. A I Math.*, **503**, 43pp.

Mäkelä, M., Nevanlinna, O. and Sipilä, A. H. (1974): Exponentially fitted multistep methods by generalized Hermite–Birkhoff interpolation, *BIT*, **14**, 437–451.

Mäkelä, M., Nevanlinna, O. and Sipilä, A. H. (1974a): On the concepts of convergence, consistency and stability in connection with some numerical methods, *Numer. Math.*, **22**, 261–274.

Makinson, G. J. (1968): Stable high order implicit methods for the numerical solution of systems of differential equations, *Comput. J.*, **11**, 305–310.

Maksimeǐ, G. A. (1971): The Runge–Kutta method and quadrature formulas (Russian), *Sibirsk. Mat. Zh.*, **12**, 1146–1150; English translation: *Siberian Math. J.*, **12**, 825–827.

Maksimīv, Ē. M., see Bodnarchuk, P. I.

Maksymīv, Ē. M. and Kutniv, M. V. (1980): A nonlinear explicit method for numerical integration of differential equations (Russian), *Vestnik L'vov. Politekhn. Inst.*, **141**, *Differentsial'nye Uravneniya i ikh Prilozhen*, 57–58.

Mannshardt, R. (1975): Runge–Kutta-Verfahren mit einem impliziten Rechenschritt, *Z. Angew. Math. Mech.*, **55**, T251–T253.

Mannshardt, R. (1978): Prädiktoren mit vorgeschriebenem Stäbilitatsverhalten, in *Numerical Treatment of Differential Equations: Proc. Oberwolfach, 1976* (Eds. R. Bulirsch, R. D. Grigorieff and J. Schröder), Lecture Notes in Mathematics No. 631, Springer, Berlin, 81–96.

Mannshardt, R. (1978a): One-step methods of any order for ordinary differential equations with discontinuous right-hand sides, *Numer. Math.*, **31**, 131–152.

Mao, Z. F., see Sun, G.

Marcowitz, U. (1975): Fehlerabschätzung bei Angangswertaufgaben für Systeme von gewöhnlichen Differentialgleichungen mit Anwendung auf das REENTRY–Problem, *Numer. Math.*, **24**, 249–275.

Marshall, R. H., see Howells, P. B.

Martin, D. W. (1958): Runge–Kutta methods for integrating differential equations on high speed digital computers, *Comput. J.*, **1**, 118–123.

Martynov, A. V., see Goncharova, I. F.

März, R. (1979): *Parametric Multistep Methods*, Seminarberichte No. 18, Humboldt University, Berlin.

März, R. (1980): Untersuchung eines parametrischen impliziten Einschrittverfahrens, *Beiträge Numer. Math.*, **8**, 85–97.

März, R. (Ed.) (1980a): *Proc. Second Conf. on Numerical Treatment of Ordinary Differential Equations, Berlin, 1980*, Seminarberichte No. 32, Humboldt University, Berlin.

März, R. (1980b): On the stability of variable methods in initial value problems and in boundary value problems, in *Proc. Second Conf. on Numerical Treatment of Ordinary Differential Equations, Berlin, 1980* (Ed. R. März), Seminarberichte No. 32, Humboldt University, Berlin, pp. 43–63.

464

März, R. (1981): Ein parametrisches Zweischrittverfahren, *Beiträge Numer. Math.*, **10**, 71–89.

März, R. (Ed.) (1982): *Proc. Third Conference on Numerical Treatment of Ordinary Differential Equations, Berlin, 1982,* Seminarberichte No. 46, Humboldt University, Berlin.

März, R, see also Alemayehu, A.

Marzulli, P. (1966): Stabilità massimale nei metodi di integrazione numerica del tipo predittore–correttore, *Calcolo*, **3**, 339–349.

Marzulli, P. (1974): Applicazione di formule A–stabili esplicite alla risoluzione numerica dell'equazione $y^{(n)} = f(x, y, y', \ldots, y^{(n-1)})$, *Boll. Un. Mat. Ital.* (4)**10**, 659–665.

Marzulli, P. (1974a): Su una classe di formule esplicite A-stabili per l'integrazione numerica di equazioni differenziali ordinarie, *Calcolo*, **11**, 403–419.

Marzulli, P. (1979): On a common feature of some Runge–Kutta and predictor–corrector methods, *Calcolo*, **16**, 181–188.

Marzulli, P., see also Gheri, G.; Lombardi, G.

Mattheij, R. M. M. (1979): On approximating smooth solutions of linear singularly perturbed ODE, in *Numerical Analysis of Singular Perturbation Problems: Proc. Conf., Math. Inst., Catholic Univ., Nijmegen, 1978* (Eds. P. W. Hemker and J. J. H. Miller), Academic Press, London, pp. 457–465.

Matthieu, P. (1951): Über die Fehlerabschätzung beim Extrapolationsverfahren von Adams I. Gleichungen 1. Ordnung, *Z. Angew. Math. Mech.*, **31**, 356–370.

Matthieu, P. (1953): Über die Fehleräbschatzung beim Extrapolationsverfahren von Adams II. Gleichungen zweiter und höherer Ordnung, *Z. Angew. Math. Mech.*, **33**, 26–41.

Mattione, R. P., Katz, I. N. and Franklin, M. A. (1978): A partitioned parallel Runge–Kutta method for weakly coupled ordinary differential equations, *Internat. J. Numer. Methods Engrg.*, **12**, 267–278.

Mayers, D. F. (1962): Methods of Runge–Kutta type, prediction and correction: deferred correction, stability of step-by-step methods, in: *Numerical Solution of Ordinary and Partial Differential Equations,* Pergamon, Oxford; Addison-Wesley, Reading, Massachusetts, pp. 19–57.

Mayers, D. F., see also Fox, L.

Mazurkiewicz, A. and Rybarski, A. (1973): A method for approximate solution of the pendulum-type differential equations, *Bull. Acad. Polon. Sci. Ser. Sci. Math. Astronom. Phys.*, **21**, 1001–1004.

McCann, M. J., see Robertson, H. H.

McCarthy, D. and Williams, H. C. (Eds.) (1978): *Proc. Seventh Manitoba Conf. Numerical Math. and Computing, Winnipeg, 1977,* Congress Numer. No. 20, Utilitas Mathematica, Winnipeg (see Bui, T. D. and Ghaderpanah, S. S., Corliss, G. F.; Swayne, D. A.)

McKee, S. and Pitcher, N. (1979): On the convergence of advanced linear multistep methods, *BIT*, **19**, 476–481

McKee, S. and Pitcher, N. (1982): Two sided error bounds for discretisation methods in ordinary differential equations, *BIT*, **22**, 314–330.

McLeod, R. J. Y., see Lambert, J. D.

McLeod, R. J. Y. and Sanz-Serna, J. M. (1982): Geometrically derived difference formulae for the numerical integration of trajectory problems, *IMA J. Numer. Anal.*, **2**, 357–370

Meister, G. (1974): Über die Integration von Differentialgleichungssystemen 1. Ordnung mit exponentiell angepaßten numerischen Methoden, *Computing*, **13**, 327–352.

Merli, C., see Dalle Rive, L.

Merluzzi, P. and Brosilow, C. (1978): Runge–Kutta integration algorithms with built-in estimates of the accumulated truncation error, *Computing*, **20**, 1–16.

Merson, R. H. (1957): An operational method for the study of integration processes, in *Proc. Symp. Data processing*, Weapons Research Establishment, Salisbury, S. Australia.

Meshaka, P. (1964): Deux méthodes d'intégration numérique pour systèmes différentiels, *Chiffres*, **7**, 135–148.

Meyer, G. E., see Keener, H. M.

Michelsen, M. L. (1976): An efficient general purpose method for the integration of stiff ordinary differential equations, *AIChE J.*, **22**, 594–597

Michelsen, M. L. (1977): Application of semi-implicit Runge–Kutta methods for integration of ordinary and partial differential equations, *Chem. Engrg J.*, **14**, 107–112.

Micula, G. (1971): Approximate integration of systems of differential equations by spline functions, *Studia Univ. Babeş-Bolyai Ser. Math.-Mech.*, **16**, No. 2, 27–39.

Micula, G. (1971a): Fonctions spline d'approximation pour les solutions des systèmes d'équations différentielles, *An. Ştiinţ. Univ. 'Al. I. Cuza' Iaşi Secţ. I a Mat. (N. S.)*, **17**, 139–155.

Micula, G. (1972): Numerical integration of the differential equation $y^{(n)} = f(x, y)$ by spline functions, *Rev. Roumaine Math. Pures Appl.* **17**, 1385–1389.

Micula, G. (1972a): Spline functions of higher degree of approximation for solutions of systems of differential equations (Romanian, English summary), *Studia Univ. Babeş-Bolyai Ser. Math.-Mech.*, **17**, No. 1. 21–32.

Micula, G. (1972b): Spline functions approximating the solution of nonlinear differential equation of nth order, *Z. Angew. Math. Mech.* **52**, 189–190.

Micula, G. (1973): Approximate solution of the differential equation $y'' = f(x, vy)$ with spline functions, *Math. Comp.*, **27**, 807–816; Corrigendum: *Math. Comp*, **29**, 673–674.

Micula, G. (1974): Die numerische Lösung nichtlinearer Differentialgleichungen unter Verwendung von Splinefunktionen, in: *Numerische Behandlung nichtlinearer Integrodifferential- und Differentialgleichungen* (Eds. R. Ansorge and W. Törnig), Lecture notes in Mathematics No. 395, Springer, Berlin, pp. 57–83.

Micula, G. (1974a): Deficient spline approximate solutions to linear differential equations of the second order, *Mathematica (Cluj)* **16**(39), 65–72.

Micula, G. (1975): Über die numerische Lösung nichtlinearer Differentialgleichungen mit Splines von niedriger Ordnung, *Internat. Ser. Numer. Math.*, **27**, 185–195.

Micula, G. (1975a): The numerical solution of nonlinear differential equations by deficient spline functions, *Z. Angew. Math. Mech.*, **55**, T254–T255.

Micula, G., (1976): Bermerkungen zur numerischen Lösung von Anfangswertproblemen mit Hilfe nichtlinearer Splinefunktionen, in *Spline Functions: Proc. Karlsruhe 1975* (Eds. K. Böhmer, G. Meinardus and W. Schempp), Lecture Notes in Mathematics No. 501, Springer, Berlin, pp. 200–209.

Micula, M., see Coţiu, A.

Mihelcić, M. (1977): Fast A-stabile Donelson–Hansensche zyklische Verfahren zur numerischen Integration von 'stiff'-Differentialgleichungssystemen, *Angew. Informat.*, **7**, 299–305.

Mihelcić, M. (1978): $A(\alpha)$-stable cyclic composite multistep methods of order 5, *Computing*, **20**, 267–272.

Mihelcić, M. and Wingerath, K. (1981): $A(\alpha)$–stable cyclic composite multistep methods of orders 6 and 7 for numerical integration of stiff ordinary differential equations, *Z. Angew. Math. Mech.*, **61**, 261–264.

Milkov, V. (1982): An iterative numerical method for solving differential equations with initial conditions (Russian), in: *Differential Equations and Applications, I, II:*

Proc., Second Conf., Ruse, 1981 (Eds. I. Dimovski and Ī. Stoyanov), Technical University Ruse, pp. 492–495.

Miller, J. C. P. (1966): The numerical solution of ordinary differential equations, in *Numerical Analysis: An Introduction* (Ed. J. Walsh), Academic Press, London, pp. 63–98.

Miller, J. C. P., see also Cash, J. R.

Miller, J. J. H. (1971): On weak stability, stability, and the type of a polynomial, in *Conf. on Applications of Numerical Analysis, Dundee, 1971* (Ed. J. L. Morris), Lecture notes in Mathematics No. **228**, Springer, Berlin, pp. 316–320.

Miller, J. J. H. (1971a): On the location of zeros of certain classes of polynomials with applications to numerical analysis, *J. Inst. Math. Appl.*, **8**, 397–406.

Miller, J. J. H. (Ed.) (1973): *Topics in Numerical Analysis,* Academic Press, London.

Miller, J. J. H. (Ed.) (1975): *Topics in Numerical Analysis II* Academic Press, London.

Miller, J. J. H. (Ed.) (1977): *Topics in Numerical Analysis III,* Academic Press, London.

Miller, J. J. H., see also Creedon, D. M.

Millman, R. S., see Klopfenstein, R. W.

Milne, W. E. (1926): Numerical integration of ordinary differential equations, *Amer. Math. Monthly,* **33**, 455–460.

Milne, W. E. (1933): On the numerical integration of certain differential equations of the second order, *Amer. Math. Monthly,* **40**, 322–327.

Milne, W. E. (1949): A note on the numerical integration of differential equations, *J. Res. Nat. Bur. Standards,* **43**, 537–542.

Milne, W. E. (1950): Note on the Runge–Kutta method, *J. Res. Nat. Bur. Standards,* **44**, 549–550.

Milne, W. E. (1953): *Numerical Solution of Differential Equations*, Wiley, New York.

Milne, W. E., see also Bennett, A. A.

Milne, W. E. and Reynolds, R. R. (1959): Stability of a numerical solution of differential equations, *J. Assoc. Comput. Mach.*, **6**, 196–203.

Milne, W. E. and Reynolds, R. R. (1960): Stability of a numerical solution of differential equations II, *J. Assoc. Comput Mach.*, **7**, 46–56.

Milne, W. E. and Reynolds, R. R. (1962): Fifth-order methods for the numerical solution of ordinary differential equations, *J. Assoc. Comput. Mach.*, **9**, 64–70.

Miranker, W. L. (1971): Matricial difference schemes for integrating stiff systems of ordinary differential equations, *Math. Comp.*, **25**, 717–728.

Miranker, W. L. (1971a): Difference schemes with best possible truncation error, *Numer. Math.*, **17**, 124–142.

Miranker, W. L. (1973): Numerical methods of boundary layer for stiff systems of differential equations, *Computing*, **11**, 221–234.

Miranker, W. L. (1981): *Numerical methods for stiff equations and singular pertubation problems,* Reidel, Dordrecht.

Miranker, W. L., see also Hoppensteadt, F. C.

Miranker, W. L. and Chern, I. L. (1981): Dichotomy and conjugate gradients in the stiff initial value problem, *Linear Algebra Appl.*, **36**, 57–77.

Miranker, W. L. and Hoppensteadt, F. (1973): Numerical methods for stiff systems of differential equations related with transistors, tunnel diodes, etc., Lecture Notes in Comput. Sci. No. 10, Springer, Berlin, pp. 416–432.

Miranker, W. L. and Liniger, W. (1967): Parallel methods for the numerical integration of ordinary differential equations, *Math. Comp.*, **21**, 303–320.

Miranker, W. L. and van Veldhuizen, M. (1978): The method of envelopes, *Math. Comp.*, **32**, 453–496.

Miranker, W. L. and Wahba, G. (1976): An averaging method for the stiff highly oscillatory problem, *Math. Comp.*, **30**, 383–399.

Mise, S., see Urabe, N.

Mises, R. von = von Mises, R.

Mitchell, A. R., see Fox, L.; Lambert, J. D.

Mitchell, A. R. and Craggs, J. W. (1953): Stability of difference relations in the solution of ordinary differential equations, *Math. Comp.*, **7**, 127–129.

Mitsui, T. (1979): The initial–value adjusting method for solving problems of the least squares type of ordinary differential equations, *Mem. Numer. Math.*, **6**, 21–37.

Mitsui, T. (1982): Runge–Kutta type integration formulas including the evaluation of the second derivative *I*, *Publ. Res. Inst. Math. Sci.*, **18**, 325–364.

Möbius, A., see Griepentrog, E.

Mohamed, J., see Coleman, J. P.

Mohr, E. (1951): Über das Verfahren von Adams zur Integration gewöhnlicher Differentialgleichungen, *Math. Nachr.*, **5**, 209–218.

Møller, O. (1965): Quasi double-precision in floating point addition, *BIT*, **5**, 37–50.

Møller, O. (1965a): Note on quasi double–precision, *BIT* **5**, 251–255.

Monastyrnyĭ, P. I., see Krylov, V. I.

Monsef, Y. (1973): Un algorithme rapide de résolution d'un système d'équations algébro–différentielles par une méthode implicite. Comparaison avec l'algorithme de Gear, *C. R. Acad. Sci. Paris,* ser. A, **277**, 1179–1181.

Monti, G. P., see Comincioli, V.

Montrone, M., see Gautschi, W.

Moore, H. (1974): Comparison of numerical integration techniques for orbital applications, in *Proc. Conf. on Numerical Solution of Ordinary Differential Equations, University of Texas at Austin, 1972* (Ed. D. G. Bettis), Lecture Notes in Mathematics No. 362, Springer, Berlin, 149–166.

Moore, R. E. (1964): The automatic analysis and control of error in digital computation based on the use of interval numbers, in *Error in Digital Computation* (Ed. L. B. Rall), vol. 1, Wiley, New York, pp. 61–130.

Moore, R. E. (1965): Automatic local coordinate transformations to reduce the growth of error bounds in the interval computation of solution of ordinary differential equations, in *Error in Digital Computation* (Ed. L. B. Rall), vol. 2, Wiley, New York, pp. 103–140.

Moore, R. E. (1966): Machine-computed error bounds for numerical solutions of differential equations, in *Information Processing 65: Proc. IFIP Congress 1965* (Ed. W. A. Kalenich), vol 2, Spartan Books, Washington, D. C., pp. 560–561.

Moore, R. E., see also Daniel, J. W.

Morel, H. (1956): Evaluation de l'erreur sur un pas dans la méthode de Runge–Kutta, *C. R. Acad. Sci. Paris*, **243**, 1999–2002.

Morino, L., Leech, J. W. and Witmer, E. A. (1974): Optimal predictor–corrector method for systems of second-order differential equations, *AIAA J.*, **12**, 1343–1347.

Morozova, Yu. I., see Vysotskiĭ, L. I.

Morris, J. L. (Ed.) (1969): *Conf. on the Numerical Solution of Differential Equations, Dundee, 1969*, Lecture Notes in Mathematics No. 109, Springer, Berlin (see Butcher, J. C.; Cooper, G. J.; Hodgkins, W. R.; Liniger, W.; Lyche, T.; Nickel, K.; Nørsett, S. P.; Piotrowski, P.; Prothero, A.; Skappel, J.; Spijker, M. N.; Stetter, H. J.; Verner, J. H.; Vitásek, E.).

Morris, J. L. (Ed.) (1971): *Conf. on Applications of Numerical Analysis, Dundee, 1971*, Lecture Notes in Mathematics No. 228, Springer, Berlin (see Cooper, G. J.; Miller, J. J. H.; Sigurdsson, S.; Stetter, H. J.; Verner, J. H.).

Morrison, D. (1962): Optimal mesh size in the numerical integration of an ordinary differential equation, *J. Assoc. Comput. Mach.*, **9**, 98–103.

Morrison, D., see also Stoller, L.

Moss, P., see Burrage, K.

Moulton, F. R. (1926): *New Methods in Exterior Ballistics,* University of Chicago.

Mukhin, I. S. (1952a): Application of the Markov–Hermite interpolation polynomials for the numerical integration of ordinary differential equations (Russian), *Akad. Nauk SSSR Prikl. Mat. Mekh.*, **16**, 231–238.

Mukhin, I. S. (1952a): On the accumulation of errors in numerical integration of differential equations (Russian), *Akad. Nauk SSSR Prikl. Mat. Mekh.*, **16**, 753–755.

Müller, D., see Friedrich, V.

Müller, K. H. (1975): Stabilitätsungleichungen für lineare Differenzenoperatoren, *Internat. Ser. Numer. Math.*, **27**, 227–253.

Mülthei, H. N. (1979): Splineapproximationen von beliebigem Defekt zur numerischen Lösung gewöhnlicher Differentialgleichungen I, *Numer. Math.*, **32**, 147–157.

Mülthei, H. N. (1979a): Splineapproximationen von beliebigem Defekt zur numerischen Lösung gewöhnlicher Differentialgleichungen, II, *Numer. Math.*, **32**, 343–358.

Mülthei, H. N. (1980): Splineapproximationen von beliebigem Defekt zur numerischen Lösung gewöhnlicher Differentialgleichungen, III, *Numer. Math.*, **34**, 143–154.

Mülthei, H. N. (1980a): Numerische Lösung gewöhnlicher Differentialgleichungen mit Splinefunktionen, *Computing*, **25**, 317–335.

Mülthei, H. N. (1980b): Zur numerischen Lösung gewöhnlicher Differential-gleichungen mit Splines in einem Sonderfall, *Math. Methods Appl. Sci.*, **2**, 419–428.

Mülthei, H. N. (1980c): Ein Konvergenzsatz für Splineapproximationen bei Anfangs-wertproblemen gewöhnlicher Differentialgleichungen, *Z. Angew. Math. Mech.*, **60**, T306–T307.

Mülthei, H. N. (1981): Numerische Behandlung von gewöhnlichen Differential-gleichungen mit Splines, *Z. Angew. Math. Mech.*, **61**, T297–T298.

Mülthei, H. N. (1982): Maximale Konvergenzordnung bei der numerischen Behandlung von gewöhnlichen Diffentialgleichungen mit Splines, *Z. Angew. Math. Mech.*, **62**, T340–T342.

Mülthei, H. N. (1982a): *A*-stabile Kollokationsverfahren mit mehrfachen Knoten, *Computing*, **29**, 51–61.

Mülthei, H. N. (1982b): Maximale Konvergenzordnung bei der numerischen Lösung von Anfangswertproblemen mit Splines, *Numer. Math.*, **39**, 449–463.

Muroya, Y. (1981): An iteration method and coefficient-wise estimates in Chebyshev series for the solutions of linear differential systems. *Mem. School Sci. Engrg. Waseda Univ.*, **45**, 123–155.

Murphy, C. P. and Evans, D. J. (1981): A flexible variable order extrapolation tech-nique for solving non-stiff ordinary differential equations, *Internat. J. Comput. Math.*, **10**, 63–75.

Murphy, W. D. (1977): Hermite interpolation and *A*-stable methods for stiff ordinary differential equations, *Appl. Math. Comput.*, **3**, 103–112.

Murray, F. J. (1950): Planning and error consideration for the numerical solution of a system of differential equations on a sequence calculator, *Math. Comp.* **4**, 133–144.

Murray, F. J., see also Brock, P.

Murthy, D. N. P., see Day, J. D.

Murthy, D. N. P., and Day, J. D. (1981): An explicit single-step multi-Jacobian method for nonstiff ordinary differential equations, *Internat. J. Comput. Math.*, **10**, 189–196.

Muszyński, J., see Litewska, K.

Mutalik, P. R. (1975): On selection of predictors for a given corrector in the solution of ordinary differential equations, *Indian J. Pure Appl. Math.*, **6**, 668–680.

Myachin, V. F. (1959): On the estimation of errors by numerical integration of the equations of celestial mechanics (Russian, English summary), *Byull. Inst. Teor. Astron.*, **7**, 257–280.

Nakashima, M. (1972): On the propagation of error in numerical integrations, *Proc.*

Japan Acad. ser. A Math. Sci., **48**, 484–488; Corrigendum: *Rep. Fac. Sci. Kagoshima Univ.*, **8**, 47–52 (1975).

Nakashima, M. (1982): On a pseudo-Runge–Kutta method of order 6, *Proc. Japan Acad. ser. A Math. Sci.*, **58**, 66–68.

Nakashima, M. (1982a): On pseudo-Runge–Kutta methods with 2 and 3 stages, *Publ. Res. Inst. Math. Sci.*, **18**, 895–909.

Naur, P. (1960): Algorithm 9, Runge–Kutta integration, *Comm. ACM*, **3**, 318 (see Thacher, H. C.).

Nazarenko, T. I., see Dombrovskaya, L. M.

Nazarenko, T. I. and Gubareva, E. B. (1978): Interval variants of Runge–Kutta type methods for solving the Cauchy problem for ordinary differential equations (Russian), in *Numerical Methods of Optimization (Applied Mathematics)* (Russian) (Eds. O. I. Artem'eva and B. A. Bel'tyukov, Adad. Nauk SSSR Sibirsk. Otdel., Ènerget. Inst., Irkutsk, pp. 153–160.

Neal, L. (1981): A simple *A*-stable method for stiff systems of differential equations, in *Proc. 1981 ACM Computer Science Conf., St. Louis, Mo.*, Assoc. for Computing Machinery, New York.

Neidhofer, G. (1959): Intégration approchée des équations différentielles lorsque la dérivée d'ordre le plus élevé ne figure que dans un terme correctif, *Chiffres*, **2**, 43–53.

Neĭmark, Yu. I. and Smirnova, V. N. (1981): Stability of difference schemes for ordinary differential equations (Russian), *Dinamika Sistem*, **1981**, 45–63.

Nevanlinna, O. (1976): On error bounds for *G*-stable methods, *BIT*, **16**, 79–84.

Nevanlinna, O. (1977): On the behaviour of global errors at infinity in the numerical integration of stable initial value problems, *Numer. Math.*, **28**, 445–454.

Nevanlinna, O. (1977a): On the numerical integration of nonlinear initial value problems by linear multistep methods, *BIT*, **17**, 58–71.

Nevanlinna, O., see also Jeltsch, R; Mäkelä, M.

Nevanlinna, O. and Liniger, W. (1978): Contractive methods for stiff differential equations, I. *BIT*, **18**, 457–474.

Nevanlinna, O. and Liniger, W. (1979): Contractive methods for stiff differential equations, II, *BIT*, **19**, 53–72.

Nevanlinna, O. and Odeh, F. (1981): Multiplier techniques for linear multistep methods, *Numer. Funct. Anal. Optim.*, **3**, 377–423.

Nevanlinna, O. and Sipilä, A. H. (1974): A nonexistence theorem for explicit *A*-stable methods, *Math. Comp.*, **28**, 1053–1055.

Nevanlinna, O. and Sipilä, A. H. (1975): Explicit *A*-stable one-step methods for ordinary differential equations, in *Proc. Fourth Manitoba Conf. Numerical Math., Winnipeg, 1974*, (Eds. B. L. Hartnell and H. C. Williams), Congress. Numer., No. 12, Utilitas Mathematica, Winnipeg, pp. 343–350.

Neves, K. W. (1975): Automatic integration of functional differential equations: an approach, *ACM Trans. Math. Software*, **1**, 357–368.

Neves, K. W. (1975a): Algorithm 497, automatic integration of functional differential equations, *ACM Trans. Math. Software*, **1**, 369–371.

Neville, E. H. (1934): Iterative interpolation, *J. Indian Math. Soc.*, **20**, 87–120.

Newbery, A. C. R. (1963): Multistep integration formulas, *Math. Comp.*, **17**, 452–455.

Newbery, A. C. R. (1967): Convergence of successive substitution starting procedures, *Math. Comp.*, **21**, 489–491.

Newbery, A. C. R. (1980): Spline-based methods for ODEs, in *Proc. Ninth Conf. Numerical Math. and Computing, Winnipeg, 1979* (Eds. G. H. J. van Rees and H. C. Williams), Congress. Numer, No. 27, Utilitas Mathematica, Winnipeg, pp. 323–327.

Newbery, A. C. R., see also Hull, T. E.

Ni, W. C., see Lee, J. S.

470

Nickel, K. (1969): The application of interval analysis to the numerical solution of differential equations, in *Conf. on the Numerical Solution of Differential Equations, Dundee, 1969* (Ed. J. L. Morris), Lecture Notes in Mathematics No. 109, Springer, Berlin, pp. 323–327.

Nickel, K. and Rieder, P. (1968): Ein neues Runge–Kutta-ähnliches Verfahren, *Internat. Ser. Numer. Math.*, **9**, 83–96.

Nicholson, D. W. (1981): On stable numerical integration methods of maximum time step, *Acta Mech.*, **38**, 191–198.

Niepage, H. D. (1980): Convergence of multistep methods for differential equations with multivalued right-hand side, in *Proc. Second Conf. on Numerical Treatment of Ordinary Differential Equations, Berlin, 1980* (Ed. R. März), Seminarberichte No. 32, Humboldt University, Berlin, pp. 72–83.

Nievergelt, J. (1964): Parallel methods for integrating ordinary differential equations, *Comm. ACM*, **7**, 731–733.

Nikityuk, Zh. M. and Slonovskiĭ, R. V. (1980): Some applications of continued fractions to the solution of first-order differential equations (Russian), *Vestnik L'vov. Politekhn. Inst.*, **141**, 69–71.

Nikol'skiĭ, È. V. and Dzhalilov, D. (1980): The method of equivalent systems and identification problems for ordinary linear differential equations (Russian), *Izv. Akad. Nauk Tadzhik. SSR Otdel. Fiz.-Mat. Khim. i Geol. Nauk*, **1980**, No. 2 (76), 3–9.

Nordsieck, A. (1962): On numerical integration of ordinary differential equations, *Math. Comp.*, **16**, 22–49.

Nordsieck, A. (1963): Automatic numerical integration of ordinary differential equations, *AMS Proc. Symp. Appl. Math.*, **15**, 241–250.

Norlund, N. E. (1924): *Vorlesungen über Differenzenrechung*, Springer, Berlin.

Norman, A. C. (1976): Expanding the solutions of implicit sets of ordinary differential equations in power series, *Comput. J.*, **19**, 63–68.

Nørsett, S. P. (1969): A criterion for $A(\alpha)$-stability of linear multistep methods, *BIT*, **9**, 259–263.

Nørsett, S. P. (1969a): An A-stable modification of the Adam–Bashforth Methods, in *Conf. on the Numerical Solution of Differential Equations, Dundee, 1969* (Ed. J. L. Morris), Lecture Notes in Mathematics No. 109, Springer, Berlin, pp. 214–219.

Nørsett, S. P. (1974): One-step methods of Hermite type for numerical integration of stiff systems, *BIT*, **14**, 63–77.

Nørsett, S. P. (1975): C-polynomials for rational approximation to the exponential function, *Numer. Math.*, **25**, 39–56.

Nørsett, S. P. (1975a): A note on local Galerkin and collocation methods for ordinary differential equations, *Utilitas Math.*, **7**, 197–209.

Nørsett, S. P. (1976): Runge–Kutta methods with a multiple real eigenvalue only, *BIT*, **16**, 388–393.

Nørsett, S. P. (1978): Restricted Padé approximations to the exponential function, *SIAM J. Numer. Anal.*, **15**, 1008–1029.

Nørsett, S. P. (1980): Collocation and perturbed collocation methods, in *Numerical Analysis: Proc. Dundee, 1979* (Ed. G. A. Watson), Lecture Notes in Mathematics, No. 773, Springer, Berlin, pp. 119–132.

Nørsett, S. P., see also Wanner, G.

Nørsett, S. P. and Wanner, G. (1979): The real-pole sandwich for rational approximations and oscillation equations, *BIT*, **19**, 79–84.

Nørsett, S. P. and Wanner, G. (1981): Perturbed collocation and Runge–Kutta methods, *Numer. Math.*, **38**, 193–208.

Nørsett, S. P. and Wolfbrandt, A. (1977): Attainable order of rational approximations to the exponential function with only real poles, *BIT*, **17**, 200–208.

Nørsett, S. P. and Wolfbrandt, A. (1979): Order conditions for Rosenbrock type methods, *Numer. Math.*, **32**, 1–15.

Nosrati, H. (1973): A modified Butcher formula for integration of stiff systems of ordinary differential equations, *Math. Comp.*, **27**, 267–272.

Novikov, V. A. and Demidov, G. V. (1972): A remark on a certain method of constructing schemes of high accuracy (Russian), *Chisl. Metody Mekh. Sploshn. Sredy*, **3**, No. 4, 89–91.

Novruzov, G. M., see Ibragimov, Sh. I.

Nugeyre, J. B. (1961): Un procédé mixte (Runge–Kutta, pas liés) d'intégration des systèmes différentiels du type $x'' = X(x, t)$, *Chiffres*, **4**, 55–68.

Nyström, E. J. (1925): Über die numerische Integration von Differentialgleichungen, *Acta Soc. Sci. Fennicae*, **50**, No. 13, 55pp.

Oberle, H. J., see Diekhoff, H. J.

Obrechkov, N. = Obreshkov, N.

Obreshkov, N. (1940): Neue Quadraturformeln, *Abh. Preuss. Akad. Wiss. Math. Nat. Kl*, **4**, 20pp.

Obreshkov, N. (1942): On mechanical quadrature (Bulgarian, French summary), *Spisanie Bulgar. Akad. Nauk.*, **65**, 191–289.

Odeh, F., see Liniger, W.; Nevanlinna, O.

Odeh, F. and Liniger, W. (1971): A note on unconditional fixed-h stability of linear multistep formulae, *Computing*, **7**, 240–253.

Odeh, F. and Liniger, W. (1977): Nonlinear fixed h-stability of linear multistep formulas, *J. Math. Anal. Appl.*, **61**, 691–712.

Odeh, F. and Liniger, W. (1980): On A-stability of second-order two-step methods for uniform variable steps, *Proc. IEEE Intl. Conf. Circuits and Computers*, **1**, 123–126.

Oesterhelt, G. (1974): Mehrschrittverfahren zur numerischen Integration von Differentialgleichungssystemen mit stark verschiedenen Zeitkonstanten, *Computing*, **13**, 279–298.

Oey, L. Y. (1982): Implicit schemes for differential equations, *J. Comput. Phys.*, **45**, 443–468.

Ohashi, T. (1970): On the conditions for convergence of one step methods for ordinary differential equations, *TRU Math.*, **6**, 59–62.

Olaefe, G. O. (1977): On the Tchebyschev method of solution of ordinary differential equations. *J. Math. Anal. Appl.*, **60**, 1–7.

Oleĭnik, A. G. (1977): A two-sided method of Runge–Kutta type, with multiple nodes, of seventh order accuracy (Russian, English summary), *Vychisl. Prikl. Mat. (Kiev)*, **32**, 117–123.

Oleĭnik, A. G. (1982): A two-sided process of Runge–Kutta type with multiple nodes of fifth order of accuracy (Russian), *Vychisl. Prikl. Mat. (Kiev)*, **46**, 3–6.

Oleĭnik, A. G., see also Lyashchenko, N. Ya.

Oliver, J. (1969): An error estimation technique for the solution of ordinary differential equations in Chebyshev series, *Comput. J.*, **12**, 57–62.

Oliver, J. (1975): A curiosity of low-order explicit Runge–Kutta methods, *Math. Comp.*, **29**, 1032–1036.

Oliver, P. (1982): a family of linear multistep methods for the solution of stiff and non-stiff ODEs, *IMA J. Numer. Anal.*, **2**, 289–301.

O'Malley, R. E. (1971): On initial value problems for nonlinear systems of differential equations with two small parameters, *Arch. Rational Mech. Anal.*, **40**, 209–222.

O'Malley, R. E. and Flaherty, J. E. (1980): Analytical and numerical methods for nonlinear singular singularly-perturbed initial value problems, *SIAM J. Appl. Math.*, **38**, No. 2, 225–247.

Oohashi, T. (1975): Predictor–corrector algorithm for reducing rounding error, *TRU Math.*, **11**, 51–56.

O'Regan, P. G. (1970): Step size adjustment at discontinuities for fourth order Runge–Kutta methods, *Comput. J.*, **13**, 401–404.

Orlov, K. (1971): Practical method for solving differential equations and their systems by means of Taylor series, *Mat. Vesnik*, **8**, 73–81.

Orlov, K. (1972): Finding of the general integral of differential equations by means of Taylor series and finding of some form of non-Cauchy's particular integrals, *Mat. Vesnik*, **9**, 273–279.

Orlov, K. (1972a): New practical methods for finding particular solutions of differential equations, *Mat. Vesnik.*, **9**, 403–408.

Orlov, K. (1977): Application of differential equations for transforming the implicit functions into the explicit ones, *Mat. Vesnik.*, **14**, 295–302.

Orlov, K. (1980): Cases when the use of the analytic method for solving differential equations is better than the use of numerical ones, *Mat. Vesnik*, **17**, 187–195.

Orlov, K. and Stojanović, M. (1974): Pseudo-integral of differential equations of the *n*-th order, *Mat. Vesnik*, **11**, 277–279.

Ortiz, E. L., see Freilich, J. H.

Osborne, M. R. (1964): A method for finite-difference approximation in ordinary differential equations, *Comput. J.*, **7**, 58–65.

Osborne, M. R. (1966): On Nordsieck's method for the numerical solution of ordinary differential equations, *BIT*, **6**, 51–57.

Osborne, M. R. (1967): Minimising truncation error in finite difference approximations to ordinary differential equations, *Math. Comp.*, **21**, 133–145.

Osborne, M. R. (1969): A new method for the integration of stiff systems of ordinary differential equations, in *Information Processing 68: Proc. IFIP Congress 1968, I-Math. Software* (Ed. A. J. H. Morrell), North Holland, Amsterdam, pp. 200–204.

O'Shea, B. B. (1978): Algorithms for the solution of systems of coupled second-order differential equations, *J. Comput. Appl. Math.*, **4**, 11–17.

Ostermann, A. (1981): *R*-Stabilität bei Zwischenschrittverfahren, *Mitt. Math. Sem. Giessen*, **148**, 26–38.

Ostermann, A., see also Filippi, S.

Pagliari, C. (1979): Un metodo del quarto ordine per l'integrazione differenziale ordinaria di secondo ordine, *Rend. Sem. Fac. Sci. Univ. Cagliari*, **49**, 541–563.

Pagliari, C. (1979a): Soluzione approssimata dell'equazione differenziale ordinaria di secondo ordine, *Rend. Sem. Fac. Sci. Univ. Cagliari*, **49**, 513–533.

Pagliari, C. (1980): Ricerca di un metodo del quinto ordine per l'integrazione numerica della equazione differenziale ordinaria di 2° ordine, *Rend. Sem. Fac. Sci. Univ. Cagliari*, **50**, 87–96.

Pakhnutov, I. A. (1975): Solution of the Cauchy problem for ordinary differential equations by means of splines (Russian), *Vychisl. Sistemy*, **65**, 96–129.

Palanisamy, K. R. and Bhattacharya, D. K. (1982): Analysis of stiff systems via single step method of block pulse functions, *Internat. J. Systems Sci.*, **13**, 961–968.

Palusinski, O. A. and Wait, J. V. (1978): Numerical techniques for partitioned dynamic system simulation, in *Proc. 1978 Summer Computer Simulation Conf. Newport Beach, Calif.*, AFIPS Press, Arlington, Virginia, pp. 113–116.

Pankova, G. D. (1980): Application of a package of programs for obtaining the exact boundaries of the solution of some equations (Russian) in *Studies in Integro-differential Equations* (Ed. M. I. Imanaliev) No. 13, 'Ilim', Frunze, pp. 368–372.

Papian, L. E. and Ball, W. E. (1970): The spectrum of numerical integration methods with computed variable stepsize, *J. Math. Anal. Appl.*, **31**, 259–284.

Papoulis, A. (1952): On the accumulation of errors in numerical solution of differential equations, *J. Appl. Phys.*, **23**, 173–176.

Paracelli, C. (1981): Alcune osservazioni su una classe de metodi lineari multistep *A*-stabili, *Note Mat.*, **1**, 261–279.

Paracelli, C., see also Guercia, L.

Parasyuk, É. M. (1980): A transformation of a matriciant (Russian), *Vestnik L'vov. Politekhn. Inst.*, No. 141, *Differentsial'nye Uravneniya i ikh Prilozhen*, pp. 73–75.

Park, K. C. (1982): An improved semi-implicit method for structural dynamics analysis, *Trans. ASME Ser. E. J. Appl. Mech.*, **49**, 589–593.

Parsaei, M., see Battye, D. J.

Pasciutti, F., see Dalle Rive, L.

Patrício, F. (1978): Cubic spline functions and initial value problems, *BIT*, **18**, 342–347.

Patrício, F. (1979): A numerical method for solving initial-value-problems with spline functions, *BIT*, **19**, 489–494.

Patrício, F. (1981): Hybrid methods for ordinary differential equations (Portuguese), *Proc. Eighth Portuguese–Spanish Conf. on Mathematics*, vol. III, University of Coimbra, Coimbra, pp. 137–142.

Patruno, V. (1972): Un metodo numerico per la risoluzione delle equazioni differenziali ordinarie, *Calcolo*, **9**, 111–132.

Pavel, P. (1981): A numerical differentiation formula and its application to the numerical integration of a third-order differential equation (Romanian, French summary), *Studia Univ. Babeş-Bolyai Math.*, **26**, No. 4, 3–8.

Pavlov, B. V., see Krestinin, A. V.

Pejović, P. (1970): The approximate solution of systems of linear differential equations by means of interval equations (Serbian, English summary), *Mat. Vesnik*, **7**, 457–472.

Pejović, P. (1972): Determination of the error in the approximate solutions of a system of differential equations—analytically and on a computer (Serbian, English summary), *Mat. Vesnik*, **9**, 383–399.

Pelekh, Ya. M., see Bodnarchuk, P. Ĭ.

Pelekh, Ya. N. (1981): Explicit A-stable methods of numerical integration of differential equations, *Mat. Metody i Fiz.-Mekh. Polya*, **13**, 30–34.

Pelekh, Ya. N. (1981a): An algorithm for constructing A-stable methods for the numerical integration of differential equations (Russian), *Mat. Metody i Fiz.-Mekh. Polya*, **14**, 12–16.

Pelekh, Ya. N. (1982): An explicit A-stable method of fourth order accuracy for the numerical integration of differential equations (Russian), *Mat. Metody i Fiz.-Mekh. Polya*, **15**, 19–23.

Pelios, A. and Klopfenstein, R. W. (1972): Minimal error constant numerical differentiation (N. D.) formulas, *Math. Comp.*, **26**, 467–475.

Peper, C., see Strehmel, K.

Perron, O. (1929): Über Stabilität und asymptotisches Verhalten der Integrale von Differentialgleichungssystemen, *Math. Z.*, **29**, 129–160.

Pesch, H. J., see Diekhoff, H. J.

Petrenko, A. I., Smirnov, A. M. and Gumen, N. B. (1981): Comparative investigation of implicit methods of integrating systems of differential equations in the solution of model problems (Russian, English summary), *Elektronnoe Modelirovanie*, **1981**, No. 3, 8–16.

Petzold, L. R. (1981): An efficient numerical method for highly oscillatory ordinary differential equations, *SIAM J. Numer. Anal.*, **18**, 455–479.

Petzold, L. R. (1982): Differential/algebraic equations are not ODEs, *SIAM J. Sci. Statist. Comput.*, **3**, 367–384.

Phillipe, B. (1982): Stabilité de la méthode des différentiations rétrogrades à ordre et pas variables (méthode de Gear), *C. R. Acad. Sci. Paris,* sér. I, **294**, 435–437.

Piazza, G. (1981): On the BN stability of the Runge–Kutta methods, *Math. Comp.*, **37**, 399–401.

Piazza, G. (1981a): *Sulla Implementazione dei Metodi Runge–Kutta Impliciti*, Istituto per le Applicazioni del Calcolo 'Mauro Picone' (IAC), Rome 12 pp.

474

Piazza, G., see also di Lena, G.

Piazza, G. and Trigiante, D. (1977): Propagazione degli errori nella integrazione numerica di equazioni differenziali ordinarie, *Pubbl. Ser.* III, No. 120.

Picel, Z. (1976): Exponentially fitted block one-step numerical methods for solution of stiff differential equations, *Delft Progress Rep.*, ser. F, **2**, 321–334.

Picel, Z., see also Bickart, T. A.; Ehle, B. L.; Tendler, J. M.

Pickard, W. F. (1964): Tables for the step-by-step integration of ordinary differential equations of the first order, *J. Assoc. Comput. Mach.*, **11**, 229–233.

Pierini, G., see Cominciolli, V.; Greco, A.

Pinilla, M. Calvo = Calvo, M.

Piotrowski, P. (1969): Stability, consistency and convergence of variable k-step methods for numerical integration of large systems of ordinary differential equations, in *Conf. on the Numerical Solution of Differential Equations, Dundee, 1969* (Ed. J. L. Morris), Lecture Notes in Mathematics No. 109, pp. 221–227.

Pirč, V. (1982): Some notes on the possibility of calculation of zero points of solutions of differential equations of second order, *Bull. Appl. Math.*, **1982**, 22–136, 7–13.

Pitcher, N. see McKee, S.

Poddar, R. S. and Trasi, M. S. (1982): Automatic integration of a stiff system using ellipsoidal norm of the error-vector for error-control, *Z. Angew. Math. Phys.*, **33**, 653–668.

Podlipenko, Yu. K. (1978): Polynomial approximation of the solution of the Cauchy problem for systems of ordinary differential equations (Russian), in *Studies in the Theory of Approximation of Functions and Their Applications* (Russian) (Ed. V. K. Dzyadyk), Akad. Nauk Ukrain. SSR, Inst. Math., Kiev, pp. 146–161.

Pokora, T. (1974): A Runge–Kutta-like method with exponential correction, *Zastos. Mat.*, **14**, 461–470.

Pokora, T. and Ząbek, Ś. (1972): On an Euler-like method with exponential correction for initial-value problems, *Zastos. Mat.*, **13**, 255–259.

Polonskaya, K. V. (1979): An analysis of round-off errors in Adams' method (Russian), in *Questions of Vector Scalar Optimization* (Ed. M. Ya. Zinger), Work Collect. Vladivostok, pp. 127–134.

Polonskaya, K. V. (1979a): A scheme for the numerical solution of a differential equation with complete feedback contour (Russian), in *Questions of Vector and Scalar Optimization* (Ed. M. Ya. Zinger), Work Collect., Vladivostok, pp. 135–141.

Poon, S. W. H., see Bui, T. D.

Popadinets, V. I. (1980): Modification of a typical computational scheme of the Runge–Kutta method (Russian), *Stochastic Methods in Problems of Optimization* (Ed. E. N. Stepanenko), Akad. Nauk Ukrain, SSR, Inst. Kibernet, Kiev, pp. 27–35.

Popadineto. V. I. (1981): A method of prediction and correction for the solution of systems of ordinary differential equations (Russian), in *Methods of Investigation of Extremal Problems* (Russian) (Eds. V. S. Mikhalevich and È. I. Nenakhov), Akad. Nauk Ukrain. SSR. Inst. Kibernet. Kiev, pp. 57–66.

Pope, D. A. (1963): An exponential method of numerical integration of ordinary differential equations, *Comm. ACM*, **6**, 491–493.

Pope, D. A., see also Danchick, R.

Popov, A. Yu. (1978): A method for the approximation of the solution of differential equations (Russian), in *Mathematical Methods in Operations Research and Reliability Theory* (Russian) (Eds. Yu. M. Ermol'ev and I. N. Kovalenko), Akad. Nauk Ukrain. SSR, Inst. Kibernet, Kiev, pp. 44–47.

Popov, V. A., see Andreev, A. S.

Pospelov, V. V. (1976): Adams' methods with an unstable operator (Russian), *Zh. Vychisl. Mat. i Mat. Fiz.*, **16**, 1467–1479; English translation: *USSR Comput. Math. and Math. Phys.*, **16**, No. 6, 83–95.

Pospelov, V. V., see also Gorbunov, A. D.

Pottle, C., see Allen, R. H.

Powell, M. J. D., see Iserles, A.

Powers, J. E. (1959): Elimination of special functions from differential equations, *Comm. ACM*, **2**, 2–3.

Práger, M., see Babuška, I.

Práger, M., Taufer, J. and Vitásek, E. (1973): Some new methods for numerical solution of initial value problems, *Proc. Equadiff III (Third Czechoslovak Conf. on Differential Equations and Their Applications, Brno, 1972)*, pp. 247–253.

Práger, M., Taufer, J. and Vitásek, E. (1973a): Overimplicit multistep methods, *Appl. Mat.*, **18**, 399–421.

Preiser, S., see Lee, D.

Priebe, K. B. (1964): Certification of algorithm 218, Kutta Merson (P. M. Lukehart, *Comm. ACM*, **6**, 737, 1963), *Comm. ACM*, **7**, 585.

Prince, P. J., see Dormand, J. R.

Prince, P. J. and Dormand, J. R. (1981): High order embedded Runge–Kutta formulae, *J. Comput. Appl. Math.*, **7**, 67–75.

Prince, P. J. and Wright, K. (1978): Runge–Kutta processes with exact principal error equations, *J. Inst. Math. Appl.*, **21**, 363–373.

Prothero, A. (1969): Local-error estimates for variable-step Runge–Kutta methods, in *Conf. on the Numerical Solution of Differential Equations, Dundee, 1969* (Ed. J. L. Morris), Lecture Notes in Mathematics No. 109, Springer, Berlin, pp. 228–233.

Prothero, A. (1980): Estimating the accuracy of numerical solutions to ordinary differential equations, in *Computational Techniques for Ordinary Differential Equations: Proc. Conf., University of Manchester, 1978* (Eds. I. Gladwell and D. K. Sayers), Academic Press, London, pp. 103–128.

Prothero, A. and Robinson, A. (1974): On the stability and accuracy of one-step methods for solving stiff systems of ordinary differential equations, *Math. Comp.*, **28**, 145–162.

Pukhov, G. E. and Ronto, N. I. (1980): On a higher precision implicit method of integrating differential equations (Russian), *Dokl. Akad. Nauk SSSR*, **251**, 554–557; English translation: *Soviet Math. Dokl.*, **21**, 484–487.

Qin, Z. F. (1982): On the numerical stability of extrapolation procedures for ordinary differential equations (Chinese, English summary), *Numer. Math. J. Chinese Univ.*, **4**, 60–67.

Quade, W. (1950): Grundsätzliches zur numerischen Integration von gewöhnlichen Differentialgleichungen, *Z. Angew. Math. Mech.*, **30**, 276–278.

Quade, W. (1957): Numerische Integration von gewöhnlichen Differentialgleichungen durch Interpolation nach Hermite, *Z. Angew. Math. Mech.*, **37**, 161–169.

Quade, W. (1959): Über die Stabilität numerischer Methoden zur Integration gewö'.nlicher Differentialgleichungen erster Ordnung, *Z. Angew. Math. Mech.*, **39**, 117–134.

Quemada, M. M., see Calvo, M.

Rademacher, G. E. (1948): On the accumulation of errors in processes of integration on high-speed calculating machines, in *Proc. of a Symposium on Large Scale Digital Calculating Machinery*, vol. 16, Annals. Comput. Labor. Harvard Univ., pp. 176–187.

Radok, J. R. M. (1959): Method of functional extrapolation for the numerical integration of ordinary equations, *J. Soc. Indust. Appl. Math.*, **7**, 425–430.

Rahme, H. S. (1969): A new look at the numerical integration of ordinary differential equations, *J. Assoc. Comput. Mach.*, **16**, 496–506.

Rahme, H. S. (1970): Stability analysis of a new algorithm used for integrating a system of ordinary differential equations, *J. Assoc. Comput. Mach.*, **17**, 284–293.

Rakitskiĭ, Yu. V. (1961): Some properties of solutions of systems of ordinary differential equations by one–step methods of numerical integration (Russian), *Zh. Vychisl.*

Mat. i Mat. Fiz., **1**, 947–962; English translation: *USSR Comput. Math. and Math. Phys.*, **1**, 1113–1128.

Rakitskiĭ, Yu. V. (1968): New methods of computing the initial tabular values in the numerical integration of ordinary differential equations, *Zh. Vychisl. Mat. i Mat. Fiz.*, **8**, 13–27; English translation: *USSR Comput. Math. and Math. Phys.*, **8**, No. 1, 14–34.

Rakitskiĭ, Yu. V. (1970): Asymptotic error formulae for solutions of systems of ordinary differential equations by functional numerical methods (Russian), *Dokl. Akad. Nauk SSSR*, **193**, 40–42; English translation: *Soviet Math. Dokl.*, **11**, 861–863.

Rakitskiĭ, Yu. V. (1972): Parametric methods of numerical integration (Russian), *Dokl. Akad. Nauk SSSR*, **207**, 544–546; English translation: *Soviet Math. Dokl.*, **13**, 1577–1580.

Rakitskiĭ, Yu. V. (1972a): Methods for successive step increase in the numerical integration of systems of ordinary differential equations, *Dokl. Akad. Nauk SSSR*, **207**, 793–795; English translation: *Soviet Math. Dokl.*, **13**, 1624–1627.

Rakitskiĭ, Yu. V., Ustinov, S. M. and Chernorutskiĭ, I. G. (1979): *Numerical Methods for Solving Stiff Systems* (Russian), 'Nauka', Moscow, 208 pp.

Ralston, A. (1960): Numerical integration methods for the solution of ordinary differential equations, in *Mathematical Methods for Digital Computers* (Eds. A. Ralston and H. S. Wilf), Wiley, New York, pp. 95–109.

Ralston, A. (1961): Some theoretical and computational matters relating to predictor–corrector methods of numerical integration, *Comput. J.*, **4**, 64–67.

Ralston, A. (1962): Runge—Kutta methods with minimum error bounds, *Math. Comp.*, **16**, 431–437; Corrigendum, *Math. Comp.*, **17**, 488.

Ralston, A. (1965): Relative stability in the numerical solution of ordinary differential equations, *SIAM Rev.*, **7**, 114–125.

Ralston, A. and Wilf, H. S. (Eds.) (1960): *Mathematical Methods for Digital Computers*, Wiley, New York (see Ralston, A.; Romanelli, M. J.).

Rao, C. P., see Iyengar, S. R. K.

Rao, C. P. and Iyengar, S. R. K. (1975): A note on optimal stiffly stable methods for ordinary differential equations, *J. Math. Phys. Sci.*, **9**, 177–184.

Rao, C. P. and Iyengar, S. R. K. (1976): Numerical differentation type formulas of higher order with an arbitrary number of offstep points, *J. Math. Phys. Sci.*, **10**, 307–318.

Rao, C. P. and Iyengar, S. R. K. (1976a): High order numerical differentiation type formulas with an off-step point for stiff ordinary differential equations, *J. Inst. Math. Appl.*, **17**, 281–293.

Rao, V. S. = Subba Rao, V.

Rapoport, I. M. (1952): A new method of approximate integration of ordinary differential equations (Russian), *Ukrain. Mat. Zh.*, **4**, 399–413.

Rebolia, L. (1972): Sulle proprietà di stabilità assoluta di una trasformazione per l'integrazione numerica dell'equazione $y^{(n)} = f(x, y, y', \ldots, y^{(n-1)})$, *Calcolo*, **9**, 59–70.

Rebolia, L. (1973): Equazioni differenziali e algebriche: un algoritmo esplicito di tipo perturbativo, *Calcolo*, **10**, 133–144.

Rebolia, L., see also Brianzi, P.

Redzhepova, Sh. (1973): Adams type formulas with a variable step for the numerical integration of ordinary differential equations (Russian, English summary), *Izv. Akad. Nauk Turkmen. SSR Ser. Fiz.-Tekhn. Khim. Geol. Nauk*, **1973**, No. 6, 3–6.

Reeves, R. F., see Conte, S. D.

Reimer, M. (1964): Optimale Verfahren zur numerischen Integration von Anfangswertproblemen, *Math. Z.*, **84**, 70–79.

Reimer, M. (1965): Eine Fehlerabschätzung für lineare Differenzenverfahren, *Numer. Math.*, **7**, 277–285.

Reimer, M. (1965a): An integration procedure including error estimation, *BIT*, **5**, 164–174.

Reimer, M. (1967): Zur Theorie der linearen Differenzenformeln, *Math. Z.*, **95**, 373–402.

Reimer, M. (1968): Finite difference forms containing derivatives of higher order, *SIAM J. Numer. Anal.*, **5**, 725–738.

Reinhardt, H. J. (1979): Stability of singularly perturbed linear difference equations, *Numerical Analysis of Singular Perturbation Problems,* Academic Press, London, pp. 485–492.

Rentrop, P., see Diekhoff, H. J.; Kaps, P.

Reutter, F. (1982): Genauigkeitsuntersuchungen bei direkten und indirekten Verfahren für gewöhnliche Differentialgleichungen *n*-ter Ordnung, *Wiss. Z. Hochsch. Archit. Bauwes. Weimar.*, **28**, 180–184.

Reutter, F. and Knapp, H. (1964): Untersuchungen über die numerische Behandlung von Anfangswertproblemen gewöhnlicher Differentialgleichungssysteme mit Hilfe von LIE-Reihen und Anwendungen auf die Berechnung von Mehrkörperproblemen, Forschungsbericht des Landes Nordrhein-Westfalen No. 1367, Westdeutscher Verlag, Cologne and Oplanden, 73pp.

Reynolds, R. R., see Milne, W. E.

Ribarič, M., see Greenspan, H.

Rice, J. R. (1960): Split Runge–Kutta methods for simultaneous equations, *J. Res. Nat. Bur. Standards*, **64B**, 151–170.

Richards, P. I., Lanning, W. D. and Torrey, M. D. (1965): Numerical integration of large, highly-damped, nonlinear systems, *SIAM Rev.*, **7**, 376–380.

Richardson, L. F. (1910): The approximate arithmetical solution by finite differences of physical problems involving differential equations, with an application to the stresses in a masonry dam, *Philos. Trans. Roy. Soc. London,* ser. A., **210**, 307–357.

Richardson, L. F. (1927): The deferred approach to the limit, *Philos. Trans. Roy. Soc. London,* ser. A., **226**, 299–361.

Richter, W. (1951): Sur l'erreur commise dans la méthode d'intégration de Milne, *C. R. Acad. Sci. Paris*, **233**, 1342–1344.

Ridder, R. de = de Ridder, R., see van Dyck, D.

Rieder, P., see Nickel, K.

Riha, W. (1972): Optimal stability polynomials, *Computing*, **9**, 37–43.

Riley, J. D., see Brown, R. R.

Rive, L. Dalle = Dalle Rive, L.

Roberts, C. E. (1979): *Ordinary Differential Equations. A Computational Approach,* Prentice-Hall, Englewood Cliffs, N.J.

Roberts, C. E. (1982): Variable order, variable stepsize, componentwise integration, *Austral. Comput. J.*, **14**, 19–25.

Robertson, H. H. (1960): Some new formulae for the numerical integration of ordinary differential equations, *Information Processing,* UNESCO, Paris, pp. 106–108.

Robertson, H. H. (1966): The solution of a set of reaction rate equations, in *Numerical Analysis: An Introduction* (Ed. J. Walsh), Academic Press, London, 178–182.

Robertson, H. H. (1976): Numerical integration of systems of stiff ordinary differential equations with special structure, *J. Inst. Math. Appl.*, **18**, 249–263.

Robertson, H. H. (1978): Some factors affecting the efficiency of stiff integration routines, in *Numerical Software—Needs and Availability: Proc. Conf. University of Sussex, 1977* (Ed. D. Jacobs), Academic Press, London, pp. 279–301.

Robertson, H. H. and McCann, M. J. (1969): A note on the numerical integration of conservative systems of first-order ordinary differential equations, *Comput. J.*, **12**, 81.

Robertson, H. H. and Williams, J. (1975): Some properties of algorithms for stiff differential equations, *J. Inst. Math. Appl.*, **16**, 23–34.

Robinson, A., see Prothero, A.

Rodabaugh, D. J. (1970): On stable correctors, *Comput. J.*, **13**, 98–100.

Rodabaugh, D. J. and Thompson, S. (1977): Corrector methods with increased ranges of stability, *Comput. Math. Appl.*, **3**, 197–201.

Rodabaugh, D. J. and Thompson, S. (1978): Adams-type methods with increased ranges of stability, *Comput. Math. Appl.*, **4**, 349–357.

Rodabaugh, D. J. and Thompson, S. (1979): Low-order A_0-stable Adams-type correctors, *J. Comput. Appl. Math.*, **5**, 225–233.

Rodabaugh, D. J. and Thompson, S. (1981): A note on the relative efficiency of Adams methods, *Comput. Math. Appl.*, **7**, 401–403.

Rökkum, L. R. and Romberg, W. (1965): Eine Runge–Kutta-Variante für $dy/dx = f(x, y)$, *Z. Angew. Math. Mech.*, **45**, T74–T75.

Roman, G. C., see Clasen, R. J.

Roman, V. M. (1980): A numerical solution of nth-order differential equations using cubic splines (Russian), *Ukrain. Mat. Zh.*, **32**, 686–693.

Romanelli, M. J. (1960): Runge–Kutta methods for the solution of ordinary differential equations, in *Mathematical Methods for Digital Computers* (Eds. A. Ralston and H. S. Wilf), Wiley, New York, pp. 110–120.

Romberg, W., see Rökkum, L. R.

Ronto, N. I., see Pukhov, G. E.

Ronveaux, A. (1962): Intégration approchée des équations et des systèmes d'équations différentielles linéaires au moyen d'opérateurs numériques, *Chiffres*, **5**, 89–107.

Roose, A. (1973): the use of Runge–Kutta methods for solving nonlinear equations (Russian), *Eesti. NSV Tead. Akad. Toimetised Füüs—Mat.*, **22**, 431–434.

Rosenbaum, J. S. (1976): Conservation properties of numerical integration methods for systems of ordinary differential equations, *J. Comput. Phys.*, **20**, 259–267.

Rosenbaum, J. S. (1978): Implicit and explicit exponential Adams methods for real time simulation, in *Numerical Methods for Differential Equations and Simulation: Proc. IMACS (AICA) Internat. Sympos., Blacksburg, Virginia, 1977* (Eds. A. W. Bennett and R. Vichnevetsky), North-Holland, Amsterdam, pp. 165–167.

Rosenbrock, H. H. (1963): Some general implicit processes for the numerical solution of differential equations, *Comput. J.*, **5**, 329–330.

Rosenbrock, H. H. and Storey, G. (1966) *Computational Techniques for Chemical Engineers,* Pergamon Press, London.

Rosser, J. B. (1967): A Runge–Kutta for all seasons, *SIAM Rev.*, **9**, 417–452.

Rowland, J. R. and Holmes, W. M. (1971): A variational approach to digital integration, *IEEE Trans. Comput.*, **20**, 894–900.

Rozenberg, D. P., see Galler, B. A.

Ruamps, F., see Crouzeix, M.

Rubin, W. B., see Bickart, T. A.

Rubin, W. B. and Bickart, T. A. (1972): A-stability of composite multistep methods, *Proc. SIAM-SIGNUM Fall Meeting, Austin, Texas, Oct., 1972.*

Ruehli, A. E., Brennan, P. A. and Liniger, W. (1980): Control of numerical stability and damping in oscillatory differential equations, in *Proc. IEEE Intl. Conf. Circuits and Computers* (Ed. N. B. Rabbat), **1**, 111–114.

Runge, C. (1895): Über die numerische Auflösung von Differentialgleichungen, *Math. Ann.*, **46**,167–178.

Runge, C. (1905): Über die numerische Auflösung totaler Differentialgleichungen, *Nachr. Gesel. Wiss.,* Göttingen, 252–257.

Russo, E., see Crisci, M. R.

Rutishauser, H. (1952): Über die Instabilität von Methoden zur Integration gewöhnlicher Differentialgleichungen, *Z. Angew. Math. Phys.*, **3** 65–74.

Rutishauser, H. (1960): Bemerkungen zur numerischen Integration gewöhnlicher Differentialgleichungen n-ter Ordnung, *Numer. Math.*, **2**, 263–279. (Preliminary version: *Z. Angew. Math. Phys.*, **6**, 497–498.)

Ryabov, Yu. A. and Tolmacheva, T. A. (1981): Construction of numerical-analytic solutions of differential equations by means of computer (Russian, English summary), *Práce Štúd. Vysokej Školy Doprav. Spojov Žiline Sér. Mat.-Fyz.*, **3**, 7–12.

Rybarski, A., see Mazurkiewicz, A.

Sacks-Davis, R. (1977): Error estimates for a stiff differential equation procedure, *Math. Comp.*, **31**, 939–953.

Sacks-Davis, R. (1977a): Solution of stiff ordinary differential equations by a second derivative method, *SIAM J. Numer. Anal.*, **14**, 1088–1100.

Sacks-Davis, R. (1980): Fixed leading coefficient implementation of SD-formulas for stiff ODEs, *ACM Trans. Math. Software*, **6**, 540–562.

Sacks-Davis, R., see also Jackson, K. R.

Sacks-Davis, R. and Shampine, L. F. (1981): A type-insensitive ODE code based on second derivative formulas, *Comput. Math. Appl.*, **7**, 487–495.

Saff, E. B. and Varga, R. S. (1975): On the zeros and poles of Padé approximants to e^z, *Numer. Math.*, **25**, 1–14.

Saldanha, J. S. V., see Jain, M. K.; Jain R. K.

Salikhov, N. P. (1962): Polar difference methods of solving Cauchy's problem for a system of ordinary differential equations (Russian), *Zh. Vychisl. Mat. i Mat. Fiz.*, **2**, 515–528; English translation, *USSR Comput. Math. and Math. Phys.*, **2**, 535–553(1963).

Sallam, S. M. (1980): On the stability of quasidouble step spline function approximations for solutions of initial value problems, in *Numerical Methods: Third Colloq., Keszthely, 1977* (Ed. P. Rózsa), Colloq. Math. Soc. János Bolyai No. 22, North-Holland, Amsterdam, pp. 517–523.

Salzer, H. E. (1956): Osculatory extrapolation and a new method for the numerical integration of differential equations, *J. Franklin Inst.*, **262**, 111–119.

Salzer, H. E. (1957): Numerical integration of $y'' = \phi(x, y, y')$ using osculatory interpolation, *J. Franklin Inst.*, **263**, 401–409.

Salzer, H. E. (1962): Trigonometric interpolation and predictor–corrector formulas for numerical integration, *Z. Angew. Math. Mech.*, **42**, 403–412.

Samara, H. (1979): The numerical solution of differential equations (Spanish), in *Formulación Operacional del Método Tau,* Cuadernos del Instituto de Matemática 'Beppo Levi', No. 10, (Edited by G. G. Garguichevich, M. B. Stampella, A. Taiana and revised by E. L. Ortiz), Universidad Nacional de Rosario, Facultad de Ciencias Exactas e Ingeniera, Rosario, 73 pp.

Sánchez López, M. (1975): Numerical solution of ordinary differential equations I (Spanish), *Gaceta Mat. ser. 1 (Madrid)*, **27**, 32–44.

Sánchez López, M. (1975a): Numerical solution of ordinary differential equations II (Spanish), *Gaceta Mat. ser. 1 (Madrid)*, **27**, 136–157.

Sandberg, I. W. (1967): Two theorems on the accuracy of numerical solutions of systems of ordinary differential equations, *Bell System Tech. J.*, **46**, 1243–1266.

Sandberg, I. W. (1967a): Some properties of a classic numerical integration formula, *Bell System Tech. J.*, **46**, 2061–2080.

Sandberg, I. W. and Shichman, H. (1968): Numerical integration of systems of stiff nonlinear differential equations, *Bell System Tech. J.*, **47**, 511–527.

Sanz-Serna, J. M. (1980): Some aspects of the boundary locus method, *BIT.*, **20**, 97–101.

Sanz-Serna, J. M. (1981): Linearly implicit variable coefficient methods of Lambert–Sigurdsson type, *IMA J. Numer. Anal.*, **1**, 39–45.

Sanz-Serna, J. M., see also McLeod, R. J. Y.

Sarafyan, D. (1968): Error estimation for Runge–Kutta methods through pseudo–iterative formulas, *Riv. Mat. Univ. Parma.*, ser. 2, **9**, 1–42.

Sarafyan, D. (1968a): Estimation of errors for the approximate solution of differential equations and their systems, *Riv. Mat. Univ. Parma*, ser. 2, **9**, 109–127.

Sarafyan, D. (1971): An investigation concerning fifth order scalar and vector Runge–Kutta processes, *Riv. Mat. Univ. Parma*, ser. 2, **12**, 41–45.

Sarafyan, D. (1972): Improved sixth-order Runge–Kutta formulas and approximate continuous solution of ordinary differential equations, *J. Math. Anal. Appl.*, **40**, 436–445.

Sarafyan, D. and Brown, R. (1967): Computer derivation of algebraic equations associated with Runge–Kutta formulas, *BIT*, **7**, 156–162.

Sarkany, E. F. and Liniger, W. (1974): Exponential fitting of matricial multistep methods for ordinary differential equations, *Math. Comp.*, **28**, 1035–1052.

Sasaki, Y. (1967): Formulae in the numerical integration of ordinary differential equations, I, *Bull. Nagoya Inst. Tech.*, **18**, 23–34.

Sasaki, Y. (1967a): Formulae in the numerical integration of ordinary differential equations, II, *Bull. Nagoya Inst. Tech.*, **19**, 105–110.

Sasaki, Y. (1968): Formulae in the numerical integration of ordinary differential equations, III, *Bull. Nagoya Inst. Tech.*, **20**, 125–129.

Saukh, S. E., see Beregovenko, G. Ya.

Savorin, A. N. = Zavorin, A. N.

Saworin, A. N. = Zavorin, A. N.

Sayers, D. K., see Gladwell, I.

Sayfy, A., see Cooper, G. J.

Schechter, E. (1957): On the error in Runge–Kutta numerical integration processes (Romanian, French summary), *Acad. R. P. Romîne Fil. Cluj. Stud. Cerc. Mat.*, **8**, 115–124.

Schechter, E. (1958): An error bound in certain numerical integration processes for differential equations (Romanian, French summary), *Acad. R. P. Romîne Fil. Cluj. Stud. Cerc. Mat.*, **9**, 343–350.

Schempp, W. (1970): Über die Theorie der linearen Diskretisierungsalgorithmen, *Z. Angew. Math. Mech.*, **50**, T75–T78.

Scherer, K., see Esser, H.

Scherer, R. (1972): Herleitung von Runge–Kutta-Verfahren der Ordnung vier mit übersichtlichen Fehlerdarstellungen, *Math. Z.*, **128**, 311–323.

Scherer, R. (1972a): Fehlerabschätzung für ein Runge–Kutta-Verfahren mit mehrfachen Knoten, *Computing*, **10**, 391–396.

Scherer, R. (1975): Exaktheitseigenschaft einiger Runge–Kutta-Formeln, *Z. Angew. Math. Mech.*, **55**, T259–T260.

Scherer, R. (1975a): Zur Stabilität halbexpliziter Runge–Kutta-Methoden, *Arch. Math. (Basel)*, **26**, 267–272.

Scherer, R. (1976): Bemerkungen zur numerischen Behandlung steifer Differentialgleichungssysteme, *Z. Angew. Math. Mech.*, **56**, T317–T319.

Scherer, R. (1977): A note on Radau and Lobatto formulae for ODEs, *BIT*, **17**, 235–238.

Scherer, R. (1978): Spiegelung von Stabilitätsbereichen, in *Numerical Treatment of Differential Equations: Proc. Oberwolfach, 1976* (Eds. R. Bulirsch, R. D. Grigorieff and J. Schröder), Lecture Notes in Mathematics No 631, Springer, Berlin, pp. 147–152.

Scherer, R. (1979): A necessary condition for *B*-stability, *BIT*, **19**, 111–115.

Scheu, G. (1977): Numerische Berechnung konvergenter Schrankenfolgen mittels Interpolationspolynomen für gewöhnliche Anfangswertaufgaben, *Z. Angew. Math. Mech.*, **57**, T301–T304.

Schiesser, W. E., see Lapidus, L.

Schilders, W. H. A., see Doolan, E. P.

Schlett, M. (1980): A class of linear multistep methods with variable matrix coefficients, in *Proc. Second Conf. on Numerical Treatment of Ordinary Differential Equations, Berlin, 1980* (Ed. R. März), Seminarberichte No. 32, Humboldt University, Berlin, pp. 88–95.

Schmidt, J. W. (1975): Bemerkungen zu einem Verfahren von H. J. Stetter, *Beiträge Numer. Math.*, **4**, 205–213.

Schneid, J., see Frank, R.

Schoen, K. (1971): Fifth and sixth order PECE algorithms with improved stability properties, *SIAM J. Numer. Anal.*, **8**, 244–248.

Schoenberg, I. J. (1974): Spline functions and differential equations—first order equations, in *Studies in Numerical Analysis* (Ed. B. K. P. Scaife), Academic Press, London, pp. 311–324.

Scholz, S. (1968): Fehlereingrenzungen bei Anfangswertproblemen gewöhnlicher Differentialgleichungen 1. Ordnung, *Z. Angew. Math. Mech.*, **48**, 203–207.

Scholz, S. (1969): Ein Runge–Kutta-Verfahren mit variablem Parameter α, *Z. Angew. Math. Mech.*, **49**, 517–524.

Scholz, S. (1975): Runge–Kutta-Verfahren mit festen und variablen Parametern zur Lösung von Systemen gewöhnlicher Differentialgleichungen 1. Ordnung, *Apl. Mat.*, **20**, 166–185.

Scholz, S. (1975a): Increase of the order of convergence by variable parameters, *Godishnik Vissh. Uchebn. Zaved. Prilozhna Mat.*, **11**, No. 3, 191–198.

Scholz, S., see also Weiss, W.

Scholz, S., Bräuer, K. and Thomas, S. (1976): Ein *A*-stabiles einstufiges Rosenbrock-Verfahren dritter Ordnung, *Beiträge Numer. Math.*, **5**, 191–199.

Schöne, F. (1982): Verallgemeinerte Mehrschrittverfahren mit variabler Schrittweite, *Mitt. Math. Sem. Giessen*, **1982**, No. 152, ii + 127 pp.

Schröder, J. (1961): Fehlerabschätzung mit Rechenanlagen bei gewöhnlichen Differentialgleichungen erster Ordnung, *Numer. Math.*, **3**, 39–61.

Schröder, J. (1961a): Verbesserung einer Fehlerabschätzung für gewöhnliche Differentialgleichungen erster Ordnung, *Numer. Math.*, **3**, 125–130.

Schröder, J., see also Bulirsch, R.

Schroll, J. (1978): Winkelstabile Mehrschrittverfahren hoher Ordnung, *Z. Angew. Math. Mech.*, **58**, T441–T442.

Schroll, J., see also Grigorieff, R. D.

Schryer, N. L. (1975): An extrapolation step-size monitor for solving ordinary differential equations, in *Proc. ACM 1974 Annual Conf.*, San Diego, California, 1974, pp. 140–148.

Schumaker, L. L. (1982): Optimal spline solutions of systems of ordinary differential equations, in *Differential Equations, Proc. First Latin American School Univ. São Paulo, 1981* (Eds. D. G. de Figueiredo and S. H. Chaim, Lecture Notes in Mathematics No. 957, Springer, Berlin–New York, pp. 272–283.

Schwermer, H. (1968): Zur Fehlererfassung bei der numerischen Integration von gewöhnlichen Differentialgleichungssystemen erster Ordnung mit speziellen Zweipunktverfahren, *Internat. Ser. Numer. Math.*, **9**, 141–155.

Sconzo, P. (1954): Formule di estrapolazione per l'integrazione numerica delle equazioni differenziali ordinarie, *Boll. Un. Mat. Ital.*, (3)**9**, 391–399.

Scraton, R. E. (1964): The numerical solution of second-order differential equations not containing the first derivative explicity, *Comput. J.*, **6**, 368–370.

Scraton, R. E. (1964a): Estimation of the truncation error in Runge–Kutta and allied processes, *Comput. J.*, **7**, 246–248.

Scraton, R. E. (1965): The solution of linear differential equations in Chebyshev series, *Comput. J.*, **8**, 57–61.

482

Scraton, R. E. (1979): Some new methods for stiff differential equations, *Internat. J. Comput. Math.*, **7**, 55–63.

Scraton, R. E. (1981): Some *L*-stable methods for stiff differential equations, *Internat. J. Comput. Math.*, **9**, 81–87.

Sedgwick, A. E., see Hull, T. E.

Seifert, P. (1980): Some results of a comparison between methods for the numerical integration of stiff systems, in *Proc. Second Conf. on Numerical Treatment of Ordinary Differential Equations, Berlin, 1980* (Ed. R. März), Seminarberichte No. 32, Humboldt University, Berlin, pp. 103–108.

Seifert, P. (1982): Verfahren und Implementierungen von Anfangswertaufgaben steifer Differentialgleichungen unter Berücksichtigung spezieller Strukturen, in *Proc. Third Conf. on Numerical Treatment of Ordinary Differential Equations, Berlin, 1982* (Ed. R. März), Seminarberichte No. 46, Humboldt University, Berlin, pp. 163–170.

Seinfeld, J. H., see Lapidus, L.

Seinfeld, J. H., Lapidus, L. and Hwang, M. (1970): Review of numerical integration techniques for stiff ordinary differential equations, *Ind. Eng. Chem. Fundam.*, **9**, 266–275.

Sen, A., see Katz, I. N.

Sendov, B., see Andreev, A. S.

Serrais, F. (1956): Sur l'estimation des erreurs dans l'intégration numérique des équations différentielles linéaires du second ordre, *Ann. Soc. Sci. Bruxelles*, sér. I, **70**, 5–8.

Seydel, R., see Diekhoff, H. J.

Shakhov, Yu. A., see Gorbunov, A. D.

Shalima, V. N. (1981): A class of numerical methods of solution of systems of second-order quasilinear equations (Russian, English summary), *Vestī Akad. Navuk BSSR Ser. Fiz.-Mat. Navuk*, **1981**, No. 6, 23–27.

Shalima, V. N. (1981a): Numerical integration of second order differential equations (Russian, English summary), *Izv. Akad. Nauk BSSR. Ser. Fiz.-Mat. Nauk*, **1981**, No. 1, 38–41.

Shalima, V. N. (1981b): Two–step methods of solution of systems of second-order quasilinear equations of a special type (Russian), *Vestnik Beloruss. Gos. Univ.*, ser. I, **1981**, No. 2, 66–68.

Shalima, V. N., see also Bobkov, V. V.

Shampine, L. F. (1973): Local extrapolation in the solution of ordinary differential equations, *Math. Comp.*, **27**, 91–97.

Shampine, L. F. (1974): Limiting precision in differential equation solvers, *Math. Comp.*, **28**, 141–144; Corrigendum, *Math. Comp.*, **28**, 1183.

Shampine, L. F. (1975): Stiffness and non-stiff differential equation solvers, *Internat. Ser. Numer. Math.*, **27**, 287–301.

Shampine, L. F. (1976): Quadrature and Runge–Kutta formulas, *Appl. Math. Comput.*, **2**, 161–171.

Shampine, L. F. (1977): Stiffness and nonstiff differential equation solvers, II: Detecting stiffness with Runge–Kutta methods, *ACM Trans. Math. Software*, **3**, 44–53.

Shampine, L. F. (1977a): Local error control in codes for ordinary differential equations, *Appl. Math. Comput.*, **3**, 189–210.

Shampine, L. F. (1978): Limiting precision in differential equation solvers, II: Sources of trouble and starting a code, *Math. Comp.*, **32**, 1115–1122.

Shampine, L. F. (1978a): Stability properties of Adams codes, *ACM Trans. Math. Software*, **4**, 323–329.

Shampine, L. F. (1978b): Solving ODEs with discrete data in SPEAKEASY, in *Recent Advances in Numerical Analysis* (Eds. C. de Boor and G. H. Golub), Academic Press, New York, pp. 177–192.

Shampine, L. F. (1978c): Solving ordinary differential equations for simulation, *Math. Comput. Simulation*, **20**, 204–207.

Shampine, L. F. (1978d): Solving ordinary differential equations for simulation, in *Numerical Methods for Differential Equations and Simulation: Proc. IMACS (AICA) Internat. Sympos., Blacksburg, Virginia, 1977* (Eds. A. W. Bennett and R. Vichnevetsky), North-Holland, Amsterdam, pp. 189–192.

Shampine, L. F. (1979): Evaluation of implicit formulas for the solution of ODEs, *BIT*, **19**, 495–502.

Shampine, L. F. (1979a): Storage reduction for Runge–Kutta codes, *ACM Trans. Math. Software*, **5**, 245–250.

Shampine, L. F. (1979b): Better software for ODEs, *Comput. Math. Appl.*, **5**, 157–161.

Shampine, L. F. (1980): What everyone solving differential equations numerically should know, in *Computational Techniques for Ordinary Differential Equations: Proc. Conf., University of Manchester, 1978* (Eds. I. Gladwell and D. K. Sayers), Academic Press, London, pp. 1–17.

Shampine, L. F. (1980a): Implementation of implicit formulas for the solution of ODEs, *SIAM J. Sci. Statist. Comput.*, **1**, 103–118.

Shampine, L. F. (1980b): Lipschitz constants and robust ODE codes, in *Computational Methods in Nonlinear Mechanics: Proc. Second Internat. Conf., University of Texas, Austin, 1979* (Ed. J. T. Oden), North-Holland, Amsterdam–New York, pp. 427–449

Shampine, L. F. (1981): Efficient use of implicit formulas with predictor–corrector error estimate, *J. Comput. Appl. Math.*, **7**, 33–35.

Shampine, L. F. (1981a): Type-insensitive ODE codes based on implicit *A*-stable formulas, *Math. Comp.*, **36**, 499–510.

Shampine, L. F. (1981b): Evaluation of a test set for stiff ODE solvers, *ACM Trans. Math. Software*, **7**, 409–420.

Shampine, L. F. (1982): Implementation of Rosenbrock methods, *ACM Trans. Math. Software*, **8**, 93–113.

Shampine, L. F. (1982a): Type-insensitive ODE codes based on implicit $A(\alpha)$-stable formulas, *Math. Comp.*, **39**, 109–123.

Shampine, L. F. (1982b): Solving ODEs in quasi steady state, in *Numerical Integration of Differential Equations and Large Linear Systems: Proc. Bielefeld, 1980* (Ed. J. Hinze), Lecture Notes in Mathematics No. 968, Springer, Berlin, pp. 234–245.

Shampine, L. F., see also Byrne, G. D.; Gordon, M. K.; Sacks-Davis, R.; Watts, H. A.

Shampine, L. F. and Gear, C. W. (1979): A user's view of solving stiff ordinary differential equations, *SIAM Rev.*, **21**, 1–17.

Shampine, L. F. and Gordon, M. K. (1972): Some numerical experiments with DIFSUB, *SIGNUM Newsletter*, **7**, 24–26.

Shampine, L. F. and Gordon, M. K. (1975): *Computer Solution of Ordinary Differential Equations: The Initial Value Problem,* Freeman, San Francisco.

Shampine, L. F. and Gordon, M. K. (1975a): Local error and variable order Adams codes, *Appl. Math. Comput.*, **1**, 47–66.

Shampine, L. F., Gordon, M. K. and Wisniewski, J. A. (1980): Variable order Runge–Kutta codes, in *Computational Techniques for Ordinary Differential Equations: Proc. Conf., University of Manchester, 1978* (Eds. I. Gladwell and D. K. Sayers), Academic Press, London, pp. 83–101.

Shampine, L. F. and Hiebert, K. L. (1977): Detecting stiffness with the Fehlberg (4,5) formulas, *Comput. Math. Appl.*, **3**, 41–46.

Shampine, L. F. and Watts, H. A. (1969): Block implicit one-step methods, *Math. Comp.*, **23**, 731–740.

Shampine, L. F. and Watts, H. A. (1971): Comparing error estimators for Runge–Kutta methods, *Math. Comp.*, **25**, 445–455.

Shampine, L. F. and Watts, H. A. (1976): Global error estimation for ordinary differential equations, *ACM Trans. Math. Software*, **2**, 172–186.

Shampine, L. F. and Watts, H. A. (1976a): Algorithm 504, GERK: global error estimation for ordinary differential equations, *ACM Trans. Math. Software*, **2**, 200–203.

Shampine, L. F. and Watts, H. A. (1977): The art of writing a Runge–Kutta code, I, in *Mathematical Software* (Ed. J. R. Rice), vol. III, Academic Press, London, pp. 257–275.

Shampine, L. F. and Watts, H. A. (1979): The art of writing a Runge–Kutta code, II, *Appl. Math. Comput.*, **5**, 93–121.

Shampine, L. F., Watts, H. A. and Davenport, S. M. (1976): Solving nonstiff ordinary differential equations—the state of the art, *SIAM Rev.*, **18**, 376–411.

Shanks, E. B. (1966): Solutions of differential equations by evaluations of functions, *Math. Comp.*, **20**, 21–38.

Shapiro, N. Z., see Clasen, R. J.

Shapkin, A. F. (1971): Stochastic version of Adams's method (Russian), *Zh. Vychisl. Mat. i Mat. Fiz.*, **11**, 766–770; English translation: *USSR Comput. Math. and Math. Phys.*, **11**, No. 3, 285–291.

Sharma, K. D. (1970): One-step methods for the numerical solution of linear differential equations based on Lobatto quadrature formulae, *J. Austral. Math. Soc.*, **11**, 115–128.

Sharma, K. D. and Gupta, R. G. (1973): A one–step method for the numerical solution of ordinary non-linear second-order differential equations based upon Lobatto four-point quadrature formula, *J. Austral. Math. Soc.*, **15**, 193–201.

Sharma, S. R., see Chawla, M. M.

Shaw, B. (1967): Modified multistep methods based upon a nonpolynomial interpolant, *J. Assoc. Comput. Mach.*, **14**, 143–154.

Shaw, B. (1967a): Some multistep formulae for special high order ordinary differential equations, *Numer. Math.*, **9**, 367–378.

Shaw, B., see also Lambert, J. D.

Sheldon, J. W., see Zondek, B.

Shepherd, J. J., see Mahony, J. J.

Sherman, A. H. and Hindmarsh, A. C. (1980): GEARS: a package for the solution of sparse, stiff ordinary differential equations, in *Electrical Power Problems: The Mathematical Challenge: Proc. Conf., Seattle, Wash., 1980* (Eds. A. M. Erisman, K. W., Neves and M. H. Dwarakanath), SIAM, Philadelphia, pp. 190–200.

Shichman, H., see Sandberg, I. W.

Shidlovskaya, N. A., see Gavurin, M. K.

Shimizu, T. (1966): Contribution to the theory of numerical integration of non-linear differential equations, I, *TRU Math.*, **2**, 55–70.

Shimizu, T. (1967): Accumulation of round-off errors in some linear multi–step methods for ordinary differential equations, *TRU Math.*, **3**, 41–47.

Shimizu, T. (1968): Contribution to the theory of numerical integration of non-linear differential equations, II, TRU Math., **4**, 80–93.

Shimizu, T. (1969): Contribution to the theory of numerical integration of non-linear differential equations, III, *TRU Math.*, **5**, 51–66.

Shimizu, T. (1970): Contribution to the theory of numerical integration of non-linear differential equations, IV, *TRU Math.*, **6**, 47–58.

Shimizu, T. (1971): Contribution to the theory of numerical integration of non-linear differential equations, V, *TRU Math.*, **7**, 45–57.

Shimizu, T. (1972): Errors in numerical integration of ordinary differential equations, *TRU Math.*, **8**, 27–32.

Shimizu, T. (1973): Numerical integration of differential equations, on $0 \le x < \infty$, *TRU Math.*, **9**, 63–67.

Shimizu, T. (1974): Contribution to the theory of numerical integration of non-linear differential equations, VI, *TRU Math.*, **10**, 83–96.

Shimizu, T. (1976): Contribution to the theory of numerical integration of non-linear differential equations, VII, *TRU Math.*, **12**, 41–49.

Shimizu, T. (1977): Contribution to the theory of numerical integration of nonlinear differential equations, VIII, *TRU Math.*, **13**, 65–75.

Shintani, H. (1965): Approximate computation of errors in numerical integration of ordinary differential equations by one-step methods, *J. Sci. Hiroshima Univ. Ser. A-I Math.*, **29**, 97–120.

Shintani, H. (1966): On a one-step method of order 4, *J. Sci. Hiroshima Univ. Ser. A-I Math.*, **30**, 91–107.

Shintani, H. (1966a): Two-step processes by one-step methods of order 3 and of order 4, *J. Sci. Hiroshima Univ. Ser. A-I Math.*, **30**, 183–195.

Shintani, H. (1971): On one-step methods utilizing the second derivative, *Hiroshima Math. J.*, **1**, 349–372.

Shintani, H. (1972): On explicit one-step methods utilizing the second derivative, *Hiroshima Math. J.*, **2**, 353–368.

Shintani, H. (1977): On one-step methods for ordinary differential equations, *Hiroshima Math. J.*, **7**, 769–786.

Shintani, H. (1980): On errors in the numerical solution of ordinary differential equations by step-by-step methods, *Hiroshima Math. J.*, **10**, 469–494.

Shintani, H. (1981): On pseudo-Runge–Kutta methods of the third kind, *Hiroshima Math. J.*, **11**, 247–254.

Shintani, H. (1982): Modified Rosenbrock methods for stiff systems, *Hiroshima Math. J.*, **12**, 543–558.

Shintani, H. (1982a): Modified Rosenbrock methods with approximate Jacobian matrices, *Hiroshima Math. J.*, **12**, 559–568.

Shintani, H. and Hayashi, Y. (1978): *A posteriori* error estimates and iterative methods in the numerical solution of systems of ordinary differential equations, *Hiroshima Math. J.*, **8**, 101–121.

Shishkin, G. I. (1981): Numerical solution of differential equations with a small parameter multiplying the highest derivatives (Russian), *Chisl. Metody Mekh. Sploshn. Sredy*, **12**, 135–147.

Shkel', V. A., see Bobkov, V. V.

Shkerstena, A. = Skerstena, A., see Klokovs, Yu. A.

Shniad, H. (1980): Global error estimation for the implicit trapezoidal rule, *BIT*, **20**, 120–121.

Shokin, Yu. I., see Kalmykov, S. A.

Shokin, Yu. I., and Kalmykov, S. A. (1980): *A two-sided Method for the Solution of the Equation $y' = f(y)$ with an Interval as Initial Value*, Frieburger Intervall–Berichte No. 80/10, University of Freiburg, Frieburg im Briesgau, pp. 22–33.

Shokin, Yu. I. and Kalmykov, S. A. (1982): *On the Interval-Analytic Method for Ordinary Differential Equations*, (Freiburger Intervall-Berichte No. 82/5, University of Freiburg, Freiburg im Briesgau, pp. 39–46.

Shouman, A. R., see Wilson, D. B.

Shura-Bura, M. R. (1952): Estimates of errors of numerical integration of ordinary differential equations (Russian), *Akad. Nauk SSSR Prikl. Mat. Mekh*, **16**, 575–588.

Shuster, A. R. and Èpshteĭn, B. S. (1973): The construction of the initial tabular values in the numerical integration of a system of ordinary differential equations (Russian), *Metody Vychisl.*, **8**, 19–22.

Siemieniuch, J. L. (1976): Properties of certain rational approximations to e^{-z}, *BIT*, **16**, 172–191.

Sierra, H. G., see Luther, H. A.

Sigurdsson, S. T. (1971): Linear multistep methods with variable matrix coefficients, in *Conf. on Applications of Numerical Analysis, Dundee, 1971*, (Ed. J. L. Morris), Lecture Notes in Mathematics No. 228, Springer, Berlin, pp. 327–331.

Sigurdsson, S. T., see also Lambert, J. D.

Silverberg, M. (1968): A new method of solving state variable equations permitting large step sizes, *Proc. IEEE*, **56**, 1352–1353.

Simon, W. E. (1965): Numerical technique for solution and error estimate for the initial value problem, *Math. Comp.*, **19**, 387–393.

Singhal, A., see Cash, J. R.

Sinha, A. S. C. (1980): High-order stiffly stable methods for analysis of nonlinear circuits and systems, *Internat. J. Systems Sci.*, **11**, 1001–1009.

Sinha, A. S. C. and Yokomoto, C. F. (1978): High order stiffly stable methods for analysis of nonlinear circuits and systems in *Sixteenth Annual Allerton Conf. on Communication, Control and Computing*, University of Illinois, Urbana-Champaign, Ill., pp. 814–823.

Sipilä, A. H., see Mäkelä, M.; Nevanlinna, O.

Sitter, J. de = de Sitter, J., see van Dyck, D.

Skappel, J. (1969): Attempts to optimize the structure of an ODE program in *Conf. on the Numerical Solution of Differential Equations, Dundee, 1969* (Ed. J. L. Morris), Lecture Notes in Mathematics No. 109), Springer, Berlin, pp. 243–248.

Skeel, R. D. (1976): Analysis of fixed-stepsize methods, *SIAM J. Numer. Anal.*, **13**, 664–685.

Skeel, R. D. (1979): Equivalent forms of multistep formulas, *Math. Comp.*, **33**, 1229–1250.

Skeel, R. D. (1982): A theoretical framework for proving accuracy results for deferred corrections, *SIAM J. Numer. Anal.*, **19**, 171–196.

Skeel, R. D. and Kong, A. K. (1977): Blended linear multistep methods, *ACM Trans. Math. Software*, **3**, 326–343.

Skelboe, S. (1977): The control of order and steplength for backward differentiation methods, *BIT*, **17**, 91–107.

Skelboe, S. (1980): Implementation of Chebyshevian linear multistep formulas, *BIT*, **20**, 356–366.

Skelboe, S. and Christensen, B. (1981): Backward differentiation formulas with extended regions of absolute stability, *BIT*, **21**, 221–231.

Skertsena, A. = Shkerstena, A., see Klokovs, Yu. A.

Sköllermo, G., see Dahlquist, G.

Sloate, H. M., see Bickart, T. A.

Sloate, H. M. and Bickart, T. A. (1970): An implicit formula for the integration of stiff network equations, in *Proc. Third Hawaii Internat. Conf. System Sciences, University of Hawaii, Honolulu, 1970*, Western Periodicals, North Hollywood, Calif., pp. 33–36.

Sloate, H. M. and Bickart, T. A. (1973): *A*-stable composite multistep methods, *J. Assoc. Comput. Mach.*, **20**, 7–26.

Slonevskiĭ, R. V. = Slon'ovs'kiĭ, R. V., see Nikityuk, Zh. M.

Smirnov, A. M. see Petrenko, A. I.

Smirnova, V. N., see Neĭmark, Yu. I.

Snider, A. D. (1976): Error analysis for a stiff system procedure, *Math. Comp.*, **30**, 216–219.

Snider, A. D. and Fleming, G. C. (1974): Approximation by aliasing with application to 'Certaine' stiff differential equations, *Math. Comp.*, **28**, 465–473.

Socea, D. (1981): Sur l'approximation des solutions des équations différentielles par des fonctions spline à déficience, *Studia Univ. Babeş-Bolyai Math.*, **26**, No. 4, 71–75.

Söderlind, G., see Dahlquist, G.

Sokolikhin, V. N., see Gorbunov, A. D.

Solak, W. W. (1973): The numerical method of solving ordinary differential equations with the Cauchy initial condition, *Ann. Polon. Math.*, **27**, 143–147.

Sommeijer, B. P., see van der Houwen, P. J.

Sommeijer, B. P. and van der Houwen, P. J. (1981): on the economization of stabilized Runge–Kutta methods with applications to parabolic initial value problems, *Z. Angew. Math. Mech.*, **61**, 105–114.

Sommer, D. (1965): Numerische Anwendung impliziter Runge–Kutta-Formeln, *Z. Angew. Math. Mech.* **45**, T77-T79.

Sommer, D., see also Glasmacher, W.

Sottas, G. (1981): Quadrature formulas with positive weights, *BIT*, **21**, 491–504.

Sottas, G. and Wanner, G. (1982): The number of positive weights of a quadrature formula, *BIT*, **22**, 339–352.

Soulé, J. C., see Guillou, A.

Southard, T. H. and Yowell, E. C. (1952): An alternative 'predictor–corrector' process, *Math. Comp.*, **6**, 253-254.

Spicher, K. (1967): Bemerkungen zur praktischen Durchführung des Verfahrens von Runge–Kutta-Fehlberg, *Elektr. Datenverarbeitung,* **2**, 79–85.

Spijker, M. N. (1966): Convergence and stability of step-by-step methods for the numerical solution of initial-value problems, *Numer. Math.*, **8**, 161–177.

Spijker, M. N. (1967): On the consistency of finite-difference methods for the solution of initial-value problems, *J. Math. Anal. Appl.*, **19**, 125–132.

Spijker, M. N. (1969): Round-off error in the numerical solution of second order differential equations, in *Conf. on the Numerical Solution of Differential Equations, Dundee, 1969* (Ed. J. L. Morris), Lecture Notes in Mathematics No. 109, Springer, Berlin, pp. 249–254.

Spijker, M. N. (1971): On the structure of error estimates for finite-difference methods, *Numer. Math.*, **18**, 73–100.

Spijker, M. N. (1971a): Reduction of roundoff error by splitting of difference formulas, *SIAM J. Numer. Anal.*, **8**, 345–357.

Spijker, M. N. (1974): Two-sided error estimates in the numerical solution of initial value problems: in: Ansorge, R. and Törnig, W. (Eds): *Numerische Behandlung nichtlinearer Integrodifferential- und Differentialgleichungen*, Lecture Notes in Mathematics No. 395, Springer, Berlin, 109–122.

Spijker, M. N. (1974a): Optimum error estimates for finite-difference methods, *Acta Univ. Carolin.—Math. Phys.*, **15**, 159–164.

Spijker, M. N. (1976): On the possibility of two-sided error bounds in the numerical solution of initial value problems, *Numer. Math.*, **26**, 271–300.

Spijker, M. N. (1977): The behaviour of error bounds in the numerical solution of initial value problems when the stepsize h tends to zero, in *Topics in Numerical Analysis III*, (Ed. J. J. H. Miller), Academic Press, London, pp. 383–400.

Spijker, M. N. (1979): Error bounds in the numerical solution of initial value problems, in *Proc. Congress Vrije Universiteit, Amsterdam* (Eds. P. C. Baayen, D. van Dulst and J. Oosterhoff), M. C. Tract No. 101, Mathematisch Centrum, Amsterdam, pp. 345–358.

Spijker, M. N. (1982): Stability in the numerical solution of stiff initial value problems, *Nieuw Arch. Wisk.*, ser. 3, **30**, 264–276.

Spijker, M. N., see also Hundsdorfer, W. H.

Squier, D. P. (1969): One-step methods for ordinary differential equations, *Numer. Math.*, **13**, 176–179.

Steihaug, T. and Wolfbrandt, A. (1979): An attempt to avoid exact Jacobian and nonlinear equations in the numerical solution of stiff differential equations, *Math. Comp.*, **33**, 521–534.

Štekauer, J. (1980): Classical Runge–Kutta formulas of the fifth order with rational coefficients for an ordinary differential equation of the fourth order, *Acta Math. Univ. Comenian*, **35**, 43–53.

Stepleman, R. S., see Chartres, B. A.

Sterne, T. E. (1953): The accuracy of numerical solutions of ordinary differential equations, *Math. Comp.*, **7**, 159–164.

Stetter, F. (1967): Eine Fehlerabschätzung für das Runge–Kutta-Verfahren der Ordnung 4, *Math. Z*, **97**, 229–237.

Stetter, F. (1968): Optimale Runge–Kutta-Verfahren der Ordnung 2 und 3, *Monatsh. Math.*, **72**, 239–244.

Stetter, F. and Zeller, K. (1967): Fehleruntersuchungen für das gewöhnliche Runge–Kutta-Verfahren, *Math. Z*, **98**, 179–184.

Stetter, H. J. (1963): Maximum bounds for the solutions of initial value problems for partial difference equations, *Numer. Math*, **5**, 399–424.

Stetter, H. J. (1965): Stabilizing predictors for weakly unstable correctors, *Math. Comp.*, **19**, 84–89.

Stetter, H. J. (1965a): A study of strong and weak stability in discretization algorithms, *SIAM J. Numer. Anal.*, **2**, 265–280.

Stetter, H. J. (1965b): Asymptotic expansions for the error of discretization algorithms for non-linear functional equations, *Numer. Math.*, **7**, 18–31.

Stetter, H. J. (1965c): The L_2-norm in the study of error propagation in initial value problems, *Apl. Mat.*, **10**, 308–311.

Stetter, H. J. (1965d): Numerische Lösung von Differentialgleichungen mit nacheilendem Argument, *Z. Angew. Math. Mech.*, **45**, T79–T80.

Stetter, H. J. (1966): Stability of non-linear discretization algorithms, in *Numerical Solution of Partial Diufferential Equations*, Academic Press, New York, pp. 111–123.

Stetter, H. J. (1968): Improved absolute stability of predictor–corrector schemes, *Computing*, **3**, 286–296.

Stetter, H. J. (1968a): Richardson-extrapolation and optimal estimation, *Apl. Mat.*, **13**, 187–190.

Stetter, H. J. (1968b): Stabilitätsbereiche bei Diskretisierungsverfahren für Systeme gewöhnlicher Differentialgleichungen, *Internat. Ser. Numer. Math.*, **9**, 157–167.

Stetter, H. J. (1968c): Stability problems in ordinary differential equations, *Coll. Math. Soc. János Bolyai*, **3**, 139–144.

Stetter, H. J. (1969): Stability properties of the extrapolation methods, in *Conf. on the Numerical Solution of Differential Equations, Dundee, 1969* (Ed. J. L. Morris), Lecture Notes in Mathematics, No. 109, Springer, Berlin, pp. 255–260.

Stetter, H. J. (1970): Symmetric two-step algorithms for ordinary differential equations, *Computing*, **5**, 267–280.

Stetter, H. J. (1971): Local estimation of the global discretization error, *SIAM J. Numer. Anal.*, **8**, 512–523.

Stetter, H. J. (1971a): Stability of discretizations on infinite intervals, in *Conf. on Applications of Numerical Analysis, Dundee, 1971*, (Ed. J. L. Morris), Lecture Notes in Mathematics No. 228, Springer, Berlin, pp. 207–222.

Stetter. H. J. (1973): *Analysis of Discretization Methods for Ordinary Differential Equations*, Springer, Berlin.

Stetter, H. J. (1973a): Discretization of differential equations on infinite intervals and applications to function minimization, in *Topics in Numerical Analysis*, (Ed. J. J. H. Miller), Academic Press, London, pp. 277–284.

Stetter, H. J. (1974): Cyclic finite-difference methods for ordinary differential equations, in *Conf. on the Numerical Solution of Differential Equations, Dundee, 1973* (Ed. G. A. Watson), Lecture Notes in Mathematics No. 363, Springer, Berlin, pp. 134–143.

Stetter, H. J. (1974a): Economic global error estimation, in *Stiff Differential Systems* (Ed. R. A. Willoughby), Plenum Press, New York, pp. 245–258.

Stetter, H. J. (1975): Recent progress in the numerical treatment of ordinary differential equations, in *Proc. International Congress of Mathematicians, Vancouver, 1974*, vol. 2, Canadian Mathematics Congress, Montreal, pp. 423–428.

Stetter, H. J. (1976): Towards a theory for discretizations of stiff differential systems, in *Numerical Analysis: Proc. Dundee Conf., 1975*, (Ed. G. A. Watson), Lecture Notes in Mathematics No. 506, Springer, Berlin, pp. 190–201.

Stetter, H. J. (1978): Global error estimation in ODE-solvers, in *Numerical Analysis: Proc. Dundee, 1977* (Ed. G. A. Watson), Lecture Notes in Mathematics No. 630, Springer, Berlin, pp. 179–189.

Stetter, H. J, (1978a): Considerations concerning a theory for ODE-solvers in *Numerical Treatment of Differential Equations: Proc. Oberwolfach, 1976* (Eds. R. Bulirsch, R. D. Griorieff and J. Schröder), Lecture Notes in Mathematics No. 631, Springer, Berlin, pp. 188–200.

Stetter, H. J. (1978b): On the maximal order in PC-codes, *Computing*, **20**, 273–278.

Stetter, H. J. (1978c): The defect correction principle and discretization methods, *Numer. Math.*, **29**, 425–443.

Stetter, H. J. (1979): Global error estimation in Adams PC-codes, *ACM Trans. Math. Software*, **5**, 415–430.

Stetter, H. J. (1979a): Interpolation and error estimation in Adams PC-codes, *SIAM J. Numer. Anal.*, **16**, 311–323.

Stetter, H. J. (1979b): Performance evaluation of ODE software through modelling, in *Performance Evaluation of Numerical Software* (Ed. L. D. Fosdick), North-Holland, Amsterdam, pp. 175–184.

Stetter, H. J. (1980): Modular analysis of numerical software, in *Numerical Analysis: Proc. Dundee, 1979* (Ed. G. A. Watson), Lecture Notes in Mathematics No. 773, Springer, Berlin, pp. 133–145.

Stetter,H. J. (1980a): Analysis and simulation in the design and performance of numerical software, in *Production and Assessment of Numerical Software* (Eds. M. A. Hennel and L. M. Delves), Academic Press, London, pp. 3-18.

Stetter, H. J. (1980b): Tolerance proportionality in ODE-codes, in *Proc. Second Conf. on Numerical Treatment of Ordinary Differential Equations, Berlin, 1980* (Ed. R. März), Seminarberichte No. 32, Humboldt University, Berlin, pp. 109–123.

Stetter, H. J. (1982): Global error estimation in ordinary initial value problems, *Numerical Integration of Differential Equations and Large Linear Systems: Proc. Bielefeld, 1980*, (Ed. J. Hinze), Lecture Notes in Mathematics No. 968, Springer, Berlin, 269–279.

Stetter, H. J., see also Feldstein, A.; Gragg, W. B.

Stetter, H. J. and Weinmüller, E. (1981): On the error control in ODE solvers with local extrapolation, *Computing*, **27**, 169–177.

Stewart, N. F. (1970): Certain equivalent requirements of approximate solutions of $x' = f(t, x)$, *SIAM J. Numer. Anal.* **7**, 256–259.

Stewart, N. F. (1971): A heuristic to reduce the wrapping effect in the numerical solution of $x' = f(t, x)$, *BIT*, **11**, 328–337.

Stewart, N. F. (1974): Computable, guaranteed local error bounds for the Adams method, *Math. Nachr.*, **60**, 145–153.

Stewart, N. F., see also Davey, D. P.

Stiefel, E., see Baumgarte, J.

Stiefel, E. and Bettis, D. G. (1969): Stabilization of Cowell's method, *Numer. Math.*, **13**, 154–175.

Stimberg, C. (1967): Vereinfachte Herleitung von Runge–Kutta-Verfahren, *Z. Angew. Math. Mech.*, **47**, 413–414.

Stineman, R. W. (1965): Digital time-domain analysis of systems with widely separated poles, *J. Assoc. Comput. Mach.*, **12**, 286–293.

Stoer, J. (1961): Über zwei Algorithmen zur Interpolation mit rationalen Funktionen, *Numer. Math.*, **3**, 285–304.

Stoer, J. (1974): Extrapolation methods for the solution of intial value problems and their practical realization, in *Proc. Conf. on the Numerical Solution of Ordinary Differential Equations, University of Texas at Austin, 1972* (Ed. D. G. Bettis), Lecture Notes in Mathematics No. 362, Springer, Berlin, pp. 1-21.

Stoer, J., see also Bulirsch, R.

Stohler, K. (1943): Eine Vereinfachung bei der numerischen Intergration gewöhnlicher Differentialgleichungen, *Z. Angew. Math. Mech.*, **23**, 120–122.

Stojanović, M. (1972): The use of a certain kind of transformation of differential equations for a practical solving of particular solutions (Serbian, English summary), *Mat. Vesnik, 9*, 281–287.

Stojanović, M., see also Orlov, K.

Stoller, L. and Morrison, D. (1958): A method for the numerical integration of ordinary differential equations, *Math. Comp.*, **12**, 269–272.

Storey, G., see Rosenbrock, H. H.

Störmer, C. (1921): Méthode d'intégration numérique des équations différentielles ordinaires, *C. R. Congress Internat. Math. Strasbourg*, pp. 243–257.

Stovall, E. J., see Lewis, H. R.

Strehmel, K. (1974): Ein neues Differenzenschemaverfahren zur Lösung von Anfangs-wertaufgaben gewöhnlicher Differentialgleichungen, *Beiträge Numer. Math.*, **3**, 157–165.

Strehmel, K. (1976): Mehrschrittverfahren mit Exponentialanpassung für Differential-gleichungssysteme 1. Ordnung, *Computing*, **17**, 247–260.

Strehmel, K. (1978): Integration von Differentialgleichungen n-ter Ordnung mit einer exponentiell angepaßten Mehrschrittmethode, *Wiss. Z. Martin-Luther-Univ. Halle-Wittenberg Math.-Natur. Reihe*, **27**, No. 4, 99–107.

Strehmel, K. (1979): Eine Klasse A-stabiler Mehrschrittverfahren für Anfangswertauf-gaben gewöhnlicher Differentialgleichungen, *Beiträge Numer. Math.*, **7**, 97–111.

Strehmel, K. (1980): Bemerkungen zur numerischen Integration gewöhnlicher Differen-tialgleichungen n-ter Ordnung mittels exponentiell angepaßter Diskretisierungsver-fahren, *Wiss. Z. Martin-Luther-Univ. Halle-Wittenberg Math. -Natur. Reihe*, **29**, No. 4, 109–115.

Strehmel, K. (1980a): Ein Extrapolationsalgorithmus für stiff Differentialgleichungen, in *Proc. Second Conf. on Numerical Treatment of Ordinary Differential Equations, Berlin, 1980* (Ed. R. März), Seminarberichte No. 32, Humboldt University, Berlin, pp. 124–135.

Strehmel, K. (1980b): Numerische Lösung großer linearer gewöhnlicher Differential-gleichungssysteme 1. Ordnung, *Wiss. Z. Martin-Luther-Univ. Halle-Wittenberg Math.-Natur. Reihe*, **29**, No. 5., 127–136.

Strehmel, K. (1980c): Bemerkungen zum Verfahren von Nickel und Rieder, *Z. Angew. Math. Mech.*, **60**, 50–53.

Strehmel, K. (1981): Stabilitätseigenschaften adaptiver Runge–Kutta-Verfahren, *Z. Angew. Math. Mech.*, **61**, 253–260.

Strehmel, K. (1981a): Numerische Lösung von steifen Differentialgleichungssystemen 1. Ordnung mittels Störansatz, *Beiträge Numer. Math.*, **10**, 153–168.

Strehmel, K. (1981b) (Ed.): *Numerische Behandlung von Differentialgleichungen*, Wiss. Beitr. Martin-Luther-Univ., Halle-Wittenberg (1981/47 (M23) 142 S).

Strehmel, K. (1982): Adaptive Runge–Kutta Methoden und ihre Stabilitäts-eigenschaften, in *Proc. Third Conf. on Numerical Treatment of Ordinary Differen-tial Equations, Berlin, 1982* (Ed. R. März), Seminarberichte No. 46, Humboldt Univ., Berlin, pp. 171–183.

Strehmel, K., see also Huťa, A.

Strehmel K. and Hoffman, V. (1975): Numerische Lösung von Anfangswertaufgaben gewöhnlicher Differentialgleichungen n-ter Ordnung, *Computing*, **14**, 225–234.

Strehmel, K. and Köhler, J. (1977): Ein neues Prediktor-Verfahren mit Exponentialanpassung für Anfangswertaufgaben gewöhnlicher Differentialgleichungssysteme 1. Ordnung, *Beiträge Numer. Math.*, **6**, 165–178.

Strehmel, K. and Peper, C. (1979): Numerische Lösung von Anfangswertaufgaben gewöhnlicher Differentialgleichungen n-ter Ordnung mittels Einschrittverfahren, *Computing*, **22**, 125–139.

Strehmel, K. and Weiner, R. (1982): Behandlung steifer Anfangswertprobleme gewöhnlicher Differentialgleichungen mit adaptiven Runge–Kutta-Methoden, *Computing*, **29**, 153–165.

Strehó, M. (1976): On the solution of ordinary differential equations of stiff type (Hungarian), *Tanulmányok—MTA Számitátstech. Automat. Kutató Int. Budapest*, **48**, 67 pp.

Strehó, M. (1981): Studies of numerical solutions of initial value problems (Hungarian, English summary), *Tanulmányok—MTA Számitátstech. Automat. Kutató Int. Budapest*, **121**, 67 pp.

Strümke, M., see Heinze, K.

Stummel, F. (1975): Biconvergence, bistability and consistency of one-step methods for the numerical solution of initial value problems in ordinary differential equations, in *Topics in Numerical Analysis II* (Ed. J. J. H. Miller), Academic Press, London, pp. 197–211.

Subba Rao, V. and Sucharitha, J. (1982): Collocation and iterated defect correction with trapezoidal scheme, *J. Math. Phys. Sci.*, **16**, 123–130.

Sucharitha, J., see Subba Rao, V.

Suleiman, M. B., see Hall, G.

Sun, G. (1979): A note on the trapezoidal rule (Chinese, English summary), *Math. Numer. Sinica*, **1**, 347–353.

Sun, G. (1980): A class of linearly implicit A-stable one-step methods (Chinese, English summary), *Math. Numer. Sinica*, **2**, 363–368.

Sun, G. (1982): A note on the extended backward differentiation formulae (Chinese, English summary), *J. Numer. Methods Comput. Appl.*, **3**, 94–99.

Sun, G. and Mao, Z. F. (1981): Two classes of linearly implicit methods (Chinese, English summary), *Math. Numer. Sinica*, **3**, 169–174.

Sun, G. and Mao, Z. F. (1981a): On a modified Gear method (Chinese, English summary), *Numer. Math. J. Chinese Univ.* **3**, 203–213.

Surla, K. (1982): Analysis of a difference method for ordinary differential equations with a small parameter (Serbo-Croation), in *Third Conf. on Applied Mathematics-Papers, Novi Sad, 1982* (Ed. D. Herceg), University Novi Sad, Novi Sad, pp. 105–116.

Swayne, D. A. (1978): Matrix operations with rational functions, in *Proc. Seventh Manitoba Conf. on Numerical Mathematics and Computing, University of Manitoba, Winnipeg, 1977* (Eds. D. McCarthy and H. C. Williams), Congress Numer. No. 20, Utilitas. Math., Winnipeg, pp. 581–589.

Sweet, R. A., see Dennis, J. E.

Szyszkowicz, M. (1982): The step-size control and the interval of absolute stability, *Proc. Third Conf. on Numerical Treatment of Ordinary Differential Equations, Berlin, 1982* (Ed. R. März), Seminarberichte No. 46, Humboldt University, Berlin, pp. 185–196.

Tagliabue, G., see Greco, A.

Tal, A. A., see Giesler, E. G.

Talbot, T. D., see Loscalzo, F. R.

Talybova, S. A. (1972): An application of functions with bounded spectrum to the estimation of the errors of methods of numerical solution of differential equations (Russian), *Izv. Vyssh. Uchebn. Zaved. Mat.*, **7**, (122), 95–99.

Tanaka, M. (1966): Runge–Kutta formulas with the ability of error estimation *Rep. Statist. Appl. Res. Un. Japan. Sci. Engrs.*, **13**, No. 3, 42–62.

Tanaka, M. (1967): On Runge–Kutta formulas using five functional values, *Information Processing in Japan*, **7**, 181–189.

Tanaka, M. (1968): On Kutta–Merson process and its allied processes (Japanese), *Joho Shori*, **9**, 18–30; English translation: *Information Processing in Japan*, **8**, 44–52 (1968).

Tanaka, M. (1968a): On Runge–Kutta type formulas with the error estimating ability (Japanese), *Joho Shori*, **9**, 261–271; English translation: *Information Processing in Japan*, **9**, 9–15 (1969).

Tanaka, M. (1969): Pseudo-Runge–Kutta methods and their application to the estimation of truncation errors in 2nd and 3rd order Runge–Kutta methods (Japanese), *Joho Shori*, **10**, 406–417.

Tanaka, M. (1971): Quadrature formulas with error estimating ability (Japanese), *Joho Shori*, **12**, 135–144; English translation: *Information Processing in Japan*, **11**, 144–151.

Tarnay, Gy., see Békéssy, A.

Taubert, K. (1974): Differenzenverfahren für gewöhnliche Anfangswertaufgaben mit unstetiger rechter Seite, in *Numerische Behandlung nichtlinearer Integrodifferential- und Differentialgleichungen* (Eds. R. Ansorge and W. Törnig), Lecture Notes in Mathematics No. 395, Springer, Berlin, pp. 137–148.

Taubert, K. (1976): Eine Erweiterung der Theorie von G. Dahlquist, *Computing*, **17**, 177–185.

Taubert, K. (1976a): Zusammenhänge zwischen Eindeutigkeitssätzen und Näherungsverfahren für gewöhnliche Anfangswertaufgaben, *Internat. Ser. Numer. Math.*, **31**, 233–240.

Taubert, K. (1981): Converging multistep methods for initial value problems involving multivalued maps, *Computing*, **27**, 123–136.

Taufer, J., see Prager, M.

Tee, G. J. (1982): Efficient solution of implicit Runge–Kutta equations, in *Computational and Asymptotic Methods for Boundary and Interior Layers: Proc. BAIL II Conf., Dublin, 1982* (Ed. J. J. H. Miller), Boole Press Conf. Ser. 4, Boole Press, Dublin, pp. 399–404.

Tendler, J. M. Bickart, T. A. and Picel, Z. (1978): A stiffly stable integration process using cyclic composite methods, *ACM Trans. Math. Software*, **4**, 339–368.

Tendler, J. M., Bickart, T. A. and Picel, Z. (1978a): Algorithm 534, STINT: STiff (differential equations) INTegrator, *ACM Trans. Math. Software*, **4**, 399–403.

Thacher, H. C. (1966): Certification of Algorithm 9, Runge–Kutta integration (P. Naur. *Comm. ACM*, **3**, 318, 1960), *Comm. ACM*, **9**, 273.

Thacher, H. C. (1972): Series solutions to differential equations by backward recurrence, in *Information Processing 71: Proc. IFIP Congress 1971* (Eds. J. E. Griffith and J. L. Rosenfeld), North Holland, Amsterdam, pp. 1287–1291.

Thomas, R. S. D. and Williams, H. C. (Eds.) (1971): *Proc. [first] Manitoba Conf. Numerical Math., Winnipeg, 1971*, Congress, Numer. No. 5, Utilitas Mathematica, Winnipeg (see Jain, R. K.; Verner, J. H.).

Thomas, R. S. D. and Williams, H. C. (Eds.) (1973): *Proc. Second Manitoba Conf. Numerical Math., Winnipeg, 1972*, Congress, Numer. No. 7, Utilitas Mathematica, Winnipeg (see Chakravati, P. C. and Worland, P. B.)

Thomas, R. S. D. and Williams, H. C. (Eds.) (1974): *Proc. Third Manitoba Conf. Numerical Math., Winnipeg, 1973*, Congress, Numer. No. 9, Utilitas Mathematica, Winnipeg (see Finden, W. F.).

Thomas, S., see Scholz, S.

Thompson, G. T., see Brush, D. G.; Kohfeld, J. J.

Thompson, S. (1982): The non-existence of certain A_0-stable Adams-type correctors, *J. Comput. Appl. Math.*, **8**, 155–157.

Thompson, S. (1982a): Stiffly stable fourth order Adams-type methods, *J. Comput. Appl. Math.*, **8**, 253–256.

Thompson, S., see also Rodabaugh, D. J.

Thomsen, P. G. (1978): Numerical solution of large systems of ordinary differential equations, *Z. Angew. Math. Mech.*, **58**, T448–T449.

Thomsen, P. G. (1982): The use of sparse matrix techniques in ODE-codes, in *Integration of Differential Equations and Large Linear Systems: Proc. Bielefeld, 1980* (Ed. J. Hinze), Lecture Notes in Mathematics No. 968, Springer, Berlin, pp. 301–309.

Thomsen, P. G., see also Zlatev, Z.

Thomsen, P. G. and Zlatev, Z. (1979): Two-parameter families of predictor–corrector methods for the solution of ordinary differential equations, *BIT*, **19**, 503–517.

Tikhonov, A. N. and Gorbunov, A. D. (1962): Asymptotic expansions of the error in the difference method of solving Cauchy's problem for systems of differential equations (Russian), *Zh. Vychisl. Mat. Fiz.*, **2**, 537–548; English translation: *USSR Comput. Math. and Math. Phys.*, **2**, 565–580, (1963).

Tikhonov, A. N. and Gorbunov, A. D. (1962a): On the optimality of implicit difference schemes of Adams type (Russian) *Zh. Vychisl. Mat. i Mat. Fiz.*, **2**, 930–933; English translation: *USSR Comput. Math. and Math. Phys.*, **3**, 1089–1093 (1963).

Tikhonov, A. N. and Gorbunov, A. D. (1963): On asymptotic estimates of the error in the Runge–Kutta type method (Russian), *Zh. Vychisl. Mat. i Mat. Fiz.*, **3**, 195–197; English translation: *USSR Comput. Math. and Math. Phys.* **3**, 257–261.

Tikhonov, A. N. and Gorbunov, A. D. (1964): Estimates of the error of a Runge–Kutta method and the choice of optimal meshes (Russian), *Zh. Vychisl. Mat. i Mat. Fiz.*, **4**, 232–241; English translation: *USSR Comput. Math. and Math. Phys.*, **4**, No. 2, 30–42.

Timlake, W. P. (1965): On an algorithm of Milne and Reynolds, *BIT*, **5**, 276–281.

Tivonchuk, V. Ĭ., see Kurpel', N. S.

Tkachenko, A. I. (1973): On 'reversing' methods of integrating systems of linear ordinary differential equations (Russian), *Ukrain. Mat. Zh.*, **25**, 843–846; English translation: *Ukrainian Math. J.*, **25**, 704–707, (1974).

Tkachenko, A. I. (1975): Increasing the accuracy of integration of a system of differential equations (Russian), *Zh. Vychisl. Mat. i Mat. Fiz.*, **15**, 509–512; English translation: *USSR Comput. Math. and Math. Phys.*, **15**, No. 2, 221–225.

Tkachenko, A. I. (1979): Some possibilities for the reduction of rounding errors in the integration of ordinary differential equations (Russian), *Zh. Vychisl. Mat. i Mat. Fiz.*, **19**, 245–248; English translation: *USSR Comput. Math. and Math. Phys.*, **19**, No. 1, 256–259.

Tkachenko, N. V. (1971): A certain method of approximate integration of a system of differential equations (Russian, English summary), *Vychisl. Prikl. Mat. (Kiev.)*, **14**, 136–147.

Todd, J. (1950): Notes on modern numerical analysis, I: Solution of differential equations by recurrence relations, *Math. Comp.*, **4**, 39–44.

Toktalaeva, S. S. (1959): Ordinate formulas for numerical integration of ordinary differential equations of first order (Russian), *Vychisl. Mat.*, **5**, 3–57.

Tollmien, W. (1938): Über die Fehlerabschätzung beim Adamsschen Verfahren zur Integration gewöhnlicher Differentialgleichungen, *Z. Angew. Math. Mech.*, **18**, 83–90.

Tollmien, W. (1953): Bemerkung zur Fehlerabschätzung beim Adamsschen Interpolationsverfahren, *Z. Angew. Math. Mech.*, **33**, 151–155.

Tolmacheva, T. A., see Ryabov, Yu. A.

Törnig, W., see Ansorge, R.

494

Torrey, M. D., see Richards, P. I.

Tortorici, M., see Cusimano, M.

Trasi, M. S., see Poddar, R. S.

Treanor, C. E. (1966): A method for the numerical integration of coupled first-order differential equations with greatly different time constants, *Math. Comp.*, **20**, 39–45.

Trigiante, D. (1977): Asymptotic stability and discretization on an infinite interval, *Computing*, **18**, 117–129.

Trigiante, D., see also Casulli, V.; Piazza, G.

Trochu, F., see Delfour, M.

Trujillo, D. M. (1975): The direct numerical integration of linear matrix differential equations using Padé approximations, *Internat. J. Numer. Methods Engrg.*, **9**, 259–270.

Tsao, H. Y., see Wilson, D. B.

Tsushima, T., see Urabe, M.

Tu, K. W., see Gear, C. W.

Tuan, P. D. (1973): An extension of Clenshaw's method for linear differential equations, *BIT*, **13**, 372–374.

Turton, F. J. (1939): The errors in the numerical solution of differential equations, *The London, Edinburgh and Dublin Philos. Mag.*, **28**, 359–363.

Turton, F. J. (1939a): Two notes on the numerical solution of differential equations, *The London, Edinburgh and Dublin Philos. Mag.*, **28**, 381–394.

Tutschke, W. (1971): Optimale Fehlerschranken für die Lösung von Differentialgleichungssystemen erster Ordnung bei Verwendung von Funktionstafeln, *Z. Angew. Math. Mech.*, **51**, 419–426.

Twizell, E. H. and Khaliq, A. Q. M. (1981): One-step multiderivative methods for first order ordinary differential equations, *BIT*, **21**, 518–527.

Überhuber, C. W. = Ueberhuber, C. W.

Ueberhuber, C. W. (1979): Implementation of defect correction methods for stiff differential equations, *Computing*, **23**, 205–232.

Ueberhuber, C. W., see also Frank, R.

Uhlmann, W. (1957): Fehlerabschätzungen bei Anfangswertaufgaben gewöhnlicher Differentialgleichungssysteme 1. Ordnung, *Z. Angew. Math. Mech.*, **37**, 88–99.

Uhlmann, W. (1957a): Fehlerabschätzungen bei Anfangswertaufgaben einer gewöhnlichen Differentialgleichung höherer Ordnung, *Z. Angew. Math. Mech.*, **37**, 99–111.

Ullman, F. D. (1972): A generalisation of the Peano–Baker method, *J. Franklin Inst.*, **293**, 137–141.

Unger, H., see Krückeberg, F.

Urabe, M. (1961): Theory of errors in numerical integration of ordinary differential equations, *J. Sci. Hiroshima Univ. ser. A-I Math.*, **25**, 3–62.

Urabe, M. (1970): An implicit one-step method of high-order accuracy for the numerical integration of ordinary differential equations, *Numer. Math.* **15**, 151–164.

Urabe, M. (1975): On the Newton method to solve problems of the least squares type for ordinary differential equations, *Mem. Fac. Sci. Kyushu Univ. ser A.* **29**, 173–183.

Urabe, M. and Mise, S. (1955): A method of numerical integration of analytic differential equations, *J. Sci. Hiroshima Univ. ser. A. Math.*, **19**, 307–320.

Urabe, M. and Tsushima, T. (1953): On numerical integration of ordinary differential equations, *J. Sci. Hiroshima Univ. ser. A. Math.*, **17**, 193–219.

Urabe M. and Yanagihara, H. (1954): On numerical integration of the differential equation $y^{(n)} = f(x, y)$, *J. Sci. Hiroshima Univ. ser. A. Math.*, **18**, 55–76.

Urbani, A. M. (1974): Accelerazione della convergenza per metodi di risoluzione di equazioni differenziali ordinarie, *Calcolo*, **11**, 509–520.

Urbani, A. M. (1976): Metodi multistep con accelerazione per la risoluzione numerica dei sistemi di equazioni differenziali ordinarie, *Calcolo*, **13**, 369–376.

Urbani, A. M., see also Gross, W.

Usmani, R. A. (1972): A multi-step method for the numerical integration of ordinary differential equations, *Z. Angew. Math. Phys.*, **23**, 465–483.

Usmani, R. A. (1976): On the implementation of Urabe's method for the integration of linear stiff equations, in *Proc. Fifth Manitoba Conf. on Numerical Mathematics, Winnipeg, 1975* (Eds. B. L. Hartnell and H. C. Williams), Congress Numer. No. 16, Utilitas Mathematica, Winnipeg, pp. 625–635

Usmani, R. A., Andres, T. H. and Walton, D. J. (1975): Error estimation in the integration of ordinary differential equations, *Internat. J. Comput. Math.*, **5**, 241–256.

Ustinov, S. M., see Rakitskiĭ, Yu. V.

Valková, A. (1982): A theoretical formula for an error of the Huťa formula of the Runge–Kutta type of the fifth order, *Acta Math. Univ. Comenian*, **40/41**, 111–128.

Valková, A. (1982a): The error of the Runge–Kutta-Huťa formula of the sixth order, *Acta Math. Univ. Comenian.*, **40/41**, 195–214.

van Bokhoven, W. M. G. (1980): Efficient higher order implicit one-step methods for integration of stiff differential equations, *BIT*, **20**, 34–43.

van der Houwen, P. J. (1971): One-step methods for linear initial value problems, *Z. Agnew. Math. Mech.*, **51**, T58–T59.

van der Houwen, P. J. (1972): Explicit Runge–Kutta formulas with increased stability boundaries, *Numer. Math.*, **20**, 149–164.

van der Houwen, P. J. (1973): One-step methods with adaptive stability functions for the integration of differential equations, in *Numerische, insbesondere approximationstheoretische Behandlung von Funktionalgleicheungen* (Eds. R. Ansorge and W. Törnig), Lecture Notes in Mathematics No. 333, Springer, Berlin, pp. 164–174.

van der Houwen, P. J. (1977): *Construction of Integration Formulas for Initial Value Problems*, North-Holland, Amsterdam.

van der Houwen, P. J. (1979): Stabilized Runge–Kutta methods for second order differential equations without first derivative, *SIAM J. Numer. Anal.*, **16**, 523–537.

van der Houwen, P. J. (1980): Multistep splitting methods of high order for initial value problems, *SIAM J. Numer. Anal.*, **17**, 410–427.

van der Houwen, P. J., see also Sommeijer, B. P.

van der Houwen, P. J. and Sommeijer, B. P. (1980): On the internal stability of explicit m-stage Runge–Kutta methods for large m-values, *Z. Angew. Math. Mech.*, **60**, 479–485.

van der Houwen, P. J. and Sommeijer, B. P. (1982): A special class of multistep Runge–Kutta methods with extended real stability interval, *IMA J. Numer. Anal.*, **2**, 183–209.

van Dooren, R. (1974): Stabilization of Cowell's classical finite difference method for numerical integration, *J. Comput. Phys.*, **16**, 186–192.

van Dyck, D., de Ridder, R. and de Sitter, J. (1980): The most storage economical Runge–Kutta methods for the solution of large systems of coupled first-order differential equations, I, *J. Comput. Appl. Math.*, **6**, 83–85.

van Kirk, J., see Leung, K. V.

van Rees, G. H. J. and Williams, H. C. (Eds.) (1980): *Proc. Ninth Manitoba Conf. Numerical Math. and Computing, Winnipeg, 1979*, Congress. Numer. No. 27, Utilitas Mathematica, Winnipeg (see Kamel, M. S.; Newbery, A. C. R.; Verner, J. H.; Worland, P. B.).

van Veldhuizen, M. (1974): Consistency and stability for one-step discretizations of stiff differential equations, in: *Stiff Differential Systems* (Ed. R. A. Willoughby), Plenum Press, New York, pp. 259–270.

van Veldhuizen, M. (1981): *D*-stability, *SIAM J. Numer. Anal.*, **18**, 45–64.

van Veldhuizen, M., see also Miranker, W. L.

van Wyk, R. (1970): Variable mesh multistep methods for ordinary differential equations, *J. Comput. Phys.*, **5**, 244–264.

Varah, J. M. (1978): Stiffly stable linear multistep methods of extended order, *SIAM J. Numer. Anal.*, **15**, 1234–1246.

Varah, J. M. (1979): On the efficient implementation of implicit Runge–Kutta methods, *Math. Comp.*, **33**, 557–561.

Varano, A., see Costabile, F.

Varga, R. S., see Saff, E. B.

Vejvoda, O. (1957): Error estimates for the Runge–Kutta formula (Czech, German summary), *Apl. Mat.*, **2**, 1–23.

Veldhuizen, M. van = van Veldhuizen, M.

Verner, J. H. (1969): The order of some implicit Runge–Kutta methods, *Numer. Math.*, **13**, 14–23.

Verner, J. H. (1969a): Implicit methods for implicit differential equations, in *Conf. on the Numerical Solution of Differential Equations, Dundee, 1969* (Ed. J. L. Morris), Lecture Notes in Mathematics No. 109, Springer, Berlin, pp. 261–266.

Verner, J. H. (1970): Quadratures for implicit differential equations, *SIAM J. Numer. Anal.*, **7**, 373–385.

Verner, J. H. (1971): On deriving explicit Runge–Kutta methods, in *Conf. on Applications of Numerical Analysis, Dundee, 1971* (Ed. J. L. Morris), Lecture Notes in Mathematics No. 228, Springer, Berlin, pp. 340–347.

Verner, J. H. (1971a): On deriving certain hybrid methods in *Proc. [first] Manitoba Conf. Numerical Math., Winnipeg, 1971* (Eds. R. S. D. Thomas and H. C. Williams), Congress Numer. No. 7, Utilitas Mathematica, Winnipeg, pp. 607–626.

Verner, J. H. (1977): Selecting a Runge–Kutta method, in *Proc. Sixth Manitoba Conf. Numerical Math., Winnipeg, 1976* (Eds. B. L. Hartnell and H. C. Williams), Congress Numer. No. 18, Utilitas Mathematica, Winnipeg, pp. 496–504.

Verner, J. H. (1978): Explicit Runge–Kutta methods with estimates of the local truncation error, *SIAM J. Numer. Anal.*, **15**, 772–790.

Verner, J. H. (1979): Families of imbedded Runge–Kutta methods, *SIAM J. Numer. Anal.*, **16**, 857–875.

Verner, J. H. (1980): John Butcher's algebraic theory: motivation for selecting simplifying conditions, in *Proc. Ninth Manitoba Conf. Numerical Math. and Computing, Winnipeg, 1979* (Eds. G. H. J. van Rees and H. C. Williams), Congress numer. No. 27, Utilitas Mathematica, Winnipeg, pp. 125–155.

Verner, J. H., see also Cooper, G. J.

Verwer, J. G. (1977): On generalized linear multistep methods with zero-parasitic roots and an adaptive principal root, *Numer. Math*, **27**, 143–155.

Verwer, J. G. (1977a): *S*-stability properties for generalized Runge–Kutta methods, *Numer. Math.*, **27**, 359–370.

Verwer, J. G. (1977b): A class of stabilized three-step Runge–Kutta methods for the numerical integration of parabolic equations, *J. Comput. Appl. Math.*, **3**, 155–166.

Verwer, J. G. (1980): An implementation of a class of stabilized explicit methods for the time integration of parabolic equations, *ACM Trans. Math. Software*, **6**, 188–205.

Verwer, J. G. (1980a): On generalized Runge–Kutta methods using an exact Jacobian at a non-step point, *Z. Angew. Math. Mech.*, **60**, 263–265.

Verwer, J. G. (1981): On the practical value of the notion of *BN*-stability, *BIT*, **21**, 355–361.

Verwer, J. G. (1982): An analysis of Rosenbrock methods for nonlinear stiff initial value problems, *SIAM J. Numer. Anal.*, **19**, 155–170.

Verwer, J. G. (1982a): Instructive experiments with some Runge–Kutta–Rosenbrock methods, *Comput. Math. Appl.*, **8**, 217–229.

Verwer, J. G. (1982b): A note on a Runge–Kutta–Chebyshev method, *Z. Angew. Math. Mech.*, **62**, 561–563.

Vichnevetsky, R., see Bennett, A. W.

Vietoris, L. (1953): Der Richtungsfehler einer durch das Adamssche Interpolationsverfahren gewonnenen Näherungslösung einer Gleichung $y' = f(x, y)$, *Österr. Akad. Wiss. Math. Nat. Kl. S-BIIa*, **162**, 157–167.

Vietoris, L. (1953a): Der Richtungsfehler einer durch das Adamssche Interpolationsverfahren gewonnenen Näherungslösung eines Systems von Gleichungen $y_k = f_k(x, y_1, y_2, \ldots, y_m)$ *Österr. Akad. Wiss. Math. Nat. Kl. S-BIIa*, **162**, 293–299.

Višňák, K. (1977): A-stable block implicit methods containing second derivatives, *Czechoslovak Math. J.*, **27**(102), 14–42.

Višňák, K. see also Kubíček, M.

Višňák, K. and Kubíček, M. (1978): A class of numerical methods of high order for stiff problems in ordinary differential equations, *J. Inst. Math. Appl.*, **21**, 251–264.

Vitásek, E. (1969): The numerical stability in solution of differential equations, in *Conf. on the Numerical Solution of Differential Eq uations, Dundee , 1969* (Ed. J. L. Morris), Lecture Notes in Mathematics No. 109, Springer, Berlin, pp. 87–111.

Vitásek, E. (1979): A-stability and numerical solution of evolution problems, *IAC 'Mauro Picone'*, Series III **186**, 42pp.

Vitásek, E., see also Babuška, I.; Práger, M.

Vlasov, I. O. and Charnyĭ, I. A. (1950): On a method of numerical integration of ordinary differential equations (Russian), *Akad. Nauk SSSR Inzhenernyi Sbornik*, **8**, 181–186.

Vogelaere, R. de = de Vogelaere, R.

Volkov, E. A. (1974): The asymptotics of an *a posteriori*, error estimate of the difference solution of an ordinary differential equation (Russian), *Differentsial'nye Uravneniya*, **10**, 2262–2266.

Voňka, P. (1972): Stability and local error of difference methods for the solution of the ordinary differential equation of the first order, *Apl. Mat.*, **17**, 18–27.

von Mises, R. (1930): Zur numerischen Integration von Differentialgleichungen, *Z. Angew. Math. Mech.*, **10**, 81–92.

Vorob'ëv, L. M. (1956): The applicability of S. A. Chaplygin's method of approximate integration to a certain class of ordinary non-linear differential equations of the second order (Russian), *Uspekhi Mat. Nauk. (new ser.)*, **11**, 181–185.

Voronovskaya, E. V. (1955): On an alteration of Chaplygin's method for differential equations of the first order, *Prikl. Mat. Mekh.*, **19**, 121–126.

Vysotskiĭ, L. I., Amel'chenko, V. V. and Morozova, Yu. I. (1970): Certain questions on the application of the Runge–Kutta method to the solution of differential equations of order higher than one (Russian), in *Comput. Methods and Programming*, vol. 3, Izdat. Saratov University, Saratov, pp. 63–65.

Waddell, E. R., see Walston, D. E.

Wahba, G., see Miranker, W. L.

Wait, J. V., see Palusinski, O. A.

Wall, D. D. (1956): Note on predictor–corrector formulas, *Math. Comp.*, **10**, 167.

Wallace, C. S., see Gupta, G. K.

Wallace, C. S. and Gupta, G. K. (1973): General linear multistep methods to solve ordinary differential equations, *Austral. Comput. J.*, **5**, 62–69.

Walsh, J. (1974): Initial and boundary value routines for ordinary differential equations, in *Software for Numerical Mathematics* (Ed. D. J. Evans), Academic Press, London, pp. 177–189.

Walston, D. E. and Waddell, E. R. (1968): Accelerating convergence of one-step methods for the numerical solution of ordinary differential equations, *Internat. J. Comput. Math.*, **2**, 23–33.

Walton, D. J, see Usmani, R. A.

Wambecq, A. (1976): Nonlinear methods in solving ordinary differential equations,*J. Comput. Appl. Math.*, **2**, No. 1, 27–33.

Wambecq, A. (1978): Rational Runge–Kutta methods for solving systems of ordinary differential equations, *Computing*, **20**, 333–342.

Wambecq, A. (1979): Some properties of rational methods for solving ordinary differential equations, in *Padé Approximation and Its Applications: Proc. Antwerp, 1979* (Ed. L. Wuytack), Lecture Notes in Mathematics No. 765, Springer, Berlin, pp. 352–365.

Wambecq, A. (1980): Solution of the equations associated with rational Runge–Kutta methods of orders up to four, *J. Comput. Appl. Math.*, **6**, 275–281.

Wang, C. Z. (1982): Formulas for the solutions of differential equations I. Formulas for the solution of higher-order linear and nonlinear ordinary differential equations with variable coefficients (Chinese, English summary), *Acta Math. Appl. Sinica*, **5**, 274–284.

Wang, M. Z. (1979): Convergence of an iterative method for stiff systems of linear algebraic equations (Chinese), *Acta Math. Appl. Sinica*, **2**, 308–315.

Wang, T. C. (1977): A modification of the Runge–Kutta procedure (Russian), *Zh. Vychisl. Mat. i Mat. Fiz.*, **17**, 769–771; English translation: *USSR Comput. Math. and Math. Pys.*, **17**, No. 3, 203–206.

Wanner, G. (1968): Theorie der Lie-Reihen in *Überblicke Mathematik* (Ed. D. Laugwitz), vol. I, B. I.-Hochschultaschenbücher. 161/161a Bibliographisches Institut, Mannheim, pp. 133–153.

Wanner, G. (1969): *Integration gewöhnlicher Differentialgleichungen. Lie-Reihen (mit Programmen), Runge–Kutta-Methoden* B. I. Hochschulskripten No. 1 831/831a. Bibligraphisches Institut, Mannheim.

Wanner, G. (1973): Runge–Kutta methods with expansion in even powers of *h*, *Computing*, **11**, 81–85.

Wanner, G. (1976): A short proof on nonlinear *A*-stability, *BIT*, **16**, 226–227.

Wanner, G. (1977): On the integration of stiff differential equations, *Internat. Ser. Numer. Math.*, **37**, 209–226.

Wanner, G. (1980): On the choice of γ for singly-implicit RK or Rosenbrock methods, *BIT*, **20**, 102–106.

Wanner, G. (1980a): Characterization of all *A*-stable methods of order $2m - 4$, *BIT*, **20**, 367–374.

Wanner, G, see also Gottwald, B. A.; Hairer, E.; Kaps, P.; Kastlunger, K. H.; Knapp, H.; Nørsett, S. P. .; Sottas, G.

Wanner, G., Hairer, E. and Nørsett, S. P. (1978): Order stars and stability theorems, *BIT*, **18**, 475–489.

Wanner, G., Hairer, E. and Nørsett, S. P. (1978a): When *I*-stability implies *A*-stability, *BIT*, **18**, 503.

Warga, J. (1953): On a class of iterative procedures for solving normal systems of ordinary differential equations, *J. Math. Phys.* **31**, 223–243.

Warner, D. D. (1976): A note on variable order strategies for differential equation solvers, *SIGNUM Newsletter*, **11**, 27–28.

Warten, R. M. (1963): Automatic step-size control for Runge–Kutta integration, *IBM J. Res. Develop.*, **1963**, 340–341.

Warten, R. M., see also Fowler, M. E.

Watanabe, D. S. (1978): Block implicit one-step methods, *Math. Comp.*, **32**, 405–414.

Watanabe, D. S., see also Gear, C. W.

Watanabe, K. and Himmelblau, D. M. (1982): Analysis of trajectory errors in integrating ordinary differential equations, *J. Franklin Inst.*, **314**, 283–321.

Watkins, D. S. (1981): Determining initial values for stiff systems of ordinary differential equations, *SIAM J. Numer. Anal.*, **18**, 13–20.

Watkins, D. S. (1981a): Efficient initialization of stiff systems with one unknown initial condition, *SIAM J. Numer. Anal.*, **18**, 794–800.

Watson, G. A. (Ed.) (1974) *Proc. Conf. on the Numerical Solution of Differential Equations, Dundee 1973*, Lecture Notes in Mathematics No. 363, Springer, Berlin (see Hull, T. E.; Lambert, J. D.; Stetter, H. J.).

Watson, G. A. (Ed.) (1976): *Numerical Analysis: Proc. Dundee Conf., 1975* Lecture Notes in Mathematics No. 506, Springer, Berlin, (see Butcher, J. C.; Dahlquist, G.; Stetter, H. J.;)

Watson, G. A. (Ed.) (1980): *Numerical Analysis: Proc. Eighth Conf., Dundee, 1979,* Lecture Notes in Mathematics no. 773, Springer, Berlin, (see Butcher, J. C.; Lambert, J. D. and McLeod, R. J. Y.; Nørsett, S. P.; Stetter, H. J.)

Watson, H. D. D. and Gourlay, A. R. (1976): Implicit integration for CSMP III and the problem of stiffness, *Simulation*, **26**, No. 2, 57–61.

Watson, I. A., see Lambert, J. D.

Watt, J. M. (1967): The asymptotic discretization error of a class of methods for solving ordinary differential equations, *Proc. Cambridge Philos. Soc.*, **63**, 461–472.

Watt, J. M. (1968): Consistency, convergence and stability of general discretizations of the initial value problem, *Numer. Math.*, **12**, 11–22.

Watt, J. M., see also Hall, G.

Watts, H. A. (1980): Survey of numerical methods for ordinary differential equations, *Electric Power Problems: The Mathematical Challenge: Proc. Conf., Seattle, Wash., 1980*, SIAM, Philadelphia, pp. 127–158.

Watts, H. A., see also Shampine, L. F.

Watts, H. A. and Shampine, L. F. (1972): *A*-stable block implicit one-step methods, *BIT*, **12**, 252–266.

Weigand, P. (1977): Splineapproximationen vom Defekt 2 und lineare Mehrschrittformeln zur numerischen Lösung gewöhnlicher Differentialgleichungen, *Beiträge Numer. Math.*, **6**, 189–195.

Weigand, P. (1978): Verallgemeinerte Splines und lineare Mehrschrittformeln zur numerischen Lösung gewöhnlicher Differentialgleichungen, *Wiss. Z. Tech. Hochsch. Karl-Marx-Stadt*, **20**, 727–733.

Weimer, A. W. and Clough, D. E. (1979): A critical evaluation of the semi-implicit Runge–Kutta methods for stiff systems, *AIChE J.*, **25**, 730–732.

Weiner, R. (1982): Nichtlineare Kontraktivität einer Klasse semiimpliziter Mehrschrittverfahren, in *Proc. Third Conf. on Numerical Treatment of Ordinary Differential Equations, Berlin, 1982* (Ed. R. März), Seminarberichte No. 46, Humboldt University, Berlin, pp. 215–220.

Weiner, R., see also Strehmel, K.

Weinmüller, E., see Stetter, H. J.

Weiss, R. (1974): The application of implicit Runge–Kutta and collocation methods to boundary-value problems, *Math. Comp.*, **28**, 449–464.

Weiss, R., see also de Hoog, F.

Weiss, W. and Scholz, S. (1974): Runge–Kutta–Nyström-Verfahren mit variablen Parametern zur numerischen Behandlung von gewöhnlichen Differentialgleichungen zweiter Ordnung, *Beiträge Numer. Math*, **2**, 211–227.

Weissinger, J. (1950): Eine verschärfte Fehlerabschätzung zum Extrapolationsverfahren von Adams, *Z. Angew. Math. Mech.*, **30**, 356–363.

Weissinger, J. (1952): Eine Felherabschätzung für die Verfahren von Adams und Störmer, *Z. Angew. Math. Mech.*, **32**, 62–67.

500

Weissinger, J. (1953): Numerische Integration impliziter Differentialgleichungen, *Z. Angew. Math. Mech.*, **33**, 63–65.

Weissinger, J. (1973): Über zulässige Schrittweiten bei den Adams-Verfahren, *Z. Angew. Math. Mech.*, **53**, 121–126.

Werner, H. (1975): Interpolation and integration of initial value problems of ordinary differential equations by regular splines, *SIAM J. Numer. Anal.*, **12**, 255–271.

Werner, H. (1979): Extrapolationsmethoden zur Bestimmung der beweglichen Singularitäten von Lösungen gewöhnlicher Differentialgleichungen, in *Numerical Mathematics: Sympos., Inst. Appl. Math. Univ. Hamburg, 1979* (Eds. R. Ansorge, K. Glashoff and B. Werner), Internat. Ser. Numer. Math. No. 49, Birkhäuser, Basel, pp 159–176.

Werner, H. (1980): Spline functions and the numerical solution of differential equations, in *Special Topics of Applied Mathematics*, North-Holland, Amsterdam, pp. 173–192.

Werner, H. (1981): Non-linear splines, some application to singular problems, in *Padé Approximation and Its Applications: Proc. Amsterdam, 1980* (Eds. M. G. de Bruin and H. van Rossum), Lecture Notes in Mathematics No. 888, Springer, Berlin, pp. 64–77.

Werschulz, A. G. (1979): Computational complexity of one-step methods for a scalar autonomous differential equation, *Computing*, **23**, 345–355.

Werschulz, A. G. (1980): Computational complexity of one-step methods for systems of differential equations, *Math. Comp.*, **34**, 155–174.

West, M. R., see Dew, P. M.

Westreich, D. (1980): An efficient predictor–corrector algorithm, *Comput. J.*, **23**, 186.

Wheeler, D. J. (1959): Note on Runge–Kutta method of integrating ordinary differential equations, *Comput. J.*, **2**, 23.

Whitney, D. E. (1966): Propagated error bounds for numerical solution of transient response. *Proc. IEEE*, **54**, 1084–1085.

Whitney, D. E. (1969): More about similarities between Runge–Kutta and matrix exponential methods for evaluation transient response, *Proc. IEEE*, **57**, 2053–2054.

Whitworth, F. C. P., see Cooper, G. J.

Widlund, O. (1967): A note on unconditionally stable linear multistep methods, *BIT*, **7**, 65–70.

Wilf, H. S. (1957): An open formula for the numerical integration of first order differential equations, *Math. Comp.*, **11**, 201–203.

Wilf, H. S. (1958): An open formula for the numerical integration of first order differential equations, II, *Math. Comp.*, **12**, 55–58.

Wilf, H. S. (1959): A stability criterion for numerical integration, *J. Assoc. Comput. Mach.*, **6**, 363–365.

Wilf, H. S. (1960): Maximally stable numerical integration, *J. Soc. Indust. Appl. Math.*, **8**, 537–540.

Wilf, H. S., see also Ralston, A.

Willems, J. C., see Glover, K.

Willers, I. M. (1974): A new integration algorithm for ordinary differential equations based on continued fraction approximations, *Comm. ACM*, **17**, 504–508.

Willers, I. M., see also Barton, D.

Williams, H. C., see Hartnell, B. L.; Thomas, R. S. D.; van Rees, G. H. J.

Williams, J., see Robertson, H. H.

William, J. and de Hoog, F. (1974): A class of A-stable advanced multistep methods, *Math. Comp.*, **28**, 163–177.

Williamson, J. H. (1980): Low-storage Runge–Kutta schemes, *J. Comput. Phys.*, **35**, 48–56.

Willoughby, R. A. (Ed.) (1974): *Stiff Differential Systems*, Plenum Press, New York.

501

Willoughby, R. A., see also Liniger, W.
Wilson, D. B., Tsao, H. Y. and Shouman, A. R. (1973): A method for numerically solving second-order non-homogeneous linear differential equations with variable coefficients, *Internat. J. Numer. Methods Engrg.*, **7**, 235–240.
Wimp, J., see Luke, Y. L.
Windeknecht, T. G. and d'Angelo, H. (1977): System theoretic implications of numerical methods applied to the solution of ordinary differential equations, *IEEE TRANS on Systems, Man, and Cybernetics*, **7**, 805–810.
Windisch, G. (1975): Über Differenzenverfahren zur Lösung nichlinearer Cauchy-Aufgaben gewöhnlicher Differentialgleichungen, *Beiträge Numer. Math.*, **3**, 173–183.
Wingerath, K. see Mihelčić, M.
Wisniewski, J. A., see Shampine, L. F.
Witmer, E. A., see Morino, L.
Witty, W. H. (1964): A new method of numerical integration of differential equations, *Math. Comp.*, **18**, 497–500.
Wojtowicz, J. (1975): On the estimation of the error of solving the system of ordinary differential equations with initial conditions and on the estimation of the error of derivatives, *Demonstratio Math.*, **8**, 123–131.
Wolfbrandt, A. (1982): Dynamic adaptive selection of integration algorithms when solving ODEs, *BIT*, **22**, 361–367.
Wolfbrandt, A., see also Nørsett, S. P.; Steihaug, T.
Wolfe, M. A. (1971): The numerical solution of implicit first order ordinary differential equations with initial conditions, *Comput. J.*, **14**, 173–178.
Worland, P. B. (1973): Local error estimates in the numerical solution of $y'' = f(x, y)$, *Proc. IEEE*, **61**, 1365–1366.
Worland, P. B. (1974): A stability and error analysis of block methods for the numerical solution of $y'' = f(x, y)$, *BIT*, **14**, 106–111.
Worland, P. B. (1976): Parallel methods for the numerical solution of ordinary differential equations, *IEEE Trans. Comput.*, **25**, 1045–1048.
Worland, P. B. (1980): Predictor–corrector methods for the parallel solution of ordinary differential equations, in *Proc. Ninth Conf. Numerical Math. and Computing, Winnipeg, 1979* (Eds. G. H. J. van Rees and H. C. Williams), Congress Numer. no. 27, Utilitas Mathematica, Winnipeg, pp. 411–423.
Worland, P. B., see also Chakravarti, P. C.
Wouk, A. (1976): Collocation for initial value problems, *BIT*, **16**, 215–222.
Wright, K. (1970): Some relationships between implicit Runge–Kutta, collocation and Lanczos τ methods, and their stability properties, *BIT*, **10**, 217–227.
Wright, K., see also Prince, P. J.
Wyk, R. van = van Wyk, R.
Yanagihara, H., see Urabe, M.
Yokomoto, C. F., see Sinha, A. S. C.
Yowell, E. C., see Southard, T. H.
Yuan, Z. D., see Zhu, C. H.
Yuldashev, Z. Kh., see Kalmykov, S. A.
Ząbek, Ś., see Pokora, T.
Zadiraka, K. V. (1951): Solution by the method of S. A. Chaplygin of linear differential equations of the second order with variable coefficients (Ukranian), *Dopovidi Akad. Nauk Ukrain. RSR*, **1951**, 163–170.
Zadiraka, K. V. (1952): The approximate integration by S. A. Chaplygin's method of linear differential equations, of second order with variable coefficients (Russian), *Ukrain. Mat. Zh.*, **4**, 299–311.
Zadunaisky, P. E. (1966): A method for the estimation of errors propagated in the

numerical solution of a system of ordinary differential equations, *International Astronomical Union, Symposium*, **1966**, No. 25, Academic Press, New York, pp. 281–287.

Zadunaisky, P. E. (1970): On the accuracy in the numerical computation of orbits in *Periodic Orbits, Stability and Resonances* (Ed. G. E. O. Giacaglia), Reidel, Dordrecht, pp. 216–227.

Zadunaisky, P. E. (1976): On the estimation of errors, propagated in the numerical integration of ordinary differential equations, *Numer. Math.*, **27**, 21–39.

Zadunaisky, P. E. and Lafferriere, G. (1980): On an iterative improvement of the approximate solution of some ordinary differential equations, *Comput. Math. Appl.*, **6**, 147–154.

Zahar, R. V. M., see Barton, D.

Zakharov, A. Yu. (1979): Comparison of numerical methods of solving stiff systems of ordinary differential equations (Russian), *Akad. Nauk SSSR Inst. Prikl. Mat. Preprint*, **124**, 25pp.

Zakharov, A. Yu. (1979a): Some results of comparison of the effectiveness of the solution of systems of ordinary differential equations (Russian), *Akad. Nauk SSSR Inst. Prikl. Mat. Preprint*, **125**, 34pp.

Zakharov, A. Yu. (1982): Some results of comparison of efficiency in solving systems of ordinary differential equations, *Proc. Third Conf. on Numerical Treatment of Ordinary Differential Equations, Berlin, 1982* (Ed. R. März), Seminarberichte, No. 46, Humboldt University, Berlin, pp. 229–239.

Zakian, V. (1975): Application of J_{MN} approximants to numerical initial-value problems in linear differential-algebraic systems, *J. Inst. Math. Appl.*, **15**, 267–272.

Zavorin, A. N. (1977): The solution of stiff ordinary differential equation systems by means of step series (Russian), *Latv. Mat. Ezhegodnik*, **21**, 123–132.

Zavorin, A. N. (1979): Stable iterative realization of some implicit methods of integration (Russian), *Zh. Vychisl. Mat. i Mat. Fiz.*, **19**, 121–128; English translation: *USSR Comput. Math. and Math. Phys.*, **19**, No. 1, 125–132.

Zavorin, A. N. (1981): The effect of step variation on the stability of some backward differentiation formulas, *Zh. Vychisl. Mat. i Mat. Fiz.*, **21**, 1582–1586; English translation: *USSR Comput. Math. and Math. Phys.*, **21**, No. 6, 223–228.

Zavorin, A. N. (1982): The stability of backward differentiation formulas for two different variable step techniques, in *Proc. Third Conf. on Numerical Treatment of Ordinary Differential Equations, Berlin, 1982* (Ed. R. März), Seminarberichte no. 46, Humboldt University, Berlin, pp. 241–246.

Zavorin, A. N. (1982a): Über die Effektivität einiger nichtlinearer Verfahren bei der numerischen Behandlung steifer Differentialgleichungssysteme, *Numer. Math.*, **40**, 169–177.

Zavorin, A. N. and Khesina, I. Ya. (1973): Some numerical methods for the solution of stiff systems of ordinary differential equations (Russian), *Zh. Vychisl. Mat. i Mat. Fiz.*, **13**, 71–79; English translation: *USSR Comput. Math. and Math. Phys.*, **13**, No. 1, 89–99.

Zayats, V. M. (1982): Stability of numerical integration algorithms with a combination of methods of different orders with a variable step (Russian, English summary), *Teoret. Elektrotekn.*, **33**, 112–119.

Zeller, K. (1969): Runge–Kutta-Approximationen, *Internat. Ser. Numer. Math.*, **10**, 365–366.

Zeller, K., see also Stetter, F.

Zhao, S. S. (1982): Numerical methods for solving initial value problems of stiff and high oscillatory systems of ODEs (Chinese, English summary), *Numer. Math. J. Chinese Univ.*, **4**, 325–337.

Zhu, C. H. (1980): A family of numerical methods for the solution of initial value problem of stiff differential equations (Chinese, English summary), *Math. Numer. Sinica*, **2**, 356–362.

Zhu, C. H. and Yuan, Z. D. (1980): The stability for numerical integration of a system of ordinary differential equations (Chinese, English summary), *Math. Numer. Sinica*, **2**, 77–89.

Zirilli, F., see Aluffi, F.

Ziv, A. and Amdursky, V. (1977): On the numerical solution of stiff linear systems of the oscillatory type, *SIAM J. Appl. Math.*, **33**, 593–606.

Zlatev, Z. (1978): Stability properties of variable stepsize variable formula methods, *Numer. Math.*, **31**, 175–182.

Zlatev, Z. (1981); Zero-stability properties of the three-ordinate variable stepsize variable formula methods, *Numer. Math.*, **37**, 157–166.

Zlatev, Z. (1981a): Modified diagonally implicit Runge–Kutta methods, *SIAM J. Sci. Statist. Comput.*, **2**, 321–334.

Zlatev, Z., see also Thomsen, P. G.

Zlatev, Z. and Thomsen, P. G. (1979): Automatic solution of differential equations based on the use of linear multistep methods, *ACM Trans. Math. Software*, **5**, 401–414.

Zlatev, Z. and Thomsen, P. G. (1980): Differential integrators based on linear multistep methods, in *Méthodes Numériques dans les Sciences de l'Ingénieur-G. A. M. N. I.* (Eds. E. Absi, R. Glowinski, P. Lascaux and H. Veysseyre), vol. 1, Dunod, Paris, pp. 221–231.

Zondek, B. and Sheldon, J. W. (1959): On the error propagation in Adams' extrapolation method, *Math. Comp.*, **13**, 52–55.

Zubov, V. I. (1975): Theory of numerical integration of differential equations (Russian), *Differentisial'nye Uravneniya*, **11**, 2269–2270.

Zurmühl, R. (1940): Zur numerischen Integration gewöhnlicher Differentialgleichungen zweiter und höherer Ordnung, *Z. Angew. Math. Mech.*, **20**, 104–116.

Zurmühl, R. (1948): Runge–Kutta-Verfahren zur numerischen Integration von Differentialgleichungen n-ter Ordnung, *Z. Angew. Math. Mech.*, **28**, 173–182.

Zurmühl, R. (1952): Runge–Kutta-Verfahren unter Verwendung höherer Ableitungen, *Z. Angew. Math. Mech.*, **32**, 153–154.

Zurmühl, R., see also Collatz, L.

Zverkina, T. S. (1968): A one-parameter analogue of Adams' formulae (Russian), *Zh. Vychisl. Mat. i Mat. Fiz.*, **8**, 797–807; English translation: *USSR Comp. Math. and Math. Phys.*, **8**, No. 4, 124–139.

Zverkina, T. S. (1972): A new class of finite-difference methods (Russian), *Zh. Vychisl. Mat. i Mat. Fiz.*, **12**, 1182–1196; English translation: *USSR Comput. Math. and Math. Phys.*, **12**, No. 5, 123–139 (1973).

Zverkina, T. S., see Gorbunov, A. D.

Additional References

Abramowitz, M. and Stegun, I. A. [1965]: *Handbook of Mathematical Functions*, Dover, New York.

Berge, C. [1962]: *The Theory of Graphs and Its Applications*, Methuen, London.

Burrage, K. and Chipman, F. H. [1985]: The stability properties of singly-implicit general linear methods, *IMA J. Numer. Anal.*, **5**, 287–295.

Butcher, J. C. [1984]: An application of the Runge–Kutta space, *BIT*, **24**, 425–440.

Butcher, J. C. [1985]: The non–existence of ten stage eighth order explicit Runge–Kutta methods, *BIT*, **25**, 521–540.

Crouzeix, M., Hundsdorfer, W. H. and Spijker, M. N. [1983]: On the existence of solutions to the algebraic equations in implicit Runge–Kutta methods, *BIT*, **23**, 84–91.

Dekker, K. [1981]: Algebraic stability of general linear methods, University of Auckland Computer Science Report No. 25.

Dekker, K. and Verwer, J. G. [1984]: *Stability of Runge–Kutta Methods for Stiff Nonlinear Differential Equations*, North-Holland, Amsterdam.

Dieudonné, J. [1969]: *Foundations of Modern Analysis*, Academic Press, New York.

Dugundji, J. [1966]: *Topology*, Allyn and Bacon, Boston.

Ehle, B. L. [1969]: On Padé approximations to the exponential function and A-stable methods for the numerical solution of initial value problems, Research report CSRR 2010, Dept. of Applied Analysis and Computer Science, University of Waterloo.

Frank, R. Schneid, J. and Ueberhuber, C. W. [1985]: Stability properties of implicit Runge–Kutta methods, *SIAM J. Numer. Anal.*, **22**, 497–514.

Frank, R., Schneid, J. and Ueberhuber, C. W. [1985a] Order results for implicit Runge–Kutta methods applied to stiff systems, *SIAM J. Numer. Anal.*, **22**, 515–534.

Gifkins, A. R. [1972]: An algebraic approach to Runge–Kutta methods, Thesis, University of Auckland.

Hardy, G. H. [1949]: *Divergent Series*, Oxford University Press.

Householder, A. S. [1970]: *The Numerical Treatment of a Single Nonlinear Equation*, McGraw-Hill, New York.

Kahan, W. [1965]: Further remarks on reducing truncation errors, *Comm. ACM*, **8**, 40.

Lang, S. [1968]: *Analysis I*, Addison-Wesley, Reading, Mass.

Ortega, J. M. and Rheinboldt, W. C. [1970]: *Iterative Solutions of Nonlinear Equations in Several Variables*, Academic Press, New York.

Ostrowski, A. M. [1966]: *Solutions of Equations and Systems of Equations*, Academic Press, New York.

Spijker, M. N. [1985]: Feasibility and contractivity in implicit Runge–Kutta methods, *J. Comput. Appl. Math.*, **12**, 563–578.

Wolfbrandt, A. [1977]: Thesis, Chalmers Institute of Technology, Göteborg.

Subject and Author Index

References are to subsections (Italic), or where it is more appropriate, to sections (Bold) or chapters (Roman). In particular, authors referred to in three or more subsections within the same section are indexed to the overall section rather than the particular subsections.

506